An Introduction to Python Programming for Scientists and Engineers

Python is one of the most popular programming languages, widely used for data analysis and modelling, and is fast becoming the leading choice for scientists and engineers. Unlike other textbooks introducing Python, typically organised by language syntax, this book uses many examples from across Biology, Chemistry, Physics, Earth science, and Engineering to teach and motivate students in science and engineering. The text is organised by the tasks and workflows students undertake day-to-day, helping them see the connections between programming tools and their disciplines. The pace of study is carefully developed for complete beginners, and a spiral pedagogy is used so concepts are introduced across multiple chapters, allowing readers to engage with topics more than once. "Try This!" exercises and online Jupyter notebooks encourage students to test their new knowledge, and further develop their programming skills. Online solutions are available for instructors, alongside discipline-specific homework problems across the sciences and engineering.

Johnny Wei-Bing Lin is an Associate Teaching Professor and Director of Undergraduate Computing Education in the Division of Computing and Software Systems at the University of Washington Bothell, and an Affiliate Professor of Physics and Engineering at North Park University. He was the founding Chair of the American Meteorological Society's annual Python Symposium.

Hannah Aizenman is a Ph.D. candidate in Computer Science at The Graduate Center, City University of New York. She studies visualization and is a core developer of the Python library Matplotlib.

Erin Manette Cartas Espinel graduated with a Ph.D. in physics from the University of California, Irvine. After more than 10 years at the University of Washington Bothell, she is now a software development engineer.

Kim Gunnerson recently retired as an Associate Teaching Professor at the University of Washington Bothell, where she taught chemistry and introductory computer programming.

Joanne Liu received her Ph.D. in Bioinformatics and Systems Biology from the University of California San Diego.

An Introduction to Python Programming for Scientists and Engineers

Johnny Wei-Bing Lin
University of Washington Bothell and North Park University

Hannah Aizenman
City College of New York

Erin Manette Cartas Espinel
Envestnet Tamarac

Kim Gunnerson
University of Washington Bothell

Joanne Liu
Novozymes A/S

CAMBRIDGE
UNIVERSITY PRESS

CAMBRIDGE
UNIVERSITY PRESS

University Printing House, Cambridge CB2 8BS, United Kingdom

One Liberty Plaza, 20th Floor, New York, NY 10006, USA

477 Williamstown Road, Port Melbourne, VIC 3207, Australia

314-321, 3rd Floor, Plot 3, Splendor Forum, Jasola District Centre,
New Delhi – 110025, India

103 Penang Road, #05–06/07, Visioncrest Commercial, Singapore 238467

Cambridge University Press is part of the University of Cambridge.

It furthers the University's mission by disseminating knowledge in the pursuit of
education, learning, and research at the highest international levels of excellence.

www.cambridge.org
Information on this title: www.cambridge.org/highereducation/isbn/9781108701129
DOI: 10.1017/9781108571531

First published 2022

Printed in the United Kingdom by TJ Books Limited, Padstow Cornwall

A catalogue record for this publication is available from the British Library.

Library of Congress Cataloging-in-Publication Data
Names: Lin, Johnny Wei-Bing, 1972– author. | Aizenman, Hannah, 1987– author. |
　　Espinel, Erin Manette Cartas, 1965– author. | Gunnerson, Kim Noreen, 1965– author. |
　　Liu, Joanne (Joanne K.), author.
Title: An introduction to Python programming for scientists and engineers /
　　Johnny Wei-Bing Lin, University of Washington, Bothell, Hannah Aizenman,
　　City College of New York, Erin Manette Cartas Espinel, Envestnet Tamarac,
　　Kim Gunnerson, University of Washington, Bothell, Joanne Liu, Biota Technology Inc.
Description: First edition. | Cambridge, United Kingdom ; New York, NY :
　　Cambridge University Press, 2022. | Includes bibliographical references and index.
Identifiers: LCCN 2022000136 | ISBN 9781108701129 (paperback)
Subjects: LCSH: Python (Computer program language) | Computer programming. |
　　Engneering–Data processing. | BISAC: SCIENCE / Earth Sciences / General
Classification: LCC QA76.73.P98 L55 2022 | DDC 005.13/3–dc23/eng/20220304
　　LC record available at https://lccn.loc.gov/2022000136

ISBN 978-1-108-70112-9 Paperback

Contents

Detailed Contents

Preface

Most introductory programming textbooks are written with the assumption that the student thinks like a computer scientist. That is, writers assume that the student best learns programming by focusing on the structure and syntax of programming languages. The result is an introductory textbook that teaches programming in a way that is accessible to future programmers and developers but not as much to scientists or engineers who mainly want to investigate scientific problems.

This textbook is written to teach programming to scientists and engineers, not to computer scientists. We assume that the reader has no background, formal or informal, in computer programming. Thus, this textbook is distinct from other introductory programming textbooks in the following ways:

- **It is organized around a scientist or engineer's workflow.** What are the tasks of a scientist or engineer that a computer can help with? Doing calculations (e.g., Chapters 2 and 6), making a plot (e.g., Chapters 4 and 5), handling missing data (e.g., Chapter 15), and saving and storing data (e.g., Chapters 9 and 18) are just a few of the tasks we address.
- **It teaches programming, not numerical methods, statistics, data analytics, or image processing.** The level of math that the reader needs is modest so the text is accessible to a first-year college student.
- **It provides examples pertinent to the natural sciences and engineering.** Jupyter notebooks associated with this textbook provide structured practice using examples from physics, chemistry, and biology, and additional notebooks for engineering are planned. For instance, the physics notebooks include problems dealing with electromagnetic fields, optics, and gravitational acceleration.
- **Syntax is secondary.** The primary goal is to teach the student how to use Python to do scientific and engineering work. Thus, we teach as much language syntax and structure as needed to do a task. Later, as we address more complex science and engineering tasks, we teach additional aspects of language syntax and structure. As a result, this textbook is not intended as a Python language reference where all (or most) of the aspects of a given feature of the language are addressed at the same time.
- **It is paced for the beginner.** This text offers many examples, explanations, and opportunities to practice. We take things slowly because learning is a step-by-step process, not a toss-into-the-deep-end process. As a result, this text is not concise, particularly in the beginning. It will seem ponderous to an expert programmer. This is intentional.

Structure of the Textbook

The textbook is divided into four parts. Parts I–III, collectively, cover most of the topics of a CS1 and CS2 sequence:

- Part I shows how to get the basic scientific and engineering workflow tasks done, including visualization, modeling, analysis, input/output, and computer administration.
- Part II shows us how to do more advanced tasks, building off of what we have seen in Part I. Throughout, the programming is taught through learning how to do the science and engineering workflow tasks, so almost every chapter in Parts I–II addresses a different science or engineering task we can use Python programming for.
- The chapters in Part III, because they cover more advanced programming concepts that are better discussed in computer science terms, are organized more traditionally, with chapters on more advanced data structures, classes and inheritance, basic searching and sorting, and recursion. Nonetheless, even in Part III, we connect those concepts to science and engineering workflow tasks.

Typical CS2 topics the textbook does not cover include linked lists, divide-and-conquer sorting algorithms, and trees.

Part IV goes beyond the topics of a sequence of introductory programming or scientific computing courses to describe how to turn our programs into something really robust.

Structure of the Chapters

The text takes a very uniform approach to each of the chapters in Parts I–III, as follows:

- The first section describes the task and gives some examples of using Python to accomplish that task. These examples and problems are general enough to be understood by most scientists and engineers.
- The second section describes why the example in the first section works the way it does: the programming concepts and Python syntax are described and demonstrated in this section.
- The third section provides exercises/examples, often similar to the examples in the first section, for the reader to try, along with solutions and discussion of why the solution works. Additional programming and Python concepts are often discussed in these solutions.
- The fourth section briefly describes the online exercises and problems that are discipline-specific (physics, chemistry, and biology), for further explication and practice.
- The final section is a chapter review containing self-test questions (with answers at the end of the section) and a chapter summary.

Chapters 1 and 11 are the exceptions in Parts I–III, having a different organization and lacking the exercises, questions, and chapter review as described above.

Because the chapters in Part IV cover techniques, packages, and utilities (some of which are still evolving) that do not lend themselves to a simple scientific or engineering workflow example nor to exercises in a Jupyter notebook, the chapters in Part IV are not organized using the five-section pattern most of the chapters in Parts I–III use. The aim of Part IV is to introduce software engineering tools and practices that help us to write better and more reliable code, in order for the reader to seek more information in works that specialize in these topics. We hope that our introduction will whet your appetite to learn more on your own about these tools and practices.

When we first start to learn something, we require more explanation and practice to get comfortable with the material, learn the vocabulary, and think in new ways. The textbook is thus structured like a pyramid:

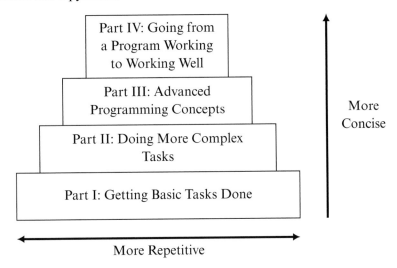

The chapters in Part I are the longest, and the chapters in Part IV are the shortest. The earlier parts are more repetitive and less concise while the later parts are less repetitive and more concise. The wider the block in the pyramid above, the more repetitive the part. The higher up the pyramid, the more concise the part.

The Logic Behind Some Decisions on Topic Coverage and Sequence

Some of the decisions about coverage and sequence may be surprising, so the rationale is explained here:

- By basing the structure of most of the chapters in Parts I–II on science and engineering workflow tasks, syntax is addressed in pieces. That is, there is no single chapter on variables, single chapter on branching, etc. Rather, those topics are covered over several chapters, in a kind of spiral approach. Pedagogically, this is better because it allows us to learn things a little at a time with subsequent treatments of a topic reinforcing what we learned earlier. Learning in this way is treated as a spiral rather than as a line. But, there are at least two costs with this method of learning. First, the first time we see a particular structure in a given chapter, because we do not fully describe that structure in that chapter, the code examples may seem awkward and more complicated than needed. Second, by spreading out the description of a single structure over multiple chapters, this textbook does not function well as a reference. What can you do to mitigate these costs? For the first, we ask for your patience as we slowly build up the description of the topic. The code examples will become more concise and Pythonic as the book progresses. For the second, we have written the textbook to provide a substantial amount of cross-referencing between sections, so those cross-references might be enough to lead you where you want to go. Skimming the Detailed Contents, **which details the subtopics included in each chapter,** or Index, may also help you find the occurrences of the topics you are interested in. Appendix C also lists the contents of the book by programming topic.

- The "how to write a program" chapter does not come early on as it does in most programming textbooks. We briefly mention how to write programs in Section 6.2.7 and provide a more detailed treatment in Chapter 11. How come? If the goal of the textbook is to teach programming by doing, rather than reading about how to "do," it should start off with using Python to do science and engineering tasks. Once the reader has some experience writing short programs, they will have more context to understand advice on how to write a longer program.

- The workflow focus influences how looping and branching are introduced in this textbook. In most textbooks, these topics are introduced separately, in separate chapters. The result, however, is to limit what can be done with such knowledge. An `if` statement by itself, without the possibility of being visited multiple times, does not accomplish much. In this textbook, we introduce both concepts together, in Chapter 6 on basic diagnostic data analysis. Neither topic is treated exhaustively in this chapter, but with the introduction of both topics in one chapter, we can vividly show how useful these structures are for using the computer to investigate a dataset.

- Because the textbook is aimed at novices rather than students with programming experience, we had to make difficult choices regarding what concepts to introduce and when. One result of these choices is that, particularly in earlier chapters, our code examples are less than Pythonic in order to make our treatment of the concepts at that point in the book clearer to novice students. We also ignore some packages and tools that can accomplish some of the tasks we address more concisely and efficiently, in order to focus on the learning goal at hand. Reasonable people will disagree regarding our choices, but we wanted instructors to know we recognize the tension, and we heartily support whatever customization will best meet the needs of your students.

Topic Sequence and Flexible Approach for Different Course Lengths

- Part I is material for approximately one quarter-long (10 week) class. The material works well for an introductory programming course where the students have no prior experience with programming.

- Parts I–II are material for approximately one semester-long (15 week) introductory programming course.
- Parts I–III are written so later chapters build on the contents of earlier chapters. They contain material for a two-quarter introductory programming sequence.
- The chapters in Part IV are supplemental and can be read independently and used for self-study.
- Going in order is one way to use the textbook. However, other sequences are possible. For instructors who want to cover the material by programming topic, the table of contents in Appendix C shows where various aspects of the programming topics are covered.

Typesetting and Coloring Conventions

Throughout the textbook, we use different forms of typesetting and coloring to provide additional clarity and functionality. Some of the special typesetting conventions we use include:

- Inline source code. These are instructions interspersed in the paragraphs of the textbook that tell Python what to do. They are typeset in a dark red, monospace font, as in `a = 4`.
- Inline commands to type on your keyboard or printed to the screen. Typeset in a dark red, monospace font, as in `print('hello')`.
- Inline generic arguments or values. These labels are placeholders to be replaced by snippets of code or concrete values. These are typeset in a dark red, italicized, proportional font, in between a less than sign and a greater than sign, as in *<condition>*.
- Blocks or listing items of source code, commands, generic arguments, or values. These are typeset in an indented block quotation structure or listing structure in a black font.
- File contents (that are not source code). These are typeset in an indented block quotation structure in a black, monospace font.
- File, directory, and app names. Typeset in a black, italicized, proportional font, as in *usr/bin*.
- Key terms. On first use, these are typeset in a dark blue, bold, proportional font, as in **program**, and can be found in the Glossary.

General references to application, library, module, and package names are typeset the same as regular text. Thus, references to the Matplotlib package are typeset just as in this sentence. As most packages have unique names, this should not be confusing. In the few cases where the package names are regular English words (e.g., the time module), references to the module will hopefully be clear from the context.

Assessment and Practice for Students

This text and the online Jupyter notebooks provide an abundance of opportunities for students to practice what they are learning:

- Try This! exercises are designed to provide practice for students to take bite-sized, incremental steps in growing their understanding the material. In a classroom setting, they are appropriate for active learning problems, group work, and as components in programming labs. The answers are publicly available, so these exercises are best used for practice and development rather than summative assessment.
- The Homework Problems are provided online as Jupyter notebooks and require more time for students to work on than the Try This!. Thus, generally speaking, they are more appropriate for students to work on outside of class. Solutions for homework notebooks are only provided to verified instructors.

The "To the Student" section following this Preface provides further description of the kinds of exercises and problems that are available.

Supporting Resources and Updates to This Textbook

The textbook's website, www.cambridge.org/core/resources/pythonforscientists, contains updates to and supporting resources for the textbook. This includes:

- Web pages describing some topics in more detail.
- Jupyter notebooks with exercises and problems.
- Copies of the larger datasets and images referenced in this textbook.
- A list of addenda and errata.
- Links to key external sites.

All the above resources (with the exception of the images, and solutions for exercises and problems) are accessible by both students and instructors.

The landscape of Python teaching resources is vast and constantly changing. We provide a web page listing some of these resources at www.cambridge.org/core/resources/pythonforscientists/instructors/.

Please let us know of any corrections by emailing us at ipyses@johnny-lin.com.

To the Student

An Introduction to Python Programming for Scientists and Engineers consists of two components:

- The print/digital book that you are now reading.
- Online resources including web pages with additional content and exercises and problems as Jupyter notebooks.

The latter are not "supplements" because they tightly link to the content and flow of the textbook. The two pieces form an integrated whole.

Applications and Exercises

The textbook provides the following kinds of questions and problems in Parts I–III:

- Try This!. These are exercises/worked examples and are found in the print/digital book. They are similar in complexity to the main chapter examples and contain solutions and discussion of the solution. The Try This! sections also extend the discussion earlier in the chapters, introducing new concepts.
- Chapter Review Self-Test Questions. These are found in the print/digital book. These are relatively short questions that give you a first-cut assessment of your understanding of the material in the chapter. The answers to these questions are found at the end of the chapter.
- Discipline-Specific Try This!. These are found online. They are similar to the exercises discussed in the print/digital book in the Try This! sections. We give you a Jupyter notebook with the Discipline-Specific Try This! and then another Jupyter notebook with answers to the Try This!.
- Discipline-Specific Homework Problems. These are found online. They are designed to be assigned by instructors for homework. We give you a blank Jupyter notebook with the Problems. These Problems are generally more involved than the Discipline-Specific Try This!.

All Discipline-Specific exercises and problems are in the form of Jupyter notebooks that can be downloaded to your own computer and run locally. The Discipline-Specific Jupyter notebooks are at www.cambridge.org/core/resources/pythonforscientists/jupyter-notebooks/. Currently, notebooks for biology, chemistry, and physics are provided. We hope to include notebooks for other disciplines in the future. Details on using Jupyter notebooks begin in Section 2.2.3.

The larger datasets referenced in this textbook and the Jupyter notebooks are found at www.cambridge.org/core/resources/pythonforscientists/datasets/. These are freely available for download.

Using the Textbook

We recommend that you approach the chapters in the following manner.

First, read the main chapter example. If possible, **type in the code** for the example and see what you get. This is more doable for earlier chapters, where the code is shorter. In later chapters, the benefits of typing in the code are probably not worth the time to key it all in. As you read the main chapter example, you might not understand everything in the example. If so, that is okay. The purpose of the main chapter example is not to teach you the concepts in the chapter. That is what the rest of the chapter is for. The main chapter example is there to give you a sense of how we can use Python to solve a particular science or engineering task and to motivate the rest of the chapter's discussion and exercises.

Second, read the Python Programming Essentials section. In most chapters, this section will have small examples you can type in. Please do so. This will help your learning.

Third, do the Try This!. Do not read read the solution that immediately follows until you have done the Try This!. If you just read the Try This! and skip immediately to the solution, you will circumvent the learning process. Doing and working problems really help us learn! Also, remember the discussion of these exercises contains additional material regarding the chapter's topics. Do not skip the Try This! as if they are optional, "mere examples."

Fourth, do as many of the Discipline-Specific Try This! you have the time to do. After you do them, look at their solutions and see if you can explain *the reason* for any differences between your solution and the solution notebook.

Fifth, to increase the sophistication of your understanding of the topics involved, do at least a few of the Discipline-Specific Homework Problems. The solutions are only available to instructors, but the practice itself will help with your learning.

The Chapter Review Self-Test Questions can be used to test your understanding of the material. You might want to use them for studying for exams, but you do not need to wait for an exam to go through them. As you are reading the textbook, these questions might also help give you a sense of your comprehension. Note, though, that these questions are not very difficult, so you should not conclude that if you are able to answer all the questions easily that you will do fine on an exam.

The Chapter Summary provides one additional opportunity for you to see the topics again. It should not be used as a substitute for taking your own notes or creating your own study sheet for an exam.

One final piece of advice. Wherever we suggest you read the text, we are implicitly assuming you will read *slowly*. Really, really slowly. When we read code, we have to go line-by-line and term-by-term. We have to ask what the state of the variables are before that line, what the state of the variables are after that line, and what did the line do to cause (or not cause) any changes. When we encounter code, we should ask how the program would behave differently if we made a change to the values and order of the code. And we should take notes (by hand, if possible, because we learn more that way) on what we have read and write down any questions we have as they come to us. If we do not write down questions immediately, we will forget them and falsely believe we understand the material. We have to *actively engage* the text.

We cannot read a programming textbook (or any math, science, or engineering textbook) as if it were a novel. When we read a novel, we can just read the words and sentences themselves, skimming if we are in a rush. The prose itself tells us about the characters, plot, and setting. But in a programming textbook, this kind of reading will not work. Skimming a programming textbook is a recipe for disaster. We have to unpack and excavate the meaning of the code and the text describing the code. If we do not, our understanding will be limited. So, read s-l-o-w-l-y! ☺

Notices and Disclaimers

Mark and Trademark Acknowledgments

Anaconda, Anaconda Navigator, Conda, and Numba are marks and/or registered trademarks of Anaconda, Inc. Apple, Mac, Mac OS, and OS X are registered trademarks of Apple Inc. Azure, Excel, Microsoft, PowerShell, Windows, and Word are registered trademarks of Microsoft Corporation in the United States and/or other countries. Chrome^{TM} browser is a trademark of Google LLC. Debian is a registered trademark of Software in the Public Interest, Inc. Django is a trademark of the Django Software Foundation. Git is either a registered trademark or trademark of Software Freedom Conservancy, Inc., corporate home of the Git Project, in the United States and/or other countries. GITHUB® is an exclusive trademark registered in the United States by GitHub, Inc. GitLab is a registered trademark of GitLab, Inc. GNOME® is a trademark of the GNOME Foundation. GNU is an operating system supported by the Free Software Foundation. Jupyter®, JupyterHub, and derivative word marks are trademarks or registered trademarks of NumFOCUS. Kubernetes® is a registered trademark in the United States and/or other countries of The Linux Foundation. LibreOffice is a registered trademark of The Document Foundation. Linux is a trademark owned by Linus Torvalds. Matlab and MathWorks are registered trademarks of The MathWorks, Inc. PyCharm[1] is a trademark of JetBrains s.r.o. Python is a registered trademark of the Python Software Foundation. Stack Overflow is a trademark of Stack Exchange Inc. Ubuntu is a registered trademark of Canonical Ltd. All other trademarks and marks mentioned in this book are the property of their respective owners. Any errors or omissions in trademark and/or other mark attributions are not meant to be assertions or denials of trademark and/or other mark rights.

Copyright Acknowledgments

Screenshots of Chrome browser sessions (e.g., Figures 2.3, 3.7, 3.8, 3.9, 4.6, and 4.7) include elements from Google LLC and are used by permission.

Images from Volkman et al. (2004) in Chapter 13 are copyright © 2004 by Volkman et al. and are used by permission under the conditions of the Creative Commons Attribution License. Volkman et al. do not specify the license version, but the Creative Commons

---

[1] www.jetbrains.com.

Attribution 4.0 International Public License is available at https://creativecommons.org/licenses/by/4.0/legalcode. The images in the present work have been modified from their original form in Volkman et al.

The screen shots in Figures 2.3, 3.7, 3.8, 3.9, 21.2, 21.3, and 21.4 are reprinted with permission from Apple Inc.

The code in Figure 24.1 is © 2010–2021, Holger Krekel and others. The code is used by permission under the conditions of the MIT License. The license agreement is available at https://github.com/tox-dev/tox/blob/master/LICENSE.[2]

Code portions and ideas referenced throughout the text are footnoted or otherwise referenced in the text. The code portions and ideas in Chapter 14, however, are built off of longer blocks of code as well as a larger variety of code sources, most prominently the cartopy manual (Met Office, 2010–2015), licensed under an Open Government License (www.nationalarchives.gov.uk/doc/open-government-licence/version/2/), and the Matplotlib documentation (matplotlib.org/contents.html; also see Hunter (2007)). Traditional footnoting and referencing does not work as well for this kind of synthetic work. Thus, we cite those references here and at www.cambridge.org/core/resources/pythonforscientists/refs/.

Use in this book of information from copyrighted sources is by permission (and is noted either in this acknowledgments section or in the respective figure captions) or is usage believed to be covered under Fair Use doctrine.

Data and Other Usage Acknowledgments

This work includes data from the Mikulski Archive for Space Telescopes (MAST) (e.g., in Chapter 1). STScI is operated by the Association of Universities for Research in Astronomy, Inc., under National Aeronautics and Space Administration (NASA) contract NAS5-26555. This includes data collected by the Kepler Mission. Funding for the Kepler Mission is provided by the NASA Science Mission directorate.

This work includes output from simulations using the molecular dynamics simulation package NAMD (Phillips et al., 2005). NAMD was developed by the Theoretical and Computational Biophysics Group in the Beckman Institute for Advanced Science and Technology at the University of Illinois at Urbana-Champaign. The official NAMD web page is at www.ks.uiuc.edu/Research/namd/.

This work includes surface/near-surface air temperature data (e.g., in Section 7.1) that is from National Centers for Environmental Prediction (NCEP) Reanalysis data provided by the NOAA/OAR/ESRL PSD, Boulder, Colorado, USA, from their website at www.esrl.noaa.gov/psd.

This work includes data from the UCI Machine Learning Repository (Dua and Graff, 2019). The repository is at https://archive.ics.uci.edu/ml.

---

[2] Accessed January 11, 2021.

Section 13.3 includes images from Patel and Dauphin (2019), published by NASA's Earth Observatory, which were created using Black Marble data from Ranjay Shrestha at NASA's Goddard Space Flight Center and Landsat data from the U.S. Geological Survey.

World Time Buddy (www.worldtimebuddy.com) provided help with time zone conversions. (See Sections 7.1 and 12.1.)

Data used for displaying natural features and political boundries (in Chapter 14) are provided by OpenStreetMap (© OpenStreetMap contributors) and Natural Earth (public domain). OpenStreetMap data are available under the Open Database Licence (www.openstreetmap.org/copyright). Free vector and raster map data are available at naturalearthdata.com.

Data in this work are also acknowledged in descriptions in the main text and footnotes. All data and images in this work are used by permission. Many of these materials are in the public domain or are otherwise freely available for republication and reuse. Please see the sources of the data for details.

Data presented in this work that are not explicitly attributed to a source should be considered fictional and created by the authors for the purposes of illustration or teaching. They must not be considered genuine or accurate descriptions of any natural or artificial phenomena or system. Most (though not all) occurrences of such data are accompanied by the term "fictitious" or "pretend."

Disclaimers

Although we have worked hard to make the text, code, and related online resources (together, "Resources") accurate and correct, the Resources are provided "as-is," without warranty of any kind, express or implied, including but not limited to the warranties of merchantability, fitness for a particular purpose and noninfringement. In no event shall the authors or copyright holders be liable for any claim, damages or other liability, whether in an action of contract, tort or otherwise, arising from, out of or in connection with the Resources or the use or other dealings in the Resources.[3]

Permission to use marks, trademarks, copyrighted materials, or any other materials by their owners does not imply an endorsement of that use or of the Resources.

---

[3] Copied and adapted from the MIT License, as listed on Opensource.org, https://opensource.org/licenses/MIT (accessed July 14, 2021).

Acknowledgments

We are grateful for the editorial help of Charles Howell, Matt Lloyd, Lisa Pinto, and Melissa Shivers at Cambridge University Press. We thank Spencer Cotkin for editorial suggestions, most we have implemented and that have made the book much better. We thank Beverley Lawrence for copyediting the text. A number of anonymous reviewers provided helpful feedback which have been incorporated into the text.

We are thankful for Cynthia Gustafson-Brown's assistance on this project, particularly in finding and providing some of the description in the Section 13.1 example.

We are appreciative of conversations with and assistance by: Bill Erdly, Michael Grossberg, Charity Flener Lovitt, Laurence Molloy, Hansel Ong, Jim Phillips, and Rob Nash. Whether through publications, workshops, conferences, or discussions, the communities we are a part of – personal, workplace, disciplinary, and, of course, the Python community – contributed ideas, encouragement, and support.

Parts of this book are based on the book, *A Hands-On Introduction to Using Python in the Atmospheric and Oceanic Sciences*,[4] slides from the 2020 American Meteorological Society's *Beginner's Course to Using Python in Climate and Meteorology*,[5] and the set of notes, *Lecture Notes on Programming Theory for Management Information Systems.*[6] These resources are by Johnny Lin and the acknowledgments made in those resources also apply to this text. We are grateful for those who gave us permission to use material they created. These are acknowledged in the Notices section, the captions of the included or adapted figures, or in the online resources.

[4] Lin (2012).
[5] Not formally published.
[6] Lin (2019).

I thank my wife Karen, and my children Timothy, James, and Christianne for their encouragement and love. *S. D. G.*

Johnny Wei-Bing Lin
Bellevue, Washington

I thank my advisor, Professor Michael Grossberg, for all his help and the AMS Python community and Matplotlib Development Team for their influence and conversations on all things code.

Hannah Aizenman
New York City, New York

I thank my husband Vicente, my family Duffy, Cheryl, and Sara, and my friends Valerine, Kaitlyn, and Alanna for all of their love and support.

Erin Manette Cartas Espinel
Kenmore, Washington

I thank Samantha Gunnerson for her help reading through notebooks to give me advice, Eric Gunnerson for his support as well as technical skills, and Dr. Paola Rodríguez Hidalgo for listening to me talk about the different Python writing I was doing during this process.

Kim Gunnerson
Bellevue, Washington

I thank my family for their support and Cynthia Gustafson-Brown for providing several of the More Discipline-Specific Practice problems in Chapters 2 and 3.

Joanne Liu
San Diego, California

Part I

Getting Basic Tasks Done

1   Prologue: Preparing to Program

The Preface and To the Student sections describe how and why we structured the book and resources the way we did and how to make the most out of them. In the present chapter, we set the stage for the study of programming as an endeavor and Python as a language. We also describe what software needs to be installed in order to make use of the rest of the book.

1.1   What Is a Program and Why Learn to Program?

A **program** is a set of instructions telling a computer what to do. Every action a computer takes, from making a calculation to displaying a graph, is ultimately controlled by a program that a human being writes. This program consists of a file with commands. The computer reads that file and executes the commands one at a time.

The problem is that the language the computer understands is different from the languages that human beings know. Natural language – the language of people – is incredibly rich and is capable of describing so much more than facts and figures. Computer languages, in contrast, are extremely simple with very limited vocabularies and capabilities. This is because a computer can only do a few things:

- Save values.
- Do calculations.
- Ask if something is true or false.
- Accept input (e.g., from a keyboard) and output (e.g., to a screen).
- Do a task over and over again.

In one sense, everything a computer does – whether sending an email, playing a cat video, or modeling the Earth's climate – is the result of many programmers breaking down whatever complex tasks they want the computer to do into some combination of the above capabilities. So, learning how to program means learning how to break down the task we want to do into (very) simple pieces and how to express those tasks in a language – a programming language – that the computer understands. As an aside, we often talk about **code** or coding when referring to the task of programming. Those terms are another way of referring to the syntax of computer languages and the task of writing programs in those languages, respectively.

For a scientist or an engineer, what is the purpose of learning to program? Few scientists or engineers are interested in becoming software developers, and the foundations of modern science were developed using pen, paper, and the human mind. Newton and Darwin did their calculations without calculators, let alone computers. Today, great science and engineering work can still be done without needing to program a computer. For work involving small datasets or analytical mathematical solutions, pen and paper (and a calculator or spreadsheet) is often enough.

Today, however, we enjoy more data than our predecessors would have believed possible: measurements from satellites, in-situ sensors, or large-scale experiments; models of fluid flow, structures, or biological systems. A spreadsheet is often inadequate to deal with such large datasets. At the same time, the sciences and engineering have been the recipients of an explosion in computational tools for calculations of all kinds. Whether we are looking for a traditional statistical analysis **routine** or want to implement the latest machine learning algorithm, someone else has written a tool we can use. The software engineering community has also developed tools and legal frameworks that enable computational tools to be easily shared and integrated into anyone's programs. The result is that a person who knows how to program can do more science and engineering.

A picture is worth a thousand words, so consider Figure 1.1: This is an image showing the light received by the National Aeronautics and Space Administration (NASA) Kepler Mission's spacecraft (a space telescope designed to search for exoplanets) from the star

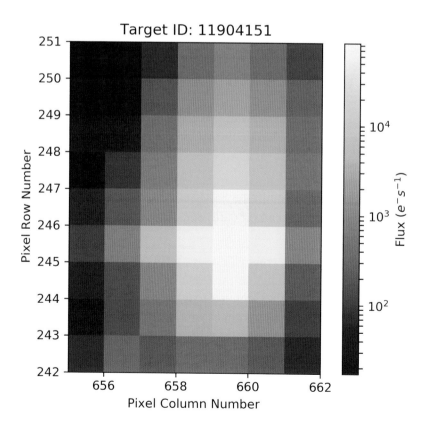

Figure 1.1 Light flux from Kepler-10 during quarter 4.

Kepler-10, around which orbits the first rocky exoplanet discovered by the spacecraft (Kepler-10b).[1] The Python code to read in the data via the Internet from the Mikulski Archive for Space Telescopes (MAST), where the archived data resides, and create the plot is as follows:

```
1   from lightkurve import KeplerTargetPixelFile
2   data = KeplerTargetPixelFile.from_archive('Kepler-10', quarter=4)
3   data.plot(scale='log')
```

That's it! Three lines of code does it! This is why scientists and engineers need to learn to program. We will not unpack these lines of code right now, but after going through the book, we will have the tools to understand their meaning and use.

1.2   What Is Python and Why Learn This Language?

There are many programming languages, and every language has its own strengths and weaknesses. But it is difficult to learn programming in the abstract. To learn programming, we have to use some particular language. Once we learn one language, other languages are more easily learned, but we have to start somewhere. The language we will use in the present work is Python.

While there is no perfect programming language, the Python language is extremely powerful and versatile while at the same time also easily understood and learned. Code written in Python is very clear, so clear that it almost reads like naturally spoken English. For scientists and engineers, Python has capabilities that make it useful for analyzing data and solving mathematically based problems. There is a large community of scientific Python users and developers that are constantly adding to the capabilities of the language to do science. As we go through the text, we will see these characteristics and features of Python. For now, just trust us that Python is a good place to start.

There is, however, one additional reason for learning Python: it is free. Python is an **open source** programming language, that is, a programming language whose underlying code is itself available and open to anyone to examine and use for their own purposes. Python is free in the sense of "freedom." Python is also free in the sense of "no cost." As a result, if we have a connection to the Internet, we can obtain a copy of everything needed to run Python programs. We can use the Python language for our own programs without worrying about whether we can purchase a (possibly) expensive license for a specialized data analysis language or whether a program we write today will be able to be run tomorrow if we distribute it to others or move to another computer ourselves. After learning Python, we can be confident

---

[1] This example is slightly altered from the one given in the documentation of the lightkurve package (see Vinícius et al. (2018)), a Python package used to access data from the Kepler Mission. That example is at http://lightkurve. keplerscience.org/tutorials/2.02-recover-a-planet.html (accessed August 15, 2018). Information about the Kepler Mission is at www.nasa.gov/mission_pages/kepler/overview/index.html (accessed August 20, 2018).

(as humanly possible) that we will be able to use the language throughout our lives for whatever scientific or engineering purpose – or business, artistic, literary, etc., purpose – we wish.

1.3   Software We Will Need

In this book, we will learn to write Python programs, but in order to get our computer to **run** (or execute the commands of) our programs, we need a little more than the program instruction files themselves. Python most usefully comes as part of a **distribution** of utility programs and tools.

We have written this text assuming version 3.x (e.g., 3.6, etc.) of Python has been installed, through a recent installation of the Anaconda distribution.[2] Installing the Anaconda distribution is preferable to installing Python from the Python Software Foundation's website, or to rely on the version of Python that is included with some operating systems, because the Anaconda distribution includes packages and utility programs that are essential for scientists and engineers.

The Anaconda distribution download page contains detailed directions on how to download and install the software. We provide an up-to-date link to the page at www.cambridge .org/core/resources/pythonforscientists/refs/, ref. 1. This edition is free to individuals and can be installed without administrator privileges. It can be installed on the same computer or account at the same time as another installation of Python and is available for all major operating systems (Windows, Mac OS X, and Linux).

When installed out-of-the-box, the Anaconda distribution provides most of what we need. Through the distribution's tools, we can also add additional utility programs and libraries (called packages) to the distribution, as desired. Use Anaconda Navigator to install additional packages and manage packages. Up-to-date links to the Navigator documentation and getting started guide are at www.cambridge.org/core/ resources/pythonforscientists/refs/, refs. 2 and 3 respectively.

Here is a list of additional packages that are used in the current text that might not come with the default installation of the Anaconda distribution. In parentheses, we note which chapters or parts of these packages are first and/or mainly used:

- cartopy (Chapter 14),
- line-profiler (Part IV),
- memory-profiler (Part IV),
- netCDF4 (Chapter 18),
- numba (Part IV),
- pytest (Section 23.3.2).

---

[2] Most of the code in this book, particularly in the earlier chapters, will also work with Python 2.7.x. However, although there is some scientific and engineering legacy code still written in Python 2.7.x, all major packages have migrated to version 3.x.

To run the code in the sections listed above, please add these packages. Note the line-profiler and memory-profiler packages are also referred to as line_profiler and memory_profiler, using underscores instead of hyphens.

If any of the packages above are already installed, Anaconda Navigator will tell us when we try to install it, so we do not have to worry about overwriting anything. If we do not know whether we have all the packages we need, no worries. If while we are running a Python program we receive a message such as:

```
ModuleNotFoundError: No module named 'netCDF4'
```

we can fix that by installing the package or module named, and then rerun our program.

That should be it for the preliminaries. We are ready now to begin our journey in learning Python programming. We start with using Python to fulfill that basic computational need of a scientist or engineer, the need to have a good calculator.

2   Python as a Basic Calculator

Pencil and paper are two of the greatest tools ever invented for studying science and engineering problems. But, there comes a point when the amount of data and complexity of the calculations require us to use more powerful approaches. The calculator is a scientist or engineer's basic workhorse for making those calculations. While Python has a wide range of applications, from graphing to modeling to data analysis, the simplest (and sometimes most useful) way is to use Python as a calculator. In fact, if our laptop is already open, it is often easier to use Python to do simple (or not-so-simple) arithmetic rather than using a calculator. In this chapter, we describe how to use Python as a basic calculator. The structure of this chapter will be the same as that in most of the chapters. See the To the Student frontmatter section for details. We center our discussion around examples of the task for the chapter and the Python code to accomplish the task. Please do not expect the examples will make complete sense when they are introduced. In the section after, we provide the explanation of what the Python code is doing. Later in the chapter, we provide Try This! examples, chapter Self-Test Questions, and a Chapter Summary.

2.1   Example of Python as a Basic Calculator

We introduce the example in this section. The detailed description of how the example works and why it does what it does is found in the next section (Section 2.2). When we read the beginning examples in most of the chapters, it is okay to not understand everything (or most things) in the examples. This is our first look at the concepts involved!

A basic calculator needs to do two tasks:

- Perform arithmetic.
- Store values for recall and reuse.

Many basic calculators also include parentheses keys to enable grouping of **operations**. Some can also store more than one value in the calculator's memory for recall, or add a current value to an existing value in memory.

Consider a cube whose sides are 2 in long. Here is a Python session to calculate the volume of this cube:

```
>>> 2*2*2
8
```

When the prompt for the Python **interpreter** appears (the interpreter is the place where instructions in Python are entered in and the computer carries them out, i.e., **executes** them), we type our arithmetic **expression**, press the [Enter] key (or [Return] key on some keyboards), and the computer will execute our expression and print the result below. Starting the Python interpreter is akin to turning on a calculator. Once the interpreter or calculator is turned on, it is ready for us to enter numbers and do calculations. In the above example, the interpreter's prompt is the three greater-than symbols (>>>) and shows us where to start typing our arithmetic expression. We do not type in the >>>; those symbols appear automatically when the interpreter starts.

In addition to using the default Python interpreter, as shown above, we can also run an interactive Python session in a different **environment**. An environment is the interface we use to interact with the Python interpreter. Below is the same example as above but with the code in a cell in the Jupyter notebook environment:

```
In [1]:   2*2*2

Out[1]:   8
```

We can group operations using parentheses. Here is an expression that calculates the combined volume of two cubes, one with 2 in sides and another with 3 in sides:

```
>>>  (2*2*2)+(3*3*3)
35
```

In a Jupyter notebook, this would be:

```
In [1]:   (2*2*2)+(3*3*3)

Out[1]:   35
```

Finally, we can save the result of an expression by assigning it a name. Once saved, we can use the saved values by referring to the name. We provide an example of calculating the volume of two cubes, saving their values as **variables**, and using those variables in another expression in Figure 2.1. In that example, the first three lines of code are each executed, one at a time, in turn. The result is the output Python provides in the last line. The top part of the figure shows how the code would be entered in and executed in the default Python interpreter. In line 1 of the top part (which is equivalent to the first line of the In [1] cell of the bottom part), we multiply three 2s and then save the result to the variable volume_cube_1. In line 2, we do the same for three 3s and save the result to the variable volume_cube_2. Finally, in line 3, we add the contents of the variable volume_cube_1 and volume_cube_2 together, and the Python interpreter prints the result, 35. The bottom part shows how the code would be entered in and executed in a Jupyter notebook. Both top and bottom are equivalent, the only difference being how the interface looks.

```
1   >>> volume_cube_1 = 2*2*2
2   >>> volume_cube_2 = 3*3*3
3   >>> volume_cube_1 + volume_cube_2
4   35
```

```
In [1]:   volume_cube_1 = 2*2*2
          volume_cube_2 = 3*3*3
          volume_cube_1 + volume_cube_2

Out[1]:   35
```

Figure 2.1 An example of evaluating expressions, saving values as variables, and using variables in other expressions, using the default Python interpreter and using a Jupyter notebook.

2.2 Python Programming Essentials

The example in Figure 2.1 illustrates the four Python programming topics we will discuss in this chapter: expressions, operators, variables, and the interpreter itself. Whereas the code can be executed using the default Python interpreter or in the Jupyter notebook environment, nearly all the interactive code examples will be given using Jupyter. For the sake of conciseness, when we show the interactive environments of the default Python interpreter, Jupyter notebooks, and other console environments described in Section 3.2.5, we will usually refer to them as the "interpreter." Section 2.2.3 discusses the default interpreter and Jupyter notebook environments in more detail.

2.2.1 Expressions and Operators

The code snippets `3*3*3` and `volume_cube_1 + volume_cube_2` are examples of expressions. Expressions are collections of values and variables (and other items, as we will see later) that are joined together and acted on by **operators**.

Values are numbers written as we would expect (e.g., `3`, `-7.2`), although to represent a number in scientific notation we enter it in the form xey, where the x represents the number multiplied by the power of 10 and y represents the exponent to the power of ten. Thus, the following two columns of numbers are equivalent:

Scientific notation	Python form
5.4×10^6	`5.4e6`
7.54×10^{-4}	`7.54e-4`
-2.4×10^{-2}	`-2.4e-2`

Some operators work on two elements (such as the addition (+) operator, which adds the numbers to the left and right of the operator), whereas others work on only one element.

Table 2.1 Arithmetic operators.

Operation	Symbol
Add	+
Subtract	−
Multiply	*
Divide	/
Integer divide	//
Exponentiation	**

The former are called **binary operators** whereas the latter are called **unary operators**. A little confusingly, we also use the term "**binary**" when talking about how data are stored in a computer. In those contexts, "binary" means a 0 or 1. This meaning of binary is described in more detail in Section 9.2.2 and Chapter 18. Table 2.1 lists some basic arithmetic operators Python uses (these are all binary operators). Later on in the text, we will encounter more operators, both unary and binary.

When Python evaluates an arithmetic expression, it follows the standard mathematical order of parenthetical blocks first, then exponentiation, then multiplication and division, and finally addition and subtraction. Operations at the same level of priority are executed left-to-right.[1] In this example:

```
In [1]:  4+3*2
```

```
Out[1]:  10
```

```
In [2]:  (4+3)*2
```

```
Out[2]:  14
```

we see that because multiplication has higher priority than addition, in cell `In [1]` the `3*2` is executed first before adding to the `4`. In the `In [2]` expression, the parentheses force the `4+3` to be executed first before multiplying the result by `2`.

The normal arithmetic operators listed in Table 2.1 act as expected. Division, however, is a little more nuanced. The division (`/`) operator acts like regular division, which returns the quotient and remainder in decimal form. Thus:

```
In [1]:  1/2
```

```
Out[1]:  0.5
```

---

[1] See the Wikibooks page on Python operators for more details. We provide an up-to-date link to the page at www.cambridge.org/core/resources/pythonforscientists/refs/, ref. 17.

```
In [2]:  7/3

Out[2]:  2.3333333333333335

In [3]:  5/5

Out[3]:  1.0

In [4]:  5.0/5.0

Out[4]:  1.0
```

It does not matter whether the numerator or denominator is an integer or decimal number, the result will be a decimal number. Also, in Out[2], it appears decimal numbers are not properly represented in Python. We discuss this in more detail in Sections 8.2.7, 9.2.2, and 9.2.3. For now, just note that decimal numbers are, in general, inexactly represented in computers.

Sometimes, however, we do not want division to give us the decimal result but instead only want the quotient (as an integer) and discard the remainder by rounding down.[2] In those cases, we want to use integer division, which is given by the // operator:

```
In [1]:  1//2

Out[1]:  0

In [2]:  7//3

Out[2]:  2

In [3]:  5//5

Out[3]:  1

In [4]:  7//3.0

Out[4]:  2.0

In [5]:  -9/2

Out[5]:  -4.5

In [6]:  -9//2

Out[6]:  -5
```

---

[2] http://stackoverflow.com/a/5365702 (accessed October 19, 2020).

If either of the **operands** (the values to the left and right of the $//$ operator) is a decimal number, integer division is still done, but the result is a decimal number.

Teaser trailer: In this section, we have been discussing integers and decimals as if they are different entities in Python, whereas in mathematics 2 and 2.0 are identical. This is a hint of the programming concept called **typing**, where we define different *kinds* of values to have different properties. We address that topic in Section 5.2.7.

2.2.2 Variables

On a calculator, we often save values in memory to use later. In Python, we save values by setting them to variable names. We can then use the values by name. In line 2 of the top part of Figure 2.1, we take the result of the expression $2*2*2$ and give it a name. In doing so, if later on in our Python calculator session we make reference to that name, it is the same as referring to the result of the expression (in this case the value 8). We do something similar in line 3 for the volume of the second cube. The act of giving a name to a value is called **assigning** a variable.

In Python, we use an equal sign to do the assignment. The name of the variable goes to the left of the equal sign and the expression whose result we are assigning that variable to is to the right of the equal sign. Thus, in line 2 of the top part of Figure 2.1, the variable name is `volume_cube_1` and the value we assign `volume_cube_1` to is 8, which is the result of the expression $2*2*2$. These tasks in line 2 happen in this order: First, the expression $2*2*2$ is evaluated to obtain 8; then, the variable `volume_cube_1` is attached to that 8.

The order of these tasks is important. In most everyday life, the equal sign connotes mathematical equality or "interchangeability" or "sameness." That is *not* what is happening in assignment. Here is an example of why this difference is important. Consider this line of Python code:

```
volume_cube_1 = volume_cube_1 + 100
```

If the above line of code were a mathematical equation, it would make no sense. We could subtract the variable `volume_cube_1` from both sides and obtain the mathematical equation:

$$0 = 100$$

which is false. Thus, what is happening in the earlier line of code is not a mathematical expression. Instead, what is happening is that the right-hand-side expression `volume_cube_1 + 100` is first evaluated using whatever the *current* value of `volume_cube_1` is. So, if `volume_cube_1` were equal to 8, the right-hand-side expression evaluates to 108. *Then*, the variable `volume_cube_1` is set to that value, overwriting the old value of `volume_cube_1`. At the end of executing the above line of code, `volume_cube_1` would be set to 108, a value 100 larger than `volume_cube_1` held before the line of code was executed.

Variables in Python differ from variables in mathematics in another way. When we have a mathematical equation such as:

$$y = mx + b$$

it does not matter whether we have values defined for any of the variables. With the equation written as-is, we can manipulate the terms using the rules of algebra, for instance rewriting the equation as:

$$x = \frac{y - b}{m}$$

Variables in Python, however, do not exist until we create them through assignment. Thus, this line of code would not work:

```
y = (m * x) + b
```

unless m, x, and b were previously defined. In contrast, while the single line of code above would not work, this set of lines would:

```
In [1]:   m = 0.5
          x = 2.1
          b = -13.0
          y = (m * x) + b
```

The m, x, and b variables do not have to be assigned to a value *immediately* before y is assigned, but they have to be assigned somewhere before y is assigned.

We call a variable that "grows," as in this line of code:

```
volume_cube_1 = volume_cube_1 + 100
```

an **accumulator**, because the variable volume_cube_1 accumulates value; more is added onto the existing value. If the variable loses value, as here:

```
volume_cube_1 = volume_cube_1 - 100
```

we call the variable a **deaccumulator**. Because the same variable can gain or lose value over multiple lines of code, sometimes it is easier to just call a variable whose value we change in this way an accumulator. Note that an accumulator or deaccumulator variable is still a regular variable. But, because we use the variable in such a way that it gains or loses value, we give it a special name to highlight that use.

Consider again Figure 2.1. Let us say we changed the code a little, to the following:

```
In [1]:   volume_cube_1 = 2*2*2
          volume_cube_2 = 3*3*3
          sum_cubes = volume_cube_1 + volume_cube_2
          volume_cube_2 = 4*4*4
          sum_cubes
```

```
Out[1]:   35
```

Instead of calculating `volume_cube_1` + `volume_cube_2` and outputting the result to the screen, we save it to `sum_cubes`. Then, we change `volume_cube_2` and output `sum_cubes` to the screen.

Why is `sum_cubes` unchanged even when `volume_cube_2` is changed in the fourth line of `In [1]`? The value of `sum_cubes` does not change when `volume_cube_2` changes, because Python does not go back and reevaluate the `volume_cube_1` + `volume_cube_2` expression using the new `volume_cube_2` value. Put another way, the line that assigned `sum_cubes` is not an algebraic expression that is true as the value of `volume_cube_2` changes. Rather, the line that assigned `sum_cubes` does the calculation and makes the assignment, and from that point on, Python only knows that `sum_cubes` has a value of 35. Python does not know, after `sum_cubes` is assigned, how `sum_cubes` was calculated.

Whereas our previous code examples have contained two variables or less to the right of the equal sign, we can have any number of variables in a line of code on the right-hand side of the equal sign. We also do not need the blank spaces on either side of the equal sign. Thus, `a = 2` and `a=2` work equally well (pun intended ☺). However, the addition of blank spaces between tokens – such as numbers, operators, and the equal sign – greatly improves the readability of longer expressions, so their use is preferred.

When using Python as a calculator, we will probably use a limited number of variables, so it is probably fine to use a single letter for our variable names. But, it makes more sense to name variables something descriptive in order to make it easier to understand what the variable contains. In later chapers, our variable names will become more descriptive.

Python variable names must begin with a letter or underscore. Letters and underscores can also be used in other parts of the variable name, as can numerals.[3] Python is also case-sensitive, so the variable name `Volume` and `volume` refer to two different variables. When we have multiple words in a Python variable name, we usually separate them by underscores (as we see in the above example). We will learn about exceptions to this rule later on.

To see what the contents of a variable are, just type in the name of the variable:

```
In [1]:  a = 2
         a
```

```
Out[1]:  2
```

This method only works for one variable in the cell, the variable we last type in the cell. If we want to see the contents of more than one variable, we will need to use the `print` command. That command is introduced in Try This! 2-4 and further discussed in Section 3.2.4.

2.2.3 The Python Interpreter

As we saw earlier, to use Python as a basic calculator, there are multiple environments we can use. The two we look at here are:

---

[3] The reference for this portion, https://docs.python.org/3/reference/lexical_analysis.html#identifiers (accessed January 2, 2021), also describes nuances to these rules.

- Using a terminal window.
- A Jupyter notebook.

In Section 3.2.5, we examine a more sophisticated Python interpreter environment.

With all this talk about different environments, it may seem as if Python is really complex and maybe that there are multiple flavors of Python. This, however, is not the case. An environment is just a different way of interacting with Python. The Python interpreter is the same. It turns out some environments are more useful than others. For the task of using Python as a basic calculator, a terminal window or Jupyter notebook is a good way of interacting with Python. As we look at other science and engineering tasks we want Python to do, we may find other interpreter environments to be more useful.

Using the Interpreter in a Terminal Window

When we log in to a Windows or Mac OS X computer, what do we see? Chances are, a whole bunch of icons. Those icons represent files or programs and most of us are used to double-clicking them in order to open a file or run a program.

That is not how people used to use computers. In the olden days, the computer presented a cursor where the user would type in a command (e.g., `mkdir NewDirectory`) that told the computer what it should do (in the previous example, making a folder or directory called *NewDirectory*). This is called a **command-line interface**.

We can still do this today. To do so, we first open up a **terminal window**, which puts a window on the desktop that has a cursor awaiting a command to be typed in. Why would we want to use this way of telling a computer what to do? That is a larger topic; for now, just think of a terminal as a quick way of getting access to the Python interpreter.

See www.cambridge.org/core/resources/pythonforscientists/term/ for operating system-specific advice for starting a terminal window. To start the Python interpreter using a terminal window, after opening that window, type in a command to start Python (as opposed to clicking a menu choice by double-clicking an icon). The www.cambridge.org/core/resources/pythonforscientists/term/ page also provides operating system-specific advice for starting Python in a terminal window and the basics of using Python in a terminal window. Details are on that page, but highlights are given below.

When successful, the window will look like Figure 2.2. The three greater-than signs (> > >) on the left of the line shows we are now in the Python interpreter. Once inside the Python interpreter, we can type in whatever we want to calculate, press Enter, and Python will process what we typed in.

When we are done with all the calculations we want to make, to exit the Python interpreter, type `quit()`. On Linux and Mac OS X computers, we can also type Ctrl + d to quit. On Windows, we can also type Ctrl + z and Return to quit.[4]

To exit from the terminal window, type `exit` once we are outside the Python interpreter (at the operating system level). Typing `exit` while we are in the Python interpreter will do nothing but tell us what we need to do to leave the Python interpreter.

---

[4] https://stackoverflow.com/a/41524896 (accessed December 24, 2020).

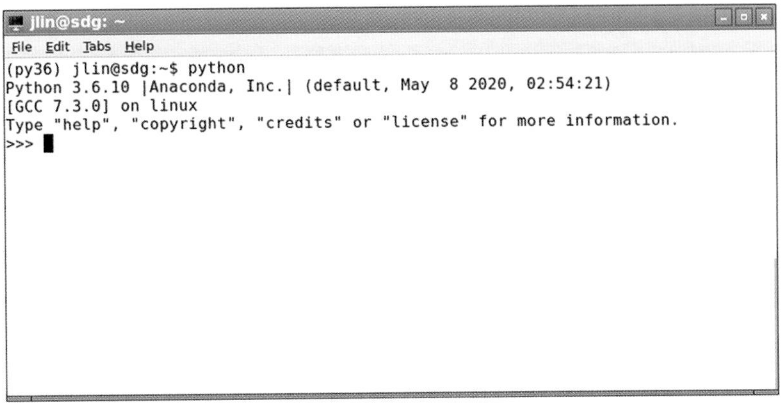

Figure 2.2 Starting the Python interpreter in a terminal window.

Using the Interpreter in a Jupyter Notebook

When we install the Anaconda distribution, we automatically install Jupyter. But, similar to the situation with starting and using Python in a terminal window, there are multiple ways of starting and using a Jupyter notebook. Generally speaking, we can open a Jupyter notebook either by using Anaconda Navigator, running the Jupyter application from a terminal window, or running the Jupyter application via the app's menu entry or icon. See www.cambridge.org/core/resources/pythonforscientists/jupyter/ for operating system-specific advice for starting Jupyter and details on the basics of using Jupyter. The present section contains highlights. Sections 3.2.4 and 3.2.5 also provide additional information on using Jupyter notebooks.

As we have seen, the standard Python interpreter run from a terminal window executes one line of Python code at a time. The Jupyter notebook allows us to process blocks of code at a time, arbitrarily edit previously executed code and rerun it, and include nicely formatted blocks of descriptive text.

Figure 2.3 shows a Jupyter notebook with the commands of Figure 2.1 typed in one cell labeled In [1]. When the notebook was initially opened, the cell next to the In [ ] prompt was empty. With the code entered in, when the cell is run, each command is executed and output from the last line of the code is output in the Out[1] cell. In multicell Jupyter notebooks, the numbering increases for each new In [ ]/Out[ ] cell. Here are some options for executing a cell(s):

- Type Ctrl + Enter : Execute the contents of the current cell.
- Type ⇧ + Enter : Execute the contents of the current cell and open an empty cell below the current cell for additional input.
- Menu command Cell ≫ Run All : Run all cells in the notebook.

Whereas basic calculators usually can only save a single value in memory, Jupyter automatically saves an open notebook. We can also manually save the current notebook via the menu command File ≫ Save and Checkpoint .

Figure 2.3 A Jupyter notebook showing the code from the example in Figure 2.1.

To make the Jupyter notebook stop executing, there are a few options. The Kernel menu contains commands to stop execution of the notebook (Kernel〉Interrupt) and restart and rerun the notebook (Kernel〉Restart & Run All). To close the window and stop the kernel, select the File〉Close and Halt menu option. Depending on the operating system, the Logout and/or Quit buttons on the notebook may close the notebook.

2.3   Try This!

In this section, we try examples similar to the ones we saw in Section 2.1. We give the questions first then provide and discuss the solutions. Try the exercises first before reading the solution and discussion ☺. This section is *not* just a rehash of the examples in Section 2.1. Through these examples (and in the Try This! sections in other chapters), we discuss additional features of using Python. Do the Try This! and carefully read through the answers and discussion.

Try This! 2-1 Calculating: Area of a Circle

What is the area of a circle with a radius of 5.2 cm? As practice, try this in a terminal-window interpreter and in a Jupyter notebook.

Try This! Answer 2-1

This terminal-window interpreter solution does not save the radius or the area as variables:

```
>>> 3.1415 * (5.2**2)
84.94616000000002
```

whereas this one does:

```
>>> radius = 5.2
>>> area = 3.1415 * (radius**2)
>>> area
84.94616000000002
```

Here are the above solutions as executed in Jupyter notebooks:

```
In [1]:  3.1415 * (5.2**2)
```

```
Out[1]:  84.94616000000002
```

```
In [2]:  radius = 5.2
         area = 3.1415 * (radius**2)
         area
```

```
Out[2]:  84.94616000000002
```

By saving the radius and area as variables, we can use those values later in the Python session, if we wish. In this example we typed in a coarse approximation to the value of π. We will see in Section 3.2.2 how to access a more accurate version of π and other common constants.

Try This! 2-2 Calculating: Kinetic Energy of a Projectile

What is the kinetic energy of a 4.3 kg projectile traveling at a speed of 1.2 m/s? Recall the kinetic energy is given as $\frac{1}{2}mv^2$ (where m is mass in kg and v is velocity in m/s).

Try This! Answer 2-2

The following Jupyter notebook solution does not save any values as variables:

```
In [1]:  1.0 / 2.0 * 4.3 * (1.2**2)
```

```
Out[1]:  3.095999999999996
```

We ensure regular division by using decimal points and zeros for the $\frac{1}{2}$ in the kinetic energy equation. Actually, we do not need to include the zeros, just the decimal points, to let Python know that the number is a decimal number:

```
In [1]:  1. / 2. * 4.3 * (1.2**2)
```

```
Out[1]:  3.095999999999996
```

In Python 3.x, because division by default does decimal division, leaving out the decimal point on the $\frac{1}{2}$ is not a problem:

```
In [1]:   1 / 2 * 4.3 * (1.2**2)
```

```
Out[1]:   3.0959999999999996
```

If we ran a Python 2.7.x Jupyter notebook (or interpreter), leaving out the decimal points can produce erroneous results:

```
In [1]:   1 / 2 * 4.3 * (1.2**2)
```

```
Out[1]:   0.0
```

because in Python 2.7.x, division of two integers using the "/" operator will default to integer division, discarding the remainder by rounding down. We only have to worry about this if we find ourselves having to run a program using this older version of Python.

Try This! 2-3 Calculating: Roots Using the Quadratic Formula

Consider the following quadratic polynomial:

$$3x^2 + 4x - 3 = 0$$

What are the roots of x? Solve for x using the quadratic formula:

$$x = \frac{-b \pm \sqrt{b^2 - 4ac}}{2a}$$

where a, b, and c are the coefficients of the quadratic polynomial and equal 3, 4, and -3, respectively, for the above polynomial.

Also use Python to save the answers and check they are correct. As practice, try this in a terminal-window interpreter and in a Jupyter notebook.

Try This! Answer 2-3

The following interpreter session will do what we want:

```
1    >>> a = 3
2    >>> b = 4
3    >>> c = -3
4    >>> x1 = (-b + (b**2 - 4*a*c)**0.5) / (2*a)
5    >>> x1
6    0.5351837584879964
7    >>> x2 = (-b - (b**2 - 4*a*c)**0.5) / (2*a)
8    >>> x2
9    -1.8685170918213299
10   >>> 3*x1**2 + 4*x1 - 3
11   -4.440892098500626e-16
12   >>> 3*x2**2 + 4*x2 - 3
13   0.0
```

We set the variables for *a*, *b*, and *c* in lines 1–3. There are two real roots of this quadratic polynomial that we calculate and save as x1 and x2 in lines 4 and 7, respectively. In lines 10 and 12, we plug x1 and x2 back into the polynomial. If the roots are correctly calculated, the result of both expressions should be zero, which they are. The very small number output in line 11 is typical of what happens with arithmetic using decimal numbers in a computer; the result will often be a little off from the analytical result.

Also, in the expressions in lines 10 and 12, no parentheses are used, but we obtain the correct results because the order of operations will ensure the exponentiation happens first, then the multiplication, then the addition and subtraction. However, using parentheses yields clearer code.

Here is the above solution as a Jupyter notebook:

```
In [1]:  a = 3
         b = 4
         c = -3
         x1 = (-b + (b**2 - 4*a*c)**0.5) / (2*a)
         x1
```

```
Out[1]:  0.5351837584879964
```

```
In [2]:  x2 = (-b - (b**2 - 4*a*c)**0.5) / (2*a)
         x2
```

```
Out[2]:  -1.8685170918213299
```

```
In [3]:  3*x1**2 + 4*x1 - 3
```

```
Out[3]:  -4.440892098500626e-16
```

```
In [4]:  3*x2**2 + 4*x2 - 3
```

```
Out[4]:  0.0
```

Try This! 2-4 Adding Up: A Moving Microbe

Pretend we have a microbe that travels the following distance each minute:

Minute	Distance (cm)
1	0
2	2
3	2.5
4	4.1
5	3.6
6	5.1

The distances given are how far the microbe has traveled during the corresponding minute. Calculate and print the total distance the microbe has traveled at each of the above times.

Try This! Answer 2-4

We could solve this by typing in an expression adding up all of the distances up to each time:

```
In [1]:  print(0)

         0

In [2]:  print(0 + 2)

         2

In [3]:  print(0 + 2 + 2.5)

         4.5

In [4]:  print(0 + 2 + 2.5 + 4.1)

         8.6

In [5]:  print(0 + 2 + 2.5 + 4.1 + 3.6)

         12.2

In [6]:  print(0 + 2 + 2.5 + 4.1 + 3.6 + 5.1)

         17.299999999999997
```

but this is a lot of repetitive typing. An easier way is to use an accumulator variable, which stores the current total distance and adds onto it. Figure 2.4 shows such a solution, which looks like (and is) more typing, but we do not have to retype any of the previous data values. Retyping breeds typos, so it is better to add on the new values to the existing total. In Chapter 6, when we discuss how to do a calculation over and over, we will also see that this feature of accumulators makes **incrementing** or growing a variable a snap.

When we execute a `print` command in a Jupyter notebook, the output does not go into an Out [ ] cell but is printed in another cell underneath the In [ ] cell. Out [ ] cells are generated when we evaluate an expression that has a result that is given back (e.g., 1 + 2 gives back 3) or when we type in the name of a variable by itself, as in:

```
In [1]:   total_distance = 0
          total_distance
```

```
Out[1]:   0
```

When we type in the name of a variable by itself, it is itself an expression whose result is the value it stores. This is why typing in the name of a variable by itself (without a `print` command) generates an Out [ ] cell.

```
In [1]:   total_distance = 0
          print(total_distance)


          0
```

```
In [2]:   total_distance = total_distance + 2
          print(total_distance)


          2
```

```
In [3]:   total_distance = total_distance + 2.5
          print(total_distance)


          4.5
```

```
In [4]:   total_distance = total_distance + 4.1
          print(total_distance)


          8.6
```

```
In [5]:   total_distance = total_distance + 3.6
          print(total_distance)


          12.2
```

```
In [6]:   total_distance = total_distance + 5.1
          print(total_distance)


          17.29999999999997
```

Figure 2.4 Solution to Try This! 2-4 using an accumulator variable.

2.4   More Discipline-Specific Practice

As mentioned in the To the Student section of the frontmatter, a key part of this book is the Discipline-Specific Try This! and Homework Problems in each chapter. Although these disciplinary-specific exercises build on the material covered in the text, they may also cover new material. Do the Try This! and Homework Problems and carefully read through the answers and discussion.

Details on obtaining these Jupyter notebooks are in the To the Student section. The notebooks for this chapter cover the following topics:

- Storing values as variables.
- Basic calculations with variables.
- Generation of output from a calculation.
- Using variables as accumulators or deaccumulators.

2.5   Chapter Review

2.5.1   Self-Test Questions

The questions and answers in Sections 2.3 and 2.4 are intended for practice and additional clarification of content. The questions in this section are designed to test our understanding of the material. Try to complete these questions without checking the book, looking at other resources, or using the Python interpreter. The answers to these Self-Test Questions are found at the end of the chapter.

Self-Test Question 2-1

What symbol is used for exponentiation in Python?

Self-Test Question 2-2

Consider the following variable names:

1. `x`
2. `2nd_area`
3. `second_area`
4. `bottom_2nd_area`
5. `speed_of_light`
6. `speedOfLight`

Are any of these names not legal in Python? If so, why?

Self-Test Question 2-3

Consider the following lines of input:

1. `3 * 5 - 7 ** 2`
2. `(3 * 5 - 7) ** 2`

3. x = 4
4. force = mass times acceleration
5. force = mass * acceleration
6. pressure = pressure + g * h

Are any of these lines of input not valid if typed in a Python interpreter? If so, why?

Self-Test Question 2-4

Is the result of the expression 3 * 5 - 7 ** 2 and (3 * 5 - 7) ** 2 the same or different? Why? Assume we are using Python 3.x.

Self-Test Question 2-5

Is the result of the expression 6 // 4 * 2 - 9 * 4 and 6 / 4 * 2 - 9 * 4 the same or different? Why? Assume we are using Python 3.x.

2.5.2 Chapter Summary

We now know how to use Python as a basic calculator! Although the material presented in this chapter is a tiny fraction of the capabilities of Python, it is still very useful. In describing how to use Python as a basic calculator, we covered these topics:

- Arithmetic operators and order of arithmetic operations

  - Look like and follow the precedence rules of algebra except multiplication and exponentiation use different symbols.
  - Division may be integer or regular decimal division, depending on the version of Python being used and/or the division symbol being used.

- Expressions

  - Consist of numbers, variables, and operators.
  - A value is returned from an expression.

- Naming, assigning, and using variables

  - Variables are words consisting of letters and numbers that are essentially names for a numerical value.
  - Variables have to be assigned before they can be used in an expression or another assignment statement.
  - Variables that gain or lose value over lines of code are called accumulators and deaccumulators, respectively. (Or, they are all called accumulators for brevity.)
  - The syntax of an accumulator has the variable name appearing on both sides of the equal sign.

- Running the Python interpreter from a terminal window.
- Running the Python interpreter from a Jupyter notebook.

2.5.3 Self-Test Answers

Self-Test Answer 2-1

The exponentiation operator is "**".

Self-Test Answer 2-2

All the names are legal except for the second one (2nd_area) which is not legal because a variable name cannot begin with a numeral.

Self-Test Answer 2-3

All these lines of input are valid in a Python interpreter except the fourth line (force = mass times acceleration), because the word "times" is not an operator and the expression as-is lacks any operators. In the fifth and sixth lines (force = mass * acceleration and pressure = pressure + g * h), these will be valid assuming the variables mass and acceleration for the fifth line and the variables pressure, g, and h for the sixth line have been previously defined.

Self-Test Answer 2-4

The expression 3 * 5 - 7 ** 2 returns -34 and the expression (3 * 5 - 7) ** 2 returns 64. The difference is because in the second expression, because of the parentheses, we first multiply 3 by 5, then subtract 7 from that, and then square the entire result. In the first expression, we first square 7, then multiply 3 times 5, and then subtract the square of 7 from the product of 3 times 5.

Self-Test Answer 2-5

The expression 6 // 4 * 2 - 9 * 4 returns -34 and the expression 6 / 4 * 2 - 9 * 4 returns -33.0. The difference is because in the first expression, we divide 6 by 4 using integer division. Integer division results in an integer and after all the rest of the operations are done, the result is an integer. In the second expression, we divide 6 by 4 using regular division. The result is a decimal number, and after all the rest of the operations are done, the result is a decimal number.

3   Python as a Scientific Calculator

In the previous chapter we saw how we can use Python as a basic calculator. Along the way, we were introduced to expressions, variables, and environments to type in and run Python code. In this chapter, we will extend our use of Python as a basic calculator to tasks associated with a scientific calculator.

3.1   Example of Python as a Scientific Calculator

A scientific calculator differs from a basic calculator in two main ways:

- A scientific calculator can use prewritten and built-in mathematical functions such as sine, logarithm, etc.
- A scientific calculator can store sets of operations and calculations, in order to perform those calculations again (often with new input values). These kinds of scientific calculators are also called programmable calculators.

Many scientific calculators today are also graphing calculators. We will discuss how to use Python for basic graphing in Chapter 4.

In biology and many other fields, exponential growth of some measurable or observable quantity with time (such as of a population of bacteria), over a period of time, is a common phenomenon. Such growth results from any process in which the rate of growth of the quantity is proportional to the quantity itself. For a population of bacteria, this kind of growth follows the equation:

$$P = Ae^{rt}$$

where P is the population, t is the time, r is a constant, and A is a constant that equals the population at $t = 0$.[1] The symbol e refers to Euler's number and is approximately equal to $2.718\,2818$.[2] If $A = 100$ (so $P = 100$ at $t = 0$ h) and $r = 0.1$ h^{-1}, what would the population be at $t = 10$ h? The following Python session would calculate this population:

---

[1] This equation for P is the solution of the differential equation:

$$\frac{dP}{dt} = rP$$

The quantity dP/dt is called a derivative. Derivatives and differential equations are described in more detail in Section 17.1.1.

[2] https://en.wikipedia.org/wiki/E_(mathematical_constant) (accessed December 26, 2020).

```
In [1]:  from numpy import exp
         A = 100.0
         r = 0.1
         t = 10.0
         A * exp(r*t)
```

```
Out[1]:  271.82818284590451
```

In the above code, the exponential function (e^x) is called exp and x is the value inside the parentheses after the exp.

In physics and other fields that deal with vector quantities, projecting a vector onto an axis is an important task. Consider the following velocity vector $\vec{v}$ that we wish to project onto the x-axis, as shown in the following figure.

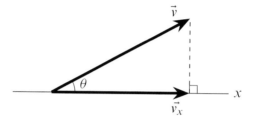

The length of $\vec{v}_x$ is a function of the cosine of the angle θ. Specifically:

$$v_x = v \cos \theta$$

The value of θ in the above figure is 28.07° or approximately 0.489 91 radians.

Figure 3.1 shows the Python code that will calculate the value of v_x. We assume $v = 2$ m/s. In Figure 3.1, we make use of the cosine function cos to calculate the cosine of the angle theta. That value is calculated by the cos(theta) expression in the fourth line of In [1]. Unlike in the bacteria population example, after we evaluate the expression v * cos(theta) in the fourth line in cell In [1], we save the result to a variable (v_x) through the use of the assignment operator (the = symbol). The last line in cell In [1], when typed in the default Python interpreter or Jupyter notebook cell, displays the contents of v_x.

3.2   Python Programming Essentials

In the examples in Section 3.1, we use Python **functions**. These functions are prewritten, and the first lines of the examples, which have the syntax:

```
from <...> import <...>
```

```
In [1]:  from numpy import cos
         theta = 0.48991
         v = 2.0
         v_x = v * cos(theta)
         v_x
```

```
Out[1]:  1.7647504227331032
```

Figure 3.1 An example of using functions to project a vector along the *x*-axis.

give us access to those functions. In this section, we will explore various programming aspects of these examples. The goal of this section is to discover how do we access and use prewritten functions and how do we write our own functions.

3.2.1 Using Prewritten Functions

Before discussing this topic in detail, we first dissect the example in Figure 3.1. We make use of the cosine function, cos, in the fourth line of In [1], by **calling** cos. What does "calling" mean? When we type the name of the function (here cos) plus parentheses after the name of the function, the Python interpreter executes the function we have named using the contents (if any) between the parentheses as input into the function. The function processes that input and then returns a result. In the fourth line in cell In [1], the input into the cos function is the value of the variable theta. The output is the result of the cosine of the value of theta. That output is multiplied by the value of v, and the result of that expression is saved as v_x.

We can think of functions as "black boxes," which accept input, process that input, and return a value that can then be set to a variable or used in an expression. Diagramatically, cos(theta) can be understood as shown.

To use a function, we do not have to know what is happening inside the box; when we call the function, we automatically make use of all the machinery in the box and obtain the output from the box, the **return value**.

In this chapter, we only look at functions that have a single item of input and a single return value. Python functions can actually take multiple items of input and return multiple items, but for now we will restrict ourselves to the one item case. We discuss multiple input items in Section 4.2.1 and multiple return values in Section 19.1 (the latter discussing lines 8–15 and 46 of Figure 19.1).

3.2.2   Importing Modules and Using Module Items

Some functions in Python are built in and are automatically made available to the programmer when the Python interpreter or Jupyter notebook is running. The `print` function from Try This! 2-4 is one such function. The `del` function is another. The `del` function deletes a variable so that Python does not recognize it, as shown in the following example:[3]

```
In [1]:   a = 3
          a

Out[1]:   3

In [2]:   del(a)
          a

          NameError    Traceback (most recent call last)
          <ipython-input-3-60b725f10c9c> in <module>()
          ----> 1 a

          NameError: name 'a' is not defined
```

The result of `In [2]` is a Python error message stating to us that the variable a is not defined. We see in cell `In [2]` that when we call the function, we enter our input in the parentheses, and the function operates on the input (in this case the variable a). The `del` method does not have a return value, as all it does is delete the variable we specify in the parentheses. Thus, setting the output of `del` to a variable (`b = del(a)`, for instance) will not work.

The vast majority of functions, however, are not built in. They are not automatically available to us as soon as we start up the Python interpreter or Jupyter notebook. This is the case for most mathematical functions. In order to access these functions, we have to first **import** the function. This is the purpose of the first line in cell `In [1]` in Figure 3.1.

What happens if we do not first import a function? If we try to execute the following block of code without first running the first line in cell `In [1]` in Figure 3.1 we get:[4]

```
In [1]:   cos(3.1415/2)

          NameError    Traceback (most recent call last)
          <ipython-input-1-16e600e1e264> in <module>()
          ----> 1 cos(3.1415/2)

          NameError: name 'cos' is not defined
```

---

[3] The error message output is edited for clarity.
[4] The error message output is edited for clarity.

The output from In [1] informs us that Python does not recognize the cos function. This is because the cos function is not built in to the interpreter or Jupyter notebook but rather is part of a **module** called NumPy (the name used in the interpreter or notebook is "numpy", without capitalization). For now, we can think of a module as a collection of functions and pre-assigned variables (such as mathematical constants). In order to use cos we first need to import it. If we add an import line before the call to cos, everything works fine:

```
In [1]:  from numpy import cos
         cos(3.1415/2)
```

```
Out[1]:  4.632679487995776e-05
```

In the cos examples above, we imported the function by typing in:

from *<module>* import *<function>*

where *<module>* is the name of the module the function resides in and *<function>* is the name of the function we are interested in being able to use from the *<module>*.

But what if we wanted to make use of more than one function in a module? Would we have to use from ... import ... every time we wanted access to the function? Thankfully, no ☺. The following import command:

import *<module>*

will make all the functions of the module available by typing *<module>*.*<function>*. That is:

```
In [1]:  import numpy
         numpy.cos(3.1415/2)
```

```
Out[1]:  4.632679487995776e-05
```

is the same as:

```
In [1]:  from numpy import cos
         cos(3.1415/2)
```

```
Out[1]:  4.632679487995776e-05
```

If we want to save a little typing, we can use:

import *<module>* as *<alias>*

to make an even shorter reference to the module. The *<alias>* can be any name that would be a valid Python variable name. That is:

```
In [1]:   import numpy as np
          np.cos(3.1415/2)

Out[1]:   4.632679487995776e-05
```

is the same as:

```
In [1]:   import numpy
          numpy.cos(3.1415/2)

Out[1]:   4.632679487995776e-05
```

Modules do not only contain functions but may also contain variables set to certain values. For instance, the NumPy module has a variable called `pi` that is set to the value of π. When we import a module using `import`, we can reference the variables defined in the module in a way similar to how we reference functions, by using the period notation:

```
In [1]:   import numpy
          numpy.pi

Out[1]:   3.141592653589793
```

Notice that when we make use of `pi`, we do not put parentheses after the variable name whereas when we use a function (like `cos`), we do put parentheses after the function name. This is because with the function, we are *calling* the function. Python conveys this through appending the pair of parentheses to the function name. In the case of `pi`, because it is a variable, we do not call it. Remember that calling a function means giving input to a function, running the function, and receiving whatever is returned.

Finally, if we want to learn more about a function, besides reading the online Python documentation, we can use the built-in function `help` for some basic information or the NumPy function `info` for more detailed information. In the above case of NumPy's `cos` function, assuming NumPy has been imported as `np`, typing in `help(np.cos)` or `np.info(np.cos)` in the Python interpreter or Jupyter notebook will provide information about the `cos` function and how to use it. Additional information on using these functions in a terminal window is found at www.cambridge.org/core/resources/pythonforscientists/term/.

3.2.3   Writing and Using Our Own Functions

So far, we have talked about how to access functions someone else has written. What if we want to use functions we have written? Here we describe two ways of doing so.

First, we can define the function in the interpreter session. Figure 3.2 shows an example. It defines and uses a function `percent_to_decimal`, which converts a percentage into its

```
In [1]:  def percent_to_decimal(input_percent):
             output = input_percent / 100.0
             return output

         percent_to_decimal(42)
```

```
Out[1]:  0.42
```

Figure 3.2 Example of defining a function in the Python interpreter.

decimal or ratio form. As we see in the first line of `In [1]`, the first line when defining a function begins with `def`, followed by the function name, then the **parameter list**. The `def` line ends with a colon.

The parameter list is the contents between the parentheses in the `def` line. In Figure 3.2, there is one parameter named `input_percent`. All references in the body of the function definition with the name `input_percent` refer to whatever value is passed into the function when it is called. That is, when a line like this is executed:

```
percent_to_decimal(42)
```

the value `42` is used in the function definition in Figure 3.2 wherever we see the variable name `input_percent`. Because we can call this function with any value in between the parentheses, and we do not know what that value is until the function is actually called, we have to give it a variable name as a placeholder in the function definition, to be filled in when the function is called. Otherwise, within this function definition, we will not be able to make reference to an input value.

Is there anything special to the name `input_percent`? No. We can use any valid Python variable name, as long as references to that parameter name in the function's body match. Thus, we can rewrite out `percent_to_decimal` in Figure 3.2 as:

```
In [1]:  def percent_to_decimal(my_value):
             output = my_value / 100.0
             return output

         percent_to_decimal(42)
```

```
Out[1]:  0.42
```

and the function would work exactly the same.

The `def` statement in Figure 3.2 (and the lines of code underneath the `def` statement that are indented by four spaces) only *defines* the function. The block of code does not *run* or use the function. That is to say, if we typed only this:

```
In [1]:  def percent_to_decimal(my_value):
             output = my_value / 100.0
             return output
```

and left out the `percent_to_decimal(42)` line in Figure 3.2, the function `percent_to_decimal` would never run and the calculation `my_value / 100.0` would never be made. The act of defining a function only means we have created this "box" of code we can refer to and run. It just defines the name of the function and says this series of commands (the lines of code in the body) are in this box with the name of the function. Only when the function is called is the code in the function definition actually run. When called, the code in the body of the function is executed one line at a time (i.e., the line after the `def` line, then the line after, in Figure 3.2), from beginning to end through the body (though beginning in Section 6.2.4, we will see how we can change the order lines of code are executed).

When the function is called (run), the variables in the body of the function are defined and used. However, these variables are not available to refer to by name outside the function. The `my_value` and `output` variables from the `def` statement immediately above only exist in the body of the function definition. If we tried to print one of those values from outside the function definition, Python would not recognize the variable.

If we want to access the contents of a variable that is defined inside the function's body from outside the function, we have to return it from our function. The item returned from a function is called the function's return value. In our example in Figure 3.2, the return value of a function is given by `return` followed by whatever is being returned. In the above example, the variable `output` is returned, but the return value can itself be an expression. For instance, the code below works exactly the same as in Figure 3.2:

```
In [1]:  def percent_to_decimal(input_percent):
             return input_percent / 100.0

         percent_to_decimal(42)
```

```
Out[1]:  0.42
```

and saves us an extra line of typing. We make the `100.0` a decimal to ensure we never have integer division operating on `input_percent`.

Typing in a function in an interpreter is easy and straightforward, but as soon as we exit the interpreter, the function definition is lost. Thus, it often makes sense to instead define a function in a file and then use `import` to give us access to the function.

For instance, we could put the function definition portion of Figure 3.2 in a file called *myfuncs.py*, as seen in Figure 3.3. Then, as long as *myfuncs.py* is in the same directory as the

```
def percent_to_decimal(input_percent):
    return input_percent / 100.0
```

Figure 3.3 Contents of *myfuncs.py*.

working directory for the Python interpreter,[5] we can import *myfuncs.py* and use its contents (i.e., use the function `percent_to_decimal`):

```
In [1]:  import myfuncs
         myfuncs.percent_to_decimal(42)
```

```
Out[1]:  0.42
```

Earlier, however, we used `import` for importing modules. Why do we use it here on our file of Python code? What this shows us is that a module is just a file containing Python commands. If we have a bunch of function, variable, etc., definitions, we can put them into a file and have a module we can import and use. We will have created a library of our own functions!

3.2.4 A Programmable Calculator

In Section 3.2.3, we saw how we could write (and use) our own functions in our Python calculator. We also saw that all we had to do was to copy-and-paste what we typed at the interpreter prompt and put it in a file, then import that file to use the functions defined in the file (i.e., module). We will now see that we can use this copy-and-paste methodology (with minor adjustments) to "record" our entire calculator session, including the setting of variables and the use or calling of functions, so that we can run the calculator session again. This record of what would have been an interactive calculator session is often referred to as a "**script**" and the writing of those lines of code "**scripting**." The neat thing about scripting is that we can use the script over and over again, without typing it in again. Note sometimes people use "script" and "program" interchangably, although programs are usually longer and more complex than scripts.

In Section 2.2.3, we learned how to access the Python interpreter using a terminal window and a Jupyter notebook. In this section, we look at how to do scripting using these same two environments.

---

[5] The more complete answer is that *myfuncs.py* has to be on the path specified by the *PYTHONPATH* environment variable, but for our purposes right now, as long as *myfuncs.py* is in the same directory we started `python` in, it will work. See the Python documentation description of *PYTHONPATH* for more information. We provide an up-to-date link to the tutorial at www.cambridge.org/core/resources/pythonforscientists/refs/, ref. 4.

A Programmable Calculator Using Terminal Windows

Assume we have four percentages that we wish to convert to decimals using our `percent_to_decimal` function. That function is defined in *myfuncs.py* (Figure 3.3). In a terminal-window interpreter session we would type this in:

```
>>> import myfuncs
>>> myfuncs.percent_to_decimal(42)
0.42
>>> myfuncs.percent_to_decimal(2)
0.02
>>> myfuncs.percent_to_decimal(83)
0.83
>>> myfuncs.percent_to_decimal(12)
0.12
```

But if we exited the interpreter and wanted to later redo these four recalculations, we would have to type it all in again. So, we copy-and-paste these lines into a file, as seen in Figure 3.4. We call this file *myscript1.py*.

When we say, "copy-and-paste these lines into a file," we do not mean into a word processor like Microsoft Word. Instead, we mean to put the lines into a plain text file. These files contain only letters, numbers, and some other symbols (e.g., punctuation, a **newline character**, etc.). They do not contain formatting (e.g., italics) of any sort and if we save Python code into a formatted file type like *.doc*, the code will not work properly. Instead of Word, we use a text editor. There are many kinds of text editors available on each operating system. We list a few at www.cambridge.org/core/resources/pythonforscientists/text-editors/.

Text files are often suffixed *.txt* but they do not have to be. Code files are text files whose suffix corresponds to the language the program is written in. Python code files are suffixed *.py*.

To run the script from a terminal window, type in `python`, a space, and the name of the file. For our file *myscript1.py* that would be:

```
python myscript1.py
```

If we do this, we find that nothing seems to happen. The results of the `percent_to_decimal` calls are not printed to the screen. This is one of the adjustments we have to make when writing a script instead of typing in commands at the interpreter prompt. In the interpreter, when we type the name of a variable or call a function, the value of the variable or the function's return value are automatically displayed to the screen. In a script, this does not occur.

Figure 3.4 Contents of *myscript1.py*.

```
import myfuncs
myfuncs.percent_to_decimal(42)
myfuncs.percent_to_decimal(2)
myfuncs.percent_to_decimal(83)
myfuncs.percent_to_decimal(12)
```

```
import myfuncs
print(myfuncs.percent_to_decimal(42))
print(myfuncs.percent_to_decimal(2))
print(myfuncs.percent_to_decimal(83))
print(myfuncs.percent_to_decimal(12))
```

Figure 3.5 Contents of *myscript2.py*.

In order to see the contents of the return value, we have to pass the function call as an input to the built-in `print` function, as seen in the *myscript2.py* file in Figure 3.5.

When we run the revised script, the expected output is produced (the `$` is the prompt showing where we type in commands in the command-line interface to the computer's operating system, such as is also seen in the first line of Figure 2.2):

```
$ python myscript2.py
0.42
0.02
0.83
0.12
```

When we run the script in the terminal as above, we automatically leave the default Python interpreter at the end of running the script. To stay in the default interpreter at the end of running the script, we add an "`-i`" between `python` and the script filename, as in:

```
$ python -i myscript2.py
0.42
0.02
0.83
0.12
>>>
```

Finally, if we do not want to keep the `percent_to_decimal` function in a separate file from our script, that is fine too. We can define functions and use them in the same script file, as well as setting variables, etc. By putting the function in our script, we also do not need to import the myfuncs module. Figure 3.6 shows this revised script, *myscript3.py*.

A Programmable Calculator Using Jupyter Notebooks

In a Jupyter notebook, the task of saving lines of Python code into a file and the task of running those lines of code in the Python interpreter are conducted in a single document, the notebook. Thus, instead of creating a .*py* file that is separate from the interpreter (as in Figure 3.6), we put those lines of code into one or more notebook cells. Figure 3.7 shows all of Figure 3.6 in a single Jupyter notebook cell. When we run the cell, the output is generated as shown.

We can also split up the contents of Figure 3.8 into multiple cells with the definition of `percent_to_decimal` in one cell and the use of `percent_to_decimal` in a second cell.

```
def percent_to_decimal(input_percent):
    return input_percent / 100.0

print(percent_to_decimal(42))
print(percent_to_decimal(2))
print(percent_to_decimal(83))
print(percent_to_decimal(12))
```

Figure 3.6 Contents of *myscript3.py*.

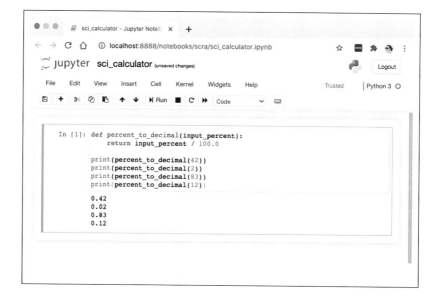

Figure 3.7 Contents of *myscript3.py* in a single Jupyter notebook cell.

As expected, the first cell produces no output because all it does is define the function (to get output we have to call the function, at the very least).

We can also import other modules, such as we defined in *myfuncs.py* in Figure 3.3 and make use of the functions and variables defined in those modules. Figure 3.9 shows an example of this use, which is a copy of what we saw in Figure 3.5.

Finally, we can export a Jupyter notebook into another format via the ⎢File⎟⟩⎢Download as⎟ menu command. The three most useful formats for that command are: as a regular Python code plain text *.py* file, as an HTML file (i.e., web page), and as a PDF file. In the first case, the result is a Python file we can run in a terminal window or any of the environments described in Section 3.2.5.

3.2.5 Python Interpreter and Code-Writing Environments for More Complex Programs

We have seen how to write and run our own programmable calculator scripts in a terminal window or in a Jupyter notebook. Both environments work fine when we are using Python

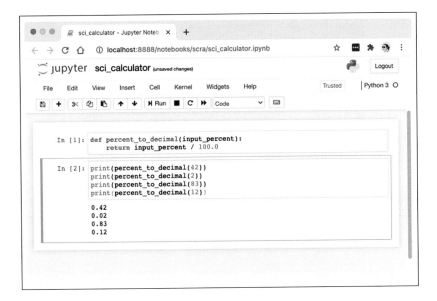

Figure 3.8 Contents of *myscript3.py* in two Jupyter notebook cells.

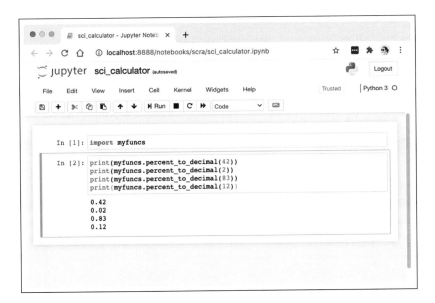

Figure 3.9 Approach of *myscript2.py* (from Figure 3.5) in two Jupyter notebook cells.

as a basic or scientific calculator. There comes a point, however, when the complexity of our programmable calculator scripts become so long and involved that writing and running them may be easier to do in an **Interactive Development Environment (IDE)**. An IDE is a program that brings together the jobs of writing and running a script into a single application. Often an IDE will have a window where we can write a script and another where the Python interpreter

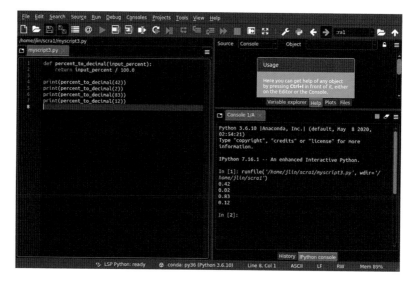

Figure 3.10 *myscript3.py* in the Spyder IDE, showing the IPython interface, after being run.

is running. IDEs also have graphical tools to help us find errors in our script or program as well as shortcut buttons or menu commands to execute common actions (e.g., running or stopping a program, etc.).

In this section, we look at the Spyder IDE which comes with the Anaconda distribution and is on by default.[6] Remember, although IDEs are useful, we do not have to use an IDE to write Python code or run a Python script. A terminal window or Jupyter notebook works fine. Ultimately, it comes down to which environment (or environments; we can use one environment for certain kinds of tasks and another environment for other tasks) makes it easier for us to get our work done. All these options are free and thus easy to try out.[7]

As with Jupyter notebooks, we can open Spyder either by running the Spyder application from a terminal window, by running the Spyder application via the app's menu entry or icon, or by using Anaconda Navigator. We provide operating system-specific instructions using these methods, as well as details on using Spyder, at www.cambridge.org/core/resources/pythonforscientists/spyder/. Here, we provide highlights.

The left-hand pane of Figure 3.10, shows an open *myscript3.py* file (the contents of which are shown in Figure 3.6) in Spyder. In the lower right-hand pane, we have an IPython interpreter interface. This behaves very similarly to the "three greater-than signs" Python interpreter, except it has the In [ ], etc., prompts similar to a Jupyter notebook.

---

[6] https://docs.continuum.io/anaconda/user-guide/tasks/integration/spyder.html (accessed June 24, 2017).

[7] Another IDE is PyCharm and we would be remiss if we did not mention it. We provide an up-to-date link to the IDE's website at www.cambridge.org/core/resources/pythonforscientists/refs/, ref. 7.

The menu command Run⟩⟩Run or the green arrow button on the toolbar will run the Python code file with the results showing in the interpreter pane. The lower right-hand pane in Figure 3.10 shows the results of running the open code file. We can also type code to execute in the IPython interpreter interface, apart from the code from the left-hand pane that was run.

If the script, for some reason, freezes while running, or we need to manually stop the script from running, type Ctrl + c while in the IPython console. More details about the Spyder IDE are available at the IDE's website. We provide an up-to-date link to the site at www.cambridge.org/core/resources/pythonforscientists/refs/, ref. 11.

3.3 Try This!

Similar to Section 2.3, this section provides additional examples that are accompanied with solutions and comments on how to program in Python and use Python as a scientific calculator. This includes using, writing, and importing functions. Do the exercises and carefully read through the answers and discussion.

Try This! 3-1 Using Functions: Projecting a Vector onto the *x*-Axis

Consider Figure 3.11, which shows the vector $\vec{v}$ (from Section 3.1) that we wish to project onto the *x*-axis. Again, the value of θ is 28.07° or approximately 0.489 91 radians. What is the value of $\vec{v}_x$? Assume $v = 4.5$ m/s.

Try This! Answer 3-1

The only thing we have to change from the solution in Section 3.1 is the value of the variable v:

```
In [1]:   from numpy import cos
          theta = 0.48991
          v = 4.5
          v_x = v * cos(theta)
          v_x
```

```
Out[1]:   3.9706884511494822
```

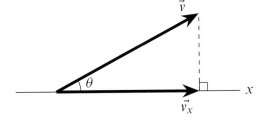

Figure 3.11 Projecting a vector onto the *x*-axis, in Try This! 3-1.

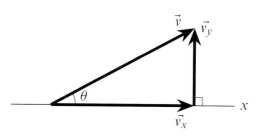

Figure 3.12 Projecting a vector onto the y-axis, in Try This! 3-2.

By default, the functions `sin`, `cos`, and `tan` in NumPy assume the angles being passed in are in radians, not degrees.

Try This! 3-2 Using Functions: Projecting a Vector onto the y-Axis

Consider Figure 3.12, which shows vector $\vec{v}$, with the same angle θ as in Try This! 3-1 (i.e., 28.07° or approximately 0.489 91 radians). What is the magnitude $\vec{v}_y$? Assume $v = 7.6$ m/s.

Try This! Answer 3-2

Recall that v_y in Figure 3.12 will be given by:

$$v_y = v \sin \theta$$

So, we need the sine function from NumPy, and thus, we need to import and use `sin`. The variable v will also change. And, as we are calculating the component v_y, we will change the name of the variable we are calculating to v_y instead of v_x:

```
In [1]:  from numpy import sin
         theta = 0.48991
         v = 7.6
         v_y = v * sin(theta)
         v_y
```

```
Out[1]:  3.5761532199404615
```

Try This! 3-3 Using Functions: Projecting a Vector onto the x- and y-Axes

Consider Figure 3.13, showing another vector $\vec{v}$, where $\alpha = 53.52°$. What are the magnitudes $\vec{v}_x$ and $\vec{v}_y$? Assume $v = 8.2$ m/s. Remember that π radians equals 180°, so to convert from degrees to radians, we multiply the number of degrees by $\frac{\pi}{180}$.

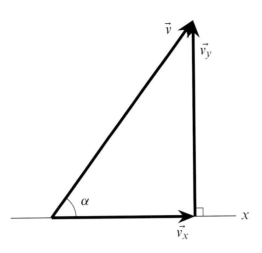

Figure 3.13 Projecting a vector onto the *x*- and *y*-axes, in Try This! 3-3.

Try This! Answer 3-3

We need both the sine and cosine functions from NumPy. We also recall from Section 3.2.2 that NumPy contains the value of π as a module variable. This code solves the problem:

```
In [1]:   import numpy as np
          alpha = 53.52 * np.pi / 180.
          v = 8.2
          v_x = v * np.cos(alpha)
          v_y = v * np.sin(alpha)
          v_x
```

```
Out[1]:   4.8752456426256483
```

```
In [2]:   v_y
```

```
Out[2]:   6.5933284404813035
```

In this solution, instead of importing the `sin` and `cos` functions by name using the `from` ... `import` ... syntax, we import NumPy by the alias `np` and then refer to the elements we want from that module (i.e., `pi`, `sin`, and `cos`) using the alias (i.e., as `np.pi`, `np.sin`, and `np.cos`, respectively). We make use of `pi` to convert α from degrees into radians, since the NumPy `sin` and `cos` functions assume `alpha` is in radians.

Try This! 3-4 Using Functions: pH

The pH of a substance is defined as:[8]

$$pH = -\log_{10}\left[H^+\right]$$

where $\left[H^+\right]$ is the hydrogen ion concentration in mol/liter. Calculate the pH of a solution whose hydrogen ion concentration is 2.1×10^{-6} mol/liter. Hint: The NumPy function to calculate the base-10 logarithm is `log10`.

Try This! Answer 3-4

This code solves the problem:

```
In [1]:   import numpy as np
          pH = -np.log10(2.1e-6)
          pH
```

```
Out[1]:   5.6777807052660805
```

In the second line of In [1] above, the value of what is being passed into the call of the function `log10` is expressed in scientific notation. This is the same as typing in the number as a decimal:

```
In [1]:   pH = -np.log10(0.0000021)
          pH
```

```
Out[1]:   5.6777807052660805
```

Note that taking the negative of the base-10 logarithm can be done by putting a minus sign (on the keyboard a hyphen) in front of the call to `log10`. While `(-1) * np.log10(2.1e-6)` also works, there is no need to do this multiplication to take the negative of the function call's result.

Try This! 3-5 Using Functions: Projectile Motion

Consider a cannon pointing upwards so its barrel makes a 45° angle with the level ground surface. If a projectile is shot out at 10 m/s, how far from the cannon will the projectile hit the ground? Assume no air resistance and the ground is always level. Also assume the gravitational acceleration g is 9.81 m/s^2.

Hints: Remember the following kinematic equations. First, for velocity at time t, assuming we know the velocity at $t = 0$ is v_0 and the acceleration is constant a:

$$v_t = v_0 + at$$

---

[8] Dickerson et al. (1984, p. 163).

Next, the position x at time t for no acceleration is:

$$x_t = x_0 + v_0 t$$

assuming x_0 is the position at $t = 0$.

Try This! Answer 3-5

We solve this problem in two parts. First, we use the first kinematic equation to calculate how long the projectile is in the air. The projectile goes up with an initial velocity equal to the y-component (assuming the y-axis is in the up–down direction) of the initial velocity vector. The time the projectile takes to reach zero y-velocity (the top of its trajectory) is half the total time it is in the air. Thus, the total time is:

```
In [1]:  import numpy as np
         angle = np.pi / 4.0
         g = -9.81
         v_y = 10.0 * np.sin(angle)
         total_time = 2.0 * ((0.0 - v_y) / g)
         total_time
```

```
Out[1]:  1.4416040391163047
```

If we assume upwards is positive y, then g is negative, because gravitational acceleration pulls down. The total time of the projectile in the air is 1.44 s. Second, we use the second kinematic equation to calculate the distance traveled, using the x-component of the initial velocity. Assuming we start at $x = 0$:

```
In [2]:  v_x = 10.0 * np.cos(angle)
         total_distance = v_x * total_time
         total_distance
```

```
Out[2]:  10.19367991845056
```

The total distance traveled by the projectile is 10.19 m. Note that a single import of np works for the entire Python interpreter session, here meaning both In [ ] cells above are in the same Jupyter notebook. We do not have to keep reimporting the module.

Try This! 3-6 Write a Function: Convert Pounds to Kilograms

Pretend we are running an experiment that processes weight measurements of wild rhesus monkeys. The values that were collected by the field staff are all in pounds, but we need the values in kilograms. Write a function `weight_to_kg` that takes a single input, the weight in pounds, and returns a single value, the weight in kilograms. Provide an example that shows the function works. Hint: There are 2.204 623 pounds in one kilogram.[9]

Try This! Answer 3-6

This Jupyter notebook cell will define the function:

```
In [1]:  def weight_to_kg(in_pounds):
             return in_pounds / 2.204623
```

and this set of cells shows the function works:

```
In [2]:  weight_to_kg(1.0)
```

```
Out[2]:  0.4535922921968971
```

Try This! 3-7 Write and Import a Function: Convert Pounds to Kilograms

Take the function written for Try This! 3-6 and place it in its own Python script. Then, write another script that imports the function for use in that second script. In that second script, use the function in the first script to convert two different values in pounds to kilograms and print out the answer.

Try This! Answer 3-7

We start out by putting the following code into the file *conversions.py*:

```
1  def weight_to_kg(in_pounds):
2      return in_pounds / 2.204623
```

In a separate file (we will call it *calculate.py*), we put the import and calls of the `weight_to_kg` function. The code in that file is:

```
1  from conversions import weight_to_kg
2  print(weight_to_kg(1.0))
3  print(weight_to_kg(1.5))
```

When *calculate.py* is run, the following values are output:

```
0.4535922921968971
0.6803884382953457
```

---

[9] Here, the term "weight" is used colloquially for mass rather than the force due to gravity. Thus, "pounds" here refers to "pounds-mass" rather than "pounds-force."

Note that we could also have imported the conversions module (recall that any Python script is a module) and used its contents in *calculate.py* the following way:

```
import conversions
print(conversions.weight_to_kg(1.0))
print(conversions.weight_to_kg(1.5))
```

We can also save the return values from the calls to `weight_to_kg` in variables and print the values of those variables:

```
from conversions import weight_to_kg
first_weight = weight_to_kg(1.0)
second_weight = weight_to_kg(1.5)
print(first_weight)
print(second_weight)
```

If we are using a text editor to write the scripts, we create the *conversions.py* and *calculate.py* files directly. If we are writing the *conversions.py* code in a Jupyter notebook, we first need to export the contents into a file called *conversions.py* using the process described in the Jupyter notebook section of Section 3.2.4.[10] Regardless of how we create *conversions.py*, that file should be in the same directory as *calculate.py*. Otherwise, *calculate.py* will have a hard time finding *conversions.py* to import it.

3.4 More Discipline-Specific Practice

The Discipline-Specific Jupyter notebooks for this chapter cover the following topics:

- Importing modules and using module items (functions and variables).
- Writing and using our own functions.
- The use of the `del` function.
- The use of the `print` function.

3.5 Chapter Review

3.5.1 Self-Test Questions

Try to do these without looking at the book or any other resources or using the Python interpreter. Answers to these Self-Test Questions are found at the end of the chapter.

---

[10] There is a mechanism to directly import a Jupyter notebook, but the process is a bit advanced. More information is available in online documentation. We provide an up-to-date link to the pertinent web page at www.cambridge.org/core/resources/pythonforscientists/refs/, ref. 5.

Self-Test Question 3-1

What is a function? Please provide a definition and describe its syntax when calling a function. How do we specify input into a function? What can we do with output from a function?

Self-Test Question 3-2

How do we import a module? A function in a module? Describe all the different ways we can gain access to a function in a module and provide examples.

Self-Test Question 3-3

What is a code file? What does it contain? How can we use it to run a program in Python? Please provide an example.

Self-Test Question 3-4

Consider the following lines of code. Describe what these lines do and what will happen if the lines for each subpart (and only those lines) are typed into the Python interpreter? If something is wrong, say why it is wrong.

1.
```
from gene_functions import countCs
```
2.
```
import numpy
value = sin(2.56)
```
3.
```
from numpy import np
```
4.
```
def kinetic_energy():
    return 0.5 * m * (v**2)
```
5.
```
def kinetic_energy():
    output = 0.5 * m * (v**2)
    return output
```
6.
```
import numpy as np
```
7.
```
def force_for_1kg_mass(acceleration):
    return 1.0 * acceleration
weight = force_for_1kg_mass(9.81)
print(weight)
```

Self-Test Question 3-5

Describe how terminal windows, Jupyter notebooks, and interactive development environments differ from one another as environments in which to use Python as a programmable calculator.

3.5.2 Chapter Summary

We now know how to use Python as a really fancy programmable scientific calculator. In doing so, we covered these topics:

- The `print` function displays the contents passed in as input onto the screen.
- The `del` function deletes a variable.
- How to use prewritten functions

    - Understanding that a function is a "black box" that accepts input value(s) and returns a value(s).
    - Calling a function in a Python interpreter session or a Python program/script.

- Importing and using a module

    - Using `import`, `import ... as ...`, and `from ... import ...` to import modules or parts of modules.
    - Referencing functions, variables, and other items that are part of a module.

- Writing our own functions

    - Using the `def` command.
    - The parameter list is the list of inputs into a function.

- Different ways of using Python as a programmable calculator

    - We can save Python code in a text file that ends in *.py*.
    - Using terminal windows to access a Python code file and run the code in the Python interpreter.
    - Using Jupyter notebooks to save and run Python code.
    - Using interactive development environments such as Spyder to save and run Python code.

3.5.3 Self-Test Answers

Self-Test Answer 3-1

A function is a "black box" that accepts an input(s) and returns a value(s). When calling a function we provide a name followed by a set of parentheses. Inside the parentheses we list the items of input into the function. If there are no items of input into the function, we keep the parentheses, but with nothing in between the delimiters. Output from a function is returned by the call: we can set that output to a variable or use it in an expression by typing the call itself. In essence, the call itself represents the return value.

Self-Test Answer 3-2

We import a module and its pieces using the `import` command, but there are a variety of kinds of `import` statements, each of which deals with a specific use case. To import a module, type `import` and then the module name. To import a module but then allow that module to be referred to by an alias, type `import`, the module name, `as`, then the alias name. To import only a single function in a module, type `from`, the module name, `import`, then the function name.

Here are some examples of importing using the NumPy module:

```
1  import numpy
2  import numpy as np
3  from numpy import exp
```

and some examples of calling the `exp` function (on a value of 1.0) in NumPy:

```
1  import numpy
2  numpy.exp(1.0)
3
4  import numpy as np
5  np.exp(1.0)
6
7  from numpy import exp
8  exp(1.0)
```

Self-Test Answer 3-3

A code file is a text file that contains Python code: variables, expressions, function definitions and calls, etc. The code file is a Python program. Put another way, for the most part, everything we can type into a Python interpreter as commands to execute we can type into a code file and have the Python interpreter execute the contents of that file. Here is an example of a code file that calculates, saves, and prints out e^1:

```
1  from numpy import exp
2  value = exp(1.0)
3  print(value)
```

Self-Test Answer 3-4

1. This line of code looks for the module `gene_functions` and finds the function (or another entity like a variable) called `countCs` and imports it, making `countCs` available to the current program.
2. This code imports the NumPy package but will return an error because it uses the `sin` function without having first imported the function. The `sin` function is not built in to the Python interpreter and so must be imported. We could fix the code by adding `numpy` and a period in front of `sin`:

```
import numpy
value = numpy.sin(2.56)
```

or by changing the import statement:

```
from numpy import sin
value = sin(2.56)
```

3. The NumPy module does not contain a function, variable, etc., named `np`, and so this line of code returns an error.
4. This function definition returns an error because the variables `m` and `v` are not defined. They either need to be passed in as parameters (the parameter list is empty in this function definition) or the variables need to be defined earlier as module variables, such as:

```
m = 10.1
v = 1.2
def kinetic_energy():
    return 0.5 * m * (v**2)
```

5. Same as in the problem above. Saving the result of `0.5 * m * (v**2)` to a variable and returning the value of that variable does not solve the problem that `m` and `v` are not defined.
6. Correctly imports the NumPy module as the alias `np`.
7. Correctly defines a function called `force_for_1kg_mass`. This function takes one input parameter (`acceleration`). The code calculates the force needed for the given acceleration on a 1 kg mass by calling the function, saves that result to the variable `weight`, and prints out that value to the screen.

Self-Test Answer 3-5

All Python programming environments do two things: enable us to write and save Python code and enable us to run the code we have written. When we use terminal windows as our environment, we do not have many "bells and whistles" in terms of the visual presentation of our code and ways of running the code, but it runs fast and is simple. The Jupyter notebook environment gives us many tools for customizing the look and feel of our code and the description of that code, and is very interactive, but it takes longer to set up and does not lend itself as easily to writing a module that will be used by other modules or programs. An interactive development environment provides tools to help a programmer (e.g., ways of presenting the code, tools to help us write code more quickly, the ability to easily step through a program) and is best for developing larger Python projects.

4 Basic Line and Scatter Plots

After making calculations, a second major task computers can help scientists and engineers with is making plots of data and calculations. In this chapter, we examine how to make a basic line and scatter plot. In Chapter 5, we discuss how to customize a basic line and scatter plot to make it more useful and readable.

4.1 Example of Making Basic Line and Scatter Plots

Here we look at how to make two basic x–y plots: a line plot (where the points are connected together) and a scatter plot (where the points are not connected together). Consider the following fictitious list of measurements of the amount of power recorded by a sensor at various points in time during the course of a day:

Decimal hour in day	Power (mW)
9.25	2.54
11	4.10
13.5	1.21
15	3.90
15.75	4.00

A decimal hour representation of 1:30 p.m. is 13.5.

Figure 4.1 shows code that will create a line plot of power versus decimal hour in day. That graph is shown in Figure 4.2. In this textbook, we follow the convention that refers to a line or scatter plot as plotting the "y-axis variable versus x-axis variable," rather than the other way around. We can either type in each line of code in a Python interpreter directly or in a file and run the script file (see Sections 3.2.4 and 3.2.5 on how to write and run a script). If we run the lines of code as a script from a terminal window, we have to keep the interpreter session open in order to see the plot.

The code to create a scatter plot of power versus decimal hour in day is given in Figure 4.3 and is very similar to the line plot code in Figure 4.1. The only difference is that we use the `scatter` function instead of the `plot` function. The scatter plot created by the code in Figure 4.3 is shown in Figure 4.4.

Assume the examples in this section are typed in to a fresh Python session (or that the current plot window is closed before the next plot window is drawn). That makes sure each plot is in its own window and simplifies our discussion in this chapter. To see how to make

```
1   import matplotlib.pyplot as plt
2   plt.plot([9.25, 11, 13.5, 15, 15.75], [2.54, 4.1, 1.21, 3.9, 4])
3   plt.xlabel('Decimal Hour in Day')
4   plt.ylabel('Power (mW)')
5   plt.show()
```

Figure 4.1 Code to create the basic line plot in Figure 4.2.

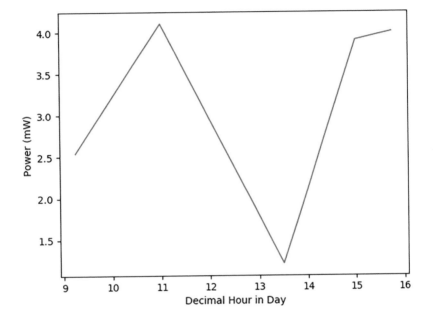

Figure 4.2 A basic line plot, generated by the code in Figure 4.1.

```
1   import matplotlib.pyplot as plt
2   plt.scatter([9.25, 11, 13.5, 15, 15.75], [2.54, 4.1, 1.21, 3.9, 4])
3   plt.xlabel('Decimal Hour in Day')
4   plt.ylabel('Power (mW)')
5   plt.show()
```

Figure 4.3 Code to create the basic scatter plot in Figure 4.4.

multiple independent plots, each in their own window, without creating a new Python session or closing the current plot window, look ahead to Section 5.2.3.

We unpack (in a broad-brushed way), this last example (Figure 4.3), line by line, to see how Python creates plots. As the code for the last example is nearly identical to the first, nearly everything we say about the last example applies to both:

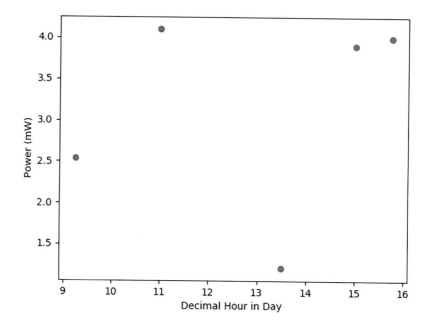

Figure 4.4 A basic scatter plot, generated by the code in Figure 4.3.

- Line 1. Import the pyplot module.
- Line 2. Create the scatter plot. The first **argument** is the list of x-values (the decimal times) and the second argument is the list of the corresponding y-values (the power values). Thus, the (x, y) pairs being plotted are: $(9.25, 2.54)$, $(11, 4.1)$, $(13.5, 1.21)$, $(15, 3.9)$, and $(15.75, 4)$.
- Line 3. Write the x-axis label.
- Line 4. Write the y-axis label.
- Line 5. Show the plot.

Our description here does not convey that much more than the Python code itself conveys. This code clarity is one of the neat features of Python. In the next section we elaborate on this description to get more insight as to what the code is doing.

4.2   Python Programming Essentials

Python's basic scientific plotting **package** (a collection of modules) is called Matplotlib. It includes a main module (called matplotlib) and a bunch of submodules or smaller collections of functions. The submodule pyplot defines a set of commands (or functions) we can use to draw graphs. As line 1 in Figure 4.3 shows, pyplot is often imported by:

```
import matplotlib.pyplot as plt
```

In Section 3.2.2, we saw that to access the functions in a module we have imported, we use a period to separate the module name and the function we want to use. This convention also applies to submodules. To reference the submodule pyplot of the package matplotlib, we reference `matplotlib`, add a period, then reference `pyplot`. Unless otherwise stated, in the examples in this book, assume that the above import has been done prior to any pyplot commands being run.

The `scatter` function (line 2) is a pyplot function we can use to create scatter plots. Unlike the other functions we looked at earlier (e.g., `sin`), there is more than one item between the parentheses that follows `scatter`. These are called "**parameters**" (or "positional input parameters") or arguments, and we describe them in more detail in Section 4.2.1. In Section 4.2.2, we also describe the sequence of values the `scatter` function in these lines accepts as input.

Once a plot is created in line 2, Matplotlib keeps track of what plot is the "current" plot. Subsequent commands, such as the making of the axis labels in lines 3 and 4, are applied to that current plot. Lines 3 and 4 also show us a value we have not seen before. Until now, all values have been decimal or integer numbers. Here we find words that are found between single quotation marks (`'`) that the `xlabel` and `ylabel` functions write to the graph as text annotation. These words are examples of **strings**, and we introduce strings in Section 4.2.3.

Finally, line 5 indicates that as we put each element of a graph onto Matplotlib's virtual canvas, Matplotlib does not render (i.e., draw onto the screen) our addition but instead waits until we call the `show` command. Usually, we care only about how a plot looks when it is all done. By waiting until `show` is called, Matplotlib avoids the extra computation involved in rendering intermediate steps. If we have more than one figure, we usually call `show` after all the plots are defined to visualize all the plots at once.

An Aside on Spyder, Jupyter, and `show`   Most of the time, Spyder and Jupyter automatically render Matplotlib plots inline (that is, in the console or in an `Out [ ]` cell). Thus, in these environments, there is often no need to call `show`. Exceptions exist, and adding `show` will not hurt. In still other cases, we have to add this line before our plot code:

```
%matplotlib inline
```

This is an additional control on how Matplotlib plots are rendered in Spyder and Jupyter. The details of how it works are not important at this point in the chapter. We can just try this if our plot is not showing in Spyder or Jupyter. This is an example of "line magic," which is discussed more in Section 22.3.

In the rest of this section, we explore the following topics in more detail: positional input parameters, lists and tuples, strings, and commenting and Jupyter markdown. As in prior chapters, after reading this chapter, go to Section 4.3 to try out the concepts.

4.2.1   Positional Input Parameters for Required Input

In Section 3.2.1, we saw that we can think of functions as "black boxes" that take in input, process it, and give us something back. We tell the function what the input is by putting it in between the parentheses after the name of the function when we call the function.

In the previous functions we have used and written, the functions only took one item of input, such as we saw in the use of the `cos` function in Figure 3.1. But, we are not limited to passing in one item of input. In principle, we can pass in as many items as we wish. The `plot` and `scatter` functions in pyplot accept multiple items of input, where each input item is listed, one after the other, in between the parentheses following the function's name when we call the function.

How does this work? Line 2 of Figure 4.3 (reproduced here for convenience):

```
plt.scatter([9.25, 11, 13.5, 15, 15.75], [2.54, 4.1, 1.21, 3.9, 4])
```

appears to have ten items, separated by commas (and inside a bunch of square brackets), between the parentheses after `scatter`. Does this mean there are ten items being input into the function? That reasoning is sound but not exactly so in this case. In actuality, there are two input items or arguments (or parameters). The first input item is this **list**:

```
[9.25, 11, 13.5, 15, 15.75]
```

and the second input item is this list:

```
[2.54, 4.1, 1.21, 3.9, 4]
```

Lists are themselves a collection of items, which we talk more about in Section 4.2.2, but for now what is important is that a list is a single entity. Thus, in line 2 of Figure 4.3, we have the left parenthesis, the first list, a comma, the second list, and the right parenthesis. This sequence of input items is called the parameter list, and for this call to `scatter`, the parameter list consists of the above two lists that are the input to the function.

But `scatter` also knows what these lists correspond to, that the first list is a list of the *x*-values of the points to plot and the second list is a list of the *y*-values of the points to plot. How does it know this? Based on the position of the input parameters in the parameter list: The first item in the parameter list is assumed by `scatter` to be the list of the *x*-values and the second item is assumed by `scatter` to be the list of the *y*-values.

Consider the following function which calculates the kinetic energy (E_k) given a mass (m) and velocity (v), using $E_k = \frac{1}{2}mv^2$:

```
def kinetic_energy(mass, velocity):
    return 0.5 * mass * (velocity**2)
```

If we wanted to use this function to calculate the kinetic energy of a mass of 4.5 kg with a velocity of 1.9 m/s, the following will make the calculation and display the result:

```
In [1]:  kinetic_energy(4.5, 1.9)
```

```
Out[1]:  8.1225
```

When we define a function, we list out the input parameters in a particular order. When we call the function and put in actual values (or variables referring to values) for those parameters, the values are mapped to the parameters one by one, based on their relative positions. Thus, in the example above, the 4.5 is mapped to the parameter `mass` while the 1.9 is mapped to the parameter `velocity`. Those values are then used in the function definition everywhere `mass` and `velocity` occur. If we change the positions of the values in the parameter list when calling `kinetic_energy`:

```
In [1]:  kinetic_energy(1.9, 4.5)
```

```
Out[1]:  19.2375
```

we obtain a totally different result because the value for `mass` and the value for `velocity` are reversed from the previous call of the function. The bottom line: When we have multiple **positional input parameters**, it is very important that we make sure the order of the arguments when we call the function exactly matches the order of the parameters in the function's definition.

A few final comments about parameter lists: First, why does it seem like there are two separate names for the input items in between the parentheses after a function's name, "parameter" and "argument"? Often, we use these two names for the same thing: to refer to the items in the parentheses after a function's name. Technically, though, these two are not the same. The term parameter refers to the items listed in the definition of the function whereas the term argument refers to the items that are substituted in for the parameters when the function is actually called. Thus, in the `kinetic_energy` example above, `mass` and `velocity` are parameters whereas 4.5 and 1.9 are arguments.

Second, why do we give names to our input parameters? Why not just leave the parameter list empty and have the function make use of variables defined outside the function? For instance, why not do the following instead?

```
mass = 4.5
velocity = 1.9
def kinetic_energy_alt():
    return 0.5 * mass * (velocity**2)
```

This code works fine and when this version of `kinetic_energy_alt` is called (with an empty parameter list), we obtain 8.1225, just as before with `kinetic_energy`.

The reason we do not do this goes back to the purpose of functions. We write functions to enable the user of the function to think of the function as a "black box." The user wants to be able to use the function without any knowledge of what is going on inside. As far as the user is concerned, as long as the user provides the proper input, the function will return the expected return value. In the `kinetic_energy_alt` example, the user needs to know that the variables in the function are named `mass` and `velocity` and that variables of that name need to be set accordingly outside the function definition. In the original `kinetic_energy` function, the user only needs to know the first parameter is the mass and the second parameter is the velocity; no knowledge of the inner workings of the function is required.

Third, do the variable names in the parameter list correspond to real variables, that is, a variable defined by that name? The short answer is no, the variables in a parameter list solely exist for us to refer to them in the body of the function definition. They are placeholders or "dummy variables" for the arguments that will be passed in when the function is called. Without variable names, the body of the function would have nothing to refer to when saying, "use the first argument's value." Once the function is called, the parameters are substituted with the arguments. The parameters only exist in the body of the function and are not accessed from outside the function.[1]

Finally, notice that because arguments are mapped to parameters based upon their position in the parameter list, all the parameters in the list are required. That is, if a function's parameter list contains two elements and we provide one argument in our call to the function, the function will not know whether that argument corresponds to the first parameter or the second parameter. Thus, we typically use positional input parameters for input to a function that is required. In Section 5.2.1 we describe how to specify optional input into a function.

4.2.2   Introduction to Lists and Tuples

We saw earlier that line 2 of Figure 4.3 consists of two lists as input. It appears that a list is a collection of numbers. In Section 9.2.3, we will find that lists can hold more than just numbers (and in fact can hold essentially anything), but for now, we do not have to think about that. We can describe the basics of lists by looking at examples that consist only of numbers.

Lists are ordered sequences. They are a collection of items or "**elements**" where the first item has the first position, the second the second position, and so on. Each of these positions has an "address" or "**index**" with which we can refer to the item stored at that position. We can think of lists as a row of cubbies or post office boxes with the items being stored inside those cubbies:

9.25	11	13.5	15	15.75
0	1	2	3	4

The diagram above represents the first list from line 2 of Figure 4.3. Each box holds one of the values from the list and the number under each box is the index for that element.

As in the example above, list element indices or addresses always start with the first element having an index of 0, the second element having an index of 1, and so on until the last element of the list, which has an index of one less than the total number of elements in the list (i.e., the last element in a list has an index of $n - 1$, where n is the total number of elements in the list).

Square brackets ([]) start and stop a list, and commas between list elements separate elements from one another. We can enter lists as an independent entity, as in the scatter function call above, but we can also save a list as a variable. Thus, line 2 of Figure 4.3 could

---

[1] The reality is a little more complex, but at this point, we can think about parameters this way.

```
x_values = [9.25, 11, 13.5, 15, 15.75]
y_values = [2.54, 4.1, 1.21, 3.9, 4]
plt.scatter(x_values, y_values)
```

Figure 4.5 Line 2 of Figure 4.3 rewritten using variables to hold the lists that are used as input into the call to `scatter`.

be rewritten as shown in Figure 4.5. Note that in Figure 4.5, there are no square brackets around the positional input parameters in the `scatter` function call (e.g., it is `x_values` not `[x_values]` in that call). The `x_values` and `y_values` variables are already lists, so we do not put square brackets around them.

To refer to an element of a list that is set to a variable, we use square brackets after the variable name, with the index in between the square brackets. Thus, the first element of list `x_values` is `x_values[0]`, the second is `x_values[1]`, etc. Because the **ordinal** value (i.e., first, second, third, etc.) of an element differs from the address of an element (i.e., 0, 1, 2, etc.), when we refer to an element by its address we append a "th" to the end of the address. That is, the "zeroth" element by address is the first element by position in the list, the "oneth" element by address is the second element by position, the "twoth" element by address is the third element by position, and so on.

Using this notation, we can change the values of an existing list or use the values of an existing list. Consider again the code in Figure 4.5. How could we change the second value of `x_values` from 11 to 11.05 (maybe we realized the time we took the data at was off from what we recorded)? We could do this with the following code, which replaces the second (i.e., index 1) element of `x_values` with 11.05:

```
x_values = [9.25, 11, 13.5, 15, 15.75]
x_values[1] = 11.05
```

or with this code, which takes the existing value of `x_values[1]` (i.e., 11) and adds 0.05 to that value, then replaces the existing value of `x_values[1]` with the resulting sum:

```
x_values = [9.25, 11, 13.5, 15, 15.75]
x_values[1] = x_values[1] + 0.05
```

Finally, the length of a list (the number of elements in the list) can be obtained using the built-in `len` function. For instance, `len(x_values)` returns the length of the list `x_values`:

```
In [1]:  x_values = [9.25, 11, 13.5, 15, 15.75]
         len(x_values)

Out[1]:  5
```

In Python, list elements can also be addressed starting from the end of the list, using a negative index. In this notation, x_values[-1] is the last element in list x_values, x_values[-2] is the next to last element, etc. We do not *have* to use negative indices, but this provides a handy way of referring to the last elements of a list without having to use the len function: For the list x_values, for instance, x_values[-1] and x_values[len(x_values)-1] refer to the same element.

We will talk more about lists in Section 9.2.3, so we finish this introduction to lists by describing how to extract parts of a list (i.e., a "**sublist**" or "**slice**") from a list, beyond a single element of that list. When specifying a range of elements to slice out of a list, we follow these rules:

- We specify a sublist range by using element index values, giving the lower and upper limits of the range, separated by a colon.
- The lower limit of the range is *inclusive*, and the upper limit of the range is *exclusive*.
- If the lower and/or upper limit is left out, the slice goes to and through the beginning and/or the end of the list, respectively.

What does slicing look like? Consider again our code from Figure 4.5. How can we plot only the first four points instead of all five points described in x_values and y_values? That is, how can we leave out the last element in the lists? We can slice only the other elements and save the result as x_values and y_values with the following code:

```
x_values = x_values[0:-1]
y_values = y_values[0:-1]
```

Here is a session that creates the Figure 4.5 version of x_values and y_values, does the above slicing, and prints out x_values and y_values, so we can see what happens:

```
In [1]:  x_values = [9.25, 11, 13.5, 15, 15.75]
         y_values = [2.54, 4.1, 1.21, 3.9, 4]
         x_values = x_values[0:-1]
         y_values = y_values[0:-1]
         x_values

Out [1]:  [9.25, 11, 13.5, 15]


In [2]:  y_values

Out [2]:  [2.54, 4.1, 1.21, 3.9]
```

How did the slicing work? The index to the left of the colon in both slicing operations is 0. The rules tell us that that index specifies the first element of the sublist we are slicing and is inclusive, i.e., that the sublist will include that element. The index to the right of the colon, however, is exclusive, i.e., is *not* included in the sublist. In the above slicing operations, that index is −1 which refers to the final element in the list. The rules tell us not to include that element so the range of elements in our sublist ends with the element before the final element.

As the slicing rules indicate, we can specify "go to the end" or "go to the beginning" of the list by leaving out one or the other of the limit indices. In the code below, we slice only the last three values of the two lists:

```
In [1]:  x_values = [9.25, 11, 13.5, 15, 15.75]
         y_values = [2.54, 4.1, 1.21, 3.9, 4]
         x_values = x_values[-3:]
         y_values = y_values[-3:]
         x_values
```

```
Out[1]:  [13.5, 15, 15.75]
```

```
In [2]:  y_values
```

```
Out[2]:  [1.21, 3.9, 4]
```

By leaving the upper limits blank in the third and fourth lines of `In [1]`, our slice goes all the way to the end of the lists. Remember a negative index means "from the end," so the −3 index means the third element from the end. In these two lines, we start with that index's element because it is the lower (and inclusive) limit of the slice. We go through more slicing cases in Section 4.3.

We have seen that a list is an ordered collections of elements whose elements we can change. What if we want a list of items that we do not want to be changed? That is, we want to store the items and read them using their indices, but we do not want to be able to replace them with different values. To do this, we store the collection of elements as a **tuple**. Defining a tuple is just like defining a list except we use parentheses around the elements instead of square brackets. Thus, if we defined the Figure 4.5 lists `x_values` and `y_values` as tuples, the code would look like:

```
x_values = (9.25, 11, 13.5, 15, 15.75)
y_values = (2.54, 4.1, 1.21, 3.9, 4)
```

We can use elements from tuples but we cannot change them:[2]

```
In [1]:  x_values = (9.25, 11, 13.5, 15, 15.75)
         x_values[1]
```

```
Out[1]:  11
```

---

[2] The error message output is edited for clarity.

```
In [2]:  x_values[1] = 11.05

         TypeError    Traceback (most recent call last)
         <ipython-input-5-64fa1aac1eef> in <module>()
         ----> 1 x_values[1] = 11.05

         TypeError: 'tuple' object does not support item assignment
```

Finally, we can convert lists into tuples using the `tuple` function and tuples into lists using the `list` function:

```
In [1]:  x_values = (9.25, 11, 13.5, 15, 15.75)
         type(x_values)

Out[1]:  tuple

In [2]:  x_values = list(x_values)
         type(x_values)

Out[2]:  list

In [3]:  x_values = tuple(x_values)
         type(x_values)

Out[3]:  tuple
```

If we create a tuple and realize we want to change the elements in it, we can convert it into a list, change the list, then convert the list back into a tuple. The built-in `type` function tells us what kind of variable a given variable is (see Section 5.2.7 for more information).

In Section 9.2.3, we describe more about what lists are and what they can do.

4.2.3   Introduction to Strings

Lines 3 and 4 of Figure 4.3 (reproduced here for convenience):

```
plt.xlabel('Decimal Hour in Day')
plt.ylabel('Power (mW)')
```

call the functions `xlabel` and `ylabel` to write out an *x*-axis and *y*-axis label, respectively. Each of these functions takes one argument, but this argument is not a number (either integer or decimal) nor a collection of numbers (a list or tuple) but rather is a new kind of variable called a string. Strings are combinations of characters that are set between matching apostrophes, quotation marks, or triple apostrophes/quotation marks. They often represent text information or words, phrases, and sentences.

A string is created by putting characters between matching single, double, or "triple" quotes. The entire sequence of characters then becomes a single value. Each of these will give us the string, `'Number of Plant Species'`:

```
'Number of Plant Species'
"Number of Plant Species"
'''Number of Plant Species'''
"""Number of Plant Species"""
```

We can set a variable to a string just as with numbers, lists, and tuples:

```
In [1]: axis_title = 'Number of Plant Species'
        axis_title
```

```
Out[1]: 'Number of Plant Species'
```

```
In [2]: type(axis_title)
```

```
Out[2]: str
```

Case matters, so `'A'` is not the same as `'a'` in a string.

Why are there multiple ways of specifying a string? Having both single quotes and quotation marks is useful when the string needs to include the other kind of punctuation mark:

```
In [1]: plot_title = "Jane Doe's Plant Experiment"
        plot_title
```

```
Out[1]: "Jane Doe's Plant Experiment"
```

```
In [2]: text = 'Calculate a "pseudo" probability distribution function.'
        text
```

```
Out[2]: 'Calculate a "pseudo" probability distribution function.'
```

The triple quotes allows us to include newline characters and spaces through typing `Enter` and pressing `Space` and having those remembered as part of the string. This makes it easier to enter in more complexly formatted strings:

```
In [1]: description = '''The force, given by:
            F = ma
        "matches" expected values.'''
        description
```

```
Out[1]: 'The force, given by:\n    F = ma\n"matches" expected values.'
```

```
In [2]:   print(description)

          The force, given by:
              F = ma
          "matches" expected values.
```

The character `'\n'` is the newline character. In a string, it is considered one character, not two. The `print` function prints out the contents of the string variable to the screen but represents formatting characters (such as newline) as the way they should look on a screen. Another common formatting character used with strings is Tab ( ), which is `'\t'`.

What if we want to specify an "**empty string**," a string value that is a string but has no characters in it (not even blank spaces)? While such a value seems useless – if we printed such a string, we would see nothing – there are times it is valuable. For instance, if we want to initialize a variable as a string, but do not want it to have any specific characters in it, we would set it to an empty string. Empty strings in Python are given by a pair of matching quotes with nothing between them, e.g., `' '` or `" "`.

Just as we can operate on numbers, we can also operate on strings. To **concatenate** two strings together, i.e., glue two strings together, we use the + operator:

```
In [1]:   phrase1 = "Number of"
          phrase2 = "Carbon Atoms"
          phrase1 + " " + phrase2
```

```
Out [1]:  'Number of Carbon Atoms'
```

Note how we had to explicitly insert a blank space between `phrase1` and `phrase2` because neither has a trailing or leading blank space, respectively.

In addition to joining two strings together, we can extracts parts of strings, including individual characters. To do so we use a syntax that is the same as the slicing of lists and tuples. Indeed, we can think of a string as a little tuple where each element is a character, with the first character having the index of 0.

If we have the following string `plot_title`:

```
plot_title = "Comparison Between Jane and Jill's Experiments"
```

we can get the following results in the Python interpreter:

```
In [1]:   plot_title[2]
```

```
Out [1]:  'm'
```

```
In [2]:   plot_title[0:5]
```

```
Out [2]:  'Compa'
```

```
In [3]:   plot_title[-5:]
```

```
Out[3]:   'ments'
```

```
In [4]:   plot_title[28:34]
```

```
Out[4]:   "Jill's"
```

```
In [5]:   plot_title[32]
```

```
Out[5]:   "'"
```

Recall that in the list slicing syntax, the index to the left of the colon is inclusive while the index to the right of the colon is exclusive. Note also that punctuation characters (such as the apostrophe) are also elements in the string and can be referenced by index.

If we want to know the number of characters in a string, we can use the `len` function:

```
In [1]:   plot_title = "Comparison Between Jane and Jill's Experiments"
          print(len(plot_title))
```

```
          46
```

We can save the results of our slicing and dicing of a string (whether one or many characters worth) as a variable:

```
In [1]:   name = plot_title[28:34]
          print(name)
```

```
          Jill's
```

When we extract one or more characters from a string, we call that subset a **substring**. Substrings are themselves strings and can be sliced and diced (and concatenated) just like the strings they were taken from.

Oftentimes in scientific or engineering work, we will want to put letters and numerals together in a string, based on numerical values that have been stored in a variable. The number `1` and the string `'1'` are not the same thing, however, and cannot be concatenated together as-is. More on this distinction is in Section 5.2.7. In order to get a version of a number as a string, we have to use the built-in `str` function:[3]

---

[3] The error message output is edited for clarity.

```
In [1]:  text1 = "Hydrogen"
         text2 = "Helium"
         i = 1
         text1 + " not " + text2 + " is number " + str(i)

Out[1]:  'Hydrogen not Helium is number 1'

In [2]:  text1 + " not " + text2 + " is number " + i

Out[2]:  TypeError    Traceback (most recent call last)
         <ipython-input-4-b4dd10772c95> in <module>()
         ----> 1 text1 + " not " + text2 + " is number " + i

         TypeError: must be str, not int
```

In cell In [2], we attempt to concatenate a string with a number (an integer) without first converting the number into a string and are told we cannot do so.

Finally, there will be times we want to create very long strings, so long that they will not fit on one line. In those cases, we can use the backslash (\) character at the very end of a line to tell Python the line of code continues on the next line. Thus:

```
text = "Experiment 1 and " + \
       "Experiment 2 are " + \
       "both finished."
```

is the same as:

```
text = "Experiment 1 and " + "Experiment 2 are " + "both finished."
```

and

```
text = "Experiment 1 and Experiment 2 are both finished."
```

The use of the backslash character to tell Python a line of code is continued on the next line works not only for defining strings but for all kinds of Python code.

This brief introduction gives us the basics of setting and slicing and dicing of strings. We discuss more about strings in Section 9.2.4.

4.2.4 Introduction to Commenting and Jupyter Markdown

We are starting to write scripts that are longer and longer. Although Python is a very clear language, at some point we need to put explanatory notes in our code to help explain why our code is written a certain way, what the code does, and what is the source of our information. In a program, these notes are called **comments**. Comments in Python are any text that begins with the hash symbol (#) and ends with the end of the line. Whether the hash is found at the beginning of a line or midway through a line, the Python interpreter considers everything after and including the hash, in that line, as a comment and does not execute its contents.

We take the code from Section 4.1 and add some commenting to it:

```
1    #- Import packages:
2    import matplotlib.pyplot as plt
3
4    #- Create plot:
5    plt.plot([9.25, 11, 13.5, 15, 15.75], [2.54, 4.1, 1.21, 3.9, 4])
6
7    #- Add labels and display:
8    plt.xlabel('Decimal Hour in Day')
9    plt.ylabel('Power (mW)')
10   plt.show()    #- Display plot on screen
```

These comment lines do not really say more than is already clear from the code, but they illustrate how the comment symbol works.

In a plain code file, comment lines are one way of providing additional information about the program to another programmer who is looking at the code (that other programmer may be the original programmer a few weeks or months later). If we are writing our Python code into a Jupyter notebook, we have an additional way of interweaving descriptions and text comments into our code, by designating certain cells as "Markdown" cells and typing our description into those cells. Figure 4.6 shows an example, as entered into the notebook (that is, before running all the cells in the notebook). Note, to tell the notebook this cell is a Markdown cell as opposed to a Python code cell, we have to select "Markdown" in the combo-box as seen in the red circled part of Figure 4.6.

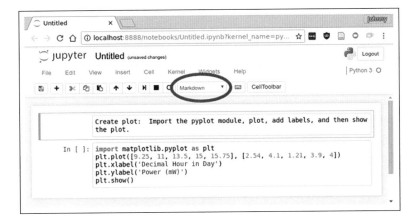

Figure 4.6 A Jupyter notebook showing plotting code with Markdown cells providing description (before running all the cells in the notebook, with "Markdown" combo-box circled in red).

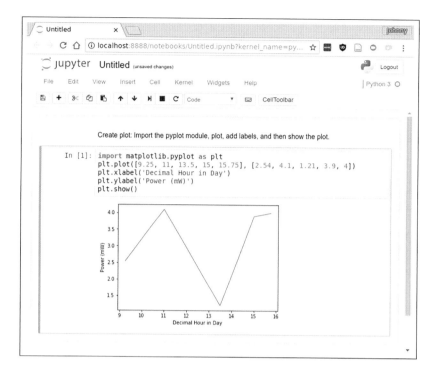

Figure 4.7 A Jupyter notebook showing plotting code with Markdown cells providing description (after running all the cells in the notebook).

Figure 4.7 shows the same notebook after we run all the cells. There is not much difference except that the Markdown cells are typeset like text in a book, i.e., in a nicer looking **proportional font**, whereas before the run, the text is displayed in a clunkier looking **nonproportional font** that looks like code. And all the cells with **executable** or runnable Python code are run and their results are given in Out [ ] cells.

For short comments, the only important feature of a Markdown cell in a Jupyter notebook is that the text in the cell is not executed. However, the benefit of Markdown is that it allows us to also include formatting into the cell, such as creating headers, making italicized text, inserting mathematical equations, etc. The Jupyter and Markdown documentation[4] describes the kinds of formatting allowed in a Markdown cell. Examples of this kind of formatting are given in the Discipline-Specific Jupyter notebooks linked in Section 4.4 and the other chapters of the book. Click in the title and question cells to see the Markdown content and run the notebooks to see how that content is typeset. Table 4.1 provides a brief summary of some basic syntax. In Section 6.2.6, we address more we can do with commenting in Python.

---

[4] http://jupyter-notebook.readthedocs.io/en/latest/examples/Notebook/Working%20With%20Markdown%20Cells. html (accessed August 9, 2017) and https://daringfireball.net/projects/markdown/basics (accessed August 23, 2018).

Table 4.1 Summary of basic Markdown syntax in a Jupyter notebook Markdown cell.

Markdown example	What it does
`# Experiment 1`	Typesets "Experiment 1" in a header font and style, something like **Experiment 1**.
`The data is *not* accurate.`	Typesets the sentence so "not" is emphasized by italics, as in, "The data is *not* accurate."
`The data is **not** accurate.`	Typesets the sentence so "not" is strongly emphasized, by bold, as in, "The data is **not** accurate."
`At [NASA](http://nasa.gov), we`	Typesets a link (in the square brackets) to a web address (in the parentheses), as in, "At NASA, we", where the gray word is a link to http://nasa.gov.

`* Frogs`
`* Fireflies`
`* Carp`
	Creates the bulleted list:
	• Frogs
	• Fireflies
	• Carp

`1. Frogs`
`2. Fireflies`
`3. Carp`
	Creates the enumerated list:
	1. Frogs
	2. Fireflies
	3. Carp

`$v_{y}$`	Typesets inline the math symbol v_y. The underscore means "subscript what is in the curly braces."
`$e^{x}$`	Typesets inline the math symbol e^x. The caret means "superscript what is in the curly braces."
`$\frac{1}{2}$`	Typesets inline the fraction $\frac{1}{2}$.
`$$F = ma$$`	Typesets, centered, on its own line, the equation $F = ma$.

4.3 Try This!

We start with graphing examples similar to ones we have seen in Section 4.1. These are followed by examples that address the main topics of Section 4.2: positional input parameters, lists and tuples, strings, and commenting. Those examples may involve using Python as a calculator or scientific calculator, not only for plotting. For some of the topics of Section 4.2, calculator questions give us some basic practice utilizing those ideas.

Try This! 4-1 Basic Scatter Plot: Animal Data

Consider the following data regarding a fictional animal:

Length (m)	Weight (kg)
0.8	1.2
1.1	1.8
0.7	0.9
1.2	2.1
1.1	0.9
0.5	0.4
1.4	2.2

Create a scatter plot of weight (on the *y*-axis) versus height (on the *x*-axis). Please label the axes appropriately.

Try This! Answer 4-1

This code solves the problem:

```
import matplotlib.pyplot as plt
plt.scatter([0.8, 1.1, 0.7, 1.2, 1.1, 0.5, 1.4], \
            [1.2, 1.8, 0.9, 2.1, 0.9, 0.4, 2.2])
plt.xlabel('Height (m)')
plt.ylabel('Weight (kg)')
plt.show()
```

The plot generated by the above code is shown in Figure 4.8.

Notice how in line 2 we use the line continuation backslash character at the end of the line to tell Python that the content of line 3 is a continuation of line 2. The leading spaces in line 3 are ignored. This way, we can format our code to look more readable.

While the use of the line continuation character in line 2 works fine, for cases such as list items and parameter lists, Python automatically knows that in line 2 we have not finished entering all the items the `plot` function is looking for (for instance, there is no close parenthesis in line 2) and automatically assumes line 3 is a continuation of line 2. Thus, we can leave out the backslash in line 2 and everything will work fine. That is:

```
plt.scatter([0.8, 1.1, 0.7, 1.2, 1.1, 0.5, 1.4],
            [1.2, 1.8, 0.9, 2.1, 0.9, 0.4, 2.2])
```

works the same as:

```
plt.scatter([0.8, 1.1, 0.7, 1.2, 1.1, 0.5, 1.4], \
            [1.2, 1.8, 0.9, 2.1, 0.9, 0.4, 2.2])
```

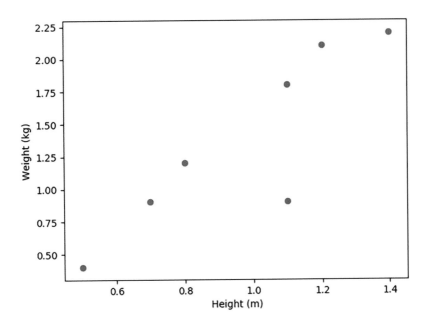

Figure 4.8 Scatter plot generated by the solution to Try This! 4-1.

Try This! 4-2 Basic Line Plot: Exponential Function

Create a table of five data points of x-values from zero to one and their correponding y-values for $y = e^x$. Create a line plot of the y-values versus the x-values.

Try This! Answer 4-2

Each of the five data points can be calculated using the `exp` function in NumPy. Here is the calculator session showing the code to do so:

```
In [1]:  from numpy import exp
         exp(0.0)
```

```
Out[1]:  1.0
```

```
In [2]:  exp(0.2)
```

```
Out[2]:  1.2214027581601699
```

```
In [3]:  exp(0.4)
```

```
Out[3]:  1.4918246976412703
```

```
In [4]:  exp(0.6)

Out[4]:  1.8221188003905089

In [5]:  exp(0.8)

Out[5]:  2.255409284924679
```

Remember to first import `exp` before using it (the first line of `In [1]` in the above code). Here is a table of the data:

x Values	$y = e^x$ Values
0.0	1.0
0.2	1.221
0.4	1.492
0.6	1.822
0.8	2.226

and this code solves the problem:

```
1   import matplotlib.pyplot as plt
2   plt.plot([0.0, 0.2, 0.4, 0.6, 0.8], \
3            [1.0, 1.221, 1.492, 1.822, 2.226])
4   plt.xlabel('x')
5   plt.ylabel('y')
6   plt.show()
```

The plot generated by the above code is shown in Figure 4.9. In later chapters, we describe ways of getting Python to make repetitive calculations, so we do not have to explicitly type in multiple function calls as we did above.

Try This! 4-3 Positional Input Parameter: Magnitude of Velocity Vector

Write a function `magnitude` that will return the magnitude of a velocity vector given the x- and y-components of the vector. That is, calling the function with an x-component of 3.0 m/s and a y-component of 4.0 m/s will produce the following:

```
In [1]:  magnitude(3.0, 4.0)

Out[1]:  5.0
```

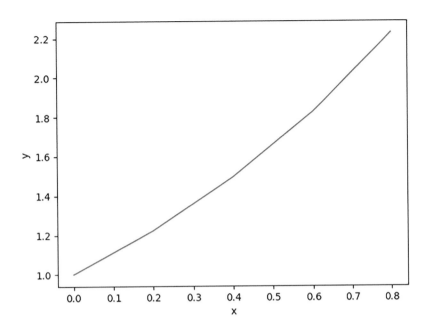

Figure 4.9 Line plot generated by the solution to Try This! 4-2.

Hint: A vector velocity $\vec{v}$ has the length given by:

$$v = \sqrt{v_x^2 + v_y^2}$$

where v_x and v_y are the x- and y-components of the velocity, respectively.

Try This! Answer 4-3

This code solves the problem:

```
def magnitude(x_component, y_component):
    return ((x_component**2) + (y_component**2))**0.5
```

We used "taking to the one-half power" in line 2 instead of using a square-root function. To do the latter, we need to import the sqrt function from the NumPy module:

```
from numpy import sqrt
def magnitude(x_component, y_component):
    return sqrt((x_component**2) + (y_component**2))
```

When we import importing modules or items from modules, we usually do this at the beginning of our Python code file (or Jupyter notebook). That way, we have access to that module or module item for the entirety of the file or notebook. In the above case, with the sqrt import line at the top of the file, the function is available for use in the magnitude definition or any function or variable definitions in that Python code file.

Try This! 4-4 Order of Input Parameters: Magnitude of Velocity Vector

Consider the solution to Try This! 4-3. What happens if we switch the order of the input arguments in the call to `magnitude`? That is, what would be different if we had:

```
magnitude(4.0, 3.0)
```

versus

```
magnitude(3.0, 4.0)
```

Does it matter? How would we demonstrate the effect of switching the order of the input arguments in the call?

Try This! Answer 4-4

Switching the order of the input arguments has no effect on the final result because in the equation for calculating the magnitude of a vector we do the same operation (i.e., squaring) to the x-component as to the y-component. Thus, both the `magnitude(4.0, 3.0)` and `magnitude(3.0, 4.0)` function calls will return 5.0. However, inside the function, in the first calling case, the parameter x_component is set to 4.0 while in the second calling case, the parameter x_component is set to 3.0. The reverse happens with the parameter y_component.

How do we show this is what happens? One way is by putting `print` statements into the body of the function. For instance, this function definition:

```
1   def magnitude(x_component, y_component):
2       print(x_component)
3       print(y_component)
4       return ((x_component**2) + (y_component**2))**0.5
```

will print out the values of x_component and y_component that are passed in when the function is called. Here is an example of this:

```
In [1]:   magnitude(3.0, 4.0)

          3.0
          4.0

Out[1]:   5.0


In [2]:   magnitude(4.0, 3.0)

          4.0
          3.0

Out[2]:   5.0
```

showing the values of x_component and y_component for each function call. The 5.0 values that are printed out in the Out [ ] cells do not come from a print statement but rather from the default behavior that calls made in the Python interpreter to functions that have a return value will print out to screen the value of the return value.

Try This! 4-5 Scatter Plot of Switched Input Parameters: Animal Data

Consider the data from Try This! 4-1. Make a scatter plot of the height versus weight instead of weight versus height. That is, place the height values on the *y*-axis and the weight values on the *x*-axis.

Try This! Answer 4-5

Recall the solution from Try This! 4-1:

```
import matplotlib.pyplot as plt
plt.scatter([0.8, 1.1, 0.7, 1.2, 1.1, 0.5, 1.4], \
            [1.2, 1.8, 0.9, 2.1, 0.9, 0.4, 2.2])
plt.xlabel('Height (m)')
plt.ylabel('Weight (kg)')
plt.show()
```

The most straightforward way of making this plot is to switch around the order of the lists passed into scatter:

```
plt.scatter([1.2, 1.8, 0.9, 2.1, 0.9, 0.4, 2.2] \
            [0.8, 1.1, 0.7, 1.2, 1.1, 0.5, 1.4])
```

If we know ahead of time, however, that we want to make these two kinds of plots, the easier way of doing so would be to save the height and weight lists as variables and make two calls to scatter with the input lists swapped. Remember that positional input parameters know what their values are based upon the order of the arguments that are passed in. This code will do what we want:

```
import matplotlib.pyplot as plt
height = [0.8, 1.1, 0.7, 1.2, 1.1, 0.5, 1.4]
weight = [1.2, 1.8, 0.9, 2.1, 0.9, 0.4, 2.2]
plt.scatter(height, weight)
plt.xlabel('Height (m)')
plt.ylabel('Weight (kg)')
plt.show()
plt.scatter(weight, height)
plt.xlabel('Weight (kg)')
plt.ylabel('Height (m)')
plt.show()
```

The switch in input argument order is in line 8. If we run the code above as-is, the first `show` call shows the first graph, and Python then waits for us to close that graph's window before showing the second graph. Normally, we prefer to see both graphs at the same time. We cover how to make multiple independent plots, each in their own window, in Section 5.2.3.

Try This! 4-6 Keeping Only the First Few Data Points: Animal Data

Consider the data from Try This! 4-1. Pretend we have now discovered only the first five data points are valid. Make a scatter plot of weight versus height using only the valid data points.

Try This! Answer 4-6

We could type in the first five data points, but doing things by hand does not harness any of the capabilities of a programming language. Instead, we use list slicing (introduced in Section 4.2.2) to keep only the first five data points. We can see this if we save the height and weight data as lists (as we saw in the solution to Try This! 4-5):

```
1   import matplotlib.pyplot as plt
2   height = [0.8, 1.1, 0.7, 1.2, 1.1, 0.5, 1.4]
3   weight = [1.2, 1.8, 0.9, 2.1, 0.9, 0.4, 2.2]
4   height = height[0:5]
5   weight = weight[0:5]
6   plt.scatter(height, weight)
7   plt.xlabel('Height (m)')
8   plt.ylabel('Weight (kg)')
9   plt.show()
```

In lines 4 and 5 we slice out the first five elements of the `height` and `weight` lists and reassign the return value of that slicing to the variables `height` and `weight`. That way, we do not have to change the argument list to the `scatter` method.

Recall that the slicing syntax for the first five elements of a list is `[0:5]` and not `[1:5]` because the first element has an index of 0. In addition, the slicing syntax is not `[0:4]` because the index value of the upper limit of the slicing range is *exclusive* of that index, not inclusive of that index. Thus, `[0:5]` extracts elements of a list with index values of 0 through 4.

With the above slicing, because we are going all the way to the beginning of the `height` and `weight` lists, we can leave out the 0. Thus, `[:5]` works the same as `[0:5]`.

Try This! 4-7 Keeping Only a Few Interior Data Points: Animal Data

Consider the data from Try This! 4-1. Pretend we have now discovered only the second, third, and fourth data points in the table are valid. Make a scatter plot of weight versus height using only those valid data points.

Try This! Answer 4-7

As in Try This! 4-6, we use list slicing to keep only the valid data points:

```
1   import matplotlib.pyplot as plt
2   height = [0.8, 1.1, 0.7, 1.2, 1.1, 0.5, 1.4]
3   weight = [1.2, 1.8, 0.9, 2.1, 0.9, 0.4, 2.2]
4   height = height[1:4]
5   weight = weight[1:4]
6   plt.scatter(height, weight)
7   plt.xlabel('Height (m)')
8   plt.ylabel('Weight (kg)')
9   plt.show()
```

The second element in a list has an index of 1, so in lines 4 and 5 we start the slicing syntax at 1. The fourth element has index 3, and the upper limit is exclusive, so the slicing syntax has a 4 to the right of the colon.

Try This! 4-8 Keeping Only the Last Few Data Points: Animal Data

Consider the data from Try This! 4-1. Pretend we have now discovered that only the last four data points in the table are valid. Make a scatter plot of weight versus height using only those valid data points.

Try This! Answer 4-8

As in Try This! 4-6, we use list slicing to keep only the valid data points:

```
1   import matplotlib.pyplot as plt
2   height = [0.8, 1.1, 0.7, 1.2, 1.1, 0.5, 1.4]
3   weight = [1.2, 1.8, 0.9, 2.1, 0.9, 0.4, 2.2]
4   height = height[-4:]
5   weight = weight[-4:]
6   plt.scatter(height, weight)
7   plt.xlabel('Height (m)')
8   plt.ylabel('Weight (kg)')
9   plt.show()
```

The fourth element from the end in a list has an index of −4, so in lines 4 and 5 we start the slicing syntax at −4. We want to go to the end of the list so there is no upper limit value to the right of the colon. If the slicing syntax was `[-4:-1]`, the last element in the list would have been left out because the slicing range is exclusive of the element specified by the index to the right of the colon.

Note that slicing with `[3:]` would have worked the same as slicing with `[-4:]` in the height and weight lists because of the total number of elements in those lists. If the total number of elements were different, `[3:]` would not have given the same result as `[-4:]`.

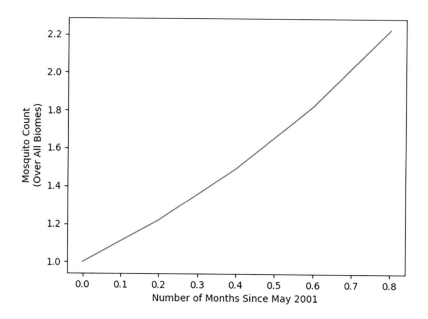

Figure 4.10 Line plot generated by the solution to Try This! 4-9.

Try This! 4-9 Label Plot Axes with a Multiple Line Label: Exponential Function

Redo the plot in Try This! 4-2 with different *x*-axis and *y*-axis labels. Make sure at least one axis label has a carriage return in it. That is, at least one label should be two lines long.

Try This! Answer 4-9

Here is code that will accomplish the task:

```
import matplotlib.pyplot as plt
plt.plot([0.0, 0.2, 0.4, 0.6, 0.8], \
        [1.0, 1.221, 1.492, 1.822, 2.226])
plt.xlabel('Number of Months Since May 2001')
plt.ylabel('Mosquito Count\n(Over All Biomes)')
plt.show()
```

This solution has a two-line-long label for the *y*-axis. The plot produced is provided in Figure 4.10. In line 5, the carriage return is introduced by including a newline character (\n) where we want the two lines in the axes label to be separated.

Try This! 4-10 Plot Labels That Include Data Values: Exponential Function

For the data in Try This! 4-2, alter the *x*-axis label to include the smallest and largest value of the data. That is, if the smallest value is 0.0 and the largest is 0.8, the label might say something like "X-Values Range from 0.0 to 0.8". In constructing the *x*-axis label, first save the data passed

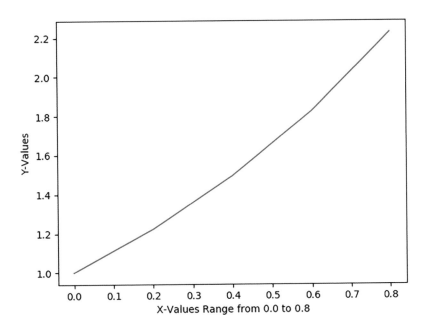

Figure 4.11 Line plot generated by the solution to Try This! 4-10.

into the plot command as list variables and then extract the smallest and largest value from the *x*-values list to use in the label.

Try This! Answer 4-10

Here is code that will accomplish the task:

```
1   import matplotlib.pyplot as plt
2   x_values = [0.0, 0.2, 0.4, 0.6, 0.8]
3   y_values = [1.0, 1.221, 1.492, 1.822, 2.226]
4   plt.plot(x_values, y_values)
5   plt.xlabel('X-Values Range from ' + str(x_values[0]) + \
6               ' to ' + str(x_values[-1]))
7   plt.ylabel('Y-Values')
8   plt.show()
```

Because the x_values list is a list of numbers, we have to use the str function in lines 5 and 6 to convert the values into strings to then append to the *x*-axis label string. The addition operator, when used on strings, concatenates the strings together. Figure 4.11 shows the plot that results from the code above.

Try This! 4-11 More Editing of Plot Labels: Exponential Function

Using the answer to Try This! 4-10, change it so xlabel accepts its input as a string variable. Before passing that variable in to xlabel, first set it to:

```
'X-Values Range from ' + str(x_values[0]) + ' to ' + str(x_values[-1])
```

then use string slicing to reassign the string variable to the above string without the "X-Values" portion.

Try This! Answer 4-11

Here is code that will accomplish the task:

```
1   import matplotlib.pyplot as plt
2   x_values = [0.0, 0.2, 0.4, 0.6, 0.8]
3   y_values = [1.0, 1.221, 1.492, 1.822, 2.226]
4   xlabel_string = 'X-Values Range from ' + str(x_values[0]) + \
5                   ' to ' + str(x_values[-1])
6   xlabel_string = xlabel_string[9:]
7   plt.plot(x_values, y_values)
8   plt.xlabel(xlabel_string)
9   plt.ylabel('Y-Values')
10  plt.show()
```

The capital-"R" in "Range" is at index position 9 (i.e., the tenth character in xlabel_string).

Try This! 4-12 Commenting Code for a Plot: Animal Data

For the solution of Try This! 4-8, provide comment lines that provide additional description of the code.

Try This! Answer 4-12

Here is the code with some comments:

```
1   #- Import packages:
2   import matplotlib.pyplot as plt
3
4   #- Data lists:
5   height = [0.8, 1.1, 0.7, 1.2, 1.1, 0.5, 1.4]
6   weight = [1.2, 1.8, 0.9, 2.1, 0.9, 0.4, 2.2]
7
8   #- Select only the last four elements from the data lists:
9   height = height[-4:]
10  weight = weight[-4:]
11
12  #- Create plot and display:
13  plt.scatter(height, weight)
14  plt.xlabel('Height (m)')
15  plt.ylabel('Weight (kg)')
16  plt.show()
```

4.4   More Discipline-Specific Practice

The Discipline-Specific Jupyter notebooks for this chapter cover the following topics:

- Writing functions with positional input parameters.
- Slicing lists and tuples.
- Creating, slicing, and concatenating strings.
- Jupyter Markdown.

4.5   Chapter Review

4.5.1   Self-Test Questions

Try to do these without looking at the book or any other resources or using the Python interpreter. Answers to these Self-Test Questions are found at the end of the chapter.

Self-Test Question 4-1

What pyplot function creates a line plot? A scatter plot? How do we import the functions to enable us to use them?

Self-Test Question 4-2

If we are given two lists of numbers (the lists are the same length), `xvalues` and `yvalues`, how would we create a line plot using these values? Please also provide any needed import lines and any code needed to display the plot on the screen.

Self-Test Question 4-3

How do we create a label for the x-axis? The y-axis?

Self-Test Question 4-4

Assume we have a function `calculate` that takes three input parameters. Assuming we have the variables x, y, and z (and, for simplicity, assume the variables are all lists of the same size but with different values), describe the difference between this call of the function:

```
calculate(x, y, z)
```

and this call:

```
calculate(z, y, x)
```

Does the first call work but the second does not? What else can we say about these two calls?

Self-Test Question 4-5

What is a list? What is a tuple? How do lists and tuples differ from each other?

Self-Test Question 4-6

Assume we have a list called `data`. How do we:

1. Determine the length of the list?
2. Access the first element of that list?
3. Access the third element of that list?
4. Access the last element of that list?
5. Create a list called `subdata` that consists of the fourth to ninth elements of the list (where fourth and ninth refer to the ordinal positions of those elements)?

Self-Test Question 4-7

What is the difference between these lines of code?

```
1   species_name = "Homo sapiens"
2   species_name = 'Homo sapiens'
3   species_name = """Homo sapiens"""
4   species_name = '''Homo sapiens'''
```

Self-Test Question 4-8

Consider the following labels of bacteria cultures: "Culture 1", "Culture 2", and "Culture 3". Each label is stored as a string in the variables `label1`, `label2`, and `label3`, respectively. What would be the code to create a string (and assigned to a variable `all_labels`) listing all the labels, separated by commas (i.e., "Culture 1, Culture 2, Culture 3")? Write this using the variables already defined, without rewriting the label text themselves.

Self-Test Question 4-9

Using the same variables from Self-Test Question 4-8, print all the labels but with just the label number listed (i.e., "Culture 1, 2, 3")? Write this using the variables already defined, without rewriting the label text themselves.

Self-Test Question 4-10

Consider the variable `sample_id` is set to the integer 23119. Create a string variable `id_label` that is set to "Sample 23119".

Self-Test Question 4-11

How does Python know when it has encountered a comment line?

Self-Test Question 4-12

What would be the Markdown entry in a Jupyter notebook for the following unenumerated (i.e., bulleted) list?

- 3 Erlenmeyer flasks.
- 10 test tubes.
- 1 Bunsen burner.

Self-Test Question 4-13

How do we continue a line of code on the next line? Please provide an example.

4.5.2   Chapter Summary

Matplotlib enables us to make x–y plots and gives intelligent defaults for the graphs we ask it to make. In looking at how to make basic line and scatter plots, we covered these topics:

- How to create a plot, label the axes, and display the plot on the screen

  - Importing the Matplotlib pyplot module.
  - The `plot` and `scatter` functions create line and scatter plots, respectively.
  - The `plot` and `scatter` functions take in two lists, one of x-values and the other of y-values.
  - The `xlabel` and `ylabel` functions write out x- and y-axis labels onto the plot.
  - When creating basic plots, first create the plot using a function like `plot` and `scatter` and then add on the labels to that plot using other functions like `xlabel` and `ylabel`.
  - The `show` function displays the plot to the screen.

- Positional input parameters

  - Functions accept input through input arguments that are mapped to input parameters.
  - The parameter list is the list of items (separated by commas) between the parentheses that directly follow the function's name. This list is defined when the function is defined.
  - The arguments are mapped to the parameters based on position in the parameter list. That is, the first argument is the value of the first parameter.
  - In the `plot` and `scatter` functions, the first argument is a list of x-values and the second argument is a list of y-values.

- Lists and tuples

  - Lists are ordered sequences.
  - Lists are comma-separated values placed inside a set of square brackets.
  - Each element of a list is referred to by its index.
  - The first element of a list has an index value of 0, the second element has an index value of 1, etc.

- The last element of a list has an index value of -1, the next to last element of a list has an index value of -2, etc.
- We refer to element(s) of a list by specifying the list variable name and then by placing the index or index range in square brackets, e.g., `data[3]`.
- We can slice a subrange of a list by specifying a range with a colon. The index value to the left of the colon is inclusive of that element while the index value to the right of the colon is exclusive of that element. Thus, `data[2:4]` will return a sublist that consists of the third (index 2) and fourth (index 3) elements of the list `data`.
- Tuples are just like lists except they cannot be changed. The values in the tuple are fixed unless the entire tuple is deleted.
- Tuples are comma-separated values placed inside a set of parentheses.

- Strings

  - Strings are made up of characters that are placed inside a matched pair of apostrophes, double quotation marks, or triple quotes.
  - Labels are a common use of strings.
  - Special characters such as a blank space, a tab, a newline, etc., are also characters in a string.
  - The newline character is `\n`.
  - Empty strings are strings with no characters in them.
  - Triple quotes can be used to easily create strings that have more complicated formatting.
  - The addition operator joins (or concatenates) two strings together.
  - The built-in `str` function converts nonstring values (such as numbers) into strings.
  - We can slice a portion of a string by using the slicing and indexing syntax of a list on the string. That is, we can think of a string as a list of characters.

- Commenting and Jupyter Markdown

  - Comment lines begin with the "#" character.
  - Jupyter notebooks can support additional documentation in Markdown cells. These cells can provide more complex formatting to their contents.

- Miscellaneous. The line continuation character is a backslash placed at the end of the line that will be continued by the next line.

As nice as the Matplotlib defaults are, we often want to customize what our graphs look like. We address such customization in Chapter 5.

By the way, the online pyplot tutorial is very good. We provide an up-to-date link to the tutorial at www.cambridge.org/core/resources/pythonforscientists/refs/, ref. 37. The online gallery of examples is also very illuminating; an up-to-date link to the gallery is at www.cambridge.org/core/resources/pythonforscientists/refs/, ref. 26.

4.5.3   Self-Test Answers

Self-Test Answer 4-1

The `plot` function creates a line plot. The `scatter` function creates a scatter plot. There are a variety of ways to import the pyplot functions. One way is by importing the pyplot submodule (usually as an alias like `plt`):

```
import matplotlib.pyplot as plt
```

Another way is by importing the functions directly:

```
from matplotlib.pyplot import plot
from matplotlib.pyplot import scatter
```

This way, there is no need to append the `plt` in front of the function names when calling the functions. We can refer to `plot` and `scatter` directly. See Section 3.2.2 for more on importing functions.

Self-Test Answer 4-2

This code would accomplish the task:

```
import matplotlib.pyplot as plt
plt.plot(xvalues, yvalues)
plt.show()
```

Self-Test Answer 4-3

Use the `xlabel` and `ylabel` functions. Each accepts a string as an input parameter, and writes that string out as the axis label, such as `xlabel("Time (sec)")`. Both functions are part of the pyplot submodule.

Self-Test Answer 4-4

In the first call, the variable x will be substituted in for the first dummy variable in the function's definition, y will be substituted in for the second dummy variable in the function's definition, and z will be substituted in for the third dummy variable in the function's definition. In the second call, the variable z will be substituted in for the first dummy variable in the function's definition, y will be substituted in for the second dummy variable in the function's definition, and x will be substituted in for the third dummy variable in the function's definition.

There is no other difference between the two function calls. The fact that "x, y, z" and "z, y, x" are in reverse order alphabetically has no impact in and of itself on what the functions do when they are called. The entire difference is in how the variable being passed in, at a given position, maps to the dummy variable at that position. In general, both function calls will work. The exception is if reordering the parameters results in an argument being passed in of a kind the function does not know how to handle. An error, if appropriate, might result.

Self-Test Answer 4-5

A list is an ordered sequence of elements. The elements of a list can be changed, reordered, etc. A tuple is also an ordered sequence of elements, but a tuple cannot be changed, reordered, etc.

Self-Test Answer 4-6

1. `len(data)`
2. `data[0]`
3. `data[2]`
4. `data[-1]` or `data[len(data)-1]`
5. `subdata = data[3:9]`

Self-Test Answer 4-7

There is no functional difference between the lines of code. Each is a valid definition of the variable `species_name` and sets the variable to the string "Homo sapiens". The triple quote versions, however, would behave differently if carriage returns were put into the definition, for instance. See Section 4.2.3 for details on triple quotes.

Self-Test Answer 4-8

Assuming we have this already defined:

```
label1 = "Culture 1"
label2 = "Culture 2"
label3 = "Culture 3"
```

Here is code that will define `all_labels`:

```
all_labels = label1 + ", " + label2 + ", " + label3
```

The "+" sign concatenates strings. The `'1'`, `'2'`, and `'3'` that are part of the contents of variables `label1`, `label2`, and `label3` are not integers but strings (i.e., they are the numerals not the numbers represented by the numerals).

Self-Test Answer 4-9

Here is code that will accomplish the task:

```
print(label1 + ", " + label2[-1] + ", " + label3[-1])
```

We use the -1 index to select the last character from label2 and label3. We can also index from the front of each string. This will do exactly the same thing as the above print call:

```
print(label1 + ", " + label2[8] + ", " + label3[8])
```

Self-Test Answer 4-10

To create id_label, we have to use the str function to convert the integer to a string before concatenating it with 'Sample ':

```
id_label = 'Sample ' + str(sample_id)
```

Note how 'Sample ' includes a space after the 'e', which separates 'Sample' from '23119'.

Self-Test Answer 4-11

When Python encounters a pound sign (#), all characters after the pound sign until the end of the line is considered a comment and is not executed.

Self-Test Answer 4-12

Type this code into the Jupyter notebook cell that has been designated a Markdown cell:

```
* 3 Erlenmeyer flasks
* 10 test tubes
* 1 Bunsen burner
```

Self-Test Answer 4-13

We put a backslash at the end of the line that will be continued in the following line, such as:

```
id_label = "Sample 23119" + \
           " from Stanford University"
```

The resulting string stored is 'Sample 23119 from Stanford University' without any line break between '23119' and 'from'.

5   Customized Line and Scatter Plots

In the previous chapter, we saw how to make a basic line and scatter plot. As useful as the default values are in Matplotlib, we often want to make our plots look different. In this chapter, we describe some of the ways of doing so. We also introduce more features of Python that enable us to do this customization.

5.1   Example of Customizing Line Plots

Let us revisit Figure 4.2. There is nothing wrong with the plot, but what if we would like to change some things? We discuss ways of making two common changes: (1) how the graph looks and what is on the graph, and (2) manipulations of the data we are plotting, so we can quickly view the data in a different way.

In Figure 5.1, we take the code that generated Figure 4.2 and make common formatting and related changes to the graph. Lines 1 and 2 are the same as in Figure 4.1. In our new version, however, the `plot` function call does not end in line 2 but continues through line 5. The **keyword input parameters** that are listed in lines 3–5 control, as the syntax suggests, how the markers and connecting lines look. Line 6 scales the plot so the *x*-axis goes from 8 to 18 and the *y*-axis goes from 0 to 5. Line 7 specifies exactly which ticks to draw on the *x*-axis. In line 10, we add a title at the top of the plot. Line 11 adds some additional text at the bottom left of the graph (the location of the text being specified in **data coordinates**). Finally, in the last line of the code snippet, we save the graph to a **Portable Network Graphics (PNG)** format image file, rather than displaying it on-screen using `show`. The plot that results from running Figure 5.1 is shown in Figure 5.2.

Scientific and engineering computations can be really complex and often involve computations on many pieces of data (hundreds, trillions, and more) instead of the simpler calculations we saw in Chapters 2 and 3. Here, we introduce the topic of doing such calculations within the context of manipulations of the data being graphed.

We take the code of Figure 5.1 above and change the *y*-values of the input data by subtracting out the mean of the *y*-values from each *y*-value. Figure 5.3 shows this changed code. The plot that results from that code is shown in Figure 5.4.

Most of the code in Figure 5.3 is the same as in Figure 5.1. The most significant difference is in lines 2–6 in Figure 5.3. Lines 10 and 13–15 in Figure 5.3 are also different, but the changes compared to similar lines in Figure 5.1 are relatively minor. In line 2, we import a module called NumPy (but whose specification in code is `numpy`). In line 3, we store the *x*-axis data

```
1    import matplotlib.pyplot as plt
2    plt.plot([9.25, 11, 13.5, 15, 15.75], [2.54, 4.1, 1.21, 3.9, 4],
3            linestyle='--', linewidth=5.0,
4            marker='*', markersize=20.0,
5            markeredgewidth=2.0, markerfacecolor='w')
6    plt.axis([8, 18, 0, 5])
7    plt.xticks([8, 10, 12, 14, 16, 18])
8    plt.xlabel('Decimal Hour in Day')
9    plt.ylabel('Power (mW)')
10   plt.title('Power vs. Time')
11   plt.text(8.2, 0.2, 'From Experiment 1')
12   plt.savefig('testplot.png', dpi=300)
```

Figure 5.1 Code to create a custom formatted version of the plot in Figure 4.2.

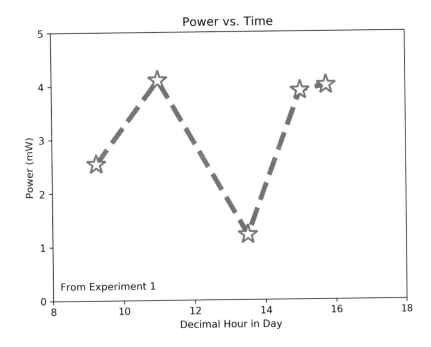

Figure 5.2 A custom formatted line plot, generated by the code in Figure 5.1.

values as a list, but in line 4, we store the *y*-axis data values as something that looks like a list but is not. It is called an **array**. Line 5 subtracts the mean of the *y*-values from each of the *y*-values and saves the result as the variable ydata_dev, which is itself an array of values.

This brief summary of the above code leaves a lot unsaid. As in previous chapters, in the next section, we go into detail as to how Python accomplishes the tasks in these two examples above.

```
1    import matplotlib.pyplot as plt
2    import numpy as np
3    xdata = [9.25, 11, 13.5, 15, 15.75]
4    ydata = np.array([2.54, 4.1, 1.21, 3.9, 4])
5    ydata_dev = ydata - np.mean(ydata)
6    plt.plot(xdata, ydata_dev,
7             linestyle='--', linewidth=5.0,
8             marker='*', markersize=20.0,
9             markeredgewidth=2.0, markerfacecolor='w')
10   plt.axis([8, 18, -3, 3])
11   plt.xticks([8, 10, 12, 14, 16, 18])
12   plt.xlabel('Decimal Hour in Day')
13   plt.ylabel('Power Deviations from the Mean (mW)')
14   plt.title('Power Deviations vs. Time')
15   plt.text(8.2, -2.8, 'From Experiment 1')
16   plt.savefig('testplot.png', dpi=300)
```

Figure 5.3 Code to create a custom formatted version of the plot in Figure 4.2, with the *y*-axis values in terms of the difference from the mean of the original *y*-values.

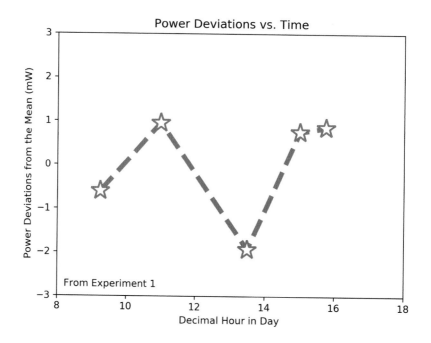

Figure 5.4 A custom formatted and transformed line plot, generated by the code in Figure 5.3.

5.2   Python Programming Essentials

In previous examples of calling functions, whether plotting functions like `scatter` or mathematical functions like `sin`, we passed in our input values to the function using positional input parameters. The calls to the function `plot` in the examples in Section 5.1 also have positional input parameters – i.e., the lists (or arrays) of x- and y-values – just as they did in Chapter 4. In Section 5.1, however, these function calls have additional input values where the values are attached to a label (e.g., `markersize`). These are **keyword input parameters** and are Python's primary way of passing optional input values into a function. As we saw above, they are also the means by which we can pass in the information to customize the formatting of our plots.

We describe keyword input parameters in detail in Section 5.2.1. In Section 5.2.2, we explore some of the keyboard input parameters and other means by which we can customize how our plots look. In Section 5.2.3 we look at some common cases where we need to handle multiple figures or curves, in Section 5.2.4 we discuss how to adjust the size of a plot, and in Section 5.2.5 we describe how to save plots to an image file.

In the second example in Section 5.1, we introduced a new kind of collection of items, the array, and saw how we can use arrays to do calculations on a collection of values. We provide a more complete introduction to array calculations in Section 5.2.6. Afterwards, in Section 5.2.7, we discuss the concept of **typing**. As of that point, we will have discussed: integers, decimal numbers, strings, lists, and arrays. Each of these, it turns out, are "types" of values. The fact that one kind of value in a computer is different from another kind of value is an important aspect of programming. Lists and arrays, as collections of items, are also called **data structures**. As we go along, we will be introduced to additional kinds of types and data structures.

Finally, recall our discussion of the `show` command in Section 4.2 and how often we do not need to call it to render a plot in the Spyder and Jupyter environments. In the examples in this present chapter, we often leave out a call to `show`. If the plot does not render in our examples, try calling `show` or using the line magic described in Section 4.2.

5.2.1   Optional Input into Functions Using Keyword Input Parameters

Keyword input parameters are the main means by which we provide optional input into the call of a function. As the examples in Section 5.1 show, the syntax of keyword input parameters consists of the keyword name or label, an equal sign, and the value the input parameter is set to. Thus, as the Figure 5.1 `plot` call shows us (reproduced here for convenience):

```
plt.plot([9.25, 11, 13.5, 15, 15.75], [2.54, 4.1, 1.21, 3.9, 4],
        linestyle='--', linewidth=5.0,
        marker='*', markersize=20.0,
        markeredgewidth=2.0, markerfacecolor='w')
```

the first keyword input parameter sets the `linestyle` input parameter to the string `'--'`, the second keyword input parameter sets the `linewidth` input parameter to the decimal number 5.0, and so on. On the inside of the `plot` function, there are (in essence) variables called `linestyle`, `linewidth`, etc., which are then set to the values assigned to the keyword in the function's calling line. The pyplot documentation lists all the keyword input parameters available to the `plot` command. An up-to-date link to this list is at www.cambridge.org/core/resources/pythonforscientists/refs/, ref. 34.

As a general rule, keyword input parameter variables are given default values in the function they are defined for. Thus, if no keyword input parameter is given in the function's calling line, the variable will still have a value. As a result, keyword input parameter arguments are optional: we do not have to set them to a value when we are calling the function unless we want the parameter to have a different value. In contrast, recall that positional input parameters are used for required input values.

Let us move from using keyword input parameters in a function call to defining keyword input parameters in functions we write. For this example, we create a function called `force_gravity` that calculates the gravitational force on an object (in N) given its mass (in kg). Assuming the object is on the surface of the Earth with a gravitational acceleration of 9.81 m/s^2, and force equals mass times acceleration ($F = ma$), the function would be the following:

```
1  def force_gravity(in_mass_kg):
2      return in_mass_kg * 9.81
```

and here is an example of using the function:

```
In [1]:  force_gravity(3)

Out [1]:  29.43
```

Say, however, that we wanted to make the function more flexible. We believe that most of the time it will be used for objects resting on the surface of the Earth, but in some cases, the objects will be resting on other planets. We want to be able to pass in the gravitational acceleration of the planet but not *require* the user to do so. By default, we want the function to assume the object is on the surface of the Earth. The following `force_gravity` definition will provide this functionality:

```
1  def force_gravity(in_mass_kg, accel=9.81):
2      return in_mass_kg * accel
```

In the above definition, the `accel` keyword (and variable in the function) is set by default to 9.81. A different value for `accel` will be used only if `force_gravity` is called with a keyword input parameter set for `accel`. Thus, both these function calls work as desired:

```
In [1]:  force_gravity(3)
```

```
Out[1]:  29.43
```

```
In [2]:  force_gravity(3, accel=3.711)
```

```
Out[2]:  11.133
```

Incidentally, 3.711 m/s^2 is the acceleration of gravity on the surface of Mars.[1]

The bottom line: Keyword input parameters enable us to pass in optional input to a function.

5.2.2 Customizing How the Plot Looks

In this section, we look at ways to customize how a plot looks. In Matplotlib, keyword input parameters are often used for this purpose, but other tools are also used to get the job done.

Controlling the Axes Ranges and Ticks

We often want to "recenter" a plot, so that the x- and y-axis ranges are exactly what we want rather than what is chosen for us by default. As we saw in the Figure 5.1 code, a way to do this is by using the `axis` function, as in line 6 of Figure 5.1 (reproduced here for convenience):

```
plt.axis([8, 18, 0, 5])
```

The single argument to the `axis` function is a list where the first two elements are the resized graph's lower and upper x-axis bounds and the third and fourth elements are the resized graph's lower and upper y-axis bounds.

To override Matplotlib's defaults and specify exactly which x- and y-values to draw and label ticks at, use the `xticks` and `yticks` functions, respectively.[2] For instance:

```
plt.xticks([8, 10, 12, 14, 16, 18])
```

puts ticks along the x-axis at values of 8, 10, 12, 14, 16, and 18. Ticks can be disabled for an axis by passing in an empty list (i.e., `[]`, a list without any values) to the `xticks` and/or `yticks` function.

Controlling Line and Marker Formatting

To control line and marker features, we can use the appropriate keyword input parameters in the `plot` function. For instance, lines 2–5 of Figure 5.1 show such a use of keyword input

---

[1] www.universetoday.com/14859/gravity-on-mars/ (accessed August 25, 2017).

[2] https://matplotlib.org/3.1.0/api/_as_gen/matplotlib.pyplot.xticks.html and https://matplotlib.org/3.1.0/api/_as_gen/matplotlib.pyplot.yticks.html (accessed June 6, 2019), https://stackoverflow.com/a/12608937 (accessed February 17, 2018).

Table 5.1 Some linestyle codes in pyplot and a plot showing the lines generated by the linestyle codes.

Linestyle	String code
Solid line	' _ '
Single dashed line	' _ _ '
Single dashed-dot line	' _ . '
Dotted line	' : '

parameters. The `linestyle`, `marker`, and `markerfacecolor` keywords use special string codes to specify the line and marker type and formatting. Linewidth, marker size, and marker edge width are in points. Thus, the lines 2–5 of Figure 5.1 `plot` call use a thick dashed line and a prominent white star for the marker. While this `plot` call does not use this, the `color` keyword sets the color for the marker and connecting line, unless overwritten by other settings.

While keyword input parameters make the code easy to read, Matplotlib provides a shorthand way of specifying a number of these format specifications without using keyword input parameters. Instead, we specify line color and type and marker color and type as a string third positional input argument, e.g.:

```
plt.plot([1, 2, 3, 4], [1, 2.1, 1.8, 4.3], 'r*--')
```

Notice that this third argument contains *all* the codes to specify line color, line type, marker color, and marker type. That is to say, all these codes can be specified in one string. In the above example, the color of the marker and connecting line is set to red, the marker is set to star, and the linestyle is set to dashed. The marker edge color is still the default, black, however. Note we can only use a third positional input argument in this way with `plot`, not `scatter`. Overall, `scatter` has fewer options for customizing plots than `plot`, so since we can use `plot` to create "scatter" plots by omitting the connecting line between markers, `plot` is generally more useful than `scatter`.

Tables 5.1 and 5.2 list some of the basic linestyles and marker codes.[3] The Matplotlib documentation provides more information on markers and lists. We provide an up-to-date link to the documentation on markers at www.cambridge.org/core/resources/pythonforscientists/refs/, ref. 31, and to the documentation on linestyles at ref. 30. Table 5.3 lists some of the preset color codes. The Matplotlib documentation contains a full list of the built-in color codes and describes ways to access other colors. We provide an up-to-date link to this documentation at www.cambridge.org/core/resources/pythonforscientists/refs/, ref. 24.

---

[3] The page http://matplotlib.sourceforge.net/api/pyplot_api.html (accessed August 13, 2012) is a reference for these tables. See http://stackoverflow.com/a/13360032 for a way of listing all the linestyle and marker codes.

Table 5.2 Some marker codes in pyplot and a plot
showing the markers generated by the marker codes.

Marker	String code
Circle	`'o'`
Diamond	`'D'`
Point	`'.'`
Plus	`'+'`
Square	`'s'`
Star	`'*'`
Up triangle	`'^'`
X	`'x'`

Table 5.3 Some color codes in pyplot.

Color	String code
Black	`'k'`
Blue	`'b'`
Green	`'g'`
Red	`'r'`
White	`'w'`

Annotation and Adjusting the Font Size of Labels

We introduced the `xlabel` and `ylabel` functions in Section 4.1 to annotate the x- and y-axes, respectively. To place a title at the top of the plot, we use the `title` function, whose basic syntax is the same as `xlabel` and `ylabel`.

General annotation in the plot uses the `text` function, whose syntax is:[4]

```
plt.text(<x-location>,<y-location>,<string to write>)
```

---

[4] For most users, plain text annotation is enough. Many scientific users, however, also want to create sophisticated mathematical annotations. Matplotlib provides the ability to use the LATEX typesetting system to render text, including mathematical symbols and equations. Using LATEX is outside the scope of the present work. We provide an up-to-date link to documentation on using LATEX at www.cambridge.org/core/resources/pythonforscientists/refs/, ref. 35.

The *x*- and *y*-locations are, by default, in terms of data coordinates. Data coordinate (x, y) locations on the plot are given in units of the plot's *x*- and *y*-values. Thus, in the Figure 5.1 `text` call:

```
plt.text(8.2, 0.2, 'From Experiment 1')
```

the text "From Experiment 1" is positioned so the left side of the text box begins at a decimal hour in day value of 8.2 and the bottom of the text box is located at a power value of 0.2 mW.

For these four functions (`xlabel`, `ylabel`, `title`, and `text`), the font size is controlled by the `size` keyword input parameter. When set to a floating-point value (that is, a number with a decimal point; see Section 5.2.7 for more on floating-point numbers), `size` specifies the size of the text in points.

5.2.3   Handling Multiple Figures or Curves

Matplotlib also handles multiple figures or curves: the creation of several plots in separate windows or the superimposing of several curves on the same plot. In this section we look at both cases as well as how to create legends for plots with multiple curves.

Plotting Multiple Figures

If we have have multiple independent figures (not multiple curves on one plot), we call the `figure` function before we call `plot`. The former creates and labels the figure accordingly. A subsequent call to that figure's number makes that figure current. For instance, consider this code:

```
1   plt.figure(3)
2   plt.plot([9.25, 11, 13.5, 15, 15.75], [2.54, 4.1, 1.21, 3.9, 4],
3           marker='o')
4   plt.figure(4)
5   plt.plot([0.1, 0.2, 0.3, 0.4], [8, -2, 5.3, 4.2], \
6           linestyle='-.')
7   plt.figure(3)
8   plt.title('Power vs. Time')
```

Line 1 creates a figure and gives it the name "3". Lines 2–3 (which we saw in Section 4.3 is a single logical line to the interpreter, even without the backslash at the end of line 2) makes a line plot with a circle as the marker to the figure named "3". Line 4 creates a figure named "4", and lines 5–6 makes a line plot with a dash-dot linestyle to that figure. Line 7 makes figure "3" the current plot again, and the final line adds a title to the plot in figure "3".

Note that the names "3" and "4" for the above figures are *not* the titles of the plots but the names of the windows. Put another way, they are the names many windowing systems (i.e., what the computer's operating system uses to render the windows it displays) will use as the label displaying in the name bar above the menu bar of the window. As in our example above, the figure name and plot title in the last figure are entirely different.

In the above example, the integers "3" and "4" were passed in as the names of the figures. Strings can also be passed in as names, rather than integers.

Plotting Multiple Curves

To plot multiple curves on a single plot, make repeated calls to `plot` without any intervening calls to `figure`. The lines for each call will be added to the same figure. Alternately, we can provide the set of three positional input arguments (x-locations, y-locations, and line/marker properties) for each plot, one right after the other. For instance, consider this code:

```
plt.plot([0, 1, 2, 3], [1, 2, 3, 4], '--o',
         [1, 3, 5, 9], [8, -2, 5.3, 4.2], '-D')
```

The first three arguments specify the x- and y-locations of the first curve, which will be plotted using a dashed line and a circle as the marker. The second three arguments specify the x- and y-locations of the second curve, which will be plotted with a solid line and a diamond as the marker. Both curves will be on the same figure.

Adding a Legend

To add a legend, with the default settings, we call the `legend` function (with no arguments) after making the plots. When we plot each curve, however, we have to set the `label` keyword input parameter to what the text for that curve should be in the legend. For instance:

```
plt.plot([0, 4, 7, 8], [1, 2, 3, 4], 'r--o', label="Sensor 1")
plt.plot([1, 3, 5, 9], [8, -2, 5.3, 4.2], 'b-D', label="Sensor 2")
plt.legend()
```

produces the plot in Figure 5.5. Note the `'r'` and `'b'` strings in the `plot` calls produce a red and blue line/marker, respectively. For more information on legends, see the Matplotlib documentation on the `legend` command and the legend guide tutorial. We provide up-to-date links to each of these resources at www.cambridge.org/core/resources/pythonforscientists/ refs/, refs. 33 and 29 respectively. This documentation is not simple, though, so reading it will likely be more fruitful after some more experience with Python.

For the related but different case of making multiple plots or panels on a single figure, see Sections 8.2.5 and 14.2.1.

5.2.4 Adjusting the Plot Size

One way of adjusting the plot size is to set the `figsize` and `dpi` keyword input parameters in the `figure` command.[5] For instance, this call to `figure`:

```
plt.figure(1, figsize=(3,1), dpi=300)
```

---

[5] http://stackoverflow.com/a/638443 (accessed August 13, 2012).

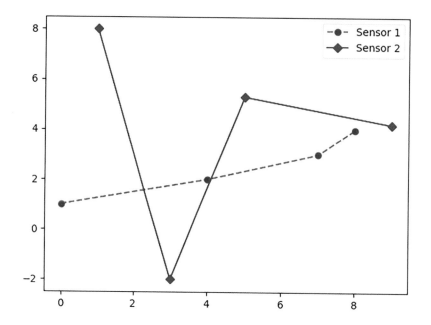

Figure 5.5 Graph created by the code in Section 5.2.3, "Adding a Legend."

before the call to the `plot` command, will make the figure named "1" 3 in wide and 1 in high, with a resolution of 300 **dots per inch (dpi)**. Edges of letters and lines look reasonably smooth when printed out at 300 dpi. As of the writing of the present work, most laser printers default to 600 dpi. The `figsize` keyword is set to a two-element tuple.

5.2.5 Saving Figures to a File

To write the plot out to a file, we can use the `savefig` function. For example, to write out the current figure to a PNG file called *testplot.png*, at 300 dpi, type:

```
plt.savefig('testplot.png', dpi=300)
```

Python knows to save the file to PNG format based on the filename suffix (the letters after the rightmost period).

In the above call, the output resolution is specified using the optional `dpi` keyword input parameter. If left out, the Matplotlib default resolution will be used. It is not enough for us to set `dpi` in our `figure` command to get an output file at a specific resolution. The `dpi` setting in `figure` will control what resolution `show` displays at while the `dpi` setting in `savefig` will control the output file's resolution. However, the `figsize` parameter in `figure` controls the figure size for both `show` and `savefig`.

We can also save figures to a file using the graphical user-interface (GUI) Save button that is part of the plot window displayed on the screen when executing the `show` function. If we save the plot using the Save button, it will save at the default resolution, even if we specify a different resolution in our figure command. Use savefig to write out the file at a specific resolution.

5.2.6 Introduction to Array Calculations

In lines 2–4 of Figure 5.3, we provided the following lines of code (reproduced here for convenience):

```
1   import numpy as np
2   ydata = np.array([2.54, 4.1, 1.21, 3.9, 4])
3   ydata_dev = ydata - np.mean(ydata)
```

The variable ydata in the above code snippet is an example of a new kind of collection of items, the array. The use of ydata to calculate ydata_dev is an example of **array syntax**, which enables us to do calculations with all or some of the elements in a group of numbers.

Like a list, an array is a collection of elements. We can also think of an array as a row of cubbies or post office boxes where each element of the array is a cubby or box and can hold something (often a value). We can create an array from a list by using the NumPy function array. Thus, ydata is an array while the argument in the array function call is a list.

Like a list, we refer to the elements in an array using array indices. The first element of an array has the index 0, the second element has the index 1, and so on. We can also index an array from the end by using negative indices, where the last element in the array is index -1, the next-to-last element is index -2, and so on. In the above array ydata, the following:

```
In [1]:  print(ydata[0])

         2.54

In [2]:  print(ydata[1])

         4.1

In [3]:  print(ydata[-1])

         4.0

In [4]:  print(ydata[-2])

         3.9
```

prints out the first, second, last, and next-to-last element in the array ydata.

Arrays also use the slicing notation for lists to extract a subarray from an existing array. Thus:

```
In [1]:  ydata_sub = ydata[0:3]
         print(ydata_sub)


         [ 2.54   4.1    1.21]
```

prints out the first three elements of ydata. In array slicing, the index to the left of the colon defines the lower bound of the range *inclusively* while the index to the right of the colon defines the upper bound of the range *exclusively*. Thus, in the above ydata[0:3] example, the slice starts with the index 0 element (i.e., 2.54) but ends with the index 2 element (i.e., 1.21). The upper bound of the range is index 3, exclusive, so the last index in the range is the index 2 element, which is one element less than the index 3 element.

Array slicing follows all the other list slicing rules given in Section 4.2.2, as these examples illustrate:

```
In [1]:  import numpy as np
         ydata = np.array([2.54, 4.1, 1.21, 3.9, 4.0])
         print(ydata[0:-1])


         [ 2.54   4.1    1.21   3.9 ]

In [2]:  print(ydata[-3:])


         [ 1.21   3.9    4.  ]

In [3]:  print(ydata[:4])


         [ 2.54   4.1    1.21   3.9 ]
```

In the third line of In [1], we select every element except the last because the -1 upper bound is exclusive, since it is to the right of the colon. In cell In [2], we start with the -3 index – i.e., the next to the next to last element – and slice to the end of the array, because there is no index to the right of the colon in the slice. In cell In [3], we start from the beginning of the array and end at the index 3 element, because the 4 index is exclusive. The lack of an index to the left of the colon tells us to slice all the way to the beginning of the array.

We can also use the slicing notation to assign multiple values at once in an array. Consider the following code operating on the array zdata:

```
In [1]:  import numpy as np
         zdata = np.array([0.43, 8.4, 2.1, 3.4, 5.9])
         zdata[0:3] = zdata[2:]
         print(zdata)


         [2.1 3.4 5.9 3.4 5.9]
```

In the third line of `In [1]`, we process the right-hand side first and extract the last three elements of `zdata`. Then, we execute the left-hand side, assigning each of those three elements to the first three locations in the `zdata` array. In the third line of `In [1]`, corresponding elements between the expressions on the left- and right-hand sides of the equal sign are assigned in the left-hand side.

So, what is the point of using an array if it is so similar to a list? Here are three reasons (a few more reasons are described later on in this book). First, operations with arrays generally execute faster than operations with lists. This is because with arrays, each element of an array must be the same *kind* of value. In Section 5.2.7, we will describe this with more precision and see that in a programming language, the "kind" of a value is described by the value's **type**. With lists, even though all the lists we have seen so far have numbers in them, we will find in Section 9.2.3 that any given list can have numbers, words, and even other lists in them. The elements in a list can hold different kinds of values, and that flexibility exacts a cost in performance. Arrays also execute faster because once we have created an array, its size cannot change. Lists, on the other hand, can grow and shrink as the program executes. The requirement that arrays have fixed size and a single element type enables them to be manipulated more efficiently.

A second reason to use arrays instead of lists is that arrays provide a myriad number of ways to operate on the values in an array. One way is through various functions. For instance, the `sum` function in the NumPy package will add up all the elements in the array that is passed into the function:

```
In [1]:   import numpy as np
          ydata = np.array([2.54, 4.1, 1.21, 3.9, 4])
          total = np.sum(ydata)
          print(total)
```

```
15.75
```

And the NumPy function `size` will tell us how many elements are in the array:

```
In [1]:   import numpy as np
          ydata = np.array([2.54, 4.1, 1.21, 3.9, 4])
          num_elements = np.size(ydata)
          print(num_elements)
```

```
5
```

Section 6.2.2 describes more functions that act on NumPy arrays and different ways those functions can behave. Section 12.2.7 provides a brief summary table of functions that work on NumPy arrays.

A second way we can operate on elements in an array is through array syntax. Array syntax means that for certain operations (or functions, which we will explore in Section 6.2.2) using arrays, the operation will execute element-wise (i.e., on each element separately). For instance, in the following code:

```
import numpy as np
data1 = np.array([1.2, 3.4, -2.1,  0.8])
data2 = np.array([9.1, 0.2,  0.1, -1.0])
data3 = data1 + data2
```

each corresponding element of `data1` and `data2` are added together and put into an array `data3`. The `data3` array has the same size as `data1` and `data2`. Thus, the 1.2 in `data1` is added to the 9.1 in `data2` to obtain the first element of `data3`, which will have the value 10.3. The index 1 elements in `data1` and `data2` are added up to obtain the index 1 element of `data3`, and so on. Thus, printing `data3` gives us:

```
In [1]:  print(data3)

         [ 10.3   3.6  -2.   -0.2]
```

In order for this to work, `data1` and `data2` have to be the same size. Otherwise, Python will not know how to match corresponding elements for the operation. When, later in this book, we examine array syntax for arrays of more than one dimension (as in the current case), there will be additional constraints.

If we add a **scalar** value (i.e., a plain number) to an array, such as:

```
import numpy as np
data1 = np.array([1.2, 3.4, -2.1,  0.8])
data_value = 0.1
data4 = data1 + data_value
```

the scalar value is added to each element, independently, in `data1` with the result being an array `data4` of the same size as `data1` with the following values:

```
In [1]:  print(data4)

         [ 1.3   3.5  -2.    0.9]
```

All the arithmetic operators (+, -, *, /, and **) operate on arrays element-wise in the way we just saw with addition. Thus, in very few lines of code, we can evaluate a complex expression on a large number of values. For instance, if we have an array of x-components of four velocity vectors called `vx` and an array of y-components of those vectors called `vy`:

```
import numpy as np
vx = np.array([ 0.2, 1.4, -1.9,  7.8])
vy = np.array([-0.3, 2.1,  3.2,  1.7])
```

we can calculate the magnitudes of all four vectors and store the result in the array `magnitudes` with just one line of code:

```
In [1]:   magnitudes = ((vx**2) + (vy**2))**0.5
          print(magnitudes)
```

```
          [ 0.36055513   2.52388589   3.72155881   7.98310716]
```

where the first line of `In [1]` calculates the magnitude of a velocity vector $\vec{v}$ using:

$$|\vec{v}| = \sqrt{v_x^2 + v_y^2}$$

for each of the four vectors encoded in vx and vy.

Note that with the code above, if vx and vy had 10 elements (describing 10 velocity vectors), rather than the 4 elements given above, the computation line for `magnitudes` (the first line of `In [1]` above) *would not change.* Python automatically makes `magnitudes` 10 elements long, does the calculations element-wise with vx and vy, and fills in the elements of `magnitudes`. This also would be exactly the same if vx and vy had 10 000 elements or 10 million. The first line of `In [1]` above would not change. Array syntax is very powerful ☺.

5.2.7 The Concept of Typing

By "typing," we do not mean pressing letters on a keyboard but the idea of a "type" or "kind" of something. In programming, typing refers to the specification that a value (or a variable set to that value) has to have certain characteristics. These characteristics restrict the kinds of values that type can be set to, how the value can be stored in memory, how the value can be manipulated, etc.

The built-in function `type` will tell us what type a variable or value is:

```
In [1]:   import numpy as np
          a = 4
          type(a)
```

```
Out [1]:  int
```

```
In [2]:   type(2.3)
```

```
Out [2]:  float
```

```
In [3]:   b = np.array([1.2, 3.1, -4.5])
          type(b)
```

```
Out [3]:  numpy.ndarray
```

The variable a in cell `In [1]` is an `int` or integer variable. The value 2.3 (`In [2]`), in contrast, is a `float` or floating-point value (roughly, a decimal value). The variable b in cell `In [3]` is a NumPy array and the `type` function tells us that the official type of a NumPy array is `numpy.ndarray`.

The variables a and b, because they are of different types, have different limits and capabilities. Because variable a has type int, we know that its value is either a positive or negative whole number (or zero) and we can use the variable in an expression that are typically defined for integer values (e.g., arithmetic). Because variable b has type numpy.ndarray, all the operations we have seen we can do to a NumPy array are available to that variable.

One special thing about Python is that it is a **dynamically typed** language, meaning that the type of a variable can change with time. In Python, the type of a variable is automatically set to the type of the value given on the right-hand side of the assignment. Thus, as we see below:

```
In [1]:   import numpy as np
          a = 4
          type(a)

Out[1]:   int


In [2]:   a = 2.3
          type(a)

Out[2]:   float


In [3]:   a = np.array([1.2, 3.1, -4.5])
          type(a)

Out[3]:   numpy.ndarray
```

the value the variable a is assigned to changes, and when it does, the type of the variable a also changes (as we see from our calls to the type function after each new assignment of a).

In contrast, in languages like Java, C++, Fortran, etc., variables are **statically typed**. In those languages, when we create (or "declare") a variable, we give it a type. That type cannot change for the life of the variable. Later, we describe some capabilities dynamic typing gives us in Python.

There are a number of built-in functions that enable us to convert one type of value or variable to another. For instance, the int function converts a value to an int type value (that is, an integer), the float function converts a value to a float type value (that is, a floating-point or decimal number), and the str function converts a value to a string:

```
In [1]:   a = 4
          type(a)

Out[1]:   int
```

```
In [2]:  b = float(a)
         print(b)
```

```
Out[2]:  4.0
```

```
In [3]:  type(b)
```

```
Out[3]:  float
```

```
In [4]:  c = int(b)
         print(c)
```

```
Out[4]:  4
```

```
In [5]:  type(c)
```

```
Out[5]:  int
```

```
In [6]:  d = str(c)
         print(d)
```

```
Out[6]:  4
```

```
In [7]:  type(d)
```

```
Out[7]:  str
```

While printing variables c and d seems to suggest they have the same values, as the calls to the type function show, the values are entirely different types. In the case of variable d, the 4 is the string representation of the numeral, not the integer value. We can see the impact of this if we try to perform the following operations on variables c and d (with continuing Jupyter notebook cells):[6]

```
In [8]:  print(c + 10)

         14
```

```
In [9]:  print(d + 10)

         TypeError     Traceback (most recent call last)
         <ipython-input-5-261d43c710bf> in <module>()
         ----> 1 print(d + 10)

         TypeError: must be str, not int
```

[6] The error messages output are edited for clarity.

```
In[10]:  print(c + " organisms")

         TypeError      Traceback (most recent call last)
         <ipython-input-3-1f551119eb43> in <module>()
         ----> 1 print(c + " organisms")

         TypeError: unsupported operand type(s) for +: 'int' and 'str'

In[11]:  print(d + " organisms")

         4 organisms
```

In cell In [8], the addition operator works arithmetically, and the result is the integer 4 plus the integer 10. The same operation on variable d in cell In [9] results in an error. Likewise, in cell In [10], we cannot concatenate the integer variable c with the string " organisms", but we can do so with variable d (In [11]).

We cannot use these built-in functions to convert an array of floating-point values to an array of integers or an array of strings to an array of floating-point values. Section 9.2.2 describes how to do such a conversion with a NumPy array. In that section, we will also go deeper into the meaning of the integer and floating-point types.

5.3   Try This!

We start with graphing examples similar to ones we have seen in Section 5.1. These are followed by examples that address the main topics of Section 5.2. We practice creating more customized line plots and using keyword input parameters. We also practice accessing elements and ranges of elements in an array and using array syntax to make calculations on a collection of values.

Try This! 5-1 Customized Line Plot: The Mouse Maze

Consider a fictional experiment where mice are given a dose of a medication and are placed in a maze which they have to complete. The following data shows the dose of the medicine given to a mouse and the time it took for the mouse to complete the maze.

Dosage (mg)	Time to complete maze (s)
0.0	105
0.08	98
0.2	54
0.37	50
0.6	65
0.84	81
1.02	182
1.2	210

Create a line plot of time to complete the maze (on the *y*-axis) versus dosage (on the *x*-axis). Make the line and marker substantially larger and make both the *x*- and *y*-axis ranges begin at zero and end at the maximum value for the respective axes. Add text in the graph frame that labels the graph as "Provisional." Write out the graph to the file *mazetimes.png* at 300 dpi. Please label the axes appropriately.

Try This! Answer 5-1

This code will solve the problem:

```
import matplotlib.pyplot as plt
import numpy as np

xdata = [0.0, 0.08, 0.2, 0.37, 0.6, 0.84, 1.02, 1.2]
ydata = [105, 98, 54, 50, 65, 81, 182, 210]
plt.plot(xdata, ydata, linewidth=5.0, marker='o', markersize=15.0)
plt.axis([0, np.max(xdata), 0, np.max(ydata)])
plt.xlabel('Dosage (mg)')
plt.ylabel('Time to Complete Maze (sec)')
plt.text(0.1, 10, 'Provisional')
plt.savefig('mazetimes.png', dpi=300)
```

The plot generated by the above code is in Figure 5.6.

In the solution, we use the default colors and linetype. In line 7, we use the return values from the calls to the NumPy max functions as the values in the list that is passed to the pyplot

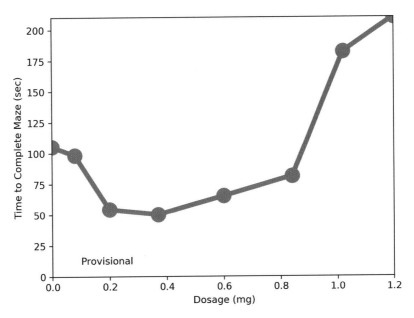

Figure 5.6 Customized line plot generated by the solution to Try This! 5-1.

axis command. Finally, in line 10, the coordinates to place the "Provisional" text are in data coordinates. In Section 14.2.1, where we introduce Matplotlib's object interface, we describe other ways of specifying the location of the text.

Try This! 5-2 Functions with Keywords: The Mouse Maze

Consider the fictional mouse maze experiment described in Try This! 5-1, where mice are given a dose of a medication and are placed in a maze which they have to complete.

Write a function that creates a plot like in Try This! 5-1 but enables the passing in of keyword values when calling the function to control the linewidth, marker size, and filename. That is, if the function is named plot_maze_times and the function is called with the following input values:

```
plot_maze_times(xdata, ydata, linewidth=2.0, markersize=10.0,
                filename='temp1.png')
```

it will produce the figure in Figure 5.7 and name it *temp1.png*. However, this call:

```
plot_maze_times(xdata, ydata, linewidth=10.0, markersize=30.0,
                filename='temp2.png')
```

will produce the figure in Figure 5.8 and name it *temp2.png*. In both calls, xdata and ydata are lists as defined in the Try This! 5-1 table for dosage and maze completion time, respectively.

Try This! Answer 5-2

This code will solve the problem:

```
 1   import matplotlib.pyplot as plt
 2   import numpy as np
 3
 4   def plot_maze_times(xdata, ydata, linewidth=1.0, markersize=3.0,
 5                       filename='default.png'):
 6       plt.figure()
 7       plt.plot(xdata, ydata, linewidth=linewidth, marker='o',
 8               markersize=markersize)
 9       plt.axis([0, np.max(xdata), 0, np.max(ydata)])
10       plt.xlabel('Dosage (mg)')
11       plt.ylabel('Time to Complete Maze (sec)')
12       plt.text(0.1, 10, 'Provisional')
13       plt.savefig(filename, dpi=300)
```

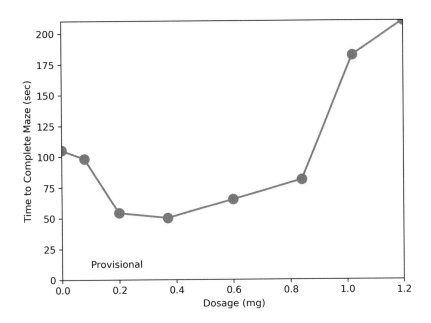

Figure 5.7 Line plot generated by a call to `plot_maze_times` in Try This! 5-2 and written to *temp1.png*.

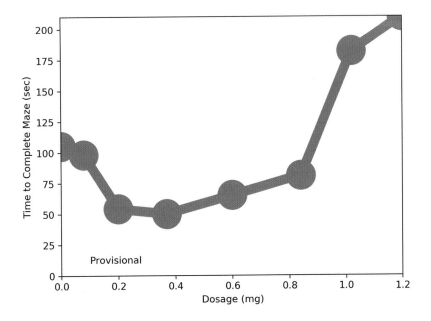

Figure 5.8 Line plot generated by a call to `plot_maze_times` in Try This! 5-2 and written to *temp2.png*.

Some items to note:

- If we have not already imported the packages we need, we need to do so (lines 1 and 2). If `plt` and `np` are already defined, there is no need to import them again.
- The body of a function has to be indented in four spaces.
- When we have lines of code that are too long but the line is made up of a sequence of items (separated by commas, as in lines 4 and 5 and 7 and 8), we can continue the line over multiple lines without using the line continuation character. See the discussion in the Try This! 4-1 solution.
- Calling pyplot's `figure` command without any arguments will create another, separate, blank figure. This prevents the later plotting commands from being applied to an existing Matplotlib figure.
- Finally, the default values for the `linewidth`, `markersize`, and `filename` keywords (as given in lines 4 and 5) mean that a call to `plot_maze_times` without any of the keyword input parameters specified will produce a plot with a line width of 1 point, marker size of 3 points, and a filename of *default.png*. Thus, this is an entirely correct use of `plot_maze_times`:

```
xdata = [0.0, 0.08, 0.2, 0.37, 0.6, 0.84, 1.02, 1.2]
ydata = [105, 98, 54, 50, 65, 81, 182, 210]
plot_maze_times(xdata, ydata)
```

Try This! 5-3 Multiple Curves: Gas Properties

Pretend we are conducting an experiment where we are compressing 4 mol of a gas into a smaller and smaller volume, and as we change the volume, we measure the pressure of the gas. Assume that we are compressing the gas so slowly that the temperature remains constant. We obtain the following measurements in this experiment:

Volume (m^3)	Pressure (Pa)
0.50	19 822
0.45	22 024
0.40	24 777
0.35	28 317
0.30	33 036
0.25	39 643

Later on, we rerun the same experiment but use 8 mol of the same gas instead of 4 mol. For the 8 mol case, we obtain the following data:

Volume (m^3)	Pressure (Pa)
0.50	39 643
0.45	44 048
0.40	49 554
0.35	56 633
0.30	66 072
0.25	79 286

Create a plot of pressure (*y*-axis) versus volume (*x*-axis) that superimposes the data from the 4 mol and 8 mol cases. Format the plot appropriately, but make sure the lines connecting the points are 3 points thick and the markers are 6 points in size. Make *x*-axis tick marks every 0.1 m^3, and include a text message in the lower-left corner in 9 point font giving the name of the experimenter. Include a legend. Save the plot to the file *gas-properties.png*. Remember to make the two curves distinguishable from one another.

Try This! Answer 5-3

This code creates the graph requested:

```
 1   import matplotlib.pyplot as plt
 2   volume = [0.5, 0.45, 0.4, 0.35, 0.3, 0.25]
 3   pressure_4mol = [19822, 22024, 24777, 28317, 33036, 39643]
 4   pressure_8mol = [39643, 44048, 49554, 56633, 66072, 79286]
 5   plt.plot(volume, pressure_4mol, 'D-r', label='4 mol',
 6            linewidth=3.0, markersize=6.0)
 7   plt.plot(volume, pressure_8mol, 'o--b', label='8 mol',
 8            linewidth=3.0, markersize=6.0)
 9   plt.xticks([0.5, 0.4, 0.3, 0.2])
10   plt.text(0.21, 20000, 'Jill Smith', size=9)
11   plt.xlabel('Volume (cubic meters)')
12   plt.ylabel('Pressure (Pa)')
13   plt.title('Pressure vs. Volume')
14   plt.legend()
15   plt.savefig('gas-properties.png', dpi=300)
```

This code uses two subsequent `plot` calls (lines 5–8), which put both curves on the same figure. For the 4 mol case (line 5), the line/marker formatting argument `'D-r'` selects a diamond marker and a solid, red line. For the 8 mol case (line 7), the line/marker formatting argument `'o--b'` selects a circle marker and a dashed, blue line. We can combine the use of the third positional input argument (`'D-r'` in line 5 and `'o--b'` in line 7) to specify the marker style, linestyle, and color with the use of keyword input parameters (in lines 6 and 8) to specify line width and the size of the marker. The `xticks` function also works with the points descending as well ascending. Figure 5.9 shows the plot generated by the above code.

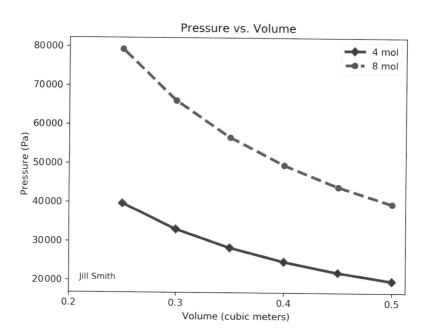

Figure 5.9 Customized line plot generated by the solution to Try This! 5-3.

Try This! 5-4 Array Indices and Slicing: A Skydiver

Consider the following fictitious values of velocity of a skydiver jumping out of an airplane at time $t = 0$ s. Remember, the skydiver feels air resistance. All velocity values are positive downward. Note the time values are irregularly sampled.

Time (s)	Velocity (m/s)
0	0
0.1	0.98
0.5	4.89
1.3	12.49
1.8	16.95
1.9	17.81
2.1	19.49
2.3	21.11
2.5	22.68
2.8	24.91
2.9	25.63
3.3	28.33

From these values, use Python to do the following subtasks:

1. Create two arrays, `time` and `velocity`.
2. Print out to the screen the velocity values at $t = 1.3$ and $t = 2.5$ s.
3. What is the average velocity using just the velocities at $t = 0$, $t = 2.1$, and $t = 3.3$ s?
4. Calculate the total number of seconds elapsed between the first and last values of velocity. We know the answer is 3.3 s by inspection, but provide the Python code to do the calculation based upon the two arrays created earlier.
5. Create an array containing only the first five velocity values.
6. Create an array containing only the last five time values.
7. Create an array containing only the velocity values from $t = 1.8$ to $t = 2.8$ s, inclusive.

Try This! Answer 5-4

1. These lines of code will accomplish these tasks:

```
import numpy as np
time = np.array([0, 0.1, 0.5, 1.3, 1.8, 1.9,
                 2.1, 2.3, 2.5, 2.8, 2.9, 3.3])
velocity = np.array([0, 0.98, 4.89, 12.49, 16.95, 17.81,
                     19.49, 21.11, 22.68, 24.91, 25.63, 28.33])
```

2. Execute the following lines of code:

```
print(velocity[3])
print(velocity[8])
```

The first element of an array has an index of zero, and each subsequent element is one larger in index value.
3. `(velocity[0] + velocity[6] + velocity[11]) / 3.0`
   We can use `velocity[-1]` instead of `velocity[11]` because we can use negative indices to reference from the back of the array (where an index value of `-1` is the last element of an array).
4. `time[-1] - time[0]`
5. `subset_velocity = velocity[:5]`
6. `subset_time = velocity[-5:]`
7. `subset_velocity = velocity[4:10]`

Try This! 5-5 Array Syntax: A Skydiver

Consider the data from Try This! 5-4 and the arrays `time` and `velocity` that were created in that Try This!. Using those arrays, do the following calculations using array syntax.

1. What are the velocity values in feet per second instead of meters per second? There are approximately 3.28 feet in a meter.
2. Create an array `time_diff` that gives the change in time from each time to the next time. That is, the first value of `time_diff` is the second value of `time` minus the first value of `time`, the second value of `time_diff` is the third value of `time` minus the second value of `time`, and so on.
3. How many elements does `time_diff` have relative to the number of elements in `time`? Why?
4. Create an array `velocity_diff` that gives the change in velocity from each time to the next time.
5. Create an array `accel` that approximates the rate of change in the velocity from one time to the next, at each time value. That is, the first value of `accel` is the change in `velocity` from the first to second element divided by the change in `time` from the first to second element, and so on.

Try This! Answer 5-5

1. `velocity_fps = velocity * 3.28`
   We obtain the following values for `velocity_fps`:

```
array([  0.    ,    3.2144,  16.0392,  40.9672,  55.596 ,  58.4168,
        63.9272,  69.2408,  74.3904,  81.7048,  84.0664,  92.9224])
```

With array syntax, when multiplying a NumPy array by a single value (i.e., a scalar value), Python goes through and does the multiplication on each of the elements. The result of the above code is an array of the same size and shape of `velocity`.

2. `time_diff = time[1:] - time[:-1]`
   The array `time_diff` is:

```
array([ 0.1,  0.4,  0.8,  0.5,  0.1,  0.2,
        0.2,  0.2,  0.3,  0.1,  0.4])
```

A carriage return and spaces are added to the above screen output so the entire line fits on a page. Similar reformatting is done with later output displays in this Try This!. All the arrays are still arrays of a single sequence of values.

The variable `time_diff` is the result of one array subtracted from another array. That is, `time[1:]` is the array:

```
array([ 0.1,  0.5,  1.3,  1.8,  1.9,  2.1,
        2.3,  2.5,  2.8,  2.9,  3.3])
```

while `time [:-1]` is this array:

```
array([ 0. ,   0.1,   0.5,   1.3,   1.8,   1.9,
        2.1,   2.3,   2.5,   2.8,   2.9])
```

When we subtract the latter array from the former array, the subtraction is done element-wise, so the first element of `time_diff` is $0.1 - 0$, the second element of `time_diff` is $0.5 - 0.1$, and so on.

The slice `time [:-1]` does not include the last element of `time` because when we slice NumPy arrays, the upper limit (the index found to the right of the colon) is exclusive (i.e., not included in the resulting slice). Because the `-1` element names the final element of `time`, `time [:-1]` will not include it. For more on using array slicing in index offset operations (including a diagram of the kind of slicing done above), see Section 12.2.5.

3. The array `time_diff` has one element less than the array `time`. This is because the final element in `time` does not have any later times to calculate a difference with.

4. `velocity_diff = velocity[1:] - velocity[:-1]`
   This is calculated the same as we calculated `time_diff` except the array we are using is the `velocity` array instead of the `time` array.

```
array([ 0.98,   3.91,   7.6 ,   4.46,   0.86,   1.68,
        1.62,   1.57,   2.23,   0.72,   2.7 ])
```

5. `accel = velocity_diff / time_diff`
   Here are the contents of `accel`:

```
array([ 9.8       ,   9.775     ,   9.5       ,
        8.92      ,   8.6       ,   8.4       ,
        8.1       ,   7.85      ,   7.43333333,
        7.2       ,   6.75      ])
```

Again, both `velocity_diff` and `time_diff` are arrays so the division is done element-wise and placed into the array `accel`.

Try This! 5-6 Plot Using Multiple Data Types: A Skydiver

Consider the data from Try This! 5-4 and the arrays `time` and `velocity` that were generated from those data in that Try This!. Using those arrays, make a plot of the velocity versus time. Label the axes appropriately and write in the title the number of data points that are being plotted. In creating the title, do not manually type in 12 to indicate there are 12 data points. Get Python to create the title using the arrays. Hint: Numbers need to be converted into strings if they are to be concatenated to other text.

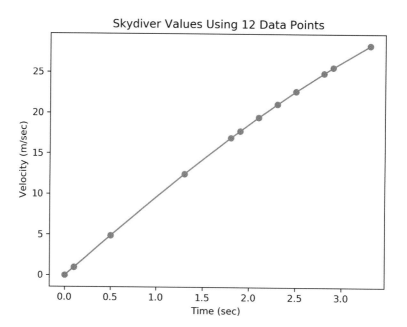

Figure 5.10 Plot of the velocity of a skydiver in Try This! 5-6.

Try This! Answer 5-6

This code will accomplish the desired tasks, resulting in the plot shown in Figure 5.10:

```
import matplotlib.pyplot as plt
import numpy as np
time = np.array([0, 0.1, 0.5, 1.3, 1.8, 1.9,
                 2.1, 2.3, 2.5, 2.8, 2.9, 3.3])
velocity = np.array([0, 0.98, 4.89, 12.49, 16.95, 17.81,
                     19.49, 21.11, 22.68, 24.91, 25.63, 28.33])
plt.plot(time, velocity, marker='o')
plt.xlabel('Time (sec)')
plt.ylabel('Velocity (m/sec)')
title = 'Skydiver Values Using ' + str(np.size(time)) \
        + ' Data Points'
plt.title(title)
```

after which we can use pyplot's show to render the plot on the screen or savefig to save the plot to an image file.

The portion of this Try This! that deals with type conversion is lines 10–11. There, we create a string variable title that contains the title of the plot and in line 12 is passed into the pyplot title function. In line 10, we obtain the number of data points from NumPy's size function. That, however, gives us an integer. We need to convert that to a string which we do using the str function.

5.4   More Discipline-Specific Practice

The Discipline-Specific Jupyter notebooks for this chapter cover the following topics:

- Optional input into functions.
- Customizing a graph. Adjusting line properties, marker properties, axes properties, labels.
- Making multiple figures and curves.
- Array syntax to making array calculations.
- Array element slicing and using both positive and negative ranges.
- Dealing with type.

5.5   Chapter Review

5.5.1 Self-Test Questions

Try to do these without looking at the book or any other resources or using the Python interpreter. Answers to these Self-Test Questions are found at the end of the chapter.

Self-Test Question 5-1

Define a keyword input parameter. How does this differ from a positional input parameter? What is a keyword input parameter used for?

Self-Test Question 5-2

Assume we have the following function `def` line (the body of the function is left out):

```
def process_data(a, b, c, size=4, label=''):
    ...
```

What will be the value of variables `a`, `b`, `c`, `size`, and `label` in the function body if the function is called as follows? Assume each code snippet is done in a separate Python session:

1. `process_data(3, 6, -1)`
2. The following code sequence:

```
1  x = 4
2  y = 2
3  z = -3
4  process_data(x, y, z)
```

3. The following code sequence:

```
x = 4
y = 2
z = -3
process_data(z, x, 9, size=z)
```

4. `process_data(3, 6, -1, label='Test 1')`

Self-Test Question 5-3

Describe what the following pyplot functions or parameters do:

1. `axis`
2. `marker`
3. `xticks`
4. `text`
5. `linestyle`

Self-Test Question 5-4

What is defined by the following string marker/line specification codes?:

1. `'o-b'`
2. `'D'`
3. `'*-.g'`
4. `'-'`

Self-Test Question 5-5

Describe the plots(s) that will result from running the following code:

```
import matplotlib.pyplot as plt
plt.figure('Chemical 1')
plt.plot(data1, data2)
plt.figure('Chemical 2')
plt.plot(data3, data4)
plt.figure('Chemical 1')
plt.xlabel('Time (sec)')
plt.show()
```

Self-Test Question 5-6

What command saves a plot to a file? How do we specify the file format to use in saving the plot?

Self-Test Question 5-7

Pretend we have a list of 1000 numbers named `data`. How do we:

1. Create an array `data_array` version of `data`?
2. Calculate the sum of all the elements in the array?
3. Calculate the average of all the elements in the array?
4. Calculate the difference between every element in the array and the first element in the array?

Assume `import numpy as np` has already been executed.

Self-Test Question 5-8

Consider the 1000-element array `data_array` from Self-Test Question 5-7:

1. Create the subarray `first_ten` that contains the first 10 elements of `data_array`.
2. Create the subarray `last_five` that contains the last five elements of `data_array`.
3. Create the subarray `from_interior` that contains the three-hundredth and first to four-hundredth and twelfth elements of `data_array`.

Assume `import numpy as np` has already been executed.

Self-Test Question 5-9

What is the type of data after each line is executed, assuming each line is executed one after the other?

```
1  import numpy as np
2  data = [2, 4, -1, 7]
3  data = np.array(data)
4  data = data[1]
```

5.5.2 Chapter Summary

There is a lot we can do in Matplotlib to make graphs look the way we want them to. In looking at how to customize line and scatter plots, we covered these topics:

- Optional input into functions

  - Keyword input parameters are used to pass in optional values into a function when the function is called.
  - Keyword input parameters are specified in a function's calling line by giving the keyword, an equal sign, and the value the parameter is being set to. That value is then used in the function at each occurrence of the variable with the keyword's name (unless that variable is redefined in the function's body).

- Keyword input parameters are often given a default value in the function's `def` definition. If the parameter is not included when the function is called, the default value is used when the lines in the function are executed.
- Functions can have both positional and keyword input parameters.

- Customizing the look of a plot

  - The `axis` command enables us to resize the two axis ranges of a plot.
  - The `xticks` and `yticks` commands control where tick marks and labels will be drawn along the *x*- and *y*-axes, respectively.
  - Line and marker formatting can be controlled by a string code passed into `plot` as a third positional input parameter. This does not work with `scatter`.
  - Many keyword input parameters exist in `plot` that enable us to customize individual features of the plot. These keywords include: `linestyle`, `linewidth`, `marker`, `markeredgewidth`, `markerfacecolor`, `markersize`, etc.
  - The `text` command places text arbitrarily on a plot. By default, the command places the text at a location given in data coordinates.

- Multiple figures and curves

  - Multiple calls to the `figure` command (either with an empty parameter list or with a unique name passed into the call to `figure`) enable us to create more than one graph in a single Python session.
  - All pyplot commands affect the current figure, which is the figure specified in the last `figure` call. To move between different figures, call `figure` with the name of the figure of interest passed in as the input parameter to make that figure the current figure.
  - To plot multiple curves on one figure, make multiple `plot` calls (or a single `plot` call) with the various sets of *x*- and *y*-values of data given one after the other in the calling line.
  - Legends are added with the `legend` command. The labels used in the legend are set by the keyword input parameter `label` passed in the call to `plot`.

- Whole plot manipulation

  - The size of the entire plot can be set in the `figure` call via the `figsize` keyword input parameter.
  - To save a plot to a file, use the `savefig` command. Generally, `savefig` will deduce what file format to use for the output file based on the suffix of the filename passed into `savefig`. The `dpi` keyword input parameter sets the resolution of the plot written out by calling `savefig`.

- An introduction to array calculations using array syntax

  - Arrays are like lists except each element of an array has the same type.
  - The NumPy `array` function creates an array from a list of values.

- The first element of an array has an index value of 0. The last element of an array has an index value equal to the total number of elements minus 1. Array elements can also be indexed from the back (where the last element has an index of -1), just like lists.
- Like lists, arrays can be sliced into subarrays. The range of the slice is given by specifying the lower bound index, putting in a colon, and then the upper bound index. The lower bound given is inclusive, and the upper bound given is exclusive.
- The `size` function returns how many elements are in an array. Other NumPy functions also operate on arrays.
- When using arrays in calculations such as addition, multiplication, etc., the operations are done element-wise, between corresponding elements of the arrays in the expression. The arrays in the expression have to be the same size, and an array of the same size as the arrays in the expression is returned. This is called array syntax. When later in this book we examine array syntax for arrays of more than one dimension, there will be additional constraints.

- Data types

  - Variables and values in Python have a type. The type of a variable or value is the "kind" of entity that variable or value is.
  - The type of a variable is set to the type of the value that variable is assigned to. Python is dynamically typed. If we reassign a variable to another value, and that variable is of a different type than the variable was before, the type of the variable will change to match the type of its new value.

5.5.3 Self-Test Answers

Self-Test Answer 5-1

A keyword input parameter is an input parameter set by an assignment syntax: the name of the keyword, followed by an equal sign, ending with the value the parameter is being set to. The variable in the function of the same name as the keyword is set to whatever the keyword input parameter is assigned to. In contrast, for a positional input parameter, the function knows what variable that parameter corresponds to based upon the order (or position) of that parameter in the parameter list. The first item or argument in the parameter list is mapped to the first variable or parameter in the function definition's parameter list, and so on. Keyword input parameters are useful for specifying optional pieces of input to a function.

Self-Test Answer 5-2

1. Variable a is 3, b is 6, c is -1, `size` is 4, and `label` is the empty string.
2. Variable a is 4, b is 2, c is -3, `size` is 4, and `label` is the empty string.

3. Variable a is −3, b is 4, c is 9, `size` is −3, and `label` is the empty string.
4. Variable a is 3, b is 6, c is −1, `size` is 4, and `label` is the string "Test 1".

Self-Test Answer 5-3

1. This function sets the limits of the *x*- and *y*-axes.
2. This keyword input parameter defines what shape to use for the data marker.
3. This function specifies what ticks to mark and annotate on the *x*-axis.
4. This function places text at a location on the plot.
5. This keyword input parameter defines what kind of line to connect the data points on the plot.

Self-Test Answer 5-4

1. Blue circle markers and solid connecting line.
2. Diamond markers with no connecting lines. A color is automatically assigned.
3. Green star markers with dashed-dot connecting line.
4. Solid connecting line with no markers. A color is automatically assigned.

Self-Test Answer 5-5

Two graphs are created, one whose window is called "Chemical 1" and the other whose window is called "Chemical 2". The "Chemical 1" plot is a graph whose points' *x*-axis values are given by `data1` and *y*-axis values are given by `data2`. The "Chemical 2" plot is a graph whose points' *x*-axis values are given by `data3` and *y*-axis values are given by `data4`. The "Chemical 2" plot has no axes labels. The "Chemical 1" plot has no *y*-axis label but does have the *x*-axis label, "Time (sec)".

Self-Test Answer 5-6

The `savefig` command. The file format is determined based on the filename suffix.

Self-Test Answer 5-7

1. `data_array = np.array(data)`
2. `np.sum(data_array)`
3. `np.average(data_array)`

    We can also calculate the average ourselves by:

    ```
    np.sum(data_array) / 1000.0
    ```

    or

    ```
    np.sum(data_array) / np.size(data_array)
    ```

Both do the same thing.

4. `data_array - data_array[0]`

Self-Test Answer 5-8

1. `first_ten = data_array[0:10]`
2. `last_five = data_array[995:]` or `last_five = data_array[-5:]`
3. `from_interior = data_array[300:412]`

Self-Test Answer 5-9

After line 2 is executed, `data` is a list. After line 3 is executed, `data` is a NumPy array. After line 4 is executed, `data` is a scalar integer.

Because Python is dynamically typed, the type of a variable can change as the program progresses. In this case, because we reassign `data` to a value of a different type at each line, the type of `data` changes. Also, it is not an issue that in lines 3 and 4 `data` appears on both the left and right sides of the equal sign. In both cases (and all cases when assigning variables), the expression to the right of the equal sign is evaluated first using what the current value of `data` is and afterwards the assignment occurs.

6 Basic Diagnostic Data Analysis

In Chapter 5, we learned about array syntax and how this enables us to use arrays not only to hold a collection of numbers but also to do calculations with those collections. In this chapter, we further explore how to analyze collections of data, such as arrays. Array syntax is powerful but there are many kinds of calculations that require the ability of examining and analyzing individual elements of an array. In this chapter, we learn about the two programming constructs that enable us to do this: **looping** (Section 6.2.3) and **branching** (Section 6.2.4).

Before we begin, we first consider what the terms in "basic diagnostic data analysis" mean. By "basic," we mainly mean "**one-dimensional.**" That is to say, we will be dealing with data that vary with a single independent variable (e.g., with distance, in time, etc.) and are stored in a single sequence of items such as the arrays we saw in Chapter 5. For instance, in this chapter's primary example (in Section 6.1), the mean precipitation data vary with month. The next term, "**diagnostic**," means that the calculations we are making describe the data as they are given, for the time(s) they are given at, as opposed to calculating or predicting the values of variables in the future. The latter is known as "**prognostic**" modeling or analysis and is described in Chapter 8. The last two terms, "**data analysis**," mean, as we would expect, the analysis of data ☺.

6.1 Example of Basic Diagnostic Data Analysis

Consider the list of mean monthly total precipitation (for the period 2000–2017) for the Seattle area, Washington, given in Table 6.1.[1] A plot of these data is given in Figure 6.2, and the Python code that generates that plot is given in Figure 6.1.

Lines 2–4 of Figure 6.1 create the plot using a format code that produces a black curve, uses a filled circle as a marker, and connects the markers using a solid line. Line 5 sets the *x*-axis and *y*-axis ranges, line 6 manually sets where to place the *x*-axis tick marks and labels, and lines 7–9 title each of the axes and the plot.

In Figure 6.3, we analyze the precipitation data given in Table 6.1. In line 1, we import the NumPy package. Once we do this import in a code file, we have access to the package using the alias np for the rest of the file. We do not have to type the import statement again. Line 3 (and elsewhere in the code) is an example of a comment line. In lines 4–6, we create an array

---

[1] Data from National Oceanic and Atmospheric Administration (NOAA) Online Weather Data, http://w2.weather.gov/climate/xmacis.php?wfo=sew (accessed February 17, 2018).

Table 6.1 Mean monthly total precipitation for the Seattle area, Washington (average over 2000–2017), in inches.

Month	Precipitation (in)
Jan.	5.58
Feb.	3.54
Mar.	4.54
Apr.	2.93
May	2.10
Jun.	1.47
Jul.	0.49
Aug.	1.03
Sep.	1.72
Oct.	4.31
Nov.	6.50
Dec.	5.41

```
1   import matplotlib.pyplot as plt
2   plt.plot([1, 2, 3, 4, 5, 6, 7, 8, 9, 10, 11, 12],
3            [5.58, 3.54, 4.54, 2.93, 2.10, 1.47,
4             0.49, 1.03, 1.72, 4.31, 6.50, 5.41], 'ko-')
5   plt.axis([1, 12, 0, 7])
6   plt.xticks([1, 2, 3, 4, 5, 6, 7, 8, 9, 10, 11, 12])
7   plt.xlabel('Month (1 = Jan)')
8   plt.ylabel('Precipitation (in)')
9   plt.title('Seattle-area, WA Mean Monthly Total Precipitation')
```

Figure 6.1 Code to create the plot of Seattle-area, Washington, mean monthly total precipitation in Figure 6.2.

for the month number of each of the precipitation values and an array for the precipitation values themselves (both tasks using the `array` function introduced in Section 5.2.6). Line 9 calculates the cumulative sum of precipitation values over the course of the 2017 calendar year and saves the 12 values in an array called `precip_cumsum`. This line makes use of a function which does the calculation and places the result into an array of the same size as the `precip` arrays.

In line 12, we create an array of all zeros called `running_mean`. In lines 13–14, we calculate and fill `running_mean` with the running mean of precipitation, defined in this case as the average between the current month and the month before, for each month. Thus, there are 11 running mean values since no "previous" month is given for the first month in the dataset.

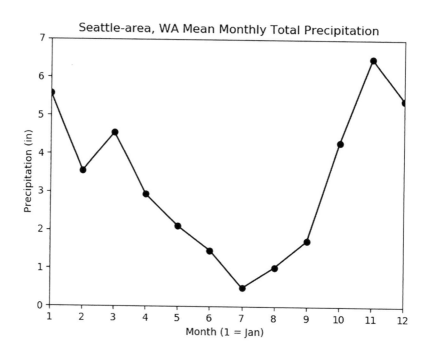

Figure 6.2 The plot of Seattle-area, Washington, mean monthly total precipitation generated by the code in Figure 6.1.

This calculation block uses a `for` loop, which enables us to go through a subset of the months and make the running mean calculation.

The last calculation block in this code involves an extra level of complexity, as the calculations include **conditional** statements. The calculation we make will be one kind of calculation in one context and another kind of calculation in a different context. Lines 17–24 calculate the maximum and minimum values of the `precip` array. This is done by going through each value in `precip` and comparing whether that value is larger or smaller than the maximum or minimum value (respectively) of the precipitation values we have already examined.

6.2   Python Programming Essentials

In Chapter 5, we were introduced to the basics of arrays: creating arrays, indexing and slicing arrays, using functions that operate on arrays, and using array syntax to do calculations on the array's elements. The analysis in the Figure 6.3 code builds on top of these array basics by showing how two new programming structures – loops and conditionals (also known as branching) – enable the code to analyze multiple pieces of data in a context-dependent way. That is, these constructs enable us to do calculations on a collection of data, a little at a time, as well as make different calculations depending on whether a certain condition is true or

```
1    import numpy as np
2
3    #- Create data arrays:
4    months = np.array([1, 2, 3, 4, 5, 6, 7, 8, 9, 10, 11, 12])
5    precip = np.array([5.58, 3.54, 4.54, 2.93, 2.10, 1.47,
6                       0.49, 1.03, 1.72, 4.31, 6.50, 5.41])
7
8    #- Calculate cumulative sum of mean precipitation:
9    precip_cumsum = np.cumsum(precip)
10
11   #- Calculate a running mean of mean precipitation:
12   running_mean = np.zeros(11, dtype='d')
13   for i in [0, 1, 2, 3, 4, 5, 6, 7, 8, 9, 10]:
14       running_mean[i] = (precip[i+1] + precip[i]) / 2.0
15
16   #- Calculate the maximum and minimum values of mean precipitation:
17   max_precip = precip[0]
18   min_precip = precip[0]
19   for iprecip in precip:
20       if iprecip > max_precip:
21           max_precip = iprecip
22
23       if iprecip < min_precip:
24           min_precip = iprecip
```

Figure 6.3 Code to analyze the Seattle-area, Washington, mean monthly total precipitation data in Figure 6.1.

not. Let us unpack Figure 6.3 in a little more detail to preview the principles and Python constructs we will expand upon in the subsections of this current section.

In lines 4–6 in Figure 6.3, we create the arrays month and precip from some lists, both using the NumPy array function. There are, however, other ways of creating arrays. Line 12 in Figure 6.3 provides another example. In Section 6.2.1, we examine additional ways of creating arrays and getting information about arrays.

The first calculation line in Figure 6.3 occurs in line 9. In this line, we use a function cumsum that operates on an array of numbers and returns an array where each element is the sum of all elements that came before. That is:[2]

```
In [1]:  import numpy as np
         precip = np.array([5.58, 3.54, 4.54, 2.93, 2.10, 1.47,
                            0.49, 1.03, 1.72, 4.31, 6.50, 5.41])
         precip_cumsum = np.cumsum(precip)
         print(precip_cumsum)

         [  5.58    9.12   13.66   16.59   18.69   20.16
           20.65   21.68   23.4    27.71   34.21   39.62]
```

[2] The result of the print command is realigned to enable the display to fit on this page.

The index 0 value of `precip_cumsum` is 5.58, because there are no values before the index 0 value. The index 1 value of `precip_cumsum` is 9.12, which is the sum of 5.58 and 3.54 from `precip`. And so on until the index 11 value of `precip_cumsum` which is the sum of all the values in `precip`. While `cumsum` is similar to the functions we looked at in Section 5.2.6 in that it acts on arrays, it differs in that the return value of the function is itself an array. We discuss more about functions that act on arrays in Section 6.2.2.

In lines 13–14 in Figure 6.3, we calculate the running mean of `precip`, going through each value of `precip` (except the last) and taking the average of the current value and the value that comes after it in `precip`. We make use of a `for` loop (line 13) which sets the variable `i` to the numbers in the list after the word `in`, one at a time. The body of the `for` loop (line 14) is executed for each of the values of `i`. We explain this in detail in Section 6.2.3.

Finally, in lines 17–24 in Figure 6.3, we calculate the maximum and minimum values of the mean monthly precipitation by examining each of the values in the `precip` array and seeing whether it is larger (for finding the maximum) or smaller (for finding the minimum) than the other values that have come before. We do this by using an `if` statement inside of a `for` loop. We discuss `if` statements in Section 6.2.4.

In the rest of the code examples in this section, assume that NumPy has already been imported as `np`.

6.2.1   More on Creating Arrays and Inquiring about Arrays

In Section 5.2.6, we saw how we can create a NumPy array from a list by using the `array` function. We use this method in lines 4–6 of Figure 6.3. In Section 5.2.6, we also saw that array syntax will return arrays. The result of adding two one-dimensional arrays that have the same number of elements is an array of the same number of dimensions and elements, where the elements are the sum of the corresponding elements in the original two arrays. In this section, we look at additional ways of creating arrays and inquiring about arrays.

Although sometimes we are interested in creating arrays from a list of values we type in, most of the time we want to have the computer create and fill our arrays for us, based upon some requirements. If we want to create a million-element array, we do not want to type in all those values by hand! To create a one-dimensional array of a given number of elements, all filled with zeros, we can use the NumPy `zeros` function. For instance, the following creates a one-dimensional array of five zeros:

```
In [1]:  import numpy as np
         data = np.zeros(5)
         print(data)

         [ 0.  0.  0.  0.  0.]
```

To create a one-dimensional array of zeros, we only need to pass in a single positional input parameter (in the above example, 5) which specifies the number of elements in the array.

Section 7.2.2 describes how we can also use `zeros` to create two-dimensional arrays, though the syntax is a little more complex. In Chapter 12, we address arrays of dimensions greater than two.

By default, `zeros` creates an array where each element is a floating-point value that occupies 8 **bytes** of memory. To create an array where each element is an integer, we can set the keyword input parameter `dtype` to the string `'l'` in the call to the function `zeros`:

```
In [1]:   data = np.zeros(5, dtype='l')
          print(data)

          [0 0 0 0 0]
```

The string `'l'` means "long integer." In Section 9.2.2, we discuss in more detail what that means. For now, just take it as meaning "an integer suitable for our needs."

What if we want a five-element, one-dimensional array of ones instead of zeros? The `ones` function accomplishes this task:

```
In [1]:   data = np.ones(5)
          print(data)

          [ 1.   1.   1.   1.   1.]
```

But, what if we wanted an array where every element was a specific value, not zero or one? We can use array syntax (as described in Section 5.2.6) to do arithmetic operations on an array of ones or zeros. For instance, if we wanted a five-element, one-dimensional array where every element equals 100, we can create an array of ones and then multiply each element by 100:

```
In [1]:   data = np.ones(5) * 100.0
          print(data)

          [ 100.   100.   100.   100.   100.]
```

With array syntax, the multiplication by 100 is automatically done on each element in the array created by the call to `ones`. The result of that multiplication is an array of the same size as the array created by the call to `ones`.

One common kind of calculation involves a collection of steadily increasing (or decreasing) values. Is there a quick way to create a one-dimensional array of steadily increasing values? Yes: the `arange` function. When called with a single positional input parameter that is a number, `arange` generates an array with that many elements where the first element is set to zero, and each subsequent element increments by one until the last element, which equals the number of elements in the array minus one. Thus, when called with the positional input parameter set to 5 we obtain a five-element array with the following values:

```
In [1]:   data = np.arange(5)
          print(data)
```

```
[0 1 2 3 4]
```

Calling arange with the integer *n* gives an array of integer values from 0 to $n - 1$.

By default, if the input parameter is an integer, the elements of the array created by arange are also integer. To obtain an array of incrementing floating-point values, pass in a floating-point input parameter or use the dtype keyword input parameter:

```
In [1]:   data = np.arange(5.0)
          print(data)
```

```
[ 0.   1.   2.   3.   4.]
```

```
In [2]:   data = np.arange(5, dtype='d')
          print(data)
```

```
[ 0.   1.   2.   3.   4.]
```

What if we want a range of values that increments by something besides one? For instance, instead of five elements incrementing from 0 to 4, what if we wanted five element from 0 to 1, incrementing by 0.25? There are two ways of doing this using arange. First, we can use array syntax and divide by 4:

```
In [1]:   data = np.arange(5.0) / 4.0
          print(data)
```

```
[ 0.     0.25  0.5   0.75  1.  ]
```

Alternately, we can call arange using three positional input parameters instead of one. When called with three input parameters, the first parameter gives the start value, the second the stop value, and the third the increment (or "step"). Note that the start is inclusive and the stop is exclusive, so if we want the last element to be 1, the stop value has to be some value greater than 1 but less than 1.25 (i.e., not more than the stop value plus the increment):

```
In [1]:   data = np.arange(0, 1.1, 0.25)
          print(data)
```

```
[ 0.     0.25  0.5   0.75  1.  ]
```

There is a built-in function called range that is similar to arange in a number of ways but is the better function to use when looping through large arrays. We discuss range in Section 6.2.3.

Having explored a few new ways of creating arrays, let us see if there are functions and other ways of taking an array that is already given and gaining information about that array. In Section 5.2.6, we learned about referencing elements or slices of an array. But we can find out more about an array than just its elements.

We saw earlier how to create a one-dimensional array of a certain size. We can do the reverse and take an array and figure out how many elements it has, using the NumPy `size` function (we saw this briefly in Section 5.2.6). Thus, for the five-element `data` array we just created:

```
In [1]:   data = np.arange(0, 1.1, 0.25)
          print(data)

          [ 0.    0.25  0.5   0.75  1.  ]

In [2]:   print(np.size(data))

          5
```

the result of calling `size` is the number of elements in `data`, which is five.

We can also find out what the type of each element (also called the **base type** of the array) in an array is. We have seen how we can set the `dtype` keyword input parameter to a single letter that tells us whether the array element is an integer, a floating-point number, etc. An existing array can also tell us what the type of each element in an array is. For instance, in:

```
In [1]:   data = np.arange(0, 1.1, 0.25)
          print(data)

          [ 0.    0.25  0.5   0.75  1.  ]

In [2]:   print(data.dtype.char)

          d
```

the cell `In [2]` prints the value of the variable `data.dtype.char`. A full explanation of what is happening requires understanding the basics of **object-oriented programming (OOP)**. We provide this in Section 9.2. For now, we can understand the syntax this way: There is a field attached to the array variable `data` called `dtype.char`. That field stores the single letter character, or **typecode**, that describes the type of each element in that array. When we print out that field, we get the letter "d", which tells us the elements of the array are all 8-byte floating-point numbers (each value occupies 8 bytes of memory in the computer). Note that we do *not* print out `np.dtype.char`. The `dtype.char` field is attached to the array variable, in this case `data`, not to the NumPy module. In Section 9.2.2, we go into more detail on bytes, `dtype`, and typecodes. A table of a number of common typecodes is found in Table 9.1.

One neat aspect of the array inquiry capabilities is that we can use them, with the array creation functions, to create customized arrays that have the same size and element types as an existing array. Consider array `data` as before, and pretend we want to create an array `newdata` that is the same size and has the same element data type as `data`, but is filled with zeros. This code accomplishes the task:

```
In [1]:  data = np.arange(0, 1.1, 0.25)
         print(data)

         [ 0.    0.25  0.5   0.75  1.  ]

In [2]:  newdata = np.zeros(np.size(data), dtype=data.dtype.char)
         print(newdata)

         [ 0.  0.  0.  0.  0.]
```

Because we used the array inquiry functions on `data`, the code in the first line of `In [2]` works the same regardless of the actual size or element types of `data`. If `data` happened to be five million elements in size rather than five elements in size, the code in the first line of `In [2]` that creates `newdata` does not change. In this way, using array inquiry functions and features rather than typing in the size of an array by hand enables us to write programs that are more flexible.

6.2.2   More on Functions on Arrays

In Section 5.2.6, we were introduced to the idea of functions acting on arrays via the `sum` function. Earlier in Chapter 3, we saw that NumPy has a variety of functions that can operate on plain numbers (i.e., scalars), such as `sin`, etc. There are many more mathematical functions in NumPy that act on arrays. A full list of NumPy functions and routines is given in the online documentation. We provide an up-to-date link to this list is at www.cambridge.org/core/resources/pythonforscientists/refs/, ref. 40. Section 12.2.7 provides a very brief summary list of important NumPy functions.

NumPy includes array creation and manipulation functions, trigonometric, transcendental, statistical, financial, sorting, and other functions. Line 9 of Figure 6.3 shows one such function, `cumsum`, which takes an array and returns another array with the cumulative sum of all values that came before in the array and including the current element:

```
In [1]:  data = np.array([3, 5, 7, -11, 2])
         print(np.cumsum(data))

         [ 3  8 15  4  6]
```

In this section, rather than list all the functions of interest to a scientist or engineer (a book in and of itself), we expand on our previous discussion by describing two key ways NumPy functions act on arrays.

The sum function from Section 5.2.6 is an example of the first way, where a function takes an array as input, does calculations with the elements of the array, and returns a scalar value. Another example would be the max and min functions, which look through the entire input array and return the largest and smallest values, respectively:

```
In [1]:  data = np.array([3, 5, 7, -11, 2])
         print(np.max(data))


         7
```

```
In [2]:  print(np.min(data))


         -11
```

A second kind of function that acts on NumPy arrays is a function that takes an array as input, does calculations element-wise on the array (that is, runs the same kind of calculation at each element of the array), and returns an array of the results. The cumsum function is an example of such a function. While each later element sums up the result of more values (because there are more "previous" values as we go through the array), the "operation" at each element is the same: add up everything up to and including the value at this index and put the result in the output array at that index.

Another example of a function that takes an array and returns an array that is the result of element-wise calculations on the input array is the sin function. When the input is a scalar, sin returns a scalar, but when the input is an array, sin returns an array, the result of calculating the sine of each element of the input array:

```
In [1]:  print(np.sin(3))


         0.14112000806
```

```
In [2]:  data = np.array([3, 5, 7, -11, 2])
         print(np.sin(data))

         [ 0.14112001 -0.95892427  0.6569866   0.99999021  0.90929743]
```

Although it appears in the example above that the sine of 3 as a scalar is different than when the 3 is an element in an array, the difference is only in how NumPy displays the result, not in the calculated value itself. If we execute print(np.sin(data)[0]), which prints out the first element of the output array generated by taking the sine of data, we obtain the same value as print(np.sin(3)).

```
In [1]:  import numpy as np
         theta = np.array([0.48991, 0.3674, 0.11229])
         v = 2.0
         v_x = v * np.cos(theta)
         print(v_x)


         [ 1.76475042  1.86652879  1.9874042 ]
```

Figure 6.4 An example of using array syntax and an array function to project a collection of vectors (with the same magnitude) along the x-axis.

An aside on how to read `np.sin(data)[0]`: The key to deciphering this code is to remember that `np.sin(data)` returns an array if `data` is an array. It is that array the `[0]` operation is applied to. When `[0]` is typed after an array – represented by a variable name or by the return value of a function that returns an array – it returns the first element of that array.

Many other mathematical functions work element-wise like `cumsum` and `sin` when the input is an array and produce an output array that is the same size as the input array. All the classical mathematical functions (e.g., `cos` for the cosine function, `tan` for the tangent function, `exp` for the exponential function) behave this way.

When combined with array syntax (as described in Section 5.2.6), functions that act element-wise on arrays are very powerful. For instance, consider the problem from Figure 3.1 where we used the `cos` function to project a vector along the x-axis. What if, instead of having a single angle `theta` to operate on we had a collection of angles? Figure 6.4 shows an adaptation of Figure 3.1 to such a situation. The variable `v_x` is now a three-element array, just as `theta` is a three-element array. In the same number of lines of code (in fact, with no change in the calculation code in the fourth line of `In [1]` in Figure 6.4), we can project any number of vectors with different angles (and the same magnitude) onto the x-axis.

6.2.3 Going Through Array Elements and an Introduction to Loops

Array slicing and array syntax are powerful ways of extracting some or all of the elements in an array and doing calculations with them. There are times, however, when we need more fine-grained control in our calculations with arrays. For instance, sometimes we need to pick-and-choose certain elements and make calculations using those elements.

Before we look at examples from our code in Figure 6.3, let us consider the following case. Pretend we have the following array of lengths (in inches):

```
lengths_in = np.array([2.3, 5.4, 12.9, 3.7, 8.8])
```

Using array syntax, the following multiplies each element of `lengths_in` by 2.54 and saves the result in the array `lengths_cm`:

```
lengths_cm = lengths_in * 2.54
```

The elements of lengths_cm are the values of lengths_in in centimeters. Both arrays have five elements.

We can also do the above inches-to-centimeters conversion calculation using the following code:

```
lengths_cm = np.zeros(5, dtype='d')
for i in [0, 1, 2, 3, 4]:
    lengths_cm[i] = lengths_in[i] * 2.54
```

Line 1 of the above code creates a one-dimensional, five-element array of floating-point zeros, set to the name lengths_cm. The next two lines are new.

Line 2, which begins with the word for, is the beginning of a loop, specifically a "for loop". The purpose of a loop is to run a block of code over and over again. The purpose of a for loop, in particular, is to run a block of code over and over again a specific number of times. In Python, we indicate the block of code to run over and over again by placing those lines of code under the line beginning with for and indenting the block of code in four spaces. This is just like how we define the body of a function by indenting in the lines of code after the statement beginning with def. Thus, in the above code, line 3 is the body of the for loop and will be repeated over and over again, in accordance to the characteristics of the for loop defined in the for statement (in line 2).

What are the characteristics of this loop? How do we interpret the line beginning with for? That line has two main entities. The first entity is between the for and the in: the variable i. The second main entity is everything between the in and the colon: the list [0, 1, 2, 3, 4]. When the Python interpreter reaches this for line, it operates using this logic:

Create a loop where I will repeat the code in the body below the for line over and over again. The first time I execute the body, the value of the variable i is set to the first value in the list [0, 1, 2, 3, 4], i.e., the value 0. The second time I execute the body, the value of the variable i is set to the second value in the list, i.e., the value 1. This continues over and over again until the last time the loop executes the body of the loop with the variable i set to 4.

The "body" of the loop is line 3, the line under the for statement that is indented in. So, the loop goes through [0, 1, 2, 3, 4] and assigns i to each of those values, and each time it does so, line 3 is executed. In line 3, the value of i is used to extract the value of index i from lengths_in, multiply that value by 2.54, and then save the result in the index i location of lengths_cm. The variable i is called an **iterator** and the list [0, 1, 2, 3, 4] is called the **iterable**.

Any valid variable name can be an iterator. It does not have to be called i. Iterables are any kind of collection of data where the idea of "go to the next item" is a meaningful concept to the computer. This does not mean the collection has to have an order, with a first, second, etc. item. We can have unordered but iterable collections. We will encounter such a collection in Section 14.2.5. Lists, tuples, and arrays can all act as iterables in a for loop. This is the meaning of line 19 in Figure 6.3:

```
for iprecip in precip:
```

The iterator in that line is `iprecip` and the iterable whose elements the iterator is being assigned to, one by one in turn, is the array `precip`. Note that iterators do not have to begin with the letter "i" either. It was just convenient to do so in the above example. Thus, this code (based on lines from Figure 6.3):

```
precip = np.array([5.58, 3.54, 4.54, 2.93, 2.10, 1.47,
                   0.49, 1.03, 1.72, 4.31, 6.50, 5.41])
for iprecip in precip:
    print(iprecip)
```

generates the following output:

```
5.58
3.54
4.54
2.93
2.1
1.47
0.49
1.03
1.72
4.31
6.5
5.41
```

Strings are also iterables and can also be looped through like the list shown above. This code:

```
for character in "Trial 1":
    print(character)
```

generates the following output:

```
T
r
i
a
l

1
```

Figure 6.5 summarizes the syntax for `for` loops.

In the Figure 6.3 code to analyze our Seattle-area precipitation data, another, more complex example of a `for` loop is given in lines 12–14 (reproduced below for convenience):

for loop syntax summary

In general, a `for` loop has the following syntax:

for *<iterator>* in *<iterable>* :
    <... body ...>

where *<iterator>* is a variable which will be set to each of the values in *<iterable>*, one in turn, as we go through each iteration of the loop. The *<... body ...>* represents one or more lines of code that are executed each iteration of the loop. The lines of the *<... body ...>* are, as a block, indented in four spaces, like we saw in Section 3.2.3 with writing our own functions.

Figure 6.5 `for` loop syntax summary.

```
running_mean = np.zeros(11, dtype='d')
for i in [0, 1, 2, 3, 4, 5, 6, 7, 8, 9, 10]:
    running_mean[i] = (precip[i+1] + precip[i]) / 2.0
```

This loop calculates the running mean of mean precipitation. In the first line, similar to the `lengths_cm` example earlier, we create an array `running_mean` that will be used to hold the results of our running mean calculation. In the second line, we loop through the integers zero to ten, meaning that the iterator `i` will be set to `0`, the body (third line) will run, the iterator will then be set to `1`, the body will run, and so on until the iterable is exhausted.

However, the third line of this code, the body of the `for` loop, is more complicated than our earlier `lengths_cm` example. How does it work? We have said that the body of the loop is executed over and over again, with the value of `i` changing each time the body is executed. The first time we go through the loop, `i` is set to `0`, and that value is used in every occurrence of `i` in the body of the loop. Thus, the first time the body of the running mean calculation loop is executed, it is the same as if we executed:

```
running_mean[0] = (precip[0+1] + precip[0]) / 2.0
```

Because `0+1` evaluates to 1, this is the same as:

```
running_mean[0] = (precip[1] + precip[0]) / 2.0
```

Thus, the first element in `running_mean` is set to the average of the first and second elements in `precip`.

The next **iteration** (or specific time we repeat the loop), `i` is set to `1`, so when the body of the loop is executed, it is the same as if we executed:

```
running_mean[1] = (precip[2] + precip[1]) / 2.0
```

The second element in `running_mean` is set to the average of the second (index 1) and third (index 2) elements in `precip`. This continues for every iteration through `i` set to `10`.

In the above example, we see that the list we are looping through ([0, 1, 2, 3, 4, 5, 6, 7, 8, 9, 10]) happens to be the indices of the 11-element array running_mean. Thus, the iterator i is set to the index values of each element of the list. This kind of task, it turns out, is quite common. Oftentimes, we want to go through each of the elements of a list and array and so will loop through all the indices of those elements. It would be nice if we could easily construct a list or array of the indices of an array.

In Section 6.2.1, we discovered a way to do so: the NumPy function arange. Thus, we could rewrite lines 12–14 from Figure 6.3 as:

```
running_mean = np.zeros(11, dtype='d')
for i in np.arange(11):
    running_mean[i] = (precip[i+1] + precip[i]) / 2.0
```

In the above code, we typed in the number of elements in running_mean (i.e., 11) in two places. We can replace the second occurrence of 11 with the output of a call to the NumPy function size, which returns the number of elements in the array passed in as the function's input parameter. Thus, we can rewrite the above as:

```
running_mean = np.zeros(11, dtype='d')
for i in np.arange(np.size(running_mean)):
    running_mean[i] = (precip[i+1] + precip[i]) / 2.0
```

The for loop in this code works even if we change the number of elements in running_mean (assuming precip also has the requisite number of elements). By using size, we have made the for loop more flexible as well as easier to maintain. Being repetitious, through writing the same value in more than one place, can lead to a program that is difficult to understand and update. As a result, when we are writing code, it is a good idea to follow the DRY principle: "don't repeat yourself."

The use of arange to enable us to loop through the indices of an array works fine for small arrays. For large arrays, however, arange does not work so well. When arange creates an array, it creates the entire thing – every element – in memory. For a loop, at any given iteration, what we really care about is the value of the iterator at that iteration. The other elements of the arange generated array do not really matter to that iteration of the loop. So, if we are creating a very large array (say millions or billions of elements), we are using up a huge amount of memory holding that entire array of indices, for no benefit.

For the cases of iterating through the indices of larger arrays, it makes more sense to use the range function (this is built in; it is not a NumPy function). The range function effectively gives us the values that arange does, which we can loop through, but without creating it all in memory all at once. Thus, we can rewrite the above code as:

```
running_mean = np.zeros(11, dtype='d')
for i in range(np.size(running_mean)):
    running_mean[i] = (precip[i+1] + precip[i]) / 2.0
```

Note that `range` only works to create a sequence of integers. It cannot create a sequence of floating-point values.

To recap: We have seen one-dimensional loops work and how we can use them to go through each element of an array to inspect the element, use the element, or change the element. We did this by looping through a list of array indices we typed in, an array of indices we generated using the NumPy function `arange`, and indices we generated using the built-in function `range`. We also did this by looping through a list of array values directly. In Section 6.2.5, we consider more complex examples of using loops on one-dimensional arrays, but before we do, we first consider how to ask questions of our data.

6.2.4   Introduction to Asking Questions of Data and Branching

In the Figure 6.3 code to analyze our Seattle precipitation data, the section to calculate the maximum and minimum values of mean precipitation, we find the following lines of code (lines 17–24 from that figure, reproduced here for convenience):

```
1    max_precip = precip[0]
2    min_precip = precip[0]
3    for i in range(np.size(precip)):
4        if precip[i] > max_precip:
5            max_precip = precip[i]
6
7        if precip[i] < min_precip:
8            min_precip = precip[i]
```

We said earlier that this block of code goes through each value in `precip` and compares whether that value is larger or smaller than the maximum or minimum value (respectively) of the precipitation values that had already been examined. We can expand that general description with more detail by looking at the code line by line. The line numbers below refer to the numbering in the reproduced snippet above, not the numbering in Figure 6.3.

In lines 1 and 2, we initialize the variables `max_precip` and `min_precip` to the first value in the `precip` array. It is important we set the initial values of `max_precip` and `min_precip` to a value in `precip`. If we set `min_value` to something else, say zero, and all the values in `precip` are positive, we would end up with the minimum value equal to zero even if the minimum value in `precip` was not zero. This is because the way we find (i.e., the **algorithm** we use) the minimum value compares the "minimum so far" with each later value in the array. Because every value in the array is larger than zero, a "minimum so far" value of zero will be the minimum returned.

Line 3 is the beginning of a `for` loop. The argument passed into the `range` function is the number of elements in `precip`, so the iterator `i` is set to 0, 1, 2, 3, ... 11, one in turn, because the number of elements in `precip` minus 1 is 11. That is to say, `i` is being set to the indices of `precip`, starting with the first index and going to the last index.

Lines 4–8 are the body of the `for` loop. We know these are the body of the loop because all these lines are indented in four spaces. This means that each one of these lines is executed one in turn for each of the values of i. Lines 4 and 7 begin with the word if, and although we have not yet discussed what a line that begins with if means, we might guess that this line is making some sort of comparison, and if that comparison is true, executing the lines of code below the if statement that are indented in.

That is exactly what is happening. In line 4, we check to see if the current value of precip is greater than the current value of max_precip. If so, we execute line 5 which sets max_precip to the current value of precip. In line 7, we check to see if the current value of precip is less than the current value of min_precip, and if so, we set min_value to the current value of min_precip.

After all the iterations of the for loop are done, the value of min_value will be equal to the smallest value in precip and the value of max_value will be equal to the largest value in precip.[3]

These lines of code illustrate what an if statement does: it asks a question, and depending on the answer to the question, Python does one thing versus another. Put another way, an if statement evaluates a test, and if the test is true, control of the computer program goes to one line (or lines) of code, while if the test is false, control of the computer program goes to another line (or lines) of code. We call these ask-a-question-then-do-something statements "if-statements" or "branching statements." With the latter name, the picture is of a computer program as a tree whose trunk we follow up, and when we reach a branch, we can either follow that branch or continue on the trunk. Whatever we call it, the purpose is the same: do something different depending on whether a certain condition is true or not.

The Basic if Statement

There are three common branching statements in Python. We have already encountered the first, a basic if statement. It begins with the word if followed by some sort of test that can be evaluated true or false, for instance, whether two values are the same or whether one is bigger than the other. A basic if statement that checks to see whether the atomic number of an element (i.e., the number of protons in the element, stored as the variable atomic_number) equals six (i.e., that the element was carbon) would look like:

```
1  if atomic_number == 6:
2      ... body ...
```

The == symbol is the operator that tests for whether two values are equal. We cannot use the single equal sign for that purpose because the single equal sign is used for assignment. Line 2 above refers to one or more lines of code that are executed if the test after the if is true. If the test is false, none of the lines in the body is executed. All of the body lines, as a block, are indented in four spaces, like we saw in Section 3.2.3.

---

[3] These lines of code have the same effect as using NumPy's max and min functions. Thus, to find the maximum, we could use max_precip = np.max(precip) instead.

What if we want to print out a message if the atomic number stored in the variable atomic_number was less than carbon's (and thus that the element has fewer protons)? The following basic if statement accomplishes this task:

```
if atomic_number < 6:
    print("Element has fewer protons than carbon")
```

If the test is true, the program enters the block indented in and under the if statement. If the test is false, the program skips all lines in the block indented in and under the if statement and goes to the line of code *after* the body of the if statement.

To illustrate this, we consider a short program using the if statement above, where we set the value of atomic_number before the if statement and place another line of code after the body of the if statement. Drawing arrows illustrating how the program progresses, the following code would look like this:

```
1   atomic_number = 2
2   if atomic_number < 6:
3       print("Element has fewer protons than carbon")
4   print("Test completed")
```

and when run, the above code would produce:

```
Element has fewer protons than carbon
Test completed
```

The program begins at line 1, goes to line 2, executes line 3 because the test in line 2 is true, then exits the body of the if statement after line 3 (because there are no more lines in the body) and goes to line 4. Line 4 will always be executed because it is outside the body of the if statement. The value of the test in the if statement plays no role in controlling whether the program will reach line 4 or not.

What happens if the value of atomic_number is greater than six? The code below shows the case where atomic_number is 17, and again the arrows show the progression of the code:

```
1   atomic_number = 17
2   if atomic_number < 6:
3       print("Element has fewer protons than carbon")
4   print("Test completed")
```

and when run, the above code would produce:

```
Test completed
```

The program begins at line 1, goes to line 2, evaluates the test in line 2 and finds out it is false, then goes directly to line 4, skipping line 3 entirely. When the condition in a basic `if` statement is false, the program skips the body of the `if` statement entirely; none of those lines of code is executed.

If we want to print out a message if the element has fewer protons than carbon or another message if the element has more protons than carbon, we can use two basic `if` statements, one after the other:

```
if atomic_number < 6:
    print("Element has fewer protons than carbon")
if atomic_number > 6:
    print("Element has more protons than carbon")
```

In this case, if `atomic_number` is greater than six, the first `if` statement condition will test false, and the `print` statement under that `if` statement will not execute. The second `if` statement's condition, however, will test true and the `print` statement after that will execute. In this example, if `atomic_number` is equal to six, neither `if` statement's conditions will be true and neither of the `print` statements will execute; nothing will be printed to the screen. To have the program print a message for the case where `atomic_number` equals six, we have to add a third `if` statement to test for that case:

```
if atomic_number < 6:
    print("Element has fewer protons than carbon")
if atomic_number > 6:
    print("Element has more protons than carbon")
if atomic_number == 6:
    print("Element is carbon")
```

If we did not have the last `if` statement and put the:

```
print("Element is carbon")
```

line after the second `if` statement like this:

```
if atomic_number < 6:
    print("Element has fewer protons than carbon")
if atomic_number > 6:
    print("Element has more protons than carbon")
print("Element is carbon")
```

every time this code ran, the message "Element is carbon" would print out, regardless of the value of `atomic_number`, because that final `print` statement is not in an `if` statement body. It is really important to make sure that whenever we write a series of `if` statements we properly cover the full range of logical options the `if` statements describe. Otherwise, the program may do something unwanted.

Table 6.2 Basic comparison operators in Python.

Operator	Code
Equal to	==
Not equal to	!=
Greater than	>
Greater than or equal to	>=
Less than or equal to	<=

In later sections, we describe other ways of testing for multiple conditions besides chaining basic if statements. For now, chaining basic if statements is one way of dealing with multiple conditions.

Table 6.2 lists basic comparison operators we can use in the test condition for an if statement. Each of these operators "connect" two operands, so there has to be something (a variable or a value) to the left and to the right of the operator. We will find in Section 8.2.2 that the kinds of questions we can ask a computer program to answer go beyond the operators listed in this table. However, this table is a good place to start.

The if-else Statement

We saw earlier in this short program:

```
atomic_number = 2
if atomic_number < 6:
    print("Element has fewer protons than carbon")
print("Test completed")
```

that the last print statement executes not only when atomic_number is less than six but also when it is not. The phrase "Test completed" will print out for all possible values of atomic_number. How then do we go about specifying Python commands to execute when the if condition is false, but *only* if the if condition is false? This is the purpose of the if-else statement.

In an if-else statement, we add an else clause after the if clause and put in a line or lines of code to execute if the if condition is false. Here is an example:

```
1  if atomic_number < 6:
2      print("Element has fewer protons than carbon")
3  else:
4      print("Element does not have fewer protons than carbon")
```

In this statement, the line or lines of code indented in four spaces after the if statement execute when the condition (atomic_number < 6) is true. If that condition is false, the line or lines of code indented in four spaces after the `else` statement execute instead. Thus, there are two bodies or blocks of code in the `if-else` statement: one under the `if` statement and one under the `else` statement. One and only one of the block of statements executes when the program reaches the `if-else` statement.

As we did for the basic `if` statement, we write two short programs, one for the case where `atomic_number` = 2 and the other for the case where `atomic_number` = 17, and see how the program progresses in each case. For the first case, the flow of execution is:

```
1   atomic_number = 2
2   if atomic_number < 6:
3       print("Element has fewer protons than carbon")
4   else:
5       print("Element does not have fewer protons than carbon")
6   print("Test completed")
```

When run, the above code produces:

```
Element has fewer protons than carbon
Test completed
```

The program goes from line 1 to 2 and then to line 3 because the condition `atomic_number < 6` is true. After line 3 executes, there are no other lines in the block of code (i.e., body) under the `if` statement, so the program goes to line 6, the line after the entire `if-else` statement. The program does not go into the block of code (i.e., body) under the `else` statement.

Let us consider the case when `atomic_number` = 17. In this case, the flow of execution is:

```
1   atomic_number = 17
2   if atomic_number < 6:
3       print("Element has fewer protons than carbon")
4   else:
5       print("Element does not have fewer protons than carbon")
6   print("Test completed")
```

When run, the above code produces:

```
Element does not have fewer protons than carbon
Test completed
```

The program goes from line 1 to 2 and then to line 4 and 5, skipping line 3, because the condition `atomic_number < 6` is false. After line 5 executes, there are no other lines in the block of code (i.e., body) under the `else` statement so the program goes to line 6, the line after the entire `if-else` statement. The program does not go into the block of code (i.e., body) under the `if` statement.

One way of understanding the basic `if` statement is as an `if-else` statement that has no lines of code under the `else` clause. That is to say, the basic `if` statement:

```
if atomic_number < 6:
    print("Element has fewer protons than carbon")
```

is the same as this `if-else` statement:

```
if atomic_number < 6:
    print("Element has fewer protons than carbon")
else:
    pass
```

The command `pass` is the way we tell the Python interpreter "do nothing." When the command is executed, nothing is done: there is no output to the screen, no calculations made, nothing. In Python, anytime we need to have a block or body of code (i.e., anytime we have a block of code indented in by four spaces), we are not allowed to just leave that blank. The code below is incorrect and will cause an error:

```
if atomic_number < 6:
    print("Element has fewer protons than carbon")
else:
```

Once we type in `else`, we have to have code after the colon for the body, even if we are not going to do anything as part of the `else` clause. Putting in `pass` fulfills this requirement.

This is also the case for other structures we have seen so far that have blocks of code (bodies) indented in four spaces: function definitions and `for` loops. In both case cases, we cannot have an empty body. These are both incorrect and will cause an error:

```
def calculate():
```

```
for i in [1, 2, 3, 4]:
```

If we want the function `calculate` to do nothing, we cannot leave the body of the function blank. Likewise, if we want the loop through the list [1, 2, 3, 4] to do nothing, we also cannot leave the body of the loop blank. We need to put something in and that something is the command `pass`:

```
def calculate():
    pass
```

```
for i in [1, 2, 3, 4]:
    pass
```

The `if-elif` Statement

We saw earlier how we can chain together multiple basic `if` statements to select one option out of a set of possible options. By making the `if` statement conditions mutually exclusive (e.g., greater than *x*, equal to *x*, and less than *x*), we guarantee only one of the basic `if` statement conditions will be true and thus only that statement's body will be executed. In such a collection of basic `if` statements, however, it is not entirely clear that these statements are all related to one another. Python has another `if` construct to make such an interrelationship clearer: the `if-elif` statement.[4]

Consider again the case where we want to test whether the value of `atomic_number` is less than, greater than, or equal to six. In an `if-elif` statement the code would look like the following:

```
1   if atomic_number < 6:
2       print("Element has fewer protons than carbon")
3   elif atomic_number > 6:
4       print("Element has more protons than carbon")
5   elif atomic_number == 6:
6       print("Element is carbon")
7   else:
8       print("Improper atomic number")
```

A Python program with this statement in it goes through each of the `if`/`elif` tests, evaluates whether the test is true, and if it is true, executes the block of code in the body underneath the `if`/`elif` line that tests that condition. After the block of code finishes executing, the program goes to the line after the entire `if-elif` statement structure. In the above example, this is whatever code is after line 8. The program does *not* consider any other of the `if`/`elif` tests.

Let us look at the case where `atomic_number` is set to 17 before the `if-elif` statement and a final `print` line is added after the `if-elif` statement. The code below would then execute in the order shown by the arrows:

```
1   atomic_number = 17
2   if atomic_number < 6:
3       print("Element has fewer protons than carbon")
4   elif atomic_number > 6:
5       print("Element has more protons than carbon")
6   elif atomic_number == 6:
7       print("Element is carbon")
8   else:
9       print("Improper atomic number")
10  print("Test completed")
```

---

[4] It should more properly be called the `if-elif-else` statement, but that is something of a mouthful.

and when run, the above code produces:

```
Element has more protons than carbon
Test completed
```

How can we be sure that once any one of the `if`/`elif` tests evaluates as true, all the other `if`/`elif` tests are skipped? We can test (ha ha, pun intended ☺) this out! We do so by creating `if`/`elif` tests that are not mutually exclusive but contain some overlap. For instance, consider the following `if-elif` statement:

```
1   if atomic_number > 2:
2       print("Element has more protons than helium")
3   elif atomic_number > 6:
4       print("Element has more protons than carbon")
5   elif atomic_number > 8:
6       print("Element has more protons than oxygen")
7   else:
8       print("Atomic number case not found")
```

If `atomic_number` is equal to seven (i.e., nitrogen), two of the `if`/`elif` tests (lines 1 and 3) could be evaluated as true, since seven is both greater than two and six. But, when we put the above `if-elif` statement into the following program:

```
1    atomic_number = 7
2    if atomic_number > 2:
3        print("Element has more protons than helium")
4    elif atomic_number > 6:
5        print("Element has more protons than carbon")
6    elif atomic_number > 8:
7        print("Element has more protons than oxygen")
8    else:
9        print("Atomic number case not found")
10   print("Test completed")
```

we obtain the following output:

```
Element has more protons than helium
Test completed
```

The second test (line 4), third test (line 6), the `else` statement are all skipped, even though the second test would have been true had it been conducted. Once one of the `if`/`elif` tests evaluates as true, all the other tests are skipped, and the program continues with the line after the `if-elif` statement.

Finally, what does the `else` clause mean in an `if-elif` statement? In an `if-elif` statement, each of the `if`/`elif` tests are conducted until one of the tests is found to be true, and the body of that true clause is executed. What if Python goes through every `if`/`elif` test and *none* of them are true? What does Python do then? If the `else` clause is missing, Python does nothing; the line following the `if`-`elif` statement is executed. But, oftentimes with `if`-`elif` statements, we want to do something if none of the `if`/`elif` tests is true. That is the purpose of the `else` clause: if none of the `if`/`elif` tests are true, the block of code in the `else` clause is executed. Here is an example, for the case where the value of `atomic_number` is 1:

```
1   atomic_number = 1
2   if atomic_number > 2:
3       print("Element has more protons than helium")
4   elif atomic_number > 6:
5       print("Element has more protons than carbon")
6   elif atomic_number > 8:
7       print("Element has more protons than oxygen")
8   else:
9       print("Atomic number case not found")
10  print("Test completed")
```

and, when run, yields the following output:

```
Atomic number case not found
Test completed
```

These three common branching statements are the workhorses for branching in Python. Figure 6.6 summarizes the syntax for branching statements. These three statements do not, however, exhaust the range of ways in Python where we can ask questions of our data or other values of variables. In Sections 8.2.2, 13.2.2, and 13.2.3, we examine more advanced ways of asking questions.

6.2.5   Examples of One-Dimensional Loops and Branching

In Sections 6.1, 6.2.3, and 6.2.4, we looked at examples of using loops and branching statement to go through elements of an array and analyze its contents. There are many more ways of using one-dimensional loops and branching statements, independently or together. We consider some of those ways as applied to basic data analysis.

Summing Up (or Subtracting) Values

In Try This! 2-4, we looked at an example of using an accumulator variable to add up over time the distances a microbe travels. Using accumulators in that solution did not save us any

Branching syntax summary

The three foundational branching statements are:

- The basic if statement:

```
if <condition>:
    <... body ...>
```

- The if-else statement:

```
if <condition>:
    <... body if the condition is true ...>
else:
    <... body if the condition is false ...>
```

- The if-elif statement:

```
if <first condition>:
    <... body if the first condition is true ...>
elif <second condition>:
    <... body if the second condition is true ...>
        ⋮
elif <nth condition>:
    <... body if the nth condition is true ...>
else:
    <body if all above conditions are false>
```

In all the above descriptions, a <... *condition* ...> is a test that evaluates to true or false. The lines in a <... *body* ...> are executed if the applicable <... *condition* ...> is true. The lines of a <... *body* ...> are, as a block, indented in four spaces. Reminder: At the end of the if, else, and elif lines is a colon. Do not forget it!

Figure 6.6 Branching syntax summary.

typing, but now with the for loop, we can make use of the fact accumulators automatically sum things up to make the calculation simpler:

```
1  distances = [0, 2, 2.5, 4.1, 3.6, 5.1]
2  total_distance = 0
3  for idistance in distances:
4      total_distance = total_distance + idistance
5      print(total_distance)
```

What is happening in the code? In line 1, we put all the distances in the Try This! 2-4 data table into a list. Next, we initialize the accumulator variable total_distance to zero. In line 3, we loop through all the values of distances and add each of those values to the accumulator (idistance is the iterator and takes on the value of each of the items in distances, each iteration of the loop). After each idistance is added to the accumulator, total_distance is printed out.

If we run this code, we obtain the following output:

```
0
2
4.5
8.6
12.2
17.299999999999997
```

which is what we obtained in Try This! 2-4, but with code that works for more than six values in distances.

Two items to point out: First, in line 2 of the code, we have to initialize total_distance to zero. If we do not, there will either be no accumulator to add each value of distances to or the sum will be incorrect. Second, lines 4 and 5 are both indented in four spaces. This tells Python that both lines constitute the body of the for loop. Thus, each iteration of the for loop (i.e., each time as we march through the distances list and set a value from that list to idistance), *both* lines 4 and 5 are executed (in that order).

Calculation Using Only a Subset of Values Using if Statements

In the code in Figure 6.1, we defined the following data arrays (reproduced here for convenience):

```
months = np.array([1, 2, 3, 4, 5, 6, 7, 8, 9, 10, 11, 12])
precip = np.array([5.58, 3.54, 4.54, 2.93, 2.10, 1.47,
                   0.49, 1.03, 1.72, 4.31, 6.50, 5.41])
```

How can we use for loops and an if statement to calculate the average Summer mean monthly precipitation in the Seattle area (where Summer is defined as June, July, and August)? Because we know the month number for each of the precipitation values, we can loop through the indices of months (which are the same as the indices of precip), check whether the value of the month is June, July, or August, and if so, add the corresponding value of precip to the accumulator variable we are using to do the summing as part of the averaging calculation. The code to make these calculations is in Figure 6.7.

Some notes regarding this code: In the if-elif block (lines 7–14), because the values of months map to the corresponding values of precip, we use the same index i to reference the appropriate values in months as well as precip. In line 14, in the else clause, we want the loop to do nothing if we are not looking at June, July, or August data, so we put in a pass statement. Finally, in lines 5 and 15, when we either initialize summer_precip or calculate summer_avg, the numbers that are part of those lines (0.0 and 3.0, respectively) are typed in as floating-point (decimal) numbers as a double-check to make sure that any operations

```
1   import numpy as np
2   months = np.array([1, 2, 3, 4, 5, 6, 7, 8, 9, 10, 11, 12])
3   precip = np.array([5.58, 3.54, 4.54, 2.93, 2.10, 1.47,
4                       0.49, 1.03, 1.72, 4.31, 6.50, 5.41])
5   summer_precip_sum = 0.0
6   for i in range(np.size(months)):
7       if months[i] == 6:
8           summer_precip_sum = summer_precip_sum + precip[i]
9       elif months[i] == 7:
10          summer_precip_sum = summer_precip_sum + precip[i]
11      elif months[i] == 8:
12          summer_precip_sum = summer_precip_sum + precip[i]
13      else:
14          pass
15  summer_avg = summer_precip_sum / 3.0
16  print(summer_avg)
```

Figure 6.7 Code to calculate the average Summer mean monthly precipitation in the Seattle area.

done with those variables do not use operators like integer division (see Section 2.2.1). The result of running the code in Figure 6.7 is:

```
0.996666666667
```

which is the average of 1.47, 0.49, and 1.03.

In Section 8.2.2, we describe another way of checking whether the month is June, July, or August rather than the if-elif statement in Figure 6.7 and that eliminates the duplicative lines 8, 10, and 12. In Section 13.2.2, we examine how array syntax can also be used with arrays of true/false "values."

We said earlier that if statements can be used inside loops or not inside loops. That is true enough as it is. From the examples we have looked at, however, we see that the use of an if statement inside a loop is generally more powerful than outside a loop. This is because by putting it inside the loop, we are repeating our question of the data as many times as we want. That is to say, we are doing the if test over and over again, without having to write the if test multiple times. This is the beauty of a loop: Whatever calculation, analysis, or prediction we want to do, a loop enables us to do those tasks over and over again, as many times as we want, without having to duplicate lines of code.

Calculation Using Only a Subset of Values Using Array Slicing or Functions

In the example above of calculating the average Summer mean monthly precipitation in the Seattle area, we used if statements to extract the precipitation values for the Summer months. We note, however, that these months are contiguous with each other. The three months June, July, and August follow one after the other. This suggests that we can calculate the mean by using array slicing to obtain just those months and then use a `for` loop to sum up those values.

In order to do this array slicing, we need the array indices corresponding to June, July, and August. From looking at the `months` array, we see that the month number (one for January, two for February, etc.) is equal to the index of that element plus one. So, the index for June is five and the index for August is seven.

The following code does what we want:

```
 1   import numpy as np
 2   months = np.array([1, 2, 3, 4, 5, 6, 7, 8, 9, 10, 11, 12])
 3   precip = np.array([5.58, 3.54, 4.54, 2.93, 2.10, 1.47,
 4                      0.49, 1.03, 1.72, 4.31, 6.50, 5.41])
 5   summer_precip = precip[5:8]
 6   summer_precip_sum = 0.0
 7   for iprecip in summer_precip:
 8       summer_precip_sum = summer_precip_sum + iprecip
 9   summer_avg = summer_precip_sum / 3.0
10   print(summer_avg)
```

In this code, instead of looping through all the elements of `precip`, we only loop through the elements of `summer_precip`. In line 5, the upper limit of the array slice uses an index value of eight, not seven, because the upper limit of an array slice is exclusive. Line 5 selects the subarray of elements with indices 5, 6, and 7.

In the above solution, we first assign the result of the array slice to a variable and loop through that variable, but that is not necessary. Because the return value of a slicing operation is an array, we can loop through that directly:

```
 1   summer_precip_sum = 0.0
 2   for iprecip in precip[5:8]:
 3       summer_precip_sum = summer_precip_sum + iprecip
 4   summer_avg = summer_precip_sum / 3.0
 5   print(summer_avg)
```

The lines importing NumPy and creating `months` and `precip` are left out for clarity.

Finally, we note that by slicing a subset of the `precip` array, the only task the `for` loop now does is to sum up all the elements in that subarray. No complicated asking questions of the data using `if` statements is needed. This suggests to us that we can accomplish all the above by replacing the `for` loop with a call to the NumPy `sum` function (introduced in Section 5.2.6). If so, the solution would become the following (again, leaving out the import and data creation lines):

```
 1   summer_precip_sum = np.sum(precip[5:8])
 2   summer_avg = summer_precip_sum / 3.0
 3   print(summer_avg)
```

Because array operations and syntax operate element by element on an array, and many NumPy array functions do the same, we often talk about array syntax as engaging in "implicit looping" as it goes about its calculations. The NumPy `sum` function above, for instance, implicitly goes through each element of `precip[5:8]` and adds it all up, just like a `for` loop would explicitly.

In this chapter, we have focused on using the `for` loop structure to go through elements (or subsets thereof) of arrays, one element at a time. Loops, however, are general structures that are used anytime we want to do something over and over again. We can, for instance, have a loop go through an iterator without ever actually using the iterator:

```
for i in range(3):
    print("Always wear safety goggles!")
```

This code generates:

```
Always wear safety goggles!
Always wear safety goggles!
Always wear safety goggles!
```

as output. The iterator `i` is never used in the body of the `for` loop. The sole purpose of this loop is to tell Python to execute the body of the loop three times.

Lastly, in this chapter we have discussed only the `for` loop. In Section 8.2.4, we discuss another kind of loop, the `while` loop.

6.2.6 Docstrings

In Section 4.2.4, we looked at how to write comments in our code. Another way of "commenting" is the use of **docstrings**.[5] Docstrings describe functions. What makes them special is that if we follow the standardized format for docstrings, Python's `help` function automatically turns the docstring into a manual page, with the proper formatting. Figure 6.8 is an example, and shows the function from Figure 3.3 with a docstring.

The docstring is the triple-quoted string (see Section 4.2.3 regarding triple quote strings) that immediately follows the `def` line of the function. The docstring is indented in four spaces, so Python knows it is part of the function's definition block. After the triple-quoted string ends, the body of the function begins (line 24 in Figure 6.8).

The first line of the docstring is a single sentence summarizing what the function does. This line is followed by a blank line and then one or more paragraphs describing the algorithm (or calculation steps) used in the function, general details about dependencies, details about the return value, etc. We would not normally say how the decimal equivalent is calculated since it is a trivial calculation, but we put such a description here for illustrative purposes. Finally, the docstring has sections that describe the input parameters, provide a summary of what the

---

[5] This section is adapted from Section 3.9 in Lin (2019).

```
def percent_to_decimal(input_percent):
    """Convert a percent to its decimal equivalent.

    The decimal equivalent is computed by dividing input_percent by 100,
    and this value is returned by the function.  The result will be
    float, regardless of whether or not input_percent is float or int.
    Input_percent is assumed to be scalar; no testing is done to see
    whether this is true.

    Positional Input Parameter:
        input_percent : float or int
            Numerical value of a percent.  Scalar.

    Keyword Input Parameters:
        None.

    Returns:
        Decimal equivalent.  Float.

    Examples:
    >>> print(percent_to_decimal(42))
    0.42
    """
    return input_percent / 100.0
```

Figure 6.8 The function in Figure 3.3 with a docstring.

function returns, gives examples of how to use the function, etc. Information is provided on the type of variables, what those variables correspond to, etc. (for more on variable type, see Section 5.2.7). If references were consulted in writing the function, citations to those should be provided. Sometimes, references are also provided to additional resources on the topic the function relates to.

In looking at Figure 6.8, it may seem a little weird that the number of lines the docstring takes up is so much larger than the number of lines of calculation (which is only one line). This is actually a good thing. As a general rule, we want our code to be concise. It is better if we can write functions that are relatively short than relatively long because it makes it more likely the code will work correctly. We can find errors more easily in short blocks of code than in long blocks of code. At the same time, when other people make use of functions we have written, ideally they should only need to look at the docstring to know how to properly use the function. Thus, we want our docstring to be clear and concise but also complete. It is quite normal for professionally written functions to have many more lines of documentation than executable lines of code.

More on docstrings and software documentation is found in Chapter 21.

6.2.7 Three Tips on Writing Code

The problems we are asking the computer to help us with are getting more complex, so we need to pay more attention to our process of writing code.[6] What can we do to help make it more likely we will write a program that works and gives the correct results? In Chapter 11, we provide a detailed and structured method of writing programs. Some students will find it helpful to read that chapter now, so by all means feel free to read ahead and use that method now. But here are three tips that can take us 80 percent of the way.

Write an Outline in Normal English

Conventional wisdom says programming = writing code. But the writing of lines of code is only a part of the process of programming. Analyzing and outlining the steps we want the computer to take to solve our problem, is, in many ways, both harder and more important than the syntax of the code. Writing a plain-English outline of what we want the computer to do, *before* we write a single line of code, can help us *tremendously* in eventually writing code that works.

In Section 11.1, we detail the steps of analyzing and outlining and give an example of an outline. Feel free to read that full explanation now. We can summarize the main message of that section, however, in two short sentences: **Do not start by writing code. Start by writing an outline of what you want the code to do.**

Write a Little Bit at a Time

What we mean by this is that we should write one or two lines of code at a time, run the code (or code snippet), print out the contents of the variables, and check to see whether those values are what we expect. If everything looks good, we write two more lines of code. If we do not get what we expect, we find what is wrong and fix it.

This sounds like a lot of work, but this "small bits" approach makes it much more likely that our code will work and work correctly. The time we spend testing the one to two lines of code we wrote will pay off in the time we save by not having so many **bugs** (errors in our code) to hunt down and fix. This "small bits" approach works particularly well with Python: we can write a code snippet, cut-and-paste it into the Python interpreter, and see if it works the way we expect.

A proficient programmer can relax this one-to-two line rule. Even proficient programmers, however, can find they make fewer mistakes and have more confidence in their code if they write a few lines of code at a time rather than tens or hundreds of lines.

---

[6] Parts of this section are adapted from Section 3.10 in Lin (2019).

Find Bugs by Simplifying the Program

Speaking of bugs, how do we find them? In Section 11.2, we describe the use of a special tool to help us find bugs called a **debugger**. And, in Chapter 23, we discuss formal **unit testing** as part of a strategy of finding bugs. For now, here is a basic step-by-step process of finding bugs.

Step 1: Run the Program and Find the Line Where the Error Appears   This line might be identified by Python when the program is run. For instance, the following code (in a file named *calc.py*):

```
1   mass = 2.0
2   velociti = 0.1
3   kinetic_energy = 0.5 * mass * (velocity ** 2)
```

when run produces this response from Python:

```
1   Traceback (most recent call last):
2     File "calc.py", line 3, in <module>
3       kinetic_energy = 0.5 * mass * (velocity ** 2)
4   NameError: name 'velocity' is not defined
```

The error appears in line 3.

Step 2: Comment Out Whatever Lines of Code Are Needed to Make the Error Go Away   In the above example, this means commenting out (i.e., putting the comment symbol at the front of the line so Python will not execute the line's commands) line 3:

```
1   mass = 2.0
2   velociti = 0.1
3   #kinetic_energy = 0.5 * mass * (velocity ** 2)
```

When we run it again, there is no error.

Step 3: Check That the Working Code Is Really Working   This means usually to check the values of the variables, but it can mean other checks. By getting the code working again, however, we can print the values of the variables to see whether the code is actually working. In the above example, we see that although our code now runs without Python complaining, when we try to print out the variables, we will find that we misspelled the `velocity` variable in line 2 as `velociti`. So, even though our code runs, it is not actually right, because we did not mean to misspell this variable.

When Python tells us the error is in a particular line, it is tempting to think that is where the error is. It often is true that is where the error is, but we cannot assume this. Sometimes the error will be elsewhere, as in the case above.

Step 4: Fix the Error and Comment Back In the Commented-Out Lines of Code "Commenting back in" means removing the comment symbol we place at the beginning of the line to "reactivate" it. Everything should now be working. Print out the values of the variables to make sure it really is working and that we did not inadvertently introduce a new error with our fix:

```
1  mass = 2.0
2  velocity = 0.1
3  kinetic_energy = 0.5 * mass * (velocity ** 2)
4  print(mass, velocity, kinetic_energy)
```

From the output from the `print`, we see everything checks out ☺.

One big warning: In Step 1, we said to find, "where the error appears." Not all errors are the result of typing the wrong Python syntax. Many times, Python will work fine and we will think everything is okay, but our code will still produce the wrong answer. For instance, consider this code:

```
1  mass = 2.0
2  velocity = 0.11
3  kinetic_energy = 0.5 * mass * (velocity ** 2)
```

We recognize this as the fixed code from above except the value of `velocity` is 0.11 m/s rather than 0.1 m/s. This code runs fine, without Python complaining, but if the `0.11` is the result of a slip of the fingers when trying to type in `0.1`, the result is absolutely and completely wrong.

But where will the error appear? Unfortunately, it will not appear by itself. Instead, we have to *make* it appear by printing out the values of the variables to make sure they are what we think they should be. In the above code, we could do this:

```
1  mass = 2.0
2  print("Mass is " + str(mass) + ": Should be 2.0")
3  velocity = 0.11
4  print("Velocity is " + str(velocity) + ": Should be 0.1")
5  kinetic_energy = 0.5 * mass * (velocity ** 2)
6  print("Kinetic energy is " + str(kinetic_energy) + ": Should be 0.01")
```

When we run this code, we get:

```
1  Mass is 2.0: Should be 2.0
2  Velocity is 0.11: Should be 0.1
3  Kinetic energy is 0.0121: Should be 0.01
```

The last two lines of output tell us there is an error in both the values of `velocity` and `kinetic_energy`. From looking at the code, we find that the latter error is the result from the former error. We found the error because we revealed the error through writing out the values of the variables. Without doing so, it is unlikely we would have detected the error let alone fixed it.

6.3   Try This!

We start with graphing examples similar to those we have seen in Section 6.1. These are followed by examples that address the main topics of Section 6.2. We practice using NumPy functions that operate element-wise, array slicing, looping, and branching. These constructs are used to accomplish analysis tasks such as calculating deviations and testing for a maximum. We also practice writing a docstring to document a function.

Try This! 6-1 Basic Arrays: A Mass–Spring System

Consider the following mass–spring system with a spring constant k and mass m:

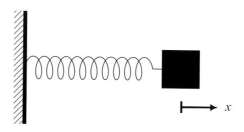

Assuming no gravity and the mass can only move in the positive or negative x-direction, the solution for the displacement x is:[7]

$$x = A\cos(\omega t + \phi)$$

where A is the amplitude of the oscillation, t is the time, ϕ is the phase angle, and the angular frequency ω is given by:

$$\omega = \sqrt{\frac{k}{m}}$$

If $A = 0.05$ m, $k = 2$ N/m, $m = 1$ kg, and $\phi = 0$ rad (which means $x = A$ at $t = 0$ s), create an array and fill it with the values of x for the first 8 s, taken every 0.1 s. Plot this array of displacement x-values versus the corresponding elements in the array of time t-values.

---

[7] Resnick and Halliday (1977, pp. 304–305).

Try This! Answer 6-1

This code solves the problem:

```
1    import matplotlib.pyplot as plt
2    import numpy as np
3
4    A = 0.05
5    k = 2.0
6    m = 1.0
7    phi = 0.0
8    omega = np.sqrt(k/m)
9    t = np.arange(0, 8.1, 0.1)
10   x = A * np.cos(omega * t + phi)
11
12   plt.plot(t, x, '-')
13   plt.xlabel('Time (sec)')
14   plt.ylabel('Displacement (m)')
15   plt.title('Displacement for a Mass-Spring System')
```

Line 10 does all the heavy lifting. In that expression, t is an array, so when t is multiplied by omega (which is a scalar), each element of t is multiplied by omega. This is part of array syntax, which we introduced in Section 5.2.6. Because omega times t is also an array, the result of that product added to phi is also an array. Thus, when we call the NumPy cosine function cos, we are calling it on an array, and the operation of the function is applied to each element of the array, one in turn (see Section 6.2.2).

In line 12, we specify the points to be connected by a solid line, with no marker. The plot has x- and y-axes labels and a title. The plot that the code generates is given in Figure 6.9.

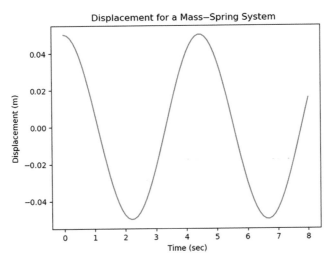

Figure 6.9 Line plot generated by the solution to Try This! 6-1.

Try This! 6-2 Calculating Diagnostics: Ideal Gas

Pretend we take a series of measurements on a gas we can treat as an ideal gas and obtain the following values of pressure at the following times:

Time (s)	Pressure (Pa)
0.0	102 000
1.0	134 000
2.0	143 100
3.0	122 050
4.0	114 180
5.0	103 060

An ideal gas is one that follows the relationship:

$$PV = nRT$$

where P is the pressure of the gas (in Pa), V its volume (in m^3), n is the number of moles of the gas, R the gas constant ($= 8.31441$ J K^{-1} mol^{-1}), and T the temperature (in K).[8] If the volume the gas is in is 0.5 m^3 and there are 25 mol of the gas, what is the temperature of the gas? While this can be solved using array syntax, try solving this using a one-dimensional `for` loop instead. Also make a plot of the temperature of the gas versus time.

Try This! Answer 6-2

This code solves the problem and plots the graph:

```
import matplotlib.pyplot as plt
import numpy as np

time = np.array([0.0, 1.0, 2.0, 3.0, 4.0, 5.0])
pressure = np.array([102000., 134000., 143100.,
                     122050., 114180., 103060.])
V = 0.5
n = 25.0
R = 8.31441

temperature = np.zeros(np.size(pressure))
for i in range(np.size(pressure)):
    temperature[i] = pressure[i] * V / (n * R)

plt.plot(time, temperature, 'o-')
plt.xlabel('Time (sec)')
plt.ylabel('Temperature (K)')
plt.title('Temperature of Gas Calculated by Ideal Gas Law')
```

[8] Dickerson et al. (1984, endsheet).

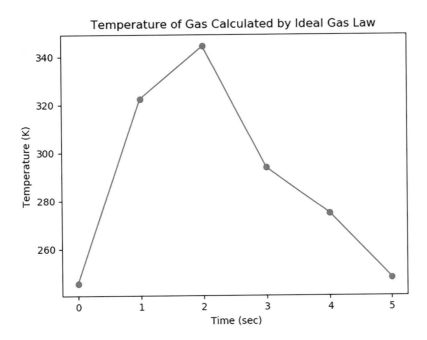

Figure 6.10 Customized line plot generated by the solution to Try This! 6-2.

after which we can use pyplot's `show` to render the plot on the screen or `savefig` to save the plot to an image file. The plot generated by the above code is in Figure 6.10.

In lines 5 and 6, when we type in the pressure values, we put a decimal point after the number. This forces the number to be stored as a floating-point value. Writing `102000.` is the same as writing `102000.0`.

In line 11, we create an array to hold the calculated temperature values. That array is prefilled with zeros. This array is the same number of elements as the `pressure` array. In the `for` loop, we loop through the index values of the array elements, setting those values to the iterator variable `i`. In the body of the loop (line 13), we calculate the temperature using the ideal gas law and the pressure at index `i` and store the result in the corresponding index `i` element in `temperature`.

Why do we loop through the index values of the `pressure` array rather than through the array elements directly? That is, why do we loop this way:

```
for i in range(np.size(pressure)):
```

rather than this way:

```
for ipress in pressure:
```

In the second way of looping, the iterator `ipress` is set to each of the values of `pressure`, one at a time. This works fine for accessing the value of each pressure measurement, but after we have done the calculation, where will we store the result? In order to store the result in the array `temperature`, we need the index where `ipress` came from (and where we will store the calculation result in `temperature`). Thus, we loop through the indices of `pressure` and `temperature` because we need to both read each element from `pressure` as well as store the result of our calcuation in the corresponding location in `temperature`.

Finally, in line 15 we use the Matplotlib letter codes to plot a solid line with circles as markers.

Try This! 6-3 Testing for a Subset: Spring Precipitation

For the mean monthly total precipitation data in Section 6.1, calculate the average Spring mean monthly precipitation in the Seattle area (where Spring is defined as March, April, and May) and print out the result to the screen. Use a `for` loop to go through the data points and an `if-elif` statement to select the right values to average.

Try This! Answer 6-3

This code accomplishes the task:

```
import numpy as np
months = np.array([1, 2, 3, 4, 5, 6, 7, 8, 9, 10, 11, 12])
precip = np.array([5.58, 3.54, 4.54, 2.93, 2.10, 1.47,
                   0.49, 1.03, 1.72, 4.31, 6.50, 5.41])
spring_precip_sum = 0.0
for i in range(np.size(months)):
    if months[i] == 3:
        spring_precip_sum = spring_precip_sum + precip[i]
    elif months[i] == 4:
        spring_precip_sum = spring_precip_sum + precip[i]
    elif months[i] == 5:
        spring_precip_sum = spring_precip_sum + precip[i]
    else:
        pass
spring_avg = spring_precip_sum / 3.0
print(spring_avg)
```

This solution is exactly the same as in Section 6.2.5 except the tests in the `if-elif` statement look for March, April, and May (`months` values of 3, 4, and 5) and the summing and averaging variable name has the word "spring" in it instead of "summer."

Try This! 6-4 Calculating Diagnostics with Flexibility: Ideal Gas

For the Try This! 6-2 scenario, create a function that accepts the pressure, volume, and number of moles as input and returns the temperature, calculated using the ideal gas law, as output. Assume that the pressure and temperature are arrays but code the function so that it works whether the volume and number of moles are scalars or arrays. If either the volume or number of moles are arrays, assume they are the same size as pressure. Use a `for` loop to go through the values in the input variables while calculating the temperature for those values. Hint: We can check whether an input parameter to the function is a multi-element array or not by using NumPy's `size` function.

Try This! Answer 6-4

The code in Figure 6.11 does what we want and results in the following arrays of temperatures output to the screen (reformatted for better display):

```
1   import numpy as np
2
3   def calculate_gas(p, V, n):
4       R = 8.31441
5
6       temperature = np.zeros(np.size(p))
7
8       for i in range(np.size(p)):
9           if np.size(V) > 1:
10              V_scalar = V[i]
11          else:
12              V_scalar = V
13
14          if np.size(n) > 1:
15              n_scalar = n[i]
16          else:
17              n_scalar = n
18
19          temperature[i] = p[i] * V_scalar / (n_scalar * R)
20
21      return temperature
22
23
24  time = np.array([0.0, 1.0, 2.0, 3.0, 4.0, 5.0])
25  pressure = np.array([102000., 134000., 143100.,
26                       122050., 114180., 103060.])
27  V = 0.5
28  n = 25.0
29  print(calculate_gas(pressure, V, n))
30
31  V = np.array([0.5, 0.55, 0.6, 0.8, 0.4, 0.75])
32  n = 25.0
33  print(calculate_gas(pressure, V, n))
```

Figure 6.11 Code to solve Try This! 6-4.

```
1    [245.35715703 322.3319514   344.22165854
2     293.58667663 274.65568814 247.9069471 ]
3    [245.35715703 354.56514653 413.06599025
4     469.7386826   219.72455051 371.86042064]
```

As we mentioned in the hint, we can support the flexibility of using scalars and arrays as input by testing to see whether the input parameters consist of more than one element/value or not. In the former case, we are dealing with an array as an input parameter and we extract the value at the index i location in that array (lines 10 and 15). If the input parameters have only one element/value, we can use just the value itself (lines 12 and 17).

If we used array syntax (as described in Section 5.2.6) to solve this problem, the function definition is much simpler:

```
1    import numpy as np
2
3    def calculate_gas(p, V, n):
4        R = 8.31441
5        return p * V / (n * R)
```

When using array syntax on NumPy arrays, Python automatically (and correctly) handles operations of arrays with other arrays, arrays with scalars, or scalars with scalars. For instance, when we multiply an array by another array, Python automatically multiplies the element in one array by the corresponding element in the other array and puts the result in the corresponding location of a results array of the same size as the two original arrays. Python also checks that the two arrays have the correct size, etc., to be multiplied together. If we multiply an array by a scalar, Python automatically multiplies each element of the array by the scalar and puts the result in the corresponding location of a results array of the same size as the original array. Thus, we do not need to test whether an input value is an array or not.

Try This! 6-5 Testing for Maximum: Spring Precipitation

For the mean monthly total precipitation data in Section 6.1, calculate the maximum mean monthly value during Spring (where Spring is defined as March, April, and May) and print out the result to the screen. Use for loops and various kinds of if statements, as needed.

Try This! Answer 6-5

Figure 6.12 gives a solution using two for loops. In the first for loop, we extract the Spring values and put them into a separate array using an algorithm similar to that used in Try This! 6-3. In the second for loop, we use the algorithm in Figure 6.3 to obtain the maximum value, that of looking at each of the Spring values and asking whether it is larger than the largest of the previously examined values. We have seen similar kinds of code earlier in lines 6–18 to select

```
1   import numpy as np
2   months = np.array([1, 2, 3, 4, 5, 6, 7, 8, 9, 10, 11, 12])
3   precip = np.array([5.58, 3.54, 4.54, 2.93, 2.10, 1.47,
4                      0.49, 1.03, 1.72, 4.31, 6.50, 5.41])
5   spring_precip = np.zeros(3, dtype='d')
6   j = 0
7   for i in range(np.size(months)):
8       if months[i] == 3:
9           spring_precip[j] = precip[i]
10          j = j + 1
11      elif months[i] == 4:
12          spring_precip[j] = precip[i]
13          j = j + 1
14      elif months[i] == 5:
15          spring_precip[j] = precip[i]
16          j = j + 1
17      else:
18          pass
19
20  max_spring_precip = spring_precip[0]
21  for j in range(np.size(spring_precip)):
22      if spring_precip[j] > max_spring_precip:
23          max_spring_precip = spring_precip[j]
24
25  print(max_spring_precip)
```

Figure 6.12 Code to solve Try This! 6-5.

the Spring values using `if-elif` statements and lines 20–23 to find the maximum of the Spring values. What is new is the use of the index variable j in lines 6–18. What is that used for?

The variable i is set to the indices of each element in the `precip` array (line 9), and we can use i to access each of the elements in that array. But there are only three Spring values versus 12 values in `precip`, so when we are filling the array `spring_precip`, we cannot use the variable i to provide index values for the elements in `spring_precip`. That is the purpose of the variable j. We set j to zero, because the first element in the array is zero, and after we come across a Spring precipitation value in `precip`, we copy that value into the array of Spring precipitation. Then, we increment the value of j by one. That way, the next time we find another Spring value in `precip`, that value will be copied and stored in the next element of `spring_precip`.

As an aside, the incrementing or decrementing operation on an accumulator or deaccumulator variable is such a common one that the `+=` and `-=` operators have been developed to limit retyping. Using these operators, we can turn:

```
j = j + 1
```

into:

```
j += 1
```

These two lines of code mean the same thing. Similarly:

```
k = k - 1
```

is the same as:

```
k -= 1
```

In these cases, the use of the += and -= does not save us much typing. It does make a noticeable difference if the variable name is longer.

There are other ways to solve this problem using branching if statements, but those require more complex if statement structures, which are not discussed until Section 8.2.

Try This! 6-6 Slicing a Subset: Spring Precipitation

For the mean monthly total precipitation data in Section 6.1, calculate the average Spring mean monthly precipitation in the Seattle area (where Spring is defined as March, April, and May) and print out the result to the screen. Use array slicing to obtain the Spring values and a for loop to help with calculating the average.

Try This! Answer 6-6

This code accomplishes the task:

```
 1   import numpy as np
 2   months = np.array([1, 2, 3, 4, 5, 6, 7, 8, 9, 10, 11, 12])
 3   precip = np.array([5.58, 3.54, 4.54, 2.93, 2.10, 1.47,
 4                      0.49, 1.03, 1.72, 4.31, 6.50, 5.41])
 5   spring_precip = precip[2:5]
 6   spring_precip_sum = 0.0
 7   for iprecip in spring_precip:
 8       spring_precip_sum += iprecip
 9   spring_avg = spring_precip_sum / 3.0
10   print(spring_avg)
```

This solution is mostly the same as in Section 6.2.5 except we slice precip to extract the index 2, 3, and 4 values (the five is exclusive in the slicing operation of line 5), which correspond to the elements for March, April, and May. Also, the summing and averaging variable has the word "spring" in it instead of "summer." Lastly, as described in the solution for Try This! 6-5, we use:

```
spring_precip_sum += iprecip
```

instead of:

```
spring_precip_sum = spring_precip_sum + iprecip
```

to save typing (and decrease the opportunity for mistyping). These two code statements mean the same thing.

Try This! 6-7 Converting Units: Spring Precipitation

For the mean monthly total precipitation data in Section 6.1, convert the precipitation values from inches into centimeters. Put the result in a new array called precip_cm. Use a for loop to do the conversion.

Try This! Answer 6-7

This solves the problem:

```
1   import numpy as np
2   precip = np.array([5.58, 3.54, 4.54, 2.93, 2.10, 1.47,
3                      0.49, 1.03, 1.72, 4.31, 6.50, 5.41])
4   precip_cm = np.zeros(np.size(precip), dtype='d')
5   for i in range(np.size(precip)):
6       precip_cm[i] = precip[i] * 2.54
```

Using array syntax, the solution to this problem is a single line of code:

```
precip_cm = precip * 2.54
```

Because precip is a NumPy array, we multiply each element by the conversion factor and store it in the corresponding location in the result array precip_cm.

Try This! 6-8 Backward/Forward Mean: Spring Precipitation

For the mean monthly total precipitation data in Section 6.1, calculate the "backward/center/forward" running mean of precipitation for all months from February to November and store the result in a separate array. For the month of February, this mean is the average of the January, February, and March precipitation values. For the month of March, this mean is the average of the February, March, and April precipitation values, and so on. Use a for loop to do the calculations.

Try This! Answer 6-8

This code solves the problem:

```
1   import numpy as np
2   running_mean = np.zeros(10, dtype='d')
3   for i in range(np.size(running_mean)):
4       running_mean[i] = (precip[i] + precip[i+1] + precip[i+2]) \
5                          / 3.0
```

In line 4, we use references relative to the value of the variable i. The value of i runs from zero to nine, that is, the indices of the output array running_mean. When i is zero, this corresponds to calculating the "backward/center/forward" running mean for February. But when applied to the `precip` array, `precip[i]` is not February's precipitation but January's (as it should be in calculating this running mean). Likewise, `precip[i+1]` for i equals zero is February's precipitation (an index value of 1) and `precip[i+2]` for i equals zero is March's precipitation. For i equal to one, `precip[i]`, `precip[i+1]`, and `precip[i+2]` are the February, March, and April precipitation values, respectively.

Try This! 6-9 Calculating Deviations: Spring Precipitation

For the mean monthly total precipitation data in Section 6.1, calculate the deviations of each mean monthly total from the period average of precipitation. Print the months when the monthly total precipitation values are above the period average. Hint: The period average is the average of mean monthly total precipitation values for the period in question (2000–2017). Use `for` loops to do the calculations.

Try This! Answer 6-9

We solve this in two stages. First, we calculate the period mean. Next, we go through each monthly value and calculate the deviation from the period mean. Lastly, we go through each monthly value and check to see if it is above or below the period mean. Depending on the results of this test, the appropriate message will print. We can put the last two steps into the same loop. Here is the solution:

```
1   import numpy as np
2   months = np.array([1, 2, 3, 4, 5, 6, 7, 8, 9, 10, 11, 12])
3   precip = np.array([5.58, 3.54, 4.54, 2.93, 2.10, 1.47,
4                      0.49, 1.03, 1.72, 4.31, 6.50, 5.41])
5
6   precip_sum = 0.0
7   for iprecip in precip:
8       precip_sum += iprecip
9   period_avg = precip_sum / 12.0
10
11  deviations = np.zeros(np.size(precip))
12  for i in range(np.size(precip)):
13      deviations[i] = precip[i] - period_avg
14      if precip[i] > period_avg:
15          print(months[i])
```

In lines 11–15, we loop through the indices of `precip` because we need to access the corresponding element in `deviations` and `months`. In lines 7–8, however, we only need the `precip` values themselves, so there is no need to loop through the indices. We can just loop through the values themselves.

Try This! 6-10 Docstrings: Ideal Gas Function

Write a docstring for the function written to solve Try This! 6-4.

Try This! Answer 6-10

In the Figure 6.13 solution, the body of the function is not given, because the focus is on the docstring. In this docstring, we specifically mention that the function only works for a one-dimensional array, because the solution to Try This! 6-4 using a `for` loop only works for such arrays. We also make the docstring's line 25 example pass in a one-element array for pressure (the first positional input parameter) because the function assumes that parameter is an array (see line 6 of the solution to Try This! 6-4).

```
1    import numpy as np
2
3    def calculate_gas(p, V, n):
4        """Calculate the temperature for an Ideal Gas.
5
6        The temperature in K is calculated using the Ideal Gas law
7        and is returned by the function.  Input variables may be
8        scalars or one-dimensional arrays, except p must be an
9        array.  The output will be an array of the same size, etc.
10       as the array p.
11
12       Positional Input Parameters:
13           p : one-dimensional array
14               Pressure in Pa.
15
16           V : scalar or one-dimensional array
17               Volume in cubic meters.
18
19           n : scalar or one-dimensional array
20               Amount of gas in mol.
21
22       Keyword Input Parameters:
23           None
24
25       Examples:
26       >>> calculate_gas(np.array([102000.0]), 0.5, 25.0)
27       [245.35715703]
28       """
29       ...
```

Figure 6.13 Code to solve Try This! 6-10.

6.4   More Discipline-Specific Practice

The Discipline-Specific Jupyter notebooks for this chapter cover the following topics:

- Creating arrays and using arrays as arguments to functions.
- Using one-dimensional loops to interact with array elements.
- Writing tests to ask questions of data.
- Using one-dimensional loops to do other repetitive tasks.
- Writing docstrings.

6.5   Chapter Review

6.5.1   Self-Test Questions

Try to do these without looking at the book or any other resources or using the Python interpreter. Answers to these Self-Test Questions are found at the end of the chapter. For all questions in this section, assume NumPy has already been imported as `np`.

Self-Test Question 6-1

What is the result of the following NumPy array creation function calls?

1. `np.arange(6)`
2. `np.zeros(3)`
3. `np.ones(4, dtype='d')`
4. `np.arange(4,9,2)`

Self-Test Question 6-2

What does the NumPy `size` function do? What kind of value does it return?

Self-Test Question 6-3

Given a one-dimensional array `data`, how would we create another one-dimensional array `data_temp` with the same number of elements, filled with zeros, and with the elements having the same type as `data`?

Self-Test Question 6-4

Assuming `data` is a one-dimensional array, what results from the following code?

```
data1 = np.sin(data / np.max(data))
```

Describe what `data1` is.

Self-Test Question 6-5

What is a `for` loop and how does it work? What is the purpose of a loop?

Self-Test Question 6-6

Using a `for` loop (and no function calls), calculate the product of all the elements of an array `data` and store that result in the value `data_product`. Assume the values in `data` are floating-point type.

Self-Test Question 6-7

If we had a variable and wanted to call a different function depending on whether that variable was equal to one of six different values, what kind of branching statement should we use and why?

Self-Test Question 6-8

Given a one-dimensional array `velocity`, write code that will take the average of all values of `velocity` that are over the scalar value stored in `threshold`. Store the result in the variable `over_threshold_avg`. Assume the values in `velocity` are floating-point type.

Self-Test Question 6-9

Describe the purpose of a docstring. How is a docstring set off from the rest of the code?

Self-Test Question 6-10

Why is it inadvisable to write many lines of code in one sitting?

6.5.2 Chapter Summary

Computers, well, compute, and thus computers are incredibly useful for analyzing data. Two constructs – looping and branching – are foundational to programming a computer to go through (lots) of data and calculating statistics and other quantities. In looking at simple diagnostic data analysis, we covered these topics:

- More on creating arrays

  - The NumPy `zeros` function creates an array of a given size (and whose elements are of a given type) that is all filled with zeros.
  - The NumPy `ones` function creates an array of a given size (and whose elements are of a given type) that is all filled with ones.
  - The NumPy `arange` function creates an array of values ascending or descending by a specified increment (or step).
  - Arithmetic operations using array syntax enable us to create an array with certain characteristics.

- Inquiring about arrays

  - The NumPy `size` function returns the total number of elements in an array.

- For an array `myarray`, the variable `myarray.dtype.char` stores a single letter character that describes what type each elements of that array are.
- The `dtype.char` typecodes can be passed into array creation functions to create arrays whose elements will be the same type as an existing array.

- Two ways of using functions with arrays

  - First kind of array function: Accepts array(s) as input, does calculations, and returns a scalar value. Example: NumPy's `min` function.
  - Second kind of array function: Accepts array(s) as input, does a calculation element-wise on the array, and returns a results array. Example: NumPy's `cumsum` function.

- Looping and going through array elements

  - Loops are the structure that describes how a computer is to do a task repeatedly.
  - Each time the task is done is called an iteration of the loop.
  - In a `for` loop, at each iteration a variable (called the iterator) is assigned to a value in a collection of items. The collection is called the iterable. The loop goes through that collection one item at a time, each time corresponding to an iteration of the loop.
  - A `for` loop is often used to iterate through the elements of an array. This can be done by iterating through the elements of an array directly.
  - A `for` loop can also be used to iterate through the elements of an array by iterating through the indices of the array (whether typed in, generated by the NumPy function `arange`, or generated by the built-in function `range`).
  - The `pass` statement is used when the body of a loop is empty.
  - Loops do not have to make use of the value stored in the iterator.
  - Figure 6.5 summarizes the syntax for `for` loops.

- Branching and asking questions of the data

  - Branching is the means by which a computer program can ask a question of data and, depending on the answer, execute one set of lines of code versus another set of lines of code.
  - The basic `if` statement asks whether a condition is true, and, if it is, executes the body of code indented in after the `if` statement.
  - The `if-else` statement asks whether a condition is true, and, if it is, executes the body of code indented in after the `if` statement. If the condition is false, Python executes the body of code indented in after the `else` statement.
  - For a list of options, the `if-elif` statement selects the first one where the condition in that option is true. If so, the body of code under that condition is executed.
  - The `pass` statement is used when the body of a a block of a branching statement (e.g., the `if` block, `else` block, etc.) is empty.
  - Figure 6.6 summarizes the syntax for branching statements.

- Examples of ways to use looping and branching to analyze data

  - Adding up (or subtracting out) a sequence of values.
  - Doing a calculation on a subset of a collection of values.

- Docstrings

  - Docstrings are triple quote strings placed right after the def statement of a function.
  - Docstrings begin with a line summarizing what the function does.
  - The docstring should describe the algorithm used, what the function returns, the nature of the input parameters, an example of the use of the function, etc.
  - Docstrings are incredibly important as they tell the user of the function how to properly use the function without having to read the code.

- Tips on how to write code

  - Write a little bit at a time.
  - To find bugs, simplify the program until the program works correctly, then add in lines of code one at a time to help in finding what is causing the problem.
  - A key part of finding bugs is getting a clear and comprehensive sense of what the values of the (key) variables are, after each line is executed. Printing out the values of those variables is one way to do this.

  With looping and branching, we can efficiently examine datasets of many elements and manipulate the data. Variables and expressions, looping, and branching are the three core constructs of a programming language. With these, the computer can do incredibly complex calculations. In the next chapter, we describe additional tools Python gives us to conduct more advanced data analysis.

6.5.3   Self-Test Answers

Self-Test Answer 6-1

```
1. array([0, 1, 2, 3, 4, 5])
2. array([ 0.,   0.,   0.])
3. array([ 1.,   1.,   1.,   1.])
4. array([4, 6, 8])
```

Self-Test Answer 6-2

The size function returns the total number of elements in the array that is passed in as input. A call to the function returns a scalar integer.

Self-Test Answer 6-3

```
data_temp = np.zeros(np.size(data), dtype=data.dtype.char)
```

Self-Test Answer 6-4

data1 is an array of the same size as data. Using array syntax, the data / np.max (data) expression takes each element of the array data and divides it by the (scalar) maximum of all the elements of data. The result of that expression is an array of the same size as data. That array is passed in as input to the sin function, which takes the sine of each element of that input array and returns the result in the corresponding element of the array data1.

Self-Test Answer 6-5

A for goes through each element of an iterable collection of items, setting the value to a variable called an iterator, one element at a time. After it sets the iterator value, Python executes the contents of the body of the for loop. The purpose of a loop is to tell the computer to do some task over and over.

Self-Test Answer 6-6

This code accomplishes the task by looping through the values of the array directly:

```
1  data_product = 1.0
2  for ivalue in data:
3      data_product = data_product * ivalue
```

whereas this code accomplishes the task by looping through the indices of the array:

```
1  data_product = 1.0
2  for i in range(np.size(data)):
3      data_product = data_product * data[i]
```

Self-Test Answer 6-7

We should use an if-elif statement. While it is possible to use basic if or if-else statements to accomplish the task, an if-elif statement will produce clearer and more understandable code for this task.

Self-Test Answer 6-8

We use a basic if statement to ensure only values greater than threshold are summed up as part of calculating the average. We also create a variable count that counts the number of elements in velocity that meet this criterion. This code will accomplish this selective averaging:

```
1   sum_over_threshold = 0.0
2   count = 0
3   for ivalue in velocity:
4       if ivalue > threshold:
5           sum_over_threshold += ivalue
6           count += 1
7   over_threshold_avg = sum_over_threshold / float(count)
```

The conversion of `count` to floating point, prior to the division operation in the calculation of the average is not needed but is included for clarity and completeness.

Self-Test Answer 6-9

A docstring provides documentation to the user as to the purpose and usage of a function. It is set off from the rest of the function definition by putting it in a set of matched triple quotes immediately after a function's `def` statement.

Self-Test Answer 6-10

The more lines of code written in one sitting, the more difficult it is to find any bugs that are included in that code. It is better to write code in small groups of lines, testing that the lines work as we go.

7   Two-Dimensional Diagnostic Data Analysis

In Chapter 6, we looked at the analysis of one-dimensional arrays of data. Along the way, we introduced the `for` loop and the branching statement. Looping enables us to make some set of calculations repeatedly. Branching enables us to ask questions of our data. In this chapter, we extend our discussion to two-dimensional arrays. Two-dimensional data scenarios exist frequently in science and engineering: the variation of stress along a cross-section of a beam, the time evolution of a plant survey transect, and the variability of radiances in the sky as the telescope points at one location and then another. In this chapter, we introduce the tools – two-dimensional array creation, manipulation, and operations and nested looping – to conduct the diagnostic analysis of such datasets. This chapter shows some contour plots. However, we do not address how to create contour plots until Chapter 14.

The larger datasets used in this chapter are available for download. See the "Supporting Resources and Updates to This Book" section in the Preface for details.

7.1   Example of Two-Dimensional Diagnostic Data Analysis

Figure 7.1 shows a filled contour plot of the surface/near-surface air temperature over the continental United States at November 11, 2017, 18:00Z.[1] The "Z" means "Zulu Time Zone." and is the same as Coordinated Universal Time (UTC). Both are the same as Greenwich Mean Time (GMT). November 11, 2017, 18:00Z is November 11, 2017, noon, Central Standard Time in the USA.

Figure 7.2 shows a "zoom-in" of Figure 7.1, centered around Southern Illinois and using only the data in Table 7.1. Tables 7.2 and 7.3 give the longitude (in degrees East) and latitude (in degrees North) for the columns and rows (respectively) of the Table 7.1 data.

Figure 7.3 shows code that analyzes the surface/near-surface air temperature over the US Midwest from Table 7.1 using a nearest-neighbor (NN) averaging algorithm. For every data point in Table 7.1, we calculate an average temperature based upon that data point plus all the values immediately adjacent to and sharing a side with that data point (i.e., not including corners). Thus, to calculate the nearest-neighbor average for the third-row, third-column data

---

[1] These are not actual measured temperatures (say from a thermometer or a satellite) but are instantaneous temperatures from the National Centers for Environmental Prediction (NCEP)/National Center for Atmospheric Research (NCAR) Reanalysis 1 project. A "reanalysis" uses a weather forecasting model coupled with weather measurements to obtain values of weather variables at regular locations in space and time. The full dataset is described in Kalnay et al. (1996).

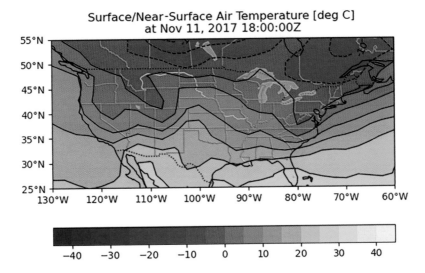

Figure 7.1 Surface/near-surface air temperature over the continental United States at November 11, 2017, 18:00Z (in °C).

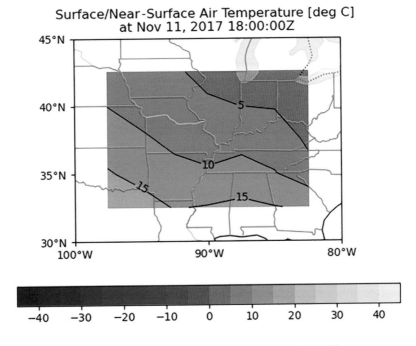

Figure 7.2 Surface/near-surface air temperature over the US Midwest at November 11, 2017, 18:00Z (in °C).

Table 7.1 Surface/near-surface air temperature over the US Midwest at November 11, 2017, 18:00Z (in °C). Only a reasonable number of decimal places are shown in this table. More digits are available in the dataset than are shown.

```
 7.9,   7.1,   5.7,   3.6,   2.3,   1.7,   1.0
 9.9,   8.5,   7.4,   5.7,   4.9,   4.5,   2.1
12.8,  10.6,   9.1,   8.1,   8.8,   8.3,   4.0
15.4,  13.5,  11.1,  10.5,  11.3,  10.1,   6.8
17.6,  17.0,  14.7,  15.3,  16.3,  15.1,  14.4
```

Table 7.2 Longitude values (in degrees East) for the columns in Table 7.1.

```
262.5, 265.0, 267.5, 270.0, 272.5, 275.0, 277.5
```

Table 7.3 Latitude values (in degrees North) for the rows in Table 7.1. The latitude of the elements in the first (top) row of Table 7.1 is the first latitude listed in the current table, and so on down the rows of Table 7.1.

```
42.5, 40.0, 37.5, 35.0, 32.5
```

point in Table 7.1 (which has the value 9.1), we would average together the five values in the boxes below:

```
 7.9,   7.1,   5.7,   3.6,   2.3,   1.7,   1.0
 9.9,   8.5,   7.4,   5.7,   4.9,   4.5,   2.1
12.8,  10.6,   9.1,   8.1,   8.8,   8.3,   4.0
15.4,  13.5,  11.1,  10.5,  11.3,  10.1,   6.8
17.6,  17.0,  14.7,  15.3,  16.3,  15.1,  14.4
```

At the boundaries of the table of data, there are fewer values available to use for the averaging. The code in Figure 7.3 generates an array with the values given in Table 7.4.

In lines 4–9 in Figure 7.3, we define the array `air_temp` which contains the values in the Table 7.1 data grid. In lines 12–13, we create an array `air_temp_avg` that is full of zeros. This array has the same number of rows and columns as `air_temp` and is used to hold the nearest-neighbor average at each (row, column) location of `air_temp`.

In lines 15–37, we execute the nearest-neighbor averaging algorithm by going through each value of `air_temp`, accumulating the values of all the nearest-neighbor points that exist for each given location in `air_temp`, and dividing that sum by the number of nearest-neighbor points to obtain the nearest-neighbor average at the location. We do the summing up for calculating the average by examining each possible neighboring value (lines 21–35), because

Table 7.4 Nearest-neighbor average of surface/near-surface air temperature over the US Midwest at November 11, 2017, 18:00Z (in °C), calculated using the code in Figure 7.3 (i.e., contents of the array `air_temp_avg`**).**

```
 8.30,   7.30,   5.95,   4.32,   3.12,   2.38,   1.60
 9.77,   8.70,   7.28,   5.94,   5.24,   4.30,   2.90
12.17,  10.90,   9.26,   8.44,   8.28,   7.14,   5.30
14.83,  13.52,  11.78,  11.26,  11.40,  10.32,   8.83
16.67,  15.70,  14.52,  14.20,  14.50,  13.98,  12.10
```

the collection of neighboring cells changes depending on whether we are in the middle of the array or at a boundary. Thus, in the corners of the domain, the average is calculated with only three points contributing to `sum_values` whereas in the interior, five points contribute to `sum_values`. The value of `num_values` also differs between these two example cases, being set to three and five, respectively.

Each of the `if` blocks in lines 21–35 checks to see whether the value above, below, to the right, and to the left of the current array element (given by the indices `i` and `j`) are within the grid domain (that is, with row and column index values that are greater than or equal to zero, and less than or equal to one less than the number of row and columns, respectively). For instance, in lines 21–23, we examine the point above the current array element, which has a row index equal to `i − 1` and a column index equal to `j`, to see if it is in the grid domain. As we know `j` is in the grid domain, we do not have to check its value. Because we are looking at the point above the current array element, its row index cannot have a value greater than the number of rows minus one, so we check only that its value is greater than or equal to zero.

As a check, we see that the third-row, third-column value in Table 7.4 (i.e., 9.26) equals the average of the boxed values surrounding and including the value 9.1 from Table 7.1, which is exactly what we expect the nearest-neighbor algorithm should give us. We also plot the Table 7.4 data in Figure 7.4. When we compare Figure 7.4 to Figure 7.2, we see that the transition in temperature – the drop going from the southwest to the northeast – is smoother and does not vary as much in space. The nearest-neighbor average makes gradients in space – the severity of changes going from one location to another – less pronounced.

Finally, in lines 40–42 in Figure 7.3, we extract a **subarray** of the just calculated nearest-neighbor average values and convert the values to kelvin (K). Those kelvin values are then cast as a fraction of the maximum value in kelvin. The values of arrays `air_temp_avg_sub` and `fraction_sub` are given in Tables 7.5 and 7.6.

The example in this section, besides illustrating two-dimensional diagnostic data analysis, will also be used to: illustrate a four-step process of writing programs (Section 11.1), describe the analysis of *n*-dimensional data (Section 12.1), and describe how to make contour plots and animation (Section 14.1).

```
1    import numpy as np
2
3    #- Create data array:
4    air_temp = np.array( \
5        [[ 7.9,    7.1,    5.7,    3.6,    2.3,    1.7,    1.0],
6         [ 9.9,    8.5,    7.4,    5.7,    4.9,    4.5,    2.1],
7         [12.8,   10.6,    9.1,    8.1,    8.8,    8.3,    4.0],
8         [15.4,   13.5,   11.1,   10.5,   11.3,   10.1,    6.8],
9         [17.6,   17.0,   14.7,   15.3,   16.3,   15.1,   14.4]] )
10
11   #- Create array of zeros to hold averages:
12   air_temp_avg = np.zeros(np.shape(air_temp),
13                           dtype=air_temp.dtype.char)
14   #- Calculate average using nearest-neighbor algorithm:
15   avg_shape = np.shape(air_temp_avg)
16   for i in range(avg_shape[0]):
17       for j in range(avg_shape[1]):
18           sum_values = air_temp[i,j]   #+ Center point
19           num_values = 1
20
21           if (i - 1) >= 0:              #+ Above point
22               sum_values += air_temp[i-1,j]
23               num_values += 1
24
25           if (i + 1) < avg_shape[0]:   #+ Below point
26               sum_values += air_temp[i+1,j]
27               num_values += 1
28
29           if (j - 1) >= 0:              #+ Left point
30               sum_values += air_temp[i,j-1]
31               num_values += 1
32
33           if (j + 1) < avg_shape[1]:   #+ Right point
34               sum_values += air_temp[i,j+1]
35               num_values += 1
36
37           air_temp_avg[i,j] = sum_values / num_values
38
39   #- Obtain subarray and Kelvin values as fraction of maximum value:
40   air_temp_avg_sub = air_temp_avg[1:4, 2:4]
41   air_temp_avg_sub_K = air_temp_avg_sub + 273.15
42   fraction_sub = air_temp_avg_sub_K / np.max(air_temp_avg_sub_K)
```

Figure 7.3 Code to analyze the surface/near-surface air temperature over the US Midwest from Table 7.1.

Table 7.5 Subarray of the nearest-neighbor average of surface/near-surface air temperature over the US Midwest at November 11, 2017, 18:00Z (in °C), calculated using the code in Figure 7.3 (i.e., contents of the array `air_temp_avg_sub`).

```
 7.28,  5.94
 9.26,  8.44
11.78, 11.26
```

Table 7.6 Table 7.5 values expressed as a fraction of the maximum temperature in the array. The values of both the numerator and denominator are converted from Celsius into kelvin before calculating the fraction. The values are calculated using the code in Figure 7.3 (i.e., contents of the array `fraction_sub`).

```
0.9842, 0.9795
0.9912, 0.9883
1.0000, 0.9982
```

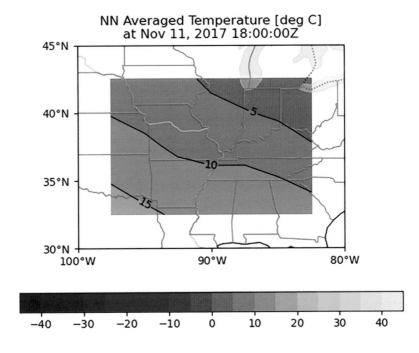

Figure 7.4 Nearest-neighbor average of surface/near-surface air temperature over the US Midwest at November 11, 2017, 18:00Z (in °C).

7.2   Python Programming Essentials

In Chapters 5 and 6, we learned how to store and do calculations on data in arrays using looping and branching. In the example in Section 7.1, we extended our understanding of arrays and looping to accommodate two-dimensional arrays. A two-dimensional array is a grid of values versus a one-dimensional array which is a list or sequence of values. A one-dimensional array would look something like Figure 7.5. A two-dimensional array, on the other hand, would look something like Figure 7.6. A two-dimensional array has both height and width, not just length. Put another way, the array is characterized by the number of rows and columns, not just the number of elements. In lines 4–9 and 12–13 of Figure 7.3, we created two different two-dimensional arrays.

After creating the two-dimensional arrays (in lines 15–37 of Figure 7.3), we use **nested loops** to go through every element of the input array `air_temp` and calculate the nearest-neighbor average for that point. In lines 40–42, we obtain a subarray of the nearest-neighbor average values, convert them to kelvin, and express those values as a fraction of the maximum value in the subarray as kelvin. In this section, we unpack how these tasks are done by examining the following regarding two-dimensional arrays: describing the **shape** of and creating these arrays, accessing and setting elements and slicing subarrays, array syntax and function calls using these arrays, and using nested loops to do computations using these arrays.

In the rest of the code examples in this section, assume that NumPy has already been imported as np.

7.9	7.1	5.7	3.6	2.3	1.7	1.0

Figure 7.5 A schematic of a one-dimensional array. These array values are the first row of Table 7.4.

7.9	7.1	5.7	3.6	2.3	1.7	1.0
9.9	8.5	7.4	5.7	4.9	4.5	2.1
12.8	10.6	9.1	8.1	8.8	8.3	4.0

Figure 7.6 A schematic of a two-dimensional array. These array values are the first three rows of Table 7.4.

7.2.1   The Shape of Two-Dimensional Arrays

Because two-dimensional arrays have both rows and columns, we need more than the total number of elements in the array to properly describe the array. In NumPy, two-dimensional arrays are described by their shape. The shape of an array is a tuple whose number of elements equals the number of dimensions of the array. The `air_temp` array defined in lines 4–9 in Figure 7.3 has a shape given by the tuple `(5, 7)`. The tuple has two elements because the array is two-dimensional.

The contents of the shape tuple tells us how many rows and columns there are in the array. The rightmost value in a shape tuple tells us how many columns there are in the array. Section 12.2.1 shows how this generalizes to *n*-dimensional arrays, where *n* is an integer 0, 1, 2, 3, and higher. *N*-dimensional arrays (for $n > 1$) are also called multidimensional arrays. For a two-dimensional array, the first value of the shape tuple tells us how many rows there are in the array. Because the `air_temp` array has a shape of `(5, 7)`, we know the array has five rows and seven columns.

If we already have an array defined and we want to know the shape of the array, the NumPy function `shape` will return the shape tuple. So, if after running the code in Figure 7.3 we typed:

```
In [1]:  print(np.shape(air_temp))

         (5, 7)
```

we would obtain the shape of `air_temp`. We can access each element of the shape tuple by saving the return of the shape tuple to a variable and accessing each tuple element by index. Because the return value of `shape` is a tuple, we can also access each tuple element by index by putting the square brackets directly after the `shape` call:

```
In [1]:  air_temp_shape = np.shape(air_temp)
         print(air_temp_shape[0])

         5
```

```
In [2]:  print(air_temp_shape[1])

         7
```

```
In [3]:  print(np.shape(air_temp)[0])

         5
```

```
In [4]:  print(np.shape(air_temp)[1])

         7
```

In cells In [3] and In [4], the call to shape is executed first, then the index 0 and index 1 (respectively) values of the shape tuple are extracted, and that value is passed to the print function to display. As another example of using shape, line 15 in Figure 7.3 saves the shape of the array air_temp_shape to the variable avg_shape.

The NumPy function size, which we introduced in Section 5.2.6 in the context of one-dimensional arrays, also works on two-dimensional arrays. The function returns the total number of elements in the array. Thus, if after running the code in Figure 7.3 we typed:

```
In [1]:  print(np.size(air_temp))


         35
```

we would obtain the total number of elements in the array, which is equal to the product of all values in the shape tuple.

The return value of shape is a tuple while the return value of size is a scalar (i.e., a plain-old number). For a two-dimensional array, this is intuitive. But this is also true for a one-dimensional array, and in this latter case it may not seem quite as intuitive:

```
In [1]:  data = np.arange(0, 1.1, 0.25)
         print(data)


         [ 0.    0.25  0.5   0.75  1.  ]

In [2]:  print(np.size(data))


         5

In [3]:  print(np.shape(data))


         (5,)
```

The call to the size function in cell In [2] returns the total number of elements in data. The call to shape in cell In [3] seems to do the same thing, but the result is inside parentheses. What is the difference? In the case of one-dimensional arrays, calls to size and shape produce the same numerical values, but in the latter case, that value is embedded inside a tuple (that is what the parentheses denote in the printed output).

7.2.2 Creating Two-Dimensional Arrays

Many of the functions we used to create one-dimensional arrays can be used to create two-dimensional arrays. As with one-dimensional arrays, we can create a two-dimensional array from a list, if the list is made up of elements that are themselves lists of the same length. Here is a small example:

```
data_list = [[ 2.3, 4.5, -1.1],
             [-0.9, 7.2,  5.0]]
```

The variable `data_list` above is a two-element list. Each element of `data_list` is a list of three elements. When the NumPy `array` function is used to turn this list into an array:

```
In [1]:   data_array = np.array(data_list)
          print(data_array)


          [[ 2.3  4.5 -1.1]
           [-0.9  7.2  5. ]]
```

the result is a two-row, three-column array. In this `print` representation, each row is grouped by a set of square brackets and all the rows (and thus the entire array itself) is grouped by the outer set of square brackets. In the code in Figure 7.3, this is how we create the array `air_temp` in lines 4–9.

Often, we want to create an array of a given shape, filled with zeros or ones. The NumPy functions `zeros` and `ones` does this for us, in a way similar to the one-dimensional case of Section 6.2.1, but in the two-dimensional case, the positional input parameter passed into the function call is a shape tuple, not the number of elements in the array. The shape tuple specifes the number of rows and columns the array of zeros or ones will have. In lines 12–13 of Figure 7.3, we use the NumPy `zeros` function to create the array `air_temp_avg` (reproduced below for convenience):

```
air_temp_avg = np.zeros(np.shape(air_temp),
                        dtype=air_temp.dtype.char)
```

The argument `np.shape(air_temp)` is the shape tuple for `air_temp`, which is the same as the shape of `air_temp_avg`. Put another way, this code produces an array `air_temp_avg`, filled with zeros, that has the same shape as `air_temp`. The elements of `air_temp_avg` also have the same data type as `air_temp`, as specified by the `dtype` keyword input parameter.

In the above lines of code, the shape tuple that is passed into the `zeros` function is the return value from calling the NumPy function `shape` on the array `air_temp`. We can also enter in a shape tuple manually. This code:

```
air_temp_avg = np.zeros((5, 7),
                        dtype=air_temp.dtype.char)
```

will give the exact same result because `air_temp` is a five-row, seven-column array. Again, the shape is entered in as a tuple (`(5, 7)`), not as just two numbers (`5, 7`). We are passing a single positional input parameter into zeros (the tuple), not two positional input parameters. To create a two-dimensional array filled with ones, we can use the NumPy function `ones` in the same way as `zeros`.

In Section 6.2.1, we created one-dimensional arrays of regularly incrementing values using the NumPy function `arange`. There is not a two-dimensional version of `arange`, but we can accomplish something similar by creating a one-dimensional array using `arange` and then using the NumPy function `reshape` to reshape that array into a two-dimensional array. For instance, in the code below, we first create a one-dimensional array `data` of the integers from 0 to 11, and then we use `reshape` to turn the array into a two-dimensional array with four rows and three columns:

```
data = np.arange(12)
data = np.reshape(data, (4, 3))
```

The result is:

```
In [1]:  print(data)

         [[ 0  1  2]
          [ 3  4  5]
          [ 6  7  8]
          [ 9 10 11]]
```

The `reshape` function creates the two-dimensional array from the one-dimensional array one row at a time. That is, it goes through the elements of the one-dimensional array until the first row is filled in the output array, then it fills the second row beginning with the next value in the one-dimensional array that has not been transferred yet, and so on.

7.2.3 Accessing, Setting, and Slicing in a Two-Dimensional Array

The syntax for accessing and setting elements and slicing subarrays in a two-dimensional array is very similar to how it is done in lists and one-dimensional arrays. The first element in a dimension has an index 0, the second element has an index 1, and so on. For slicing, a range of values is described by giving the lower bound index, a colon, then the upper bound index, with the lower bound being inclusive but the upper bound being exclusive. The difference in all these cases is that instead of specifying index values for one dimension, we have to specify them for two dimensions. The index values for each dimension are separated by a comma in the square brackets after the name of the array.

Accessing and Setting Elements   Consider the following four-row, three-column array of some (other) temperatures `temp` (in °C):

```
temp = np.array([[ 5.1,  1.4, -2.1],
                 [-0.9,  7.2, -1.7],
                 [ 2.3, -9.1, -2.0],
                 [-0.7,  8.4,  6.9]])
```

To access the value -1.7, the second-row, third-column value, we type `temp[1,2]`:

```
In [1]: print(temp[1,2])


         -1.7
```

The first value in the square brackets after the array name gives the row index of the element. The second value in the square brackets gives the column index. The second row has index 1 while the third column has index 2. When an array is displayed in Python, whether by typing it in or through printing its contents, the top row is the first row, with a row index of 0, and the leftmost column is the first column, with a column index of 0. The order of dimensions in the square brackets follows the order in the two-dimensional array's shape tuple: the first value specifies row and the second value specifies column.

To set the value of a two-dimensional array, we use the same syntax except that the reference to the array element being changed is to the left of the equal sign in an assignment statement. Thus, to set the second-row, third-column value of the `temp` array to 4.9, we type:

```
temp[1,2] = 4.9
```

If the following code is executed:

```
temp = np.array([[ 5.1,   1.4,  -2.1],
                 [-0.9,   7.2,  -1.7],
                 [ 2.3,  -9.1,  -2.0],
                 [-0.7,   8.4,   6.9]])
temp[1,2] = 4.9
```

the result is:

```
In [1]: print(temp)

         [[ 5.1   1.4 -2.1]
          [-0.9   7.2   4.9]
          [ 2.3  -9.1 -2. ]
          [-0.7   8.4   6.9]]
```

The second-row, third element in `temp` has been changed from -1.7 to 4.9.

Where do we access and set elements in the code in Figure 7.3? In that code, we do not specify the elements we are accessing or setting by typing in numerals but rather through the values of the variables `i` and `j` and operations using those variables. Thus, in line 18, we initialize the array `sum_values` to the value of `air_temp` at the element given at row index `i` and column index `j` (i.e., the point at the center of the averaging neighborhood). In other lines, such as line 22, we refer to the element in `air_temp` that is one row above the index `i` row and at the index `j` column by the syntax `air_temp[i-1,j]`. In line 37, we set the value of `air_temp_avg` at row index `i` and column index `j` to the return value of the expression `sum_values / num_values`.

Slicing Subarrays   Consider the original, unchanged array `temp`:

```
temp = np.array([[ 5.1,   1.4,  -2.1],
                 [-0.9,   7.2,  -1.7],
                 [ 2.3,  -9.1,  -2.0],
                 [-0.7,   8.4,   6.9]])
```

How can we create a subarray that consists of the values in the second and third rows and the first and second columns? The following extracts such a subarray from `temp` and saves it in `temp_sub`:

```
temp_sub = temp[1:3, 0:2]
```

and thus:

```
In [1]:  print(temp_sub)

         [[-0.9   7.2]
          [ 2.3  -9.1]]
```

The `1:3` specifies the range of rows starting from the index 1 row (inclusive) and ending with the index 3 row (exclusive). The `0:2` specifies the range of columns starting from the index 0 column (inclusive) and ending with the index 2 column (exclusive).

To slice to the beginning or to the end of a dimension, leave off the index value either before the colon (for the "to the beginning" case) or after the colon (for the "to the end" case). Thus, this code:

```
temp = np.array([[ 5.1,   1.4,  -2.1],
                 [-0.9,   7.2,  -1.7],
                 [ 2.3,  -9.1,  -2.0],
                 [-0.7,   8.4,   6.9]])
temp_sub = temp[:3, 0:]
```

yields this result:

```
In [1]:  print(temp_sub)

         [[ 5.1   1.4  -2.1]
          [-0.9   7.2  -1.7]
          [ 2.3  -9.1  -2. ]]
```

The `:3` syntax slices all rows from the first row through the index 2 row (i.e., index 3 is the upper bound, exclusive). The `0:` syntax slices all columns because index 0 is the first column and, as a lower bound of the slice in the dimension, is inclusive.

If we wanted to slice all values along a dimension, we would use a colon with no indices specified. Thus, the slicing syntax `0:` is the same as `:`.

We can also use negative indices to specify either the lower bound (inclusive) or upper bound (exclusive) of a slice along a dimension. Thus:

```
temp = np.array([[ 5.1,   1.4,  -2.1],
                 [-0.9,   7.2,  -1.7],
                 [ 2.3,  -9.1,  -2.0],
                 [-0.7,   8.4,   6.9]])
temp_sub = temp[:3, 0:-1]
```

yields:

```
In [1]:  print(temp_sub)

         [[ 5.1   1.4]
          [-0.9   7.2]
          [ 2.3  -9.1]]
```

The `0:-1` syntax does *not* extend the slice through all columns, including the last element. Instead, because the `-1` index refers to the last column, and that value is to the right of the colon, that index is considered the upper bound (exclusive). Thus, the last column in `temp_sub` is the index 1 column, not the index 2 (or last) column.

As with one-dimensional arrays, we can use array slicing to assign values in a two-dimensional array all at once. Thus:

```
temp = np.array([[ 5.1,   1.4,  -2.1],
                 [-0.9,   7.2,  -1.7],
                 [ 2.3,  -9.1,  -2.0],
                 [-0.7,   8.4,   6.9]])
temp[2:, :2] = temp[:2, 1:]
```

puts the first two-rows, last two-column subarray elements into the last two-rows, first two-column subarray element locations. The result is `temp` becomes:

```
In [1]:  print(temp)

         [[ 5.1   1.4  -2.1]
          [-0.9   7.2  -1.7]
          [ 1.4  -2.1  -2. ]
          [ 7.2  -1.7   6.9]]
```

With this as background, we see that the line 40 code in Figure 7.3 extracts a subarray of `air_temp_avg` running from the index 1 row through the index 3 row and the index 2 column through the index 3 column. The result is the subarray in Table 7.5.

7.2.4   Array Syntax and Functions in Two-Dimensional Arrays

As we saw in Sections 5.2.6, 6.2.1, and 6.2.2, array syntax and many NumPy functions on arrays allow us to: (1) operate element-wise on one-dimensional arrays, (2) apply a function element-wise on one-dimensional arrays, and (3) use a function to make a calculation that considers all the elements in an array, without using loops. The same works with two-dimensional arrays. Thus, the following code that uses the array `temp`:

```
temp = np.array([[ 5.1,   1.4,  -2.1],
                 [-0.9,   7.2,  -1.7],
                 [ 2.3,  -9.1,  -2.0],
                 [-0.7,   8.4,   6.9]])
double_temp = temp * 2.0
```

creates an array `double_temp` of the same shape as `temp` whose contents are:

```
In [1]:  print(double_temp)

         [[ 10.2    2.8   -4.2]
          [ -1.8   14.4   -3.4]
          [  4.6  -18.2   -4. ]
          [ -1.4   16.8   13.8]]
```

where each element is twice the value of the corresponding value in the `temp` array. The multiplication by 2.0 is done element-wise on `temp`.

Likewise, if we combine two two-dimensional arrays in some way, the operation is done element-wise between corresponding elements. If we add `temp` and `double_temp` above, each corresponding element of the two arrays are added together and the result is an array of the same shape:

```
In [1]:  combined_temp = temp + double_temp
         print(combined_temp)

         [[ 15.3    4.2   -6.3]
          [ -2.7   21.6   -5.1]
          [  6.9  -27.3   -6. ]
          [ -2.1   25.2   20.7]]
```

If we pass the `temp` array into a NumPy function that accepts arrays and operates on them element-wise – the `sin` function is such a function – the result is another array of the same shape as the input `temp`, where each element is the sine of the corresponding element in `temp`:

```
In [1]:  sine_temp = np.sin(temp)
         print(sine_temp)
```

```
[[-0.92581468   0.98544973  -0.86320937]
 [-0.78332691   0.79366786  -0.99166481]
 [ 0.74570521  -0.31909836  -0.90929743]
 [-0.64421769   0.85459891   0.57843976]]
```

Finally, some NumPy functions accept arrays as input arguments and considers all the elements in the array in making its calculations, without explicitly using loops. This works not only on one-dimensional arrays but also two-dimensional arrays. The NumPy `min` function, for instance, looks through all elements in a two-dimensional array and returns the element with the lowest value. If applied to the `sine_temp` array above, we obtain:

```
In [1]:  print(np.min(sine_temp))
```

```
-0.9916648104524686
```

Line 41 in the code in Figure 7.3 illustrates the use of array syntax on two-dimensional arrays. Line 42 in the code illustrates the use of both array syntax and a function that consider all elements in a two-dimensional array for its calculation. In line 41, the addition of 273.15 is done on every element of `air_temp_avg_sub` and the result is an array of the same shape that holds those values called `air_temp_avg_sub_K`. The `np.max(air_temp_avg_sub_K)` expression returns the maximum value of all the elements in `air_temp_avg_sub_K`. In the rest of the right-hand side of line 42, each element in `air_temp_avg_sub_K` is divided by the maximum value in that array, to transform the "units" of `air_temp_avg_sub_K` into fraction of the maximum value of all the elements in `air_temp_avg_sub_K`.

As an aside, in Section 12.2.3, we will find that the syntax for array syntax and NumPy functions for higher-dimensional arrays is, for the most part, exactly the same as in the one-dimensional and two-dimensional cases. Thus, the code examples in this section showing operations and function calls using arrays would not change even if the arrays were three-dimensional or of more dimensions.

7.2.5 Nested `for` Loops

Section 6.2.3 introduced the use of loops to go through each element in an array to access or set the elements. This can be done by looping through the values of the array elements or the indices of the array elements. For two-dimensional arrays, a single `for` loop cannot directly loop through all elements of the array. But, by using two `for` loops, one inside the other – by nesting the loops – we can loop through all the row indices and, for each row index, all of the column indices, to access all possible combinations of row and column indices in the array.

Consider the following code snippet that defines the array `temp` and then loops through each element, writing out the values to the screen:

```
temp = np.array([[ 5.1,   1.4,  -2.1],
                 [-0.9,   7.2,  -1.7],
                 [ 2.3,  -9.1,  -2.0],
                 [-0.7,   8.4,   6.9]])
for i in range(4):
    for j in range(3):
        print("(i,j) = (" + str(i) + "," + str(j) + \
            ") value:  " + str(temp[i,j]))
```

When run, the following is printed out to the screen:

```
(i,j) = (0,0) value:   5.1
(i,j) = (0,1) value:   1.4
(i,j) = (0,2) value:   -2.1
(i,j) = (1,0) value:   -0.9
(i,j) = (1,1) value:   7.2
(i,j) = (1,2) value:   -1.7
(i,j) = (2,0) value:   2.3
(i,j) = (2,1) value:   -9.1
(i,j) = (2,2) value:   -2.0
(i,j) = (3,0) value:   -0.7
(i,j) = (3,1) value:   8.4
(i,j) = (3,2) value:   6.9
```

What is happening? The first `for` loop goes through the values 0, 1, 2, and 3 (the result of the `range(4)` call), setting them to the iterator `i`, each in turn, with each iteration. So, in the first iteration of the first (or outer) `for` loop, `i` is set to 0 and the body of the loop is executed. But the first line of the body of the outer loop is the beginning of the second (or inner) `for` loop. The inner loop goes through the values 0, 1, and 2 (the result of the `range(3)` call), setting them to the iterator `j`, each in turn, with each iteration of the second loop. The value of `i` is unchanged – i.e., 0 – for all these iterations of the inner loop.

Put another way, for the first iteration of the outer loop, we run through all iterations of the inner loop, because that inner loop is the body of the outer loop. For all those iterations of the inner loop, we execute the body of the inner loop (the `print` statement of lines 7–8). The value of `i` is unchanged – i.e., one – for all these iterations of the inner loop.

Thus, the code first sets `i` to 0, then sets `j` to 0, 1, and 2, and for each of those values of `j`, we print out the values of `i` and `j` as well as the value of `temp` at row index `i` and column index `j`. That finishes the first iteration of the outer loop.

Next, we do the second iteration of the outer loop: `i` is set to one, the `j` is set to 0, 1, and 2, and for each of those values of `j`, we print out the values of `i` and `j` as well as the value of `temp` at row index `i` and column index `j`. That finishes the second iteration of the outer loop.

The third iteration begins with setting i to two, and then the rest occurs just as in the first two iterations. The fourth iteration begins with setting i to three, and then the rest occurs just as in the other iterations. As the i equals three iteration is the last iteration of the outer loop, when that iteration ends, no more looping is done.

The values the outer and inner loops run through are the indices for the rows and columns, respectively, of temp. Since there are four rows in temp, the outer loop goes through 0, 1, 2, and 3. Since there are three columns in temp, the inner loop goes through 0, 1, and 2. While it is possible to loop through columns then rows, rather than rows then columns as in our example above, it is generally more efficient to loop through rows in the outer loop and columns in the inner loop. Section 12.2.6 describes why this is the case.

In our example above, the row and column indices are generated using calls to range, passing in the number of rows and columns, respectively. We can use the shape function to obtain the number of rows and columns, thus enabling us to write the nested for loop so it will work on any array, not just those whose rows and columns we can visually count. This code saves the shape tuple of the array temp and uses those values in the range calls in the for loops:

```
temp_shape = np.shape(temp)
for i in range(temp_shape[0]):
    for j in range(temp_shape[1]):
        print("(i,j) = (" + str(i) + "," + str(j) + \
            ") value:  " + str(temp[i,j]))
```

This code directly uses the return value from the call to shape and extracts the appropriate element for use in the for loop range call:

```
for i in range(np.shape(temp)[0]):
    for j in range(np.shape(temp)[1]):
        print("(i,j) = (" + str(i) + "," + str(j) + \
            ") value:  " + str(temp[i,j]))
```

All the above nested for loops in this section do the same thing: they loop through all the rows in temp and then, for each row, all the columns in that row.

Lines 15–37 in Figure 7.3 illustrate the use of nested loops to go through all elements of air_temp, calculate the nearest-neighbor average for each element, and save the average value in the corresponding element of air_temp_avg. In that example, the body of the inner loop is more than a single print statement but includes lines 18–37. All those lines are executed for each iteration of the inner loop, and all iterations of the inner loop occur for each iteration of the outer loop.

In Section 6.2.3 and the current section, we limit our use of for loops to iterating through arrays of numbers. In Section 13.2.6, we will find that for loops can loop through more than arrays of numbers. And, in Section 8.2.6, we examine more complex nested loops.

7.3   Try This!

We consider examples similar to those we have seen in Section 7.1 and that address the main topics of Section 7.2. We practice creating and manipulating two-dimensional arrays. We also practice writing nested loops to operate on two-dimensional arrays.

For all the exercises in this section, assume NumPy has already been imported as np.

Try This! 7-1 Create Arrays: Air Quality

Table 7.7 lists one day of hourly $PM_{2.5}$ concentration (in $\mu g/m^3$) measurements at four locations in Beijing, China (in relatively close proximity to each other). $PM_{2.5}$ refers to airborne particles with diameters less than 2.5 μm. High concentrations of these particles are indicative of poor air quality.[2] Using Python, create a two-dimensional array called pm25 that holds the data in Table 7.7. Then, create an array of zeros called pm25_new that has the same shape as pm25.

Try This! Answer 7-1

The code in Figure 7.7 accomplishes the above tasks.

In defining the array pm25, we specify the 'd' typecode in line 25 of Figure 7.7. The same is true in line 26 when we define pm25_new. This makes sure all the elements of the array are stored as floating-point numbers rather than integers. If we left out the dtype='d' keyword input parameter setting, the elements of pm25 might be defined as integer (in earlier versions of NumPy) because the values in the input list are all integers. Generally, this is not a problem, because if later we used pm25 with arithmetic operators, Python would automatically convert the array as needed to obtain results that retained as much information as possible. Nonetheless, if we know ahead of time we want the elements of pm25 to be floating point, we should make the array elements floating point to begin with.

In line 26 of Figure 7.7, we use the NumPy shape function to obtain the shape of the pm25 array and use that shape as the shape for pm25_new. We could also have manually entered in the shape of the new array:

```
pm25_new = np.zeros((24,4), dtype='d')
```

This gives the same result as line 26 as pm25 has 24 rows and 4 columns.

Remember that "(24,4)" as well as the return value of "np.shape(pm25)" are tuples. We can, however, use a list or an array instead of a tuple in the input parameter list to zeros; the zeros function does not care whether the input is mutable or immutable, etc. Thus:

```
pm25_new = np.zeros([24,4], dtype='d')
```

---

[2] Liang et al. (2016).

and

```
pm25_new = np.zeros(np.array([24,4]), dtype='d')
```

work fine too.

Table 7.7 Hourly PM$_{2.5}$ concentrations (in $\mu g/m^3$) for June 8, 2013, in Beijing, China, at four different but nearby locations. The first row is hour 0 and the last row is hour 23. The first column are measurements from the Dongsi site, the second column are measurements from the Dongsihuan site, the third column are measurements from the Nongzhanguan site, and the last column is from the US Embassy site (Liang et al., 2016; Dua and Graff, 2019). Data are described in Try This! 7-1.

118,	132,	128,	139
114,	129,	125,	138
128,	119,	130,	139
122,	118,	119,	135
124,	114,	121,	134
108,	119,	125,	139
112,	138,	130,	140
148,	141,	140,	147
163,	162,	124,	161
195,	169,	154,	191
216,	198,	199,	212
226,	220,	213,	220
234,	218,	218,	221
219,	197,	187,	206
204,	193,	172,	189
191,	190,	181,	187
185,	194,	186,	187
193,	193,	185,	191
195,	193,	194,	191
188,	191,	174,	196
200,	176,	154,	194
186,	166,	128,	184
177,	162,	134,	182
184,	153,	104,	172

```
pm25 = np.array( \
    [[118, 132, 128, 139],
     [114, 129, 125, 138],
     [128, 119, 130, 139],
     [122, 118, 119, 135],
     [124, 114, 121, 134],
     [108, 119, 125, 139],
     [112, 138, 130, 140],
     [148, 141, 140, 147],
     [163, 162, 124, 161],
     [195, 169, 154, 191],
     [216, 198, 199, 212],
     [226, 220, 213, 220],
     [234, 218, 218, 221],
     [219, 197, 187, 206],
     [204, 193, 172, 189],
     [191, 190, 181, 187],
     [185, 194, 186, 187],
     [193, 193, 185, 191],
     [195, 193, 194, 191],
     [188, 191, 174, 196],
     [200, 176, 154, 194],
     [186, 166, 128, 184],
     [177, 162, 134, 182],
     [184, 153, 104, 172]], dtype='d')
pm25_new = np.zeros(np.shape(pm25), dtype='d')
```

Figure 7.7 Code to solve Try This! 7-1.

Try This! 7-2 Access and Slice Arrays: Surface Temperature

Consider the data in Table 7.1, given at the locations whose longitude and latitude coordinates are given in Tables 7.2 and 7.3, respectively. Do the following tasks:

1. Save the subarray of the temperatures of all locations whose longitude is less than (i.e., west of) 268° E. Call this subarray `air_temp_1`.
2. Save the value of the temperature at 275° E and 35° N as `point_1`.
3. Say we discover the last column of values is actually all supposed to be the same as the values one column to the left. Change `air_temp` to reflect this discovery.

Do the accessing and slicing by explicitly entering in the indices. The computer does not need to figure out which rows and columns correspond to the conditions above.

Try This! Answer 7-2

The solutions below assume the array `air_temp` has been defined as by lines 1–9 in Figure 7.3:

```
1. air_temp_1 = air_temp[:, :3]
2. point_1 = air_temp[3,5]
3. air_temp[:, -1] = air_temp[:, -2]
```

In the slicing operation in #1 above, the first dimension is sliced by a colon with no lower or upper bounds because all rows are desired. The second dimension is sliced by `:3` because we want the columns from the first column through the third, which is the last column whose longitude value is less than 268° E. The third column has an index of 2, and in the slicing syntax, the upper-bound values are exclusive. Thus, an upper-bound column index of 3 will run through and include the index 2 colunn.

The top row of `air_temp` (row index 0) locations have a latitude value of 42.5° N (the index 0 value if the list of values in Table 7.3 were saved as an array), not a latitude value of 32.5° N (the index 4 value if the list of values in Table 7.3 were saved as an array). That is to say, the top row of `air_temp` is not the "northernmost" row merely because it is printed out first.

Try This! 7-3 Array Syntax: Surface Temperature

Using the data in Table 7.1, convert the values from Celsius to Fahrenheit. Recall that

$$°F = \frac{9}{5}°C + 32$$

Use array syntax in the solution.

Try This! Answer 7-3

This code solves the problem. It assumes the array `air_temp` has been defined as by lines 1–9 in Figure 7.3:

```
air_temp_F = ((9.0 / 5.0) * air_temp) + 32.0
```

Because `air_temp` is an array, the arithmetic operations are applied to each element independently and the resulting array `air_temp_F` is the same shape as `air_temp`.

The code above would have worked if we had used 9 instead of 9.0, 5 instead of 5.0, and 32 instead of 32.0. By putting in the decimal, however, we guarantee that the result will be floating point.[3]

---

[3] This is only needed if we are using Python 2.7.x, but by adding in the decimal version, we guarantee our code will work with that older version of Python.

As a check, here are the values of `air_temp_F`:

```
1    46.22, 44.78, 42.26, 38.48, 36.14, 35.06, 33.80
2    49.82, 47.30, 45.32, 42.26, 40.82, 40.10, 35.78
3    55.04, 51.08, 48.38, 46.58, 47.84, 46.94, 39.20
4    59.72, 56.30, 51.98, 50.90, 52.34, 50.18, 44.24
5    63.68, 62.60, 58.46, 59.54, 61.34, 59.18, 57.92
```

Try This! 7-4 Array Slicing and Syntax: Air Quality

Given the array `pm25`, defined in the Figure 7.7 solution to Try This! 7-1, calculate the $PM_{2.5}$ concentration difference between each hour and 12 hours later, for each monitoring site, for the first 12 hours. For instance, for the first column measurements from the Dongsi site, we would calculate hour 12 minus hour 0, hour 13 minus hour 1, hour 14 minus hour 2, and so on until hour 23 minus hour 11. These same calculations would be repeated for the measurements at the other sites given in the other three columns. The resulting array will have 12 rows and 4 columns. Call the array `diff`. Do this using array slicing and array syntax.

Try This! Answer 7-4

Assuming `pm25` is already defined, using array slicing and array syntax, we can write this solution:

```
1    last_half = pm25[12:, :]
2    first_half = pm25[:12, :]
3    diff = last_half - first_half
```

The contents of `diff` are:

```
In [1]:  diff

Out[1]:  array([[ 116.,    86.,    90.,    82.],
               [ 105.,    68.,    62.,    68.],
               [  76.,    74.,    42.,    50.],
               [  69.,    72.,    62.,    52.],
               [  61.,    80.,    65.,    53.],
               [  85.,    74.,    60.,    52.],
               [  83.,    55.,    64.,    51.],
               [  40.,    50.,    34.,    49.],
               [  37.,    14.,    30.,    33.],
               [  -9.,    -3.,   -26.,    -7.],
               [ -39.,   -36.,   -65.,   -30.],
               [ -42.,   -67.,  -109.,   -48.]])
```

Let us unpack the solution. In line 1, the first slicing operation, pm25[12:, :], extracts the last 12 rows of the pm25 array. It starts at row index 12 and runs all the way to the end of the rows (row index 23). The slice selects all columns, as is specified by the lone colon in the column dimension position. This slice is saved to the array last_half and holds the measurements from the last half of the day.

In line 2 of the solution, the second slicing operation, pm25[:12, :], extracts the first 12 rows of the pm25 array. It starts at the first row (row index 0) and runs all the way to and including row index 11 (the 12 in the row slice is an upper bound and so it is exclusive, not inclusive). The slice selects all columns, as is specified by the lone colon in the column dimension position. This slice is saved to the array first_half and holds the measurements from the first half of the day.

In line 3, we subtract first_half from last_half. Because both of these arrays have 12 rows and 4 columns, the subtraction is done on corresponding elements. That is, the first-row, first-column element in first_half is subtracted from the first-row, first-column element in last_half, and so on for all the elements in each array at the same locations in their respective arrays. The result of this operation is placed into an output array (that is automatically created) called diff that also has 12 rows and 4 columns. The result of the subtraction on the first-row, first-column element of first_half and last_half is placed in the first-row, first-column location of diff, and so on for all other locations in the arrays.

In the above solution, first_half and last_half are arrays whose only purpose is to hold the measurements from the slicing operations. If we do not plan to use those arrays again, we can shorten the solution to a single line of code:

```
diff = pm25[12:, :] - pm25[:12, :]
```

The result of each slice is an array, and the subtraction operation automatically works element-wise on those arrays and places the difference in diff.

Try This! 7-5 Array Functions: Surface Temperature

Using the data in Table 7.1, calculate the range of the values in the array and the mean of all the temperatures in the array (in °C). The range of the values is the difference between the largest and smallest values.

Try This! Answer 7-5

This code solves the problem. It assumes the array air_temp has been defined as by lines 1–9 in Figure 7.3:

```
1   range_air_temp = np.max(air_temp) - np.min(air_temp)
2   mean_air_temp = np.average(air_temp)
```

The result is:

```
In [1]:  print(range_air_temp)

         16.6

In [2]:  print(mean_air_temp)

         9.231427873883929
```

Here is another way of calculating the mean of the temperatures. The `sum` function adds up all values in the array, and using that with the `size` function we obtain:

```
In [1]:  print(np.sum(air_temp) / np.size(air_temp))

         9.231427873883929
```

which gives us the same answer. In the code above, instead of storing the result of the expression in a variable, we directly pass the result of that expression to the `print` call and print it to the screen.

Try This! 7-6 Nested Looping: Surface Temperature

Using the data in Table 7.1, convert the values from Celsius to Fahrenheit. The equation for the conversion is in Try This! 7-3. Use nested looping instead of array syntax.

Try This! Answer 7-6

This code solves the problem. It assumes the array `air_temp` has been defined as by lines 1–9 in Figure 7.3:

```
1    air_temp_shape = np.shape(air_temp)
2    air_temp_F = np.zeros(air_temp_shape, dtype=air_temp.dtype.char)
3    for i in range(air_temp_shape[0]):
4        for j in range(air_temp_shape[1]):
5            air_temp_F[i,j] = ((9.0) / 5.0 * air_temp[i,j]) + 32.0
```

Unlike the array syntax solution for Try This! 7-3, this solution requires that we first create the array `air_temp_F` before we can go through the elements of `air_temp` using the nested `for` loops and do the Fahrenheit conversion calculations. Otherwise, there will be nowhere to store the converted value.

The elements of the `air_temp_F` array will have the same datatype as `air_temp`. We can also force the `air_temp_F` array to have a floating-point datatype by passing in `'d'` via the `dtype` keyword input parameter, rather than passing in `air_temp.dtype.char`.

Try This! 7-7 Array Syntax and Nested Looping: Wind Speed

Pretend we have two two-dimensional arrays, `v_x` and `v_y`. The `v_x` array contains the East–West components of surface wind velocity (i.e., the *x*-components of wind velocity, where the *x*-direction is defined as East–West) at regular longitude and latitude locations. The `v_y` array contains the North–South components of wind velocity (i.e., the *y*-components of wind velocity, where the *y*-direction is defined as North–South) at the same regular longitude and latitude locations as `v_x`. Calculate the surface wind speed *v* at the regular longitude and latitude locations, given:

$$v = |\vec{v}| = \sqrt{v_x^2 + v_y^2}$$

Do this calculation once using nested looping and once using array syntax.

Try This! Answer 7-7

This code solves the problem using nested looping:

```
1   wind_shape = np.shape(v_x)
2   v = np.zeros(wind_shape, dtype=v_x.dtype.char)
3   for i in range(wind_shape[0]):
4       for j in range(wind_shape[1]):
5           v[i,j] = ((v_x[i,j] ** 2) + (v_y[i,j] ** 2)) ** 0.5
```

The above code assumes the shape of `v_x` is the same as the shape of `v_y`. It is probably better to explicitly check this is the case before doing the calculations, which is not done above.

The array syntax solution takes only one line:

```
v = ((v_x ** 2) + (v_y ** 2)) ** 0.5
```

When doing operations with arrays, the shapes of `v_x` and `v_y` are automatically checked to make sure they are compatible, in order to apply the exponentiation operations on elements at corresponding locations. If the shapes are not compatible, Python automatically throws an error and tells us it cannot perform the operation.

In comparing the array syntax and nested looping solutions for calculating `v`, we might wonder, "why not use array syntax all the time?" Array syntax is a great tool. Code is more concise and (as Section 13.2.4 describes), array syntax is faster than explicit looping. However, array syntax is not always the best tool. It usually works when all the calculations would occur in the body of the inner loop and the same syntax is applied to all the points in the array, with no `if` statements. Although, in Section 13.2.3, we discuss ways to apply array syntax even with `if`-like and other complicating situations. In Section 8.2.6, we examine more complex cases where nested loops may be the better option. The bottom line: As a general rule, array syntax works better in noncomplex situations.

Try This! 7-8 Nested Looping: Air Quality

Repeat Try This! 7-4, but use nested looping instead of array syntax.

Try This! Answer 7-8

Assuming `pm25` is already defined, using nested looping, we can write this solution:

```
1   diff = np.zeros((12,4), dtype='d')
2   for i in range(np.shape(diff)[0]):
3       for j in range(np.shape(diff)[1]):
4           diff[i,j] = pm25[i+12,j] - pm25[i,j]
```

The main difference in the above solution from the other examples of nested loops we have seen in the Try This! in this section is the use of the offset indexing of rows in the `pm25[i+12,j]` term in line 4. The looping variable `i` runs through the row index values 0, 1, 2, ..., 11. As a result, the `pm25[i+12,j]` term will run through the row index values 12, 13, 14, ..., 23. Those are the values 12 hours from those at row index `i`. This use of an offset to the index value obtained from a `for` loop's iterator mirrors the use of offsets in the `if` statements in lines 21–35 of Figure 7.3.

In the above solution, no slicing to extract a subarray is needed. In the array syntax solution to Try This! 7-4, we created subarrays to align the portions of the original array that are offset from each other by 12 hours, to enable us to use array syntax to do the subtraction calculations element-wise. In our above nested loop solution, the offset is explicitly done one element at a time in the nested loop, using the `i+12` syntax.

Try This! 7-9 Nested Looping: NN Averaging

In the example in Section 7.1, we calculated the nearest-neighbor average of a two-dimensional array of surface temperature using the code in Figure 7.3. How would we change that code to include only the adjacent neighboring points at the same latitude of the location that is the center of our average?

Try This! Answer 7-9

The portion of code we will change is lines 15–37 of Figure 7.3. To calculate an average using only the adjacent neighboring points at the same latitude as the center of the average, we can

```
1    avg_shape = np.shape(air_temp_avg)
2    for i in range(avg_shape[0]):
3        for j in range(avg_shape[1]):
4            sum_values = air_temp[i,j]      #+ Center point
5            num_values = 1
6
7            if (j - 1) >= 0:                    #+ Left point
8                sum_values += air_temp[i,j-1]
9                num_values += 1
10
11            if (j + 1) < avg_shape[1]:   #+ Right point
12                sum_values += air_temp[i,j+1]
13                num_values += 1
14
15            air_temp_avg[i,j] = sum_values / num_values
```

Figure 7.8 Code snippet to calculate the nearest-neighbor average using only the adjacent neighboring points at the same latitude as the center of the average, in Try This! 7-9.

delete the code that considers the neighbors above and below the center of the average. Our solution replaces lines 15–37 of Figure 7.3 with the lines in Figure 7.8.

7.4 More Discipline-Specific Practice

The Discipline-Specific Jupyter notebooks for this chapter cover the following topics:

- Creating two-dimensional arrays.
- Accessing and slicing two-dimensional arrays.
- Array syntax using two-dimensional arrays.
- Using two-dimensional arrays as arguments to functions.
- Using nested loops to analyze data stored in two-dimensional arrays.

7.5 Chapter Review

7.5.1 Self-Test Questions

Try to do these without looking at the book or any other resources or using the Python interpreter. Answers to these Self-Test Questions are found at the end of the chapter. For all questions in this section, assume NumPy has already been imported as np.

Self-Test Question 7-1

What is the shape tuple of a two-dimensional array? What does it describe? How many elements does it have?

Self-Test Question 7-2

Pretend there is a number of gas samples taken at various regular latitude and longitude locations. The array ppm holds the concentration in parts per million of a certain chemical in the sample.

```
ppm = np.array([[ 20.80, 20.71, 21.72, 18.81],
                [ 19.91, 18.42, 20.73, 20.32],
                [ 17.32, 21.74, 23.19, 19.98]])
```

What is the code to obtain the shape of ppm? What is the shape tuple equal to?

Self-Test Question 7-3

Create a three-row, two-column integer array of zeros called `values`.

Self-Test Question 7-4

Create a two-row, four-column floating-point array called `factors` where the first row's values are 0, 1, 2, and 3, and the second row's values are 4, 5, 6, and 7.

Self-Test Question 7-5

Create a three-row, five-column integer array called `sevens` where every element is 7.

Self-Test Question 7-6

Given the array `factors` from Self-Test Question 7-4:

1. Save the second-row, third-column element of `factors` as the variable `value`.
2. Create a subarray `first_row` that contains only the first-row elements of `factors`.
3. Create a subarray `subset` that contains all the rows and the second and third columns of `factors`.
4. Make the first three elements in the first row of `factors` equal to the last three elements in the last row of `factors`.

Self-Test Question 7-7

For the concentrations in the array ppm from Self-Test Question 7-2, calculate the values of each element if the concentrations are to be expressed in parts per billion. Save the results as the array ppb.

Self-Test Question 7-8

Given a two-dimensional array `data1` and another two-dimensional array `data2` (both of the same shape), calculate the average between corresponding elements of the two arrays. Save the result as `avg_between_both`.

Self-Test Question 7-9

Consider the two-dimensional array `data3`. Find the maximum value in the array, first using the NumPy `max` function, and second using nested looping. Save the maximum value as a variable `max_data3`.

7.5.2   Chapter Summary

In looking at two-dimensional arrays and how to do analysis of data stored in these arrays, we covered these topics:

- What is the shape of an array?

  - Two-dimensional arrays are described in terms of the number of rows and number of columns they have. These values are stored in a two-element tuple called the shape tuple. The first element in that tuple stores the number of rows and the second element stores the number of columns.
  - One-dimensional arrays also have a shape tuple containing one element, the number of elements in that array. We can think of this as the value given by the `size` function, but stored as an element in a tuple.

- Creating two-dimensional arrays

  - The NumPy `array` function takes a list whose elements are lists and converts that into a two-dimensional array.
  - The NumPy `zeros` function takes a shape tuple as input and creates an array filled with zeros that has that shape. The `ones` function does the same thing as `zeros`, but all the elements are ones.
  - The NumPy `reshape` function can take a one-dimensional array and reshape it into a two-dimensional array.

- Accessing and setting elements in a two-dimensional array

  - Accessing and setting elements in a two-dimensional array works the same as in a one-dimensional array except two indices are provided in the square brackets after the array name (e.g., `data[2,1]`). The first number gives the row index and the second number gives the column index for the element of interest. The two indices are separated by a comma.
  - Both row and column indices are zero-based: the first row in the array has a row index of 0 and the first column in the array has a column index of 0.
  - Negative index values can be used in accessing and setting elements in two-dimensional arrays. The last element of a dimension has an index of `-1`.

- Slicing elements in a two-dimensional array

  - Slicing elements to extract subarrays to use or to assign values works the same as in a one-dimensional array except two index ranges are provided in the square brackets after the array name (e.g., `data[2:4, 1:6]`).

- The first range in the square bracket defines the lower and upper bounds of the row indices, where the lower bound is the index to the left of the colon and the upper bound is the index to the right of the colon. In that range, the lower bound is inclusive of that index but the upper bound is exclusive of that index.
- The second range in the square bracket defines the lower and upper bounds of the column indices. The values of the column index range behave the same way as with the row index range.
- If either the lower- and/or upper-bound index value is left off of the range, the range goes through the beginning and/or end element in that dimension. This is inclusive for both beginning and ending elements. Thus, for the rows range 2:4:
    * `data[2:4, :6]` slices all columns from the index 0 column (inclusive) to the index 5 column (inclusive),
    * `data[2:4, 6:]` slices all columns from the index 6 column (inclusive) to the last column (inclusive),
    * `data[2:4, :]` slices all columns.
- Negative index values can be used in specifying lower and/or upper bounds for a slicing operation.

- Array syntax and functions in two-dimensional arrays:

  - When two-dimensional arrays are part of arithmetic operations with scalar values (e.g., plain numbers), the operation is conducted between each element of the array and the scalar value. The result of each operation is stored in the corresponding row, column location of the output array. That output array is the return value from the operation and is an array of the same shape of the original array.
  - When two two-dimensional arrays are chained together by an arithmetic operation, the operation is conducted on the values in corresponding locations between the two arrays. The return value is an array of the same shape as the original arrays. The two arrays have to have the same shape in order for the operation to work.
  - Many NumPy functions (e.g., `cos`) accept two-dimensional (and other) arrays as input, do their calculations on each element of the input array, and put the result in the corresponding location in an output array of the same shape as the input array.
  - Other NumPy functions (e.g., `max`) operate on the array as a whole and return a single number as a result.

- Nested `for` loops

  - A doubly nested `for` loop consists of two `for` loops, one inside the body of the other.
  - For every iteration of the outer `for` loop, the inner `for` loop goes through all the iterations the inner loop defines.
  - Use the values of the different elements of the shape tuple as the values passed into the `range` calls that are part of the `for` loops in the nested loop structure. That way, the loop will work even if the array we are looping through changes size.

   – Nested loops are more flexible than array syntax and give us more granular control over what elements in the array we are accessing. Array syntax is more concise and usually runs faster than nested loops.

7.5.3 Self-Test Answers

Self-Test Answer 7-1

The shape tuple of a two-dimensional array is a two-element tuple. The first element describes the number of rows in the array, and the second element describes the number of columns in the array.

Self-Test Answer 7-2

The `shape` function will return the shape of ppm. If the array is defined as above, then:

```
In [1]:  print(np.shape(ppm))

         (3, 4)
```

The shape tuple indicates there are three rows and four columns in the ppm array.

Self-Test Answer 7-3

```
values = np.zeros((3,2), dtype='l')
```

Self-Test Answer 7-4

This code solves the problem:

```
factors = np.reshape(np.arange(8.0), (2, 4))
```

Passing in 8.0 into the `arange` call makes the array a floating-point array rather than an integer array. Section 6.2.1 discusses this further.

Self-Test Answer 7-5

This works:

```
sevens = np.zeros((3, 5)) + 7
```

as will this:

```
sevens = np.ones((3, 5)) * 7
```

Self-Test Answer 7-6

1. `value = factors[1,2]`
2. `first_row = factors[0, :]`
3. `subset = factors[:, 1:3]`
4. `factors[0, 0:3] = factors[1, 1:]   or   factors[0, 0:3] = factors[1, -3:]`

As can be seen above, we can put a blank space after the comma between the dimension index slices in the square brackets, or not. The examples in this chapter include a space to make the slicing description for each dimension visually more separate from the other dimension slicing descriptions.

Self-Test Answer 7-7

```
ppb = ppm * 1000.0
```

Self-Test Answer 7-8

```
avg_between_both = (data1 + data2) / 2.0
```

Self-Test Answer 7-9

Finding the maximum using a function is a single line:

```
max_data3 = np.max(data3)
```

whereas using nested looping requires a few more lines:

```
max_data3 = data3[0,0]
for i in range(np.shape(data3)[0]):
    for j in range(np.shape(data3)[1]):
        if data3[i,j] > max_data3:
            max_data3 = data3[i,j]
```

8   Basic Prognostic Modeling

We have looked at calculations involving diagnostic analysis, calculations of various quantities at a given moment in time based on other values at that time. In this chapter, we consider a case of prognostic modeling, where we examine how a quantity changes with time, given an equation that relates that quantity to time (mathematically, where $y = f(t)$). In a prognostic model, the quantity we are modeling at some future time is somehow dependent on values at the present time.

We can categorize prognostic models into **deterministic** and **nondeterministic** models. Deterministic models are those where the state of the variables being modeled are entirely calculated using the state of the model at an earlier time. Thus, given the same initial conditions – the values of the variables being modeled at time zero – the future states calculated by the model will always be the same. A nondeterministic model is a model where the earlier (or initial) state of the variables being modeled does not completely define (or determine) the later state of those variables. There is some element of randomness in the model so that each time the model is run, even though the initial conditions may be the same, the calculated values of the model variables will be different.

In this chapter, we examine a nondeterministic prognostic model, the random walk model, which has applications in everything from Brownian motion to stock market prices. We examine a simplified form of a random walk model that can be used as part of a model of the movement of a colony of bacteria in a petri dish. Using this example, we introduce additional computational constructs, of use in prognostic modeling and also data analysis: boolean values and expressions, **nested branching**, conditionals and floating-point numbers, and looping an indefinite number of times.

Although, in science and engineering, deterministic prognostic models are incredibly important, and used in applications ranging from modeling the aerodynamics of airplanes to the ecological dynamics of predators and prey, we postpone our discussion of such models until Chapter 17. The mathematics of such models is more complex than the random walk model we examine in the present chapter, and so the latter is a better application to illustrate the programming concepts of this chapter.

8.1   Example of a Basic Prognostic Model

A random walk model models the movement of an entity or agent as if the direction each step the agent takes is randomly decided.[1] Thus, if the agent is on a two-dimensional surface

---

[1] Shiflet and Shiflet (2014, pp. 405–414) has a very nice and accessible description of the random walk model.

Figure 8.1 A schematic of a two-dimensional array showing the neighboring elements (shaded lighter gray) to the element shaded darker gray.

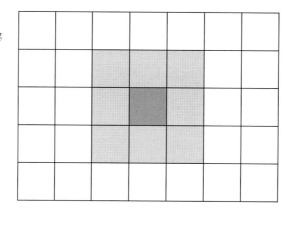

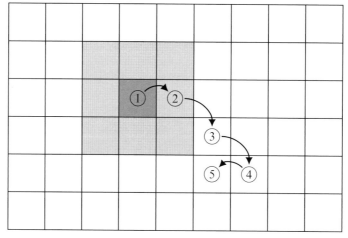

Figure 8.2 A schematic of a two-dimensional array showing a random walk through five positions from cells labeled ① to ⑤, in order. Cell ① and its neighboring elements are shown shaded darker and lighter gray, respectively.

(such as a table-top or petri dish), represented as a two-dimensional array, at each moment in time the agent can move to one of the array locations adjoining the current location. If the agent's current location is the darker gray square shown in Figure 8.1, the possible locations the agent could move into are shown by the lighter gray squares. Once the agent has moved to a new location, the next step is chosen from that new location's neighbors. In that way, the path of the agent is built up one step at a time. At a minimum, the agent's location is a time-evolving prognostic variable whose current location depends on the past locations. Figure 8.2 shows a random walk through five locations, starting from the cell labeled ① and ending with the cell labeled ⑤. Each step of the walk is to one of the neighbors of the current location. The walker can move in any direction given by these cells. Although Figure 8.2 does not show this, in a basic random walk, the walker can also move to a previously visited location.

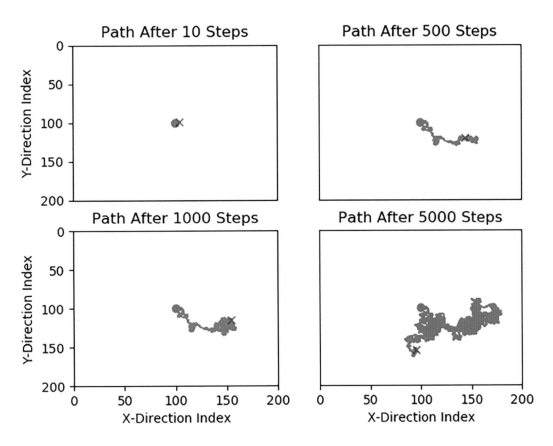

Figure 8.3 Path for a random walker whose behavior is modeled by the code in Figure 8.4. A green circle marks the beginning of the path and a red "×" marks the end of the path. In this model, $N_{crit} = 0$.

At any step, which location the agent chooses is determined randomly. For a classic random walk, each possible direction is equally probable. Other random walk models may have different rules governing the direction the agent moves in. For instance, a self-avoiding random walk does not permit the agent to return to any previously visited location on its path. There are also random walk models that are confined to just left or right (a one-dimensional random walk model) and those that are free to move in all spatial dimensions (a three-dimensional random walk model).

Figure 8.3 shows the path for a random walker whose behavior is modeled by the code in Figure 8.4. The structure of this model is based upon the bacterial colony propagation model of Ben-Jacob et al. (1994), with some differences.[2] In this model, each random walker represents a "chunk" of the bacterial colony rather than an individual bacterium, and in

---

[2] At www.cambridge.org/core/resources/pythonforscientists/refs/, ref. 93, we provide an up-to-date link to an online description of concentration gradients and how random walks can be used in studies similar to Ben-Jacob et al. (1994).

```
1    #- Module imports:
2
3    import numpy as np
4    import matplotlib.pyplot as plt
5
6
7    #- Set random seed, number of positions, and domain parameters:
8
9    np.random.seed(31493293)
10   n_positions = 5001
11   domain_shape = (200, 200)
12   food = np.zeros(domain_shape) + 5.0
13   N = np.zeros(domain_shape)
14   N_crit = 0
15
16
17   #- Set parameters and arrays for walker:
18
19   max_eat_rate = 0.1
20   use_energy = 0.05
21
22   energy = np.zeros(n_positions, dtype='d')
23   x = np.zeros(n_positions, dtype='l')
24   y = np.zeros(n_positions, dtype='l')
25
26   energy[0] = 0.3
27   x[0] = domain_shape[1] // 2
28   y[0] = domain_shape[0] // 2
29
30   step_x = np.array([ 0,  1, 1, 1, 0, -1, -1, -1])
31   step_y = np.array([-1, -1, 0, 1, 1,  1,  0, -1])
```

Figure 8.4 Code modeling a random walker whose behavior is displayed in the plots in Figure 8.3. In this model, $N_{crit} = 0$. The code to generate the plots in Figure 8.3 is given in Figure 8.5.

Ben-Jacob et al.'s study, approximately 100 000 walkers were represented. In our Figure 8.4 model, we show the behavior of only a single walker. Also, in our model (unlike Ben-Jacob et al.'s), the bacteria does not reproduce, so we model only one walker for the entire simulation.

In our model, the domain of the walker is a 200×200 grid that contains food. Each element of the grid starts with 5.0 units of food. Units of food and energy are arbitrary, time is measured in units of walker steps, and length is in units of grid cells. Food in this model is distributed uniformly, does not diffuse (i.e., move from one cell to another), and is never replenished once consumed by bacteria.[3] At each step of the walker, the walker eats an

---

[3] In contrast, Ben-Jacob et al.'s (1994) model includes diffusion of food, so food will move from higher-concentration locations to lower-concentration locations. We set the initial value of food to a high enough value so that for the number of steps calculated by the model, it is unlikely for the walker to remain stationary because of the lack of food. This is to make the demonstration of this model more interesting.

```
32    #- Take steps:
33
34    i = 1
35    while i < n_positions:
36        step_idx = np.random.randint(0, np.size(step_x))
37        xtrial = x[i-1] + step_x[step_idx]
38        ytrial = y[i-1] + step_y[step_idx]
39
40        if (xtrial >= domain_shape[1]) or (xtrial < 0):
41            continue
42        if (ytrial >= domain_shape[0]) or (ytrial < 0):
43            continue
44
45        if N[ytrial,xtrial] < N_crit:
46            N[ytrial,xtrial] += 1
47            x[i] = x[i-1]
48            y[i] = y[i-1]
49        else:
50            if energy[i-1] > 0.0:
51                x[i] = xtrial
52                y[i] = ytrial
53            else:
54                x[i] = x[i-1]
55                y[i] = y[i-1]
56
57        energy_from_eat = np.min([ max_eat_rate, food[y[i-1],x[i-1]] ])
58        energy[i] = energy[i-1] + energy_from_eat - use_energy
59        food[y[i-1],x[i-1]] -= energy_from_eat
60
61        if energy[i] < 0.0:
62            energy[i] = 0.0
63
64        if food[y[i-1],x[i-1]] < 0.0:
65            food[y[i-1],x[i-1]] = 0.0
66
67        i += 1
```

Figure 8.4 (continued)

amount of food (a maximum of 0.1 units worth), uses some energy (0.05 units worth) and changes its store of energy. The walker starts with 0.3 units of energy.

As the walker is a random walker, it chooses which direction to move in randomly, from amongst the neighboring cells to its current location (as shown in Figure 8.1). However, following Ben-Jacob et al., we simulate the resistance of the agar (the medium of the petri dish through which the walker is moving) by an N_{crit} value. Although the choice of the direction the walker moves in is random, we do not let the walker move into a cell unless the cell has previously been visited N_{crit} times. If the walker tries to move into a cell that has not been visited N_{crit} times, we increment the memory of how many times that cell has been visited and the walker remains stationary (a step distance of zero). If the cell has been visited at least

```
1    plot_t_index = [11, 501, 1001, 5001]
2    plt.figure()
3
4    for j in range(4):
5        plt.subplot(2, 2, j+1)
6        plt.plot(x[:plot_t_index[j]], y[:plot_t_index[j]], '-')
7        plt.plot(x[0], y[0], 'og')
8        plt.plot(x[plot_t_index[j]-1], y[plot_t_index[j]-1], 'xr')
9        plt.title("Path After " + str(plot_t_index[j]-1) + " Steps")
10       plt.axis([0, domain_shape[1], domain_shape[0], 0])
11
12       if j >= 2:
13           plt.xlabel("X-Direction Index")
14       else:
15           plt.xticks([])
16
17       if (j % 2) == 0:
18           plt.ylabel("Y-Direction Index")
19       else:
20           plt.yticks([])
21
22   plt.savefig("random_walk_ncrit0.png")
```

Figure 8.5 Code to create the graphs in Figures 8.3, 8.6, and 8.7. Assume the code in Figure 8.4 (with changes as described in Figures 8.6 and 8.7, as applicable) is run prior to running this plotting code. The argument for line 22 in the present figure also changes depending on the name of the graph being created.

N_{crit} times, and the walker has a positive store of energy, the walker moves into that random cell. Whether the walker moves or is stationary, the walker consumes some food each step. Figure 8.3 shows the case when $N_{crit} = 0$ whereas Figure 8.6 shows the case when $N_{crit} = 1$. Figure 8.7 shows a "zoomed-in" view for the case when $N_{crit} = 1$. The path of the walker for N_{crit} greater than zero is more compressed and local than when the agar offers no resistance to the walker (i.e., $N_{crit} = 0$).

Lines 3–4 in Figure 8.4 import the needed packages. Line 9 sets the random number **generator** "seed," which allows the sequence of random numbers generated by the computer to be repeatable. Line 10 sets the number of steps or new "positions" (i.e., movements to a new location or to stay stationary) the walker will take in this simulation. Lines 11–14 set the shape of the domain, the initial values of food at each grid cell, an array to store the number of times the walker tries to enter each cell, and the value of N_{crit}. Lines 19–20 set the maximum amount the walker will try to eat during a step and the amount of energy the walker uses during that time. Lines 22–24 create arrays that will hold the amount of energy the walker has and the x- and y-locations of the walker at each step the walker takes. The x-location at the left side of the domain is zero, and the y-location at the top of the domain is zero.

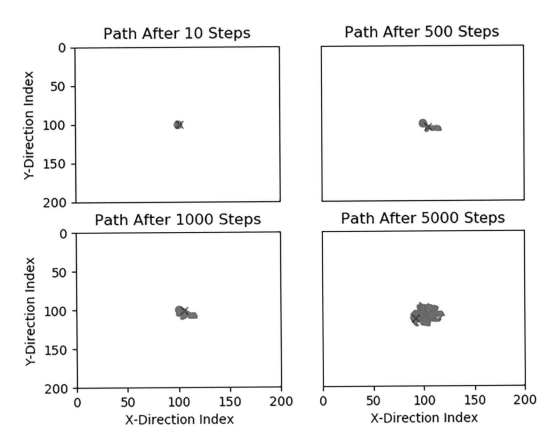

Figure 8.6 Path for a random walker whose behavior is modeled by the code in Figure 8.4 but with $N_{crit} = 1$. A green circle marks the beginning of the path and a red "×" marks the end of the path.

Lines 26–28 set the initial amount of energy the walker has and the x- and y-location of the walker. The walker's initial position is set to the middle of the domain. Recall the $//$ symbol refers to integer division (see Table 2.1).

Lines 30–31 define the relative x- and y-locations for the eight neighborhood points shown in Figure 8.1, starting at the point located at 12 o'clock and moving clockwise from there. Thus, the first elements of `step_x` and `step_y` refer to the x- and y-locations, respectively, of the neighborhood point directly above the grid point in question (i.e., 0 cells to the left or right and -1 cell in the vertical direction, that is, up 1 cell). The second elements of `step_x` and `step_y` refer to the x- and y-locations, respectively, of the neighborhood point northwest of the grid point in question (i.e., right 1 cell and up 1 cell), and so on.

Lines 32–67 in Figure 8.4 loop through and calculate the energy and positions for each of the steps the walker takes. Lines 36–38 select a random direction amongst the Figure 8.1 neighbors. The effect of lines 40–43 is to redo the selection of a random direction (without the walker moving or consuming food) if the proposed random step is outside the domain.

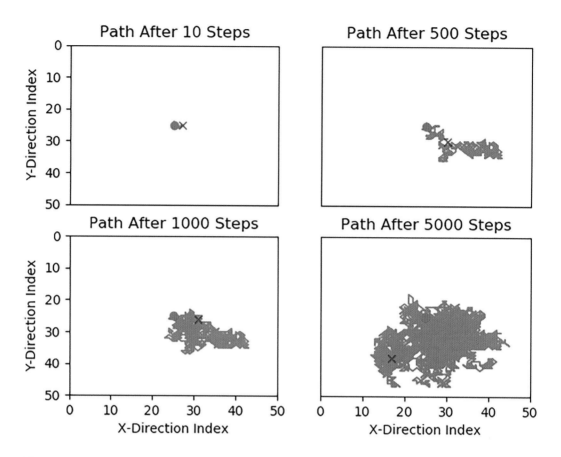

Figure 8.7 A "zoom-in" of Figure 8.6. This means we use the code that generated Figure 8.4 but with $N_{crit} = 1$, and we set `domain_shape` to `(50, 50)` instead of `(200, 200)`. A green circle marks the beginning of the path and a red "×" marks the end of the path.

Lines 45–48 keep the walker stationary if the cell the walker is trying to step into has not been visited at least N_{crit} times and increments the count of attempts to visit that cell. Lines 49–55 allow the walker to move into the new cell if the cell has been visited at least N_{crit} times and the walker has nonzero amounts of energy. Otherwise, the walker is kept stationary.

The walker eats food in lines 57–59. The amount of food eaten is the minimum of the maximum amount the walker will try to eat during a step and the amount of food at the current location. NumPy's `min` function acts on arrays and lists, so the two items we are finding the minimum of in line 57 are given in a list that is the input argument to the `min` function. The amount of food the walker eats is added to the walker's energy level, and that amount of food is subtracted from the amount of food at the current location. Lines 61–65 adjust the walker's energy level and food level at the current location to make sure they are not less than zero.

Figure 8.5 shows the code to plot Figure 8.3 and, with small changes, Figures 8.6 and 8.7. Line 1 in Figure 8.5 specifies a list of the indices for the four panels for each figure to be plotted. Line 2 creates a Matplotlib figure, and the loop in lines 4–20 creates each one of the four panels (or subplots). Line 5 creates a panel. Lines 6–8 plot the line and symbol elements in the panel: Line 6 plots the random walker path up to the number of steps given by `plot_t_index[j]`, line 7 plots the starting location of the path as a green dot, and line 8 plots the ending location of the path as a red "×". Line 9 adds the title for the panel. Line 10 sets the axis limits for the panel. The call to the `axis` function is similar to what is described in Section 5.2.2, except we flip the y-axis so the "maximum" value of the y-axis is 0 and the "minimum" value of the y-axis is 200 (or 50, depending on what size domain we use).[4] Lines 12–20 write the x- and y-axis labels and tick marks for the axes at the left of the figure and the bottom of the figure, but nowhere else. The `%` is the modulus operator. See Section 8.2.4 for details on how it is used to test for whether a number is even. Finally, line 22 saves the figure as a Portable Network Graphics (PNG) file.

8.2 Python Programming Essentials

Computationally speaking, a prognostic model is a set of calculations like the data analysis we saw in Chapters 6 and 7. The only difference with a prognostic model is that certain variables evolve with time. Thus, many of the computational constructs and structures we used for data analysis – function calls, arrays, branching, looping – are also used in prognostic models, whether deterministic or nondeterministic. We use all of these constructs in our example in Section 8.1. In that example, however, we also utilize the following: random numbers, **boolean expressions**, nested branching, looping an indefinite number of times, and making multiple subplots. In this section, we unpack these constructs and tasks, provide further examples of nested loops, and address how to handle conditionals using floating-point numbers. The latter two topics are not illustrated by the example in Section 8.1, but we discuss them because they go well with the topic of boolean expressions, the nesting of statements (whether looping or branching statements), and indefinite looping structures.

In the rest of the code examples in this section, assume that NumPy has already been imported as `np`.

8.2.1 Random Numbers in Computers

We said earlier that nondeterministic models have some element of randomness to them. To create a nondeterministic model in a computer, we need a way of generating random numbers. In nature, physical phenomena such as quantum states exhibit random behavior. A computer, however, by its very nature, is deterministic. Every time we execute a computer program, we should obtain the same result. Every time we execute `value = 2 + 3`, the variable `value` should always equal 5.

---

[4] https://stackoverflow.com/a/2052203 (accessed June 10, 2019). See also the Matplotlib documentation of `axis`. An up-to-date link to that page is at www.cambridge.org/core/resources/pythonforscientists/refs/, ref. 32.

Whereas computers are deterministic, certain mathematical algorithms can create a sequence of numbers that appears to be random, although the algorithms are creating a periodic sequence of numbers (that is, a sequence that will eventually repeat). The functions that utilize these algorithms are called **random number generators** (or, more properly, "pseudorandom number generators," because the sequences they generate are not true random numbers, which never repeat). The repeat period for a modern computer's random number generator is so long, however, that for many applications, we can treat the random number generator as a true random number generator. For instance, Python's random module's `random` function uses an algorithm called the Mersenne Twister which has a repeat period of $2^{19937-1}$, which, while inadequate for cryptography, is long enough for many scientific and engineering applications.[5]

By saying a sequence of random numbers does not repeat, we are not saying that the same random number will never show up again in the sequence but that the full sequence of random numbers will not show up again as a sequence until the number of elements in the sequence has reached the repeat period. That is, if we have a random number generator whose repeat period is 6 elements long, we are not saying "1, 3, −4, 2, 1, 7" (where the 1 repeats) cannot occur but that "1, 3, −4, 2, 1, 7, 1, 3, −4, 2, 1, 7" will occur, because the sequence is two times the repeat period in length.

Sequences of random numbers are controlled by the **seed** value of the generator. By changing the seed value, we can create different sequences of random numbers. Thus, by setting the seed to a specific value prior to calling a random number generator function, we can obtain the same sequence of random numbers each time we run a program. Note that the seed does not control the properties of the random values, such as the probability distribution function the values are chosen from, the mean of the values, etc. The seed only controls the rules by which the computer chooses the next random value in a sequence and in doing so controls what values appear, one after the other, from the random number generator. This may seem to be a strange property for a random number generator to have. After all, is not the point of a random number generator to create a sequence of random numbers whose values appear to be uncorrelated to prior values? It turns out, however, that even for nondeterministic models, we want to be able to produce the same sequence of random values every time the program is run. We can then analyze how the model works without wondering whether the behavior we are seeing is due to differences in the random numbers we generate each time we run the model. We can also create tests of the model results that can be run repeatedly (and will produce the same results every time) as we develop the model.

To set the seed of the random number generator, we call the NumPy random submodule function `seed`, as in this example:[6]

```
np.random.seed(194373)
```

---

[5] https://docs.python.org/3.6/library/random.html (accessed June 18, 2019).
[6] https://docs.scipy.org/doc/numpy/reference/generated/numpy.random.seed.html (accessed June 19, 2019).

There is essentially no return value for the function. The integer that is passed in is the seed for the random number generator. In the example random walk model code in Figure 8.4, the seed is set in line 9. The sequence of random numbers calls to the random number generator function will be exactly the same each time for this seed, as seen in this example using the random number generator function `uniform` (explained in more detail later on in this section):

```
In [1]:   np.random.seed(194373)
          np.random.uniform(size=10)
```

```
Out[1]:   array([0.52288852, 0.44023728, 0.39897908, 0.39568769, 0.24734779,
                 0.96315086, 0.8762412 , 0.29162933, 0.70948303, 0.12636644])
```

```
In [2]:   np.random.uniform(size=10)
```

```
Out[2]:   array([0.76319279, 0.31555945, 0.68727639, 0.15578094, 0.91665176,
                 0.95272056, 0.36920565, 0.03566116, 0.74779538, 0.96306654])
```

```
In [3]:   np.random.seed(194373)
          np.random.uniform(size=10)
```

```
Out[3]:   array([0.52288852, 0.44023728, 0.39897908, 0.39568769, 0.24734779,
                 0.96315086, 0.8762412 , 0.29162933, 0.70948303, 0.12636644])
```

Each of the above calls to `uniform` generates 10 random numbers from the range 0 (inclusive) to 1 (exclusive). The first call (second line of `In [1]`) generates 10 values and then 10 more values are generated in the second call (`In [2]`). When we reset the seed value in the first line of `In [3]`, the next call to `uniform` (second line of `In [3]`) gives the exact same 10 random numbers that we obtained from the `In [1]` call to `uniform`.

A different seed value will give a different set of random numbers, but using the same seed value will give the same set of random numbers. The actual value of the seed does not matter, except for this caveat: For reasons we will not get into, it still is a good rule of thumb to avoid very small seed values and to avoid seed values that are very close to each other, if we are changing the seed.

The random submodule of NumPy contains a number of different functions that will produce random values following various statistical distributions.[7] Here we mention only two functions from that module: `randint` and `uniform`.

---

[7] See the NumPy documentation regarding the random submodule and the `Generator` class for details. Up-to-date links to these pages are at www.cambridge.org/core/resources/pythonforscientists/refs/, refs. 96 and 95 respectively. Shiflet and Shiflet (2014, pp. 382–383, 390–403) also has an excellent introductory description of how random number generators work and how to generate values following any arbitrary probability distribution.

The `randint` Function   This function returns a random integer from the range specified by two positional (or keywords `low` and `high`) input parameters.[8] The range is inclusive at the lower bound and exclusive at the upper bound. The keyword input parameter `size` enables us to create a NumPy array of that size or shape. Here are a few examples:

```
In [1]:  np.random.randint(2, 5, size=10)

Out[1]:  array([2, 4, 3, 2, 4, 3, 2, 2, 2, 2])

In [2]:  np.random.randint(low=2, high=5, size=10)

Out[2]:  array([4, 2, 3, 2, 3, 4, 3, 4, 4, 2])

In [3]:  np.random.randint(-3, 2, size=(4,3))

Out[3]:  array([[ 1, -3,  1],
                [-3,  1, -3],
                [-3, -2, -2],
                [ 0,  0,  1]])

In [4]:  np.random.randint(-3, 2)

Out[4]:  0
```

Note that although the `In [1]` and `In [2]` calls to `randint` specify the same range to pull the random variable out of (i.e., 2 inclusive and 5 exclusive), the numbers that are obtained are not the same because the seed is not set to the same value prior to the function calls. If the `size` keyword input parameter is left out, the call to `randint` returns a scalar random value rather than an array of random values.

The `uniform` Function   As we have already seen, this function returns a uniformly distributed random number between the values zero (inclusive) and one (exclusive).[9] The uniform distribution means that the probability of any subrange of values between zero and one is the same as another subrange of the same width. Thus, the probability a random value is between 0.1 and 0.2 is the same as the probability it is between 0.5 and 0.6, 0.87 and 0.97, etc. The values returned are floating-point numbers.

If we want a uniform random number from a different range than zero to one, we can specify the range as positional (or keywords `low` and `high`) input parameters. As with `randint`, the keyword input parameter `size` enables us to create a NumPy array of that size or shape. Here are a few examples:

---

[8] https://docs.scipy.org/doc/numpy/reference/generated/numpy.random.randint.html#numpy.random.randint (accessed June 19, 2019).

[9] https://docs.scipy.org/doc/numpy/reference/generated/numpy.random.uniform.html#numpy.random.uniform (accessed June 19, 2019).

```
In [1]:  np.random.uniform(2, 5, size=10)

Out[1]:  array([4.28957838, 2.94667836, 4.06182917, 2.46734282, 4.74995529,
                4.85816167, 3.10761694, 2.10698349, 4.24338615, 4.88919962])

In [2]:  np.random.uniform(low=2, high=5, size=10)

Out[2]:  array([4.96628997, 4.6234993 , 2.3004134 , 4.18769449, 3.65203165,
                3.25920672, 3.73427537, 4.33592053, 3.50078218, 2.95364009])

In [3]:  np.random.uniform(-3, 2, size=(4,3))

Out[3]:  array([[-2.84002099, -1.61651669,  0.22639601],
                [ 1.99281631,  0.56290567, -1.66896522],
                [-0.50796508, -1.75067486,  0.10367237],
                [-2.99249145, -1.76806082,  0.82341562]])

In [4]:  np.random.uniform(-3, 2)

Out[4]:  -0.19310014570797307
```

These functions of the submodule random only require the parent module NumPy to be imported. That is, to access, `seed`, `randint`, etc., as shown above, we only need to have this import line:

```
import numpy as np
```

There is no need to explicitly import the random submodule.

We can now describe lines 36–38 in Figure 8.4 in more detail. In line 36, a random integer between zero (inclusive) and eight (exclusive) is chosen and set to `step_idx`. This corresponds to a randomly chosen index for elements from the `step_x` and `step_y` arrays, that is, a randomly chosen neighboring point relative to the current location. The variables `xtrial` and `ytrial` store the next step we are considering for the random walker to take. The values are determined by the current x- and y-locations (i.e., `x[i-1]` and `y[i-1]`) and the relative distance from the current point to the neighboring point of index `step_idx`.

8.2.2 Scalar Boolean Type and Expressions

In Section 6.2.4, we introduced the idea of using branching to ask questions of data. We found that our questions could be written as conditionals and that these conditionals are used in `if` statements to execute different blocks of code depending on how the conditionals evaluate (as true or false).

The conditionals we have looked at so far are relatively simple: tests to see whether a variable is greater than, less than, or equal to a certain value. It turns out these basic examples are a subset of a system of storing and manipulating true/false values. Entities that are either true or false are **boolean** entities.[10] We can manipulate boolean variables using a variety of operators and create boolean values by using operators on boolean and/or nonboolean values. Expressions that evaluate to a true or false value are called boolean expressions, and the mathematical rules that govern the manipulations of boolean variables is known as **boolean algebra**.

In Python, just as the `int` type represents integers and the `float` type represents floating-point (decimal) values, the `bool` type represents boolean values. A boolean value in Python can have one of two values: `True` or `False` (note that the capitalization matters). These are special values that we can set variables to, just as we can set integer variables to integer values like `-2` and `183`. The values `True` and `False` are not strings: They are values in and of themselves. That is to say, `True` and `'True'` are two entirely different kinds of values, the former being a boolean and the latter being a string. Another way to think of it is that a boolean variable is a variable that is either `True` or `False`, just as an integer variable is a variable that is set to a value that is a whole number.

Here is an example of setting a boolean variable `test`:

```
In [1]:   test = True
          type(test)
```

```
Out [1]:  bool
```

```
In [2]:   test = False
          type(test)
```

```
Out [2]:  bool
```

In the above case, the variable `test` is assigned first to `True` and then to `False`. In both cases, `test` is a boolean type variable. Notice there are no quotation marks around `True` and `False` as neither entity (or **token**) are strings. They are boolean values.

Besides directly assigning variables to one of the boolean values `True` or `False`, we can also create boolean values (and assign boolean variables) through boolean expressions. Consider the following `if` statement from Section 6.2.4 (reproduced here for convenience):

```
1   atomic_number = 2
2   if atomic_number < 6:
3       print("Element has fewer protons than carbon")
4   print("Test completed")
```

Because `atomic_number` is less than six, the body of the `if` statement will execute and the message will print out informing us that the element has fewer protons than carbon. But, what if we printed out the *condition* of the `if` statement?

---

[10] The term "boolean" is named after mathematician George Boole.

```
In [1]:  print(atomic_number < 6)
```

```
         True
```

The condition expression, when evaluated, produces a boolean value. This is to say, it produces either `True` or `False` (for this value of `atomic_number`, it produces `True`). The condition expression, then, is a boolean expression, because it evaluates to `True` or `False`.

This also means that an `if` statement's condition is nothing more than a boolean value. Put another way, what the `if` statement is doing is asking whether whatever is in the condition part of the statement evaluates to `True` or `False`. If the condition is equal to `True`, the body of the `if` statement executes. If the condition is equal to `False`, the body of the `if` statement does not execute. Thus, we can replace the condition of an `if` statement with a boolean variable:

```
1    atomic_number = 2
2    is_fewer_protons = atomic_number < 6
3    if is_fewer_protons:
4        print("Element has fewer protons than carbon")
5    print("Test completed")
```

And the value of `is_fewer_protons`, in this case, is `True`:

```
In [1]:  print(is_fewer_protons)
```

```
         True
```

Again, to create a boolean value, the boolean expression does *not* need to have a boolean variable in it. While we can create boolean values directly, any expression that returns a `True` or `False`, regardless of what variable types go into the expression, is a boolean expression and creates a boolean value. In the code above, `atomic_number` and the value 6 are both integers, but when combined with a less than operator, the result is a boolean value.

An aside about naming boolean variables: Although we can name boolean variables any valid variable name we want, usually we want a name that naturally conveys the idea of "true" and "false." For instance, by naming the boolean variable above `is_fewer_protons`, we are naturally connecting the value of the variable to whether or not the item in question has fewer protons. If the boolean variable is `True`, it means "yes, there are fewer protons." If the boolean variable is `False`, it means "there are not fewer protons." This can make `if` statements using such boolean variables read even more like an English phrase: `if is_fewer_protons` is not all that far from the English phrase, "if it has fewer protons." Hopefully, this makes the code more readable.

Although the restrictedness of boolean values – they can only be either `True` or `False` – may make it seem that there is not much we can do with boolean values, this is not the case. There are a variety of operations we can conduct with boolean values, and these operations enable us to express complex logical tests. The three most basic are the "and," "or," and "not"

Table 8.1 Basic boolean algebra operators in Python. *<Expr>*, *<Expr1>*, and *<Expr2>* are boolean expressions.

Operator	Code
And	*<Expr1>* and *<Expr2>*
Or	*<Expr1>* or *<Expr2>*
Not	not *<Expr>*

operators, which in Python are specified by `and`, `or`, and `not`. Table 8.1 summarizes the syntax for these operators, but let us describe them in more detail.

The and Operator This operator operates on two boolean values (just like the multiplication operator operates on two numbers). It checks to see whether both the operand values are `True`. If so, the result of the expression is `True`. For all other possible combinations of values around the `and`, the result of the expression is `False`. Consider these examples:

```
In [1]:  test1 = True
         test2 = True
         test1 and test2
```

```
Out[1]:  True
```

```
In [2]:  True and True
```

```
Out[2]:  True
```

```
In [3]:  test1 = False
         test2 = True
         test1 and test2
```

```
Out[3]:  False
```

```
In [4]:  True and False
```

```
Out[4]:  False
```

```
In [5]:  False and False
```

```
Out[5]:  False
```

In cell `In [1]`, we create two boolean variables `test1` and `test2` and set their values to `True`. The result of the `and` operator on these two boolean variables is `True`, because both values are `True`. In cell `In [2]`, we put the `True` values in the `and` expression directly and find the result is the same (`True`). In cell `In [3]`, we change the value of `test1` to `False` and find the `and` operation on `test1` and `test2` results in `False`. Cells `In [4]` and `In [5]` show how any other combination of the two operands (the values on either side of the `and`) results in the return of `False`, because at least one of the operands is `False`.

As Table 8.1 describes, the values on either side of the `and` operator can be expressions themselves. All that has to be true for the `and` statement to work is that the operands on either side of the `and` operator have to be boolean expressions, i.e., expressions that result in a boolean value. Consider a revision to the number of protons test we gave earlier. In this example, we want to see whether `atomic_number` is between the atomic numbers for helium and carbon:

```
1   atomic_number = 2
2   is_between = (atomic_number > 2) and (atomic_number < 6)
3   if is_between:
4       print("Element has more protons than helium and fewer than carbon")
5   print("Test completed")
```

The result of running this code is:

```
Test completed
```

because `is_between` evaluates to `False`, since `atomic_number > 2` evaluates to `False`. Because `is_between` is `False`, we never execute the body of the `if` statement. If we set `atomic_number` to three (lithium), the body of the `if` statement will execute and the output will be:

```
Element has more protons than helium and fewer than carbon
Test completed
```

The or Operator   Like the `and` operator, the `or` operator operates on two boolean values. It checks to see whether at least one of the values is `True`. If so, the result of the expression is `True`. For all other possible combinations of values around the `or`, the result of the expression is `False`. Consider these examples:

```
In [1]:   test1 = True
          test2 = True
          test1 or test2

Out[1]:   True

In [2]:   False or True

Out[2]:   True
```

```
In [3]:   True or False

Out[3]:   True

In [4]:   False or False

Out[4]:   False
```

The only time the `or` operator yields `False` is if both operands are `False` (`In [4]`). In all other cases, because at least one of the operands is `True`, the result of the `or` operator is `True`.

Let us revise our above number of protons test to check whether the `atomic_number` is greater than or less than six (the number of protons in carbon):

```
1    atomic_number = 2
2    is_more_or_less = (atomic_number > 6) or (atomic_number < 6)
3    if is_more_or_less:
4        print("Element has more or fewer protons than carbon")
5    print("Test completed")
```

Running the above code will result in this output:

```
Element has more or fewer protons than carbon
Test completed
```

as two is less than six, so `atomic_number < 6` in line 2 evaluates to `True` and `is_more_or_less` is `True`.

Because `atomic_number` is an integer, the boolean expression:

```
(atomic_number > 6) or (atomic_number < 6)
```

gives the same value as:

```
atomic_number != 6
```

because an integer that is greater than or less than six is one that is not equal to six.

The not Operator   This operator operates on only a single boolean value, which comes after the `not`. The result is a boolean value that is "opposite" the operand after the `not`. That is, if the operand is `True`, the result of applying `not` to that operand is `False`, and vice versa. Consider the following examples:

```
In [1]:   test = True
          not test

Out[1]:   False
```

```
In [2]:  not True
```

```
Out[2]:  False
```

```
In [3]:  not False
```

```
Out[3]:  True
```

Let us revise our above number of protons test to check whether `atomic_number` is not six (i.e., not carbon):

```
1   atomic_number = 2
2   is_carbon = atomic_number == 6
3   if not is_carbon:
4       print("Element is not carbon")
5   print("Test completed")
```

In line 2, the first equal sign assigns the result of the `atomic_number == 6` boolean expression to the boolean variable `is_carbon`. As with all assignment statements, we have to first read the expression to the right of the equal sign, evaluate it, then assign that result to the variable given on the left side of the equal sign. If we are not careful, we might miss the second equal sign in the double-equals operator between `atomic_number` and 6.

Running the above code will result in this output:

```
Element is not carbon
Test completed
```

because `atomic_number` is 2, `atomic_number == 6` will evaluate to `False`. The value of `is_carbon` is `False`, so `not is_carbon` will be `True`, and the body of the `if` statement will execute.

One common way of concisely describing all the possible outcomes of a boolean algebra operation (or combination of boolean algebra operations) is to create a **truth table**. Each column of a truth table shows the values for one of the input boolean variables involved in the operations, except for the last column which shows the output boolean value from the boolean algebra operation. Each row of a truth table (except the first one, which is a header) represents a single possible combination of values for the input boolean variables. All rows together represent all possible combinations of values for the input boolean variables. Tables 8.2–8.4 are the truth tables for the `and`, `or`, and `not` operators. Try This! 8-3 gives an example of a truth table for a combination of boolean algebra operations that utilizes three input boolean variables.

Finally, the boolean type in Python has a useful feature in that when a boolean value is used in an arithmetic expression, Python will automatically translate a `True` to one

Table 8.2 Truth table for the `and` operator. The first two columns (Var1 and Var2) are input boolean variables and the last column is the output of the expression given above. The first row is a header row.

Var1	Var2	Var1 and Var2
True	True	True
False	True	False
True	False	False
False	False	False

Table 8.3 Truth table for the `or` operator. The first two columns (Var1 and Var2) are input boolean variables and the last column is the output of the expression given above. The first row is a header row.

Var1	Var2	Var1 or Var2
True	True	True
False	True	True
True	False	True
False	False	False

Table 8.4 Truth table for the `not` operator. The first columns (Var) is the input boolean variable and the last column is the output of the expression given above. The first row is a header row.

Var	not Var
True	False
False	True

and `False` to zero. This can be useful if we want to assign different numerical values to a variable based upon the value of a boolean variable. For instance, pretend we have a scale that cannot accurately measure anything weighing less than 0.01 g. In our analysis, we decide we will consider any measurement less than 0.01 g as 0 g. Utilizing boolean expressions and arithmetic, the following code takes the variable `weight` and correctly assigns `corrected_weight` using that rule:

```
corrected_weight = (weight >= 0.01) * weight
```

If `weight` is set to 0.003, we obtain the following:

```
In [1]:  weight = 0.003
         corrected_weight = (weight >= 0.01) * weight
         print(corrected_weight)


         0.0
```

while if `weight` is set to 0.7, we obtain the following:

```
In [1]:  weight = 0.7
         corrected_weight = (weight >= 0.01) * weight
         print(corrected_weight)


         0.7
```

In the second line of each of these two executed code snippets, the `weight >= 0.01` boolean expression returns `True` if the weight is equal to or above the 0.01 g threshold and `False` otherwise. When that boolean value is multiplied by `weight`, if the weight is less than 0.01 g, we are multiplying `weight` by 0, which gives us 0 as the result saved to `corrected_weight`. If the weight is greater than or equal to 0.01 g, we are multiplying `weight` by 1, which gives whatever the value of `weight` is as the result saved to `corrected_weight`. By utilizing this property of boolean values, we have reduced to one line what is accomplished by this four-line `if` statement:

```
1   if weight >= 0.01:
2       corrected_weight = weight
3   else:
4       corrected_weight = 0.0
```

In Section 13.2.3, we examine this property in more detail and its use in selectively processing some elements but not others in an array. In the present section, we limited ourselves to scalar boolean values (i.e., a single boolean value or an expression that returns a single boolean value). In Section 13.2.2, we examine arrays of boolean values and how array syntax can be used on such a collection of boolean values.

One final comment regarding boolean algebra operator precedence: Comparison operators have higher precedence than boolean algebra operators, so the former are evaluated before the latter, and operators at the same level of precedence are evaluated left-to-right. Thus both of these statements evaluate the same way:

```
In [1]:   4 < 5 and 8 > 2

Out[1]:   True

In [2]:   (4 < 5) and (8 > 2)

Out[2]:   True
```

Likewise, arithmetic operators have higher precedence than comparison operators, so all of these statements evaluate the same way:

```
In [1]:   14 - 10 < 5 and 8 - 1 > 2

Out[1]:   True

In [2]:   (14 - 10) < 5 and (8 - 1) > 2

Out[2]:   True

In [3]:   ((14 - 10) < 5) and ((8 - 1) > 2)

Out[3]:   True
```

The boolean operators or, and, and not, however, do *not* have the same precedence. The list or, and, and not is in order of increasing precedence. The Python operator precedence table describes the precedence order for these and other operators. We provide an up-to-date link to the table at www.cambridge.org/core/resources/pythonforscientists/refs/, ref. 16. That being said, parentheses should almost always be used to make the order of operations explicit and clear. Relying on operator precedence order can result in code that is diffcult to understand.

8.2.3   Nested Branching

In Section 7.2.5, we described the nested for loop, that is, a for loop that is part of the body of another for loop. With a nested loop, the inner for loop executes in its entirety (i.e., all of its iterations) for each iteration of the outer loop.

Branching statements can also be nested. In a nested branching statement, a second (or more) branching statement is placed in the body of the outer branching statement. In such a case, the inner statement will be reached only if the outer statement is evaluated as True. Consider the following code:

```
1   atomic_number = 2
2   if atomic_number > 2:
3       if atomic_number < 6:
4           print("Element has more protons than helium " + \
5                   "and fewer than carbon")
6   print("Test completed")
```

In line 2, we check to see if `atomic_number` is greater than two. If it is, we enter in the body of the `if` statement, the first line of which is a check to see whether `atomic_number` is less than six. If that condition is true, the `print` statement is executed. We will never reach line 3 if the line 2 condition evaluates to `False`.

The result of running the above code is:

```
Test completed
```

Nesting the `if` statements in this way results in the same behavior as if we had a single `if` statement with this boolean expression as the condition:

```
(atomic_number > 2) and (atomic_number < 6)
```

Because the `and` operator requires both operands to be `True` in order for the expression to evaluate as `True`, the fact that the first operand (`atomic_number > 2`) evaluates to `False` is enough to ensure the program never enters the body of the `if` statement with this condition.

Lines 45–55 of Figure 8.4 show another example of nested branching statements. In that example, the line 50 `if` test only executes if the line 45 `if` condition evaluates as `False`. As a result, this nested branching statement enables us to describe the following three-choice scenario (where one of the following three choices will be executed):

1. Execute lines 46–48 if `N[ytrial,xtrial]` < `N_crit` is True.
2. Execute lines 51–52 if `N[ytrial,xtrial]` < `N_crit` is False and `energy[i-1]` > `0.0` is True.
3. Execute lines 54–55 if `N[ytrial,xtrial]` < `N_crit` is False and `energy[i-1]` > `0.0` is False.

We can also describe the above three-choice logic using an `if-elif` statement. An `if-elif` statement, however, would be messier, as we would have to duplicate the:

```
N[ytrial,xtrial] < N_crit
```

test in each of the `if-elif` conditions and the `energy[i-1]` > `0.0` test in two of the `if-elif` conditions. As we saw in Section 6.2.3, it is better to avoid duplicating code (the "don't repeat yourself" or DRY principle). Repeating ourselves makes it easy to inadvertently introduce errors, if one occurrence of the code is slightly different from another occurrence. Note, though, the DRY principle has to also be balanced with overall code clarity. In some cases the repetitive way of writing the program is logically clearer. For the above three-choice logic, the nested branching approach avoids the duplicate code the `if-elif` statement would result in. And, in general, nested branching enables us to express more complex branching logic than a single `if`, `if-else`, or `if-elif` statement alone can provide.

Lastly, in the examples we have seen above, there have been only two levels of nesting (an outer branching statement and an inner branching statement). There can, however, be more than two levels of nesting. Deeper levels of nesting are created by adding another level of indentation. And there can be more than one branching statement at the inner level(s).

8.2.4 Looping an Indefinite Number of Times Using `while`

With the introduction of boolean variables in Section 8.2.2, we can discuss another way of looping that enables us to do a task without knowing ahead of time how many times we will do that task. If we want to loop through the elements of an array of numbers, we know ahead of time how many times we need to run the loop. As a result, a `for` loop is the best way to go. When we do not know ahead of time how many times we will need to go through the loop, the `while` loop is the better approach.

Although the `while` loop is best used for cases where we do not know ahead of time how many iterations we will want to go through, the most basic use of `while` is to iterate through a preset number of times, just like a `for` loop. So, for our first short example of a `while` loop, we duplicate one of the `for` loops from Section 6.2.3. Recall the following `for` loop from near the beginning of that section (reproduced here for convenience):

```
1   precip = np.array([5.58, 3.54, 4.54, 2.93, 2.10, 1.47,
2                      0.49, 1.03, 1.72, 4.31, 6.50, 5.41])
3   for iprecip in precip:
4       print(iprecip)
```

Each iteration of the loop, the iterator `iprecip` is set to each element in `precip`, one at a time in turn. The body of the loop has a single line and prints out the current value of `iprecip` for that iteration of the loop.

Unlike a `for` loop, a `while` loop cannot automatically advance through an array and set the elements of that array to an iterator. Instead, we have to create a variable that will increment by one each iteration and use the value of that variable to access the elements in the array. This code will do the job:

```
1   precip = np.array([5.58, 3.54, 4.54, 2.93, 2.10, 1.47,
2                      0.49, 1.03, 1.72, 4.31, 6.50, 5.41])
3   i = 0
4   while i < np.size(precip):
5       print(precip[i])
6       i += 1
```

Lines 1 and 2 are the same as in the `for` loop version. In line 3, we initialize the variable `i` to zero. We will use `i` as the index to reference elements in `precip`.

In line 4, we begin the loop. A `while` loop works in the following way:

1. Evaluate the condition after the `while`. This is similar to how when the program hits an `if` statement, it evaluates the condition after the `if`.
   (a) If the condition after the `while` evaluates to `False`, do not execute the body (the lines below the `while` line and that are indented in four spaces) but instead continue the program with the line after the body of the `while` loop.
   (b) If the condition after the `while` evaluates to `True`, execute the code in the body.
2. After the body finishes executing, return to step 1 above and repeat.

From this description of a `while` loop, we can infer the following:

- The body in the `while` loop does not have to execute, even once. If the initial evaluation of the condition after the `while` results in `False`, the program does not enter the body of the loop.
- The condition after the `while` is only evaluated when control of the program returns to the line with the `while`, that is, at the end of the execution of the body. If the variables used in the condition change while the loop body is running, it will not affect whether the loop will repeat until control returns to the `while` line and the condition is reevaluated.
- If the `while` loop is to ever stop, the body of the loop (generally) has to change the variables in the condition after the `while` such that the condition will, at some point, evaluate to `False`. Otherwise, the loop will continue on forever.

In line 4, the condition is `i < np.size(precip)`. For the 12-element array `precip`, this condition evaluates to `True` for the initial value of `i` equal to 0.

The body of the loop is lines 5 and 6. In line 5, we print out the value of `precip` at index `i`. For the first iteration, `i` is zero so this prints out the first element of `precip`. In line 6, we increment `i` by one. Thus, when `i` becomes 12 due to the incrementing, the loop will stop, because after the incrementing, control will go back to the `while` line and `i < np.size(precip)` will be evaluated and yield `False`. Thus, this loop runs the value of `i` from 0 to 11, both inclusive. Notice that if line 6 is left out, if `i` remains unchanged from the value 0, this loop will never terminate. It will be an **infinite loop** – a loop that never stops running. Figure 8.8 summarizes the syntax for a `while` loop.

The `while` loop will terminate when control returns to the `while` line and the condition is reevaluated to `False`. There are, however, additional ways of altering how a `while` loop runs:

- The `continue` statement. If executed, the rest of the body of the `while` loop is skipped and control of the program is returned to the `while` line to reevaluate the condition.
- The `break` statement. If executed, the `while` loop is terminated and control of the program is moved to the line after the `while` loop's body.

---

while loop syntax summary

A while loop has the following syntax:

while *<condition>* :
     <... body ...>

where *<condition>* is a boolean expression such that, if the expression evaluates as True, we enter the loop and execute *<... body ...>*. After *<... body ...>* is executed, control returns to the while line and *<condition>* is reevaluated. If *<condition>* evaluates as True, we enter the loop for another iteration and execute *<... body ...>*. The lines of the *<... body ...>* are, as a block, indented in four spaces.

Figure 8.8 while loop syntax summary.

Let us modify the earlier while loop example and include in continue and break:

```
precip = np.array([5.58, 3.54, 4.54, 2.93, 2.10, 1.47,
                   0.49, 1.03, 1.72, 4.31, 6.50, 5.41])
i = 0
while i < np.size(precip):
    if i == 6:
        break
    if (i % 2) == 0:
        i += 1
        continue
    else:
        print(precip[i])
        i += 1
```

This code will produce the following output:

```
3.54
2.93
1.47
```

The break statement in line 6, when executed, terminates the loop. Because that statement is inside an if statement that only executes when i equals six, the loop will ignore this statement for the first six iterations (for values of i from zero to five, inclusive for both). The effect of line 6 is to prevent printing of the last six elements of precip.

The continue statement in line 9 is also behind an if statement which tests to see whether i is even. The percentage sign is the "modulus" operator which returns the remainder from a division operation. Thus, $3 \% 2$ will return 1 (i.e., the remainder of 3 divided by 2) whereas $4 \% 2$ will return 0 (i.e., the remainder of 4 divided by 2). Because continue skips the rest of the body of the loop and moves control to the line with the while, line 9 ensures that only odd values of i will be printed. We also use this method of testing whether a number is even in line 17 of Figure 8.5.

In the above example, we have to include the `i += 1` incrementing statement in two places (lines 8 and 12). What would happen if we left out line 8? In that case, for even values of `i`, `i` would not increment and the rest of the body would be skipped without changing the value of `i`. Because the initial value of `i` is 0, and 0 is even, the lack of the line 8 incrementing statement would mean that `i` remains 0 and control would return to line 4. The result would be an infinite loop.

A few random items about `continue` and `break`: First, `continue` and `break` statements in a `while` loop are often found inside `if` statements in the body of the loop. If they were not behind an `if` statement, they would execute every iteration of the loop, including the first. This would make later iterations of the loop and/or the body after the `continue` or `break` statement meaningless, at which point we have to ask why we need a `while` loop in the first place. Second, the `continue` and `break` statements also work in other constructs in Python, but perhaps they are most useful in `while` loops. Last, some programmers like writing `while` statements with this syntax:

```
while True:
    <... body ...>
```

and rely on `continue` or `break` statements to exit the loop. While there are instances where such a loop results in the clearest code, in general it is better to avoid such statements. When `continue` and `break` statements proliferate, a program becomes confusing to read.

Earlier, we said a `for` loop is used when we know ahead of time how many times to iterate through the body of the loop, and a `while` loop is used when we do not know ahead of time how many times to iterate through the body of the loop. Besides these use cases, there is another major difference between these two loops: the `iterator` variable in a `for` loop should *not* be changed in the body of the loop, but the nearest analogue in a `while` loop, the test condition, should be changed in the body of that loop. In the case of the `for` loop, every time the loop moves to a new iteration, the iterator is overwritten to the next value in the iterable (i.e., the list, array, or other collection of items between the `in` and the colon). Thus:

```
for i in [1, 2, 3, 4]:
    print(i)
    i = i + 1
```

will produce the following output:

```
1
2
3
4
```

The `i = i + 1` statement has no effect on the `print` call. In contrast, in this `while` loop (which produces the same output as the `for` loop above):

```
i = 1
while i <= 4:
    print(i)
    i = i + 1
```

the `i = i + 1` statement changes `i` each iteration of the loop and affects the `print` call. The `i = i + 1` statement is needed for the loop to work properly. If absent, the loop will never terminate. The bottom line: Use a `for` loop only if there is no reason to change the iterator in the body of the loop (or there is no reason for changes to the iterator to be propagated to the next iteration). Otherwise, use a `while` loop.

We are now in a position to unpack the `while` loop in Figure 8.4 (lines 34–67). Although the body in this loop is substantially longer than the examples in the current section, the structure is the same. The variable `i` is also used as an array index to reference elements in the `x`, `y`, `energy`, and `food` arrays, which provide those quantities for each step of the walker. The variable `i` is initialized to one (in line 34) because the initial values (i.e., at index 0) of `x`, `y`, `energy`, and `food` are set in lines 12 and 26–28.

As described earlier in Section 8.1, in lines 36–38 we randomly select a neighboring cell to consider moving the walker into. Why is this a "trial" movement? Because the point we choose might not be legal to move into. In lines 40–43, we check to see if `xtrial` and `ytrial` are outside of the domain. If so, the rest of the body is skipped (by calling `continue`) for the iteration and control of the loop is returned to line 35. Put another way, if `xtrial` and `ytrial` are not legal choices, we ignore all the other computations regarding food eaten and energy changes in the walker and take another random sample from the neighboring points.

In lines 45–55, as described earlier in Section 8.1, we move the walker depending on how many times the walker has tried to go into the (`ytrial`, `xtrial`) space and the amount of energy the walker currently has. Energy and food levels are adjusted in lines 57–65. In line 67, we increment `i` by one, and control of the loop is returned to the `while` in line 35.

8.2.5 Making Multiple Subplots

The plots we learned to create in Chapters 4 and 5 are "singletons": one single plot in a Matplotlib figure window. Many times, we are interested in putting multiple plots (or panels) on a single figure, in order to compare snapshots at different times or locations. Figures 8.3, 8.6, and 8.7 are examples of such plots. Figure 8.5 shows the code to plot these figures. We have seen many of the commands in Figure 8.5 before: `plot`, `title`, `axis`, etc. However, the function `subplot`, to create subplots on a Matplotlib figure window, is new.

The function `subplot` creates a subplot (or panel) in a grid of subplots in a figure window. Just as the `figure` command creates a figure window and subsequent plotting commands apply to the figure defined by the last `figure` call (see Section 5.2.3 for more on `figure`), the `subplot` function activates a subplot for subsequent plotting commands to draw on. The `subplot` function accepts three positional input parameters (as line 5 in Figure 8.5

shows).[11] The first two arguments specify the number of rows and number of columns in the grid of subplots, respectively. The third argument gives the index number of the subplot on the grid of subplots. This index number, unlike list and array indices, starts with one (inclusive) and ends at *n* (inclusive), where *n* is the total number of subplots in the grid of subplots.

Thus, line 5 in Figure 8.5 creates a 2×2 grid of subplots on the current figure window. For each iteration of the line 4 `for` loop, the `subplot` call in line 5 creates one of the subplots, with the index incrementing from one (inclusive) to four (inclusive). As the loop draws each subplot, it draws from left to right in a single row, then up to down, a row at a time.

8.2.6 More on Nested Loops

Having discussed nested branching in Section 8.2.3, we can add on to our description of nested loops from Section 7.2.5. In Section 7.2.5, we used nested `for` loops to go through each of the index pairs for the elements in a two-dimensional array. In the current section, we show more complex examples of nested loops.

The first example is of a nested loop that does calculations (or other actions) on more complicated scenarios than applying the same calculation to every single element. For instance, in Try This! 7-7, we calculated surface wind speed given a two-dimensional array of the *x*-component of wind velocity `v_x` and a two-dimensional array of the *y*-component of wind velocity `v_y` at regular latitude and longitude locations on a grid, using both nested loops and array syntax. Here is the nested loops solution (reproduced here for convenience):

```
1   wind_shape = np.shape(v_x)
2   v = np.zeros(wind_shape, dtype=v_x.dtype.char)
3   for i in range(wind_shape[0]):
4       for j in range(wind_shape[1]):
5           v[i,j] = ((v_x[i,j] ** 2) + (v_y[i,j] ** 2)) ** 0.5
```

This calculation is done on/for all points in `v`, `v_x`, and `v_y` and the code for the calculations is found only in the innermost loop.

Let us change the problem to the following: For the latitude points at the top (northernmost row) of the grid and the bottom (southernmost row) of the grid, we calculate the wind speed at each longitude point as the average of all longitude points at that latitude and assign each element of `v` in that row to this average value.[12] For all other locations, we calculate and assign the wind speed as before in Try This! 7-7. Here is code that will solve this revised problem:

---

[11] https://matplotlib.org/3.1.0/api/_as_gen/matplotlib.pyplot.subplot.html (accessed June 25, 2019), throughout this section.

[12] When might this scenario occur? Perhaps the northernmost and southernmost rows are nearly at the North and South Poles, respectively, so the wind speeds are better represented by a single value as they are so close together (due to the convergence of longitude lines as we approach the poles).

```
1    wind_shape = np.shape(v_x)
2    v = np.zeros(wind_shape, dtype=v_x.dtype.char)
3    for i in range(wind_shape[0]):
4        for j in range(wind_shape[1]):
5            v[i,j] = ((v_x[i,j] ** 2) + (v_y[i,j] ** 2)) ** 0.5
6        if (i == 0) or (i == wind_shape[0]-1):
7            avg = 0.0
8            for j in range(wind_shape[1]):
9                avg += v[i,j]
10           avg /= wind_shape[1]
11           for j in range(wind_shape[1]):
12               v[i,j] = avg
```

Lines 1–5 are the original solution to Try This! 7-7. To solve the revised problem, we "added" lines 6–12. But we did not add those lines anywhere. What is important here is the indentation of those lines, which consists of a single `if` block. Because that block is indented in four spaces (one indentation level), it is at the same level of indentation as the inner `for` loop. This means that the `if` block is executed after the entire inner loop is executed, *for each index i row*. That is to say, the `if` block is in the body of the *outer* loop, not the inner loop, and thus is executed only `wind_shape[0]` number of times (the number of `i` values). Line 5, in contrast, is in the inner loop and thus is executed `wind_shape[0]` × `wind_shape[1]` number of times.

Line 6 tests to see whether or not the value of `i` is either at the northernmost boundary (`i = 0`) or the southern boundary (`i == wind_shape[0]-1`). If so, we calculate the average wind speed over all points at that latitude (lines 7–10). This requires adding up the value through all longitudes (lines 7–9), then dividing by the number of longitude values (line 10). In lines 11–12, we go through all longitude values at this latitude (again) and assign `v` to the average. Thus, in the times the `if` condition is true, this solution contains three `for` loops, all at the "inner" loop level, as well as calculations (the line 6 `if` test) outside an inner loop, at the level of the body of the outer loop.

Our second example of a more complex nested loop is to mix `for` and `while` loops together. In the code below, we rewrite the Try This! 7-7 solution by replacing the outer loop with a `while` loop:

```
1    wind_shape = np.shape(v_x)
2    v = np.zeros(wind_shape, dtype=v_x.dtype.char)
3    i = 0
4    while i < wind_shape[0]:
5        for j in range(wind_shape[1]):
6            v[i,j] = ((v_x[i,j] ** 2) + (v_y[i,j] ** 2)) ** 0.5
7        i += 1
```

Note that the final line (line 7) has to be at the same indentation level as the `for` statement in line 5. If the `i += 1` statement is indented in four more spaces, it will be in the body of the `for` statement and will execute not `wind_shape[0]` times but `wind_shape[0]` × `wind_shape[1]` times. The value of `i` will then likely exceed the maximum index value for the rows of the two-dimensional arrays.

Functionally, the code has not changed. The `while` loop iterates `i` through all integers from 0 (inclusive) to the value of `wind_shape[0]` (exclusive). So, this solution does not offer anything new to this particular task. But, if we chose a different condition for the `while` statement, we could introduce looping behavior that does not occur a predetermined number of times. And our solution again shows that nested loops are individual loops within loops, and can be edited accordingly.

The take-home message from these examples: Remember to think of complex nested loops as one loop inside of the other, instead of as an entity in and of itself. That is to say, when analyzing code with nested loops, break apart each of the loops and analyze what each loop does. While it is common to see the pattern of having both nested loops be `for` loops and the inner loop start at the line immediately following the outer loop (as we have seen in the Chapter 7 examples), this pattern is not the only way of structuring nested loops. To understand these more complex loops, we need to look at the loops individually.

8.2.7   Conditionals Using Floating-Point Numbers

So far, we have used numerical conditionals (i.e., comparisons that involve numbers) without asking whether there are any issues when they are implemented on a computer (as opposed to on paper in a math problem). It turns out, there are. In Section 5.2.7, we introduced the concept of typing and described how decimal numbers are represented by the `float` type while integers were represented by the `int` type. Although we also discussed how values of different types have different capabilities, with regards to decimals and integers, we did not go much beyond saying these are different categories of values. However, in computers, unlike mathematics, this difference in type makes a real difference. In mathematics, because decimals and integers are in reality the same kind of thing – real numbers – this distinction does not arise.

All variables in a computer are represented by a finite amount of memory. Practically, this means a computer cannot store an arbitrary number of digits for any number. For integers, although this means there is a limit (even if it is the total amount of memory available on the computer) to the size of the integer that can be stored, each integer is still unique. Thus, the integer 3 will always be different from 7.

For floating-point numbers, most decimal numbers cannot be fully represented by the floating-point type. The set of real numbers is ultimately infinite in any finite subrange. We can always create a new real number by adding another decimal place. It is impossible to represent all possible decimal numbers even in a finite subrange by a finite amount of memory. Practically, this (plus other considerations mentioned below) means that two *different* floating-point numbers can be the *same*. Consider the following:

```
In [1]:  3.1 == 3.1001
```

```
Out[1]:  False
```

```
In [2]:   3.1 == 3.10000000000001
```

```
Out[2]:   False
```

```
In [3]:   3.1 == 3.1000000000000000000001
```

```
Out[3]:   True
```

Mathematically, the three equality tests in all three cells should evaluate to `False`, because in each case, the numbers to the left of the `==` are not the same as the numbers to the right of the `==`. But in the last test (`In [3]`), the test returns `True`. To the Python interpreter, the two values are the same.

Worse, this behavior is not necessarily consistent from computer to computer. Exactly when two different floating-point numbers become the "same" depends on the specific scheme the computer uses to represent floating-point numbers (there is more than one way to do so), the amount of memory allocated to a floating-point number, etc., and these can be different depending on the operating system, programming language, etc.

Even floating-point numbers that appear to be "integers" are not necessarily exactly represented by the floating-point type. Consider this example:

```
In [1]:   123456789012345678901.0
```

```
Out[1]:   1.234567890123456e+20
```

The number entered in cell `In [1]` is an "integer" in the sense that there is no fractional component, but its type is floating point, because of the `.0` portion. When Python echoes back the value, some of the digits are not shown. In fact, if we add `1` to the number as first entered in, and compare it with the number entered in without the addition, we see the addition makes no difference in changing the value:

```
In [1]:   (123456789012345678901.0 + 1) == 123456789012345678901.0
```

```
Out[1]:   True
```

If we left off the `.0` portion so the values are of integer type, adding `1` would make a difference:

```
In [1]:   (123456789012345678901 + 1) == 123456789012345678901
```

```
Out[1]:   False
```

Section 9.2.2 contains additional discussion about memory and type.

What is the practical result of the difference between how integers and floating-point numbers behave? Although there is no problem in comparing two integers to one another with the equality operator, it is dangerous to do so when one or both of the values being compared is a floating-point number. Instead, we should do what Bruce Bush calls "safe comparisons,"[13] where what we compare is whether two floating-point numbers are "close enough" to each other relative to some small amount of tolerance. The NumPy function isclose implements this algorithm and is a better way of checking whether two floating-point numbers are the "same":[14]

```
In [1]:  3.1 == 3.10000000000001

Out[1]:  False

In [2]:  np.isclose(3.1, 3.10000000000001)

Out[2]:  True

In [3]:  np.isclose(3.1, 3.1001)

Out[3]:  False
```

The function returns True or False depending on whether or not the two arguments passed in are "close" to each other. The function gives an interpretation of "equality" that is more consistent across different computers and that allows us to adjust how many digits are significant in the comparison. In Section 13.2.2, we revisit isclose for the case of an array of numbers.

In the Figure 8.4 code, we do not use safe comparisons because none of the comparisons test for equality. The only floating-point value comparisons made (e.g., line 50) are inequality comparions (e.g., >). However, because in lines 61–65 we made sure values of energy and food will never be negative, we could substitute line 50 with:

```
if not np.isclose(energy[i-1], 0.0):
```

and the code would work essentially the same (though very small positive values of energy would not pass the if statement).

8.3 Try This!

Although this section includes some cases similar to ones we have seen in Section 8.1, many of the Try This! in this chapter are unrelated to prognostic modeling. We practice obtaining and using random numbers, nested branching, boolean algebra operators, and using the while loop. We also practice how to compare floating-point numbers.

For all the exercises in this section, assume NumPy has already been imported as np and the Matplotlib pyplot submodule has already been imported as plt.

---

[13] Bush (1996).
[14] https://docs.scipy.org/doc/numpy/reference/generated/numpy.isclose.html (accessed June 26, 2019).

Try This! 8-1 Random Numbers: Mouse Path

Pretend we are simulating a mouse walking through a maze where at every branch (or intersection) in the path, the mouse can go either left or right. Assuming the mouse only walks forward and makes its choice of direction at each intersection randomly, write a program that prints out the simulated path decisions through six intersections. Here is an example of the output we are looking for:

```
Right
Left
Right
Right
Right
Left
```

Try This! Answer 8-1

Here are two ways of doing this. In both ways, we use a `for` loop to loop through six times. In this first solution, we use the `randint` function to return a zero or one, randomly, and an `if` test that results in printing the message `Left` if the random number is one and `Right` otherwise:

```
1   np.random.seed(973491)
2   for i in range(6):
3       if np.random.randint(0, 2) == 0:
4           print("Left")
5       else:
6           print("Right")
```

In this second solution, we use the `uniform` function to return a floating-point number between zero (inclusive) and one (exclusive). We then check to see whether the value is greater than or equal to 0.5 or not, and print `Left` and `Right` respectively:

```
1   np.random.seed(973491)
2   for i in range(6):
3       if np.random.uniform() >= 0.5:
4           print("Left")
5       else:
6           print("Right")
```

Because the uniform probability distribution is "flat" from 0 to 1 – i.e., the probability a value is in a subrange from 0 to 1 is the same in any subrange from 0 to 1 of the same width – the probability the value is between 0 (inclusive) and 0.5 (exclusive) is the same as the probability the value is between 0.5 (inclusive) and 1 (exclusive). Thus, this code simulates an equal probability of printing `Left` or `Right`.

Try This! 8-2 Nested Branching and and: Atomic Number

In Section 8.2.2's discussion of the and operator, we found the following assignment to the result of a boolean expression:

```
is_between = (atomic_number > 2) and (atomic_number < 6)
```

where `atomic_number` is defined earlier. Using nested branching, write a code that sets the value of `is_between` using the logic as given above but does *not* use the and operator. That is, write a nested branching statement that duplicates the behavior of the and operator.

Try This! Answer 8-2

This code solves the problem, assuming `atomic_number` is already defined:

```
1   if atomic_number > 2:
2       if atomic_number < 6:
3           is_between = True
4       else:
5           is_between = False
6   else:
7       is_between = False
```

The two boolean values on either side of the and operator both have to be `True` in order for the result of the and expression to return `True`. In all other cases, the result of the expression is `False`. This is why the line 7 else clause (to the outer `if-else` statement) does not need to have an inner `if-else` statement similar to the one in lines 2–5. If `atomic_number > 2` is `False`, it does not matter what `atomic_number < 6` evaluates to; `is_between` has to be `False`.

This Try This! also shows us that the and operator is equivalent to asking, "is the first boolean operand `True` or `False`," and for each one of those cases, asking, "is the second boolean operand `True` or `False`?" In that way, we can think of nested branching statements as the successive consideration of each of the alternatives (as needed) for each of the input boolean variables in a truth table, one in turn.

Try This! 8-3 Truth Tables: Mouse Study

Pretend we are designing a simulation of a mouse whose behavior is affected by the state of these three inputs:

- Has the mouse just eaten?
- Has the mouse been given a dose of the test drug?
- Is the mouse alone or with a group of other mice?

The behavior that results is governed by these rules:

- A mouse that has not eaten and has been given a dose of the test drug is lethargic (or sleepy, inactive), regardless of the presence or absence of other mice.
- A mouse that has not eaten and has not been given a dose of the test drug is lethargic if alone but not lethargic if with other mice.
- A mouse that has eaten is not lethargic, regardless of whether or not the mouse is alone or not or has taken a dose of the test drug or not.

Make a truth table that shows, for each possible value of the three inputs, whether the mouse will be lethargic or not (the output). In creating the truth table, choose descriptive names for the input and output boolean variables.

Try This! Answer 8-3

Here is a truth table that describes this scenario:

has_eaten	has_dose	is_alone	is_lethargic
True	True	True	False
True	True	False	False
True	False	True	False
True	False	False	False
False	True	True	True
False	True	False	True
False	False	True	True
False	False	False	False

Because there are three input boolean variables, there are eight possible permutations of the values of these variables. In general, given two possible values for a variable (True and False), and three such variables, there are $2^3 = 8$ possible permutations and thus eight rows in the truth table.

One benefit of a truth table is that it can be directly transformed into an if-elif statement where each row is one if or elif clause, and the condition for each of these clauses is formed by joining the input boolean variables together with and statements and negating an input variable with a not if the input variable in that row is False. Here is code that translates the above truth table:

```
if has_eaten and has_done and is_alone:
    is_lethargic = False
elif has_eaten and has_done and (not is_alone):
    is_lethargic = False
elif has_eaten and (not has_done) and is_alone:
    is_lethargic = False
elif has_eaten and (not has_done) and (not is_alone):
    is_lethargic = False
```

```
9    elif (not has_eaten) and has_done and is_alone:
10       is_lethargic = True
11   elif (not has_eaten) and has_done and (not is_alone):
12       is_lethargic = True
13   elif (not has_eaten) and (not has_done) and is_alone:
14       is_lethargic = True
15   elif (not has_eaten) and (not has_done) and (not is_alone):
16       is_lethargic = False
17   else:
18       is_lethargic = "Something is wrong"
```

where has_eaten, has_dose, and is_alone are boolean variables that have been set before the above if-elif block. In the else clause at the end of the if-elif block, we set the output variable is_lethargic to a string value rather than a boolean value which will show something is wrong with the code if the value is printed out. The better way of dealing with such errors is described in Section 9.2.7. The code above can also be expressed more concisely, which we will see in Try This! 8-4.

When expressed in regular English sentences, the possibilities the truth table describes take up only three bullet points rather than eight rows. Often, narrative descriptions of logical reasoning read differently than the truth table that summarizes all the possible states being described. Part of the challenge of programming is the proper translation of some complex logic into a truth table (and, ultimately, into a series of branching statements and/or a boolean expression). Techniques and theorems we can use for that translation are studied in boolean algebra, and their use is beyond the scope of the current text.

Try This! 8-4 Nested Branching: Mouse Study

Rewrite the solution to Try This! 8-3 using nested branching instead of if-elif statements.

Try This! Answer 8-4

Here is a solution:

```
1    if has_eaten:
2        is_lethargic = False
3    else:
4        if has_dose:
5            is_lethargic = True
6        else:
7            if is_alone:
8                is_lethargic = True
9            else:
10               is_lethargic = False
```

The nested branching solution takes fewer lines than the `if-elif` solution, because we can eliminate unnecessary tests. For instance, if `has_eaten` is `True`, it does not matter what the values of `has_dose` or `is_alone` are, `is_lethargic` will be `False`. We can ignore the values of `has_dose` and `is_alone` entirely, for that case (in line 2 above).

Try This! 8-5 Boolean Operators: Summer Precipitation

In Figure 6.7, we described a program to calculate the average Summer mean monthly precipitation in the Seattle area. Replace the `if-elif` statement of lines 7–14 with a single `if-else` statement, making use of the appropriate boolean operators in the condition so the program calculates the same average as before.

Try This! Answer 8-5

Here is the replacement `if-else` statement:

```
1        if (months[i] == 6) or (months[i] == 7) or (months[i] == 8):
2            summer_precip_sum = summer_precip_sum + precip[i]
3        else:
4            pass
```

We use the `or` operator because `months` values of six, seven, and eight all designate Summer months. Precipitation for any one of those months are part of the averaging calculation. The "extra" indentation above is included to match the indentation of the original `if-elif` statement in Figure 6.7. Finally, the use of the boolean expression with `or` statements enables us to remove the duplicative lines (8, 10, and 12) in Figure 6.7.

Try This! 8-6 Boolean Operators: Nearest-Neighbor Averaging

In the Section 7.1 example, we calculated the nearest-neighbor average of a two-dimensional array of surface temperature using the code in Figure 7.3. How would we change that code to use the "corner neighbors" instead of the adjacent neighbors for the average?

Try This! Answer 8-6

Because the corner neighbors are offset by both row and column from the center element of the average, to calculate the nearest-neighbor average using the corner neighbors, we have to use both row and index offsets to specify the corner neighbors. We also have to check that the row and column values of the corners are within the grid domain. The solution: Replace lines 17–37 of Figure 7.3 with the lines in Figure 8.9. Hint: The `and` operator is our friend.

```
1    avg_shape = np.shape(air_temp_avg)
2    for i in range(avg_shape[0]):
3        for j in range(avg_shape[1]):
4            sum_values = air_temp[i,j]     #+ Center point
5            num_values = 1
6
7            #+ Upper-left point
8            if ((i - 1) >= 0) and ((j - 1) >= 0):
9                sum_values += air_temp[i-1,j-1]
10               num_values += 1
11
12           #+ Upper-right point
13           if ((i - 1) >= 0) and ((j + 1) < avg_shape[1]):
14               sum_values += air_temp[i-1,j+1]
15               num_values += 1
16
17           #+ Lower-left point
18           if ((i + 1) < avg_shape[0]) and ((j - 1) >= 0):
19               sum_values += air_temp[i+1,j-1]
20               num_values += 1
21
22           #+ Lower-right point
23           if ((i + 1) < avg_shape[0]) and \
24              ((j + 1) < avg_shape[1]):
25               sum_values += air_temp[i+1,j+1]
26               num_values += 1
27
28           air_temp_avg[i,j] = sum_values / num_values
```

Figure 8.9 Code snippet to calculate the nearest-neighbor average using the corner neighboring points, in Try This! 8-6.

Try This! 8-7 Using `while`: Mouse Path

In Try This! 8-1, we pretended we were simulating a mouse walking through a maze where at every branch (or intersection) in the path, the mouse randomly chooses to go either left or right. We wrote a program that printed out the simulated path decisions through six intersections. Change that program so the mouse will stop walking only after making a "circle," where a circle is four consecutive right turns or four consecutive left turns. Again, assume the mouse only walks forward.

Try This! Answer 8-7

Here is a solution using `uniform` for the random number generation:

```
1    np.random.seed(973491)
2
3    num_lefts_in_a_row = 0
4    num_rights_in_a_row = 0
5    has_gone_around = False
6
7    while not has_gone_around:
8        if np.random.uniform() > 0.5:
9            print("Left")
10           num_lefts_in_a_row += 1
11           num_rights_in_a_row = 0
12       else:
13           print("Right")
14           num_rights_in_a_row += 1
15           num_lefts_in_a_row = 0
16
17       if (num_lefts_in_a_row == 4) or (num_rights_in_a_row == 4):
18           has_gone_around = True
```

We use a boolean variable to control the while loop rather than a boolean expression. As long as has_gone_around is False, the loop will continue iterating. Why False? In line 7, the condition after the while is not has_gone_around instead of merely has_gone_around. Thus, when has_gone_around is True, not has_gone_around evaluates to False. When has_gone_around is False, not has_gone_around evaluates to True. In lines 17–18, we test to see whether the number of consecutive left or right turns equals four, and if so, we set has_gone_around to True.

```
Right
Left
Right
Right
Right
Left
Left
Right
Left
Right
Right
Left
Left
Right
Right
Right
Right
```

Figure 8.10 Output from the answer to Try This! 8-7.

The tricky part of the program is recognizing that whenever the mouse takes a left turn, the **counter** for the number of *right* turns the mouse has taken (i.e., num_rights_in_a_row) is reset to 0, and vice versa whenever the mouse takes a right turn. If we forget to do so, the incrementing of the num_lefts_in_a_row and num_rights_in_a_row counters (in lines 10 and 14) will occur every time a left or right turn occurs (respectively), regardless of whether the previous turn was also left or right (respectively). Figure 8.10 shows output from the above program.

Try This! 8-8 Multiple Subplots: A Mass–Spring System

In Try This! 6-1, we plotted the displacement of a mass–spring system. Create a grid of four subplots, each of which graphs the displacement for masses of 1, 2, 3, and 4 kg. The values of A, k, and ϕ are the same as in Try This! 6-1.

Try This! Answer 8-8

The solution code is in Figure 8.11. The calculation and individual plotting lines (lines 2–5, 12–14) are nearly the same as in Try This! 6-1. The main adaptation in this solution is to embed

```
1   mass = [1, 2, 3, 4]
2   A = 0.05
3   k = 2.0
4   phi = 0.0
5   t = np.arange(0, 8.1, 0.1)
6
7   plt.figure()
8
9   for j in range(4):
10      plt.subplot(2, 2, j+1)
11      m = mass[j]
12      omega = np.sqrt(k/m)
13      x = A * np.cos(omega * t + phi)
14      plt.plot(t, x, '-')
15      plt.title('Displacement for a ' + str(m) + ' kg Mass')
16      plt.axis([t[0], t[-1], -0.06, 0.06])
17
18      if j >= 2:
19          plt.xlabel('Time (sec)')
20      else:
21          plt.xticks([])
22      if (j % 2) == 0:
23          plt.ylabel('Displacement (m)')
24      else:
25          plt.yticks([])
26
27  plt.savefig("mass-spring_subplots.png")
```

Figure 8.11 Code snippet to create multipanel plot of displacement for a mass–spring system for four different mass values, for Try This! 8-8. The plot generated by this code is in Figure 8.12.

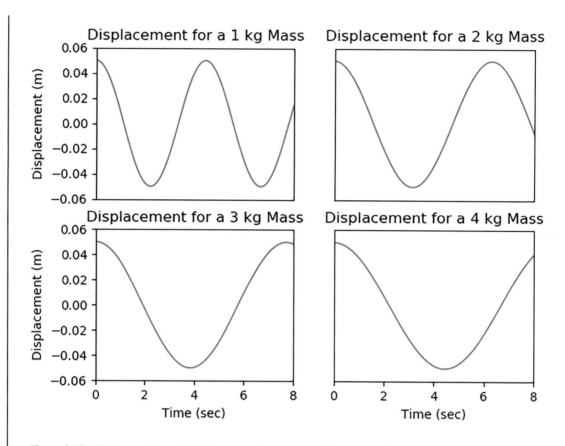

Figure 8.12 Multipanel plot of displacement for a mass–spring system for four different mass values, generated by the code in Figure 8.11, as part of the solution for Try This! 8-8.

those lines in a `for` loop with a `subplot` function call and to change the value of m to a different mass value (given in the list `mass`) for each iteration of the loop. The plot generated by the solution code is in Figure 8.12.

Try This! 8-9 More Complex Nested Loops: Air Quality

Consider the air quality data in Beijing, China, given in Table 7.7. Try This! 7-1 provides additional information regarding these data. Calculate the average amongst all four locations at each hour in the day. That is, create a 24-element array where each element is the average of the data from the four air quality monitoring locations. Use nested loops (`for` and/or `while`) instead of array syntax or array functions.

Try This! Answer 8-9

Assume the data array is called `pm25`, as given in Figure 7.7. Here is a solution using `for` loops:

```
1    avg_over_loc = np.zeros(24, dtype='d')
2    for i in range(np.shape(pm25)[0]):
3        sum_over_loc = 0.0
4        for j in range(np.shape(pm25)[1]):
5            sum_over_loc += pm25[i,j]
6        avg_over_loc[i] = sum_over_loc / np.shape(pm25)[1]
```

Line 1 creates the array that will hold the averages over the four locations at each hour of the day. In line 2, we loop through the rows, so `i` is the index for each of the 24 rows in the data. The body of this outer loop calculates the average for one of the rows of data. In line 3, we initialize a counter to sum up all the PM$_{2.5}$ concentration values at each measurement station at a given hour. The inner loop (lines 4–5) does this summation. Line 6 calculates the average for that hour and stores it as an element in `avg_over_loc`. Note that `np.shape(pm25)[0]` gives the number of rows and `np.shape(pm25)[1]` gives the number of columns in the data array.

This solution looks like other nested loops we have seen in Chapter 7, except for lines 3 and 6. What is the importance of line 3? What if we, for instance, put line 3 outside the outer loop (i.e., right after line 1)? The result would be that the averages calculated and stored in `avg_over_loc` would be based on the sum of all stations at the current hour *and all previous hours* rather than just the current hour. In order to obtain the average for just a given hour, we have to reinitialize `sum_over_loc` before looping through the stations *for every hour* (i.e., row of data). The place for that reinitialization is where line 3 is.

What about line 6? What if we, for instance, indented line 6 in one more level so that the statement was placed inside the inner `for` loop? In that event, the calculation would occur not only every iteration over `i` (i.e., over each row) but also every iteration over `j` (i.e., over each column value), and that calculation would write over the `avg_over_loc[i]` element each iteration over `j`. Interestingly, in this specific instance, this would not affect the final answer generated by the program, because line 6 does not affect `sum_over_loc`, but for the first 3 iterations of the inner loop, the values saved in `avg_over_loc[i]` would be incorrect. The value saved in `avg_over_loc[i]` in the final, fourth iteration would be correct, which is why the program will still give the correct answer. This is, however, confusing, and in another program, things may not end up so well. Putting line 6 outside the inner loop makes it clear the average is calculated using the value of `sum_over_loc` that sums over all column values for a row.

The bottom line: Lines 3 and 6 are placed where they are at (including their indentation levels) because the work they are engaged in should happen before looping over all the columns (the inner loop) for a row or after looping over all the columns for a row.

Try This! 8-10 Comparing Floating Point: Temperatures

Pretend we have the following measurements from a highly accurate thermometer inside an ice water bath (a beaker filled with ice and water). All values are in °C and are taken at various, consecutive times:

```
1   bath_temp1 = 0.008490942
2   bath_temp2 = 0.000000005
3   bath_temp3 = 0.00000023
```

Which measurement(s) are "close to" the freezing point of water, 0°C?

Try This! Answer 8-10

We use the NumPy `isclose` function's default definition of what constitutes "close to." Using that function to test where `bath_temp` is close to 0.0 yields the following results:

```
In [1]:  bath_temp1 = 0.008490942
         bath_temp2 = 0.000000005
         bath_temp3 = 0.00000023

         print(np.isclose(bath_temp1, 0.0))
         print(np.isclose(bath_temp2, 0.0))
         print(np.isclose(bath_temp3, 0.0))


         False
         True
         False
```

The value of `bath_temp2` is "close to" the freezing point of water, 0°C.

8.4   More Discipline-Specific Practice

The Discipline-Specific Jupyter notebooks for this chapter cover the following topics:

- Random numbers in computers.
- Scalar boolean type and expressions.
- Nested branching.
- Looping an indefinite number of times using `while`.
- Making multiple subplots.
- More on nested loops.
- Conditionals using floating-point numbers.

8.5   Chapter Review

8.5.1   Self-Test Questions

Try to do these without looking at the book or any other resources or using the Python interpreter. Answers to these Self-Test Questions are found at the end of the chapter. For all questions in this section, assume NumPy has already been imported as np and Matplotlib's pyplot module is imported as plt.

Self-Test Question 8-1

Describe the purpose of the seed for a random number generator. What happens if we set the seed to the same value in different parts of a program?

Self-Test Question 8-2

Describe the difference between the NumPy random uniform and randint functions.

Self-Test Question 8-3

Describe the difference between the following variables:

```
1   is_large = True
2   is_small = true
3   is_short = "False"
4   is_tall = False
```

Self-Test Question 8-4

Assuming the variables data equals −17 and count equals 3, what is the result of:

```
1. data < 10
2. (data < 10) and (data == -17)
3. ((data > 0) or (count > data)) and (count == 3)
4. (data > 0) or (count > data) and (count == 3)
5. data + count > 0
```

Self-Test Question 8-5

Describe the difference between an if-elif statement and a two-level nested branching statement (i.e., one if statement inside another if statement).

Self-Test Question 8-6

Consider this code:

```
1   data = 42
2   if (data > -3) and (data < 50):
3       factor = 2
4   else:
5       factor = 34
```

Write a nested branching statement that has the same behavior as the code above.

Self-Test Question 8-7

What will be the output of the following code?:

```
1   age = 19
2   while True or age < 21:
3       print("Age: " + str(age))
4       age += 1
```

Please describe why the output is what it is.

Self-Test Question 8-8

What is the Matplotlib syntax to make the second-row, first-column plot in a four-row, two-column grid of panels the current subplot (i.e., where all subsequent plotting commands affect that panel)?

Self-Test Question 8-9

What is the output of the following code?:

```
1   print("Decade   Age")
2   print("-----------")
3   for idecade in range(2):
4       age =  idecade * 10
5       while age < (idecade + 1) * 10:
6           print(str(idecade + 1) + "          " + str(age))
7           age += 1
```

What would happen if we replaced the code in line 7 by:

```
age += 2
```

at the same indentation level as the current line 7?

Self-Test Question 8-10

Describe how to safely compare two floating-point numbers to see whether they are equal to each other.

8.5.2   Chapter Summary

In Chapter 6, we introduced branching and looping. In Chapter 7, we introduced nested for loops that accessed two-dimensional arrays. In this chapter, we deepened the complexity of the branching and looping operations we can conduct. We covered these topics by looking at basic prognostic modeling, though these same concepts and structures are also used in data analysis:

- Random numbers in computers

    - Random number generators create sequences of pseudorandom numbers, that is, numbers that appear random but whose sequences ultimately repeat after many values.
    - The random number generator's seed (set by the NumPy random submodule function `seed`) determines the sequence of random numbers that are generated. We obtain the same sequence of random numbers on a given computer if the value of the seed is set to the same value prior to calculating the random numbers.
    - The `randint` function returns one integer or an array of integers chosen randomly within a range of values, described by an inclusive lower limit and an exclusive upper limit. Each integer in the range has the same probability of being drawn.
    - The `uniform` function returns one or an array of uniformly distributed floating-point random numbers between zero (inclusive) and one (exclusive).

- Scalar boolean type and expressions

    - `True` and `False` are special, reserved values in Python that represent the situation "true" and "false." These are values, not variables, and must be capitalized and written as shown, without quotation marks or other characters.
    - Boolean type variables have the value of either `True` or `False`.
    - Give boolean variables a name or identifier that automatically connects with a true or false answer, e.g., `is_missing`.
    - Boolean expressions are any Python expression that results in a boolean value. The elements of a boolean expression can be boolean values or they can be other types of variables (e.g., integers, strings, etc.) and operators that operate on those values (e.g., the greater than operator, the logical equality testing operator, etc.).
    - A boolean algebra is defined by the mathematical rules that boolean variables follow. Boolean algebra expressions consist of boolean variables (which themselves can be the result of boolean expressions) and boolean algebraic operators (e.g., and, etc.).
    - The `and` operator checks to see if two boolean operands are both `True` and returns `True` if so. `False` is returned for all other combinations of the two boolean operands.
    - The `or` operator checks to see if two boolean operands are both `False` and returns `False` if so. `True` is returned for all other combinations of the two boolean operands.
    - The `not` operator takes a single boolean operand and returns the opposite of that operand's value.
    - Truth tables summarize the relationship between the input boolean variables in a boolean expression and the output boolean variable that is the result of the expression. Each row represents one possible set of values for the input boolean variables. The set of all rows in the table represents all possible combinations of values of the input boolean variables.

- Nested branching

    - A nested branching statement is a branching statement placed inside another branching statement.

- If the inner branching statement is placed in the body of an outer `if` or `elif` condition, that inner branching statement will only be executed if the outer condition is `True`. If the inner branching statement is placed in the `else` body of a branching statement, that inner branching statement will only be executed if the outer condition is `False`.
- We can have more than two levels of nesting. Deeper levels of nesting are created by adding another level of indentation.

- Looping an indefinite number of times

  - The `for` loop is best used when we know ahead of time how many times we want to iterate through the loop.
  - The `while` loop is best used when we do not know ahead of time how many times we want to iterate through the loop.
  - The `for` loop should not be used if the value of the iterator will be changed in the body of the loop and we want to use that changed iterator in a later iteration of the loop.
  - The condition in the `while` line is an expression that evaluates to `True` or `False`.
  - The body of the `while` loop will execute when the condition in the line with the `while` evaluates to `True`. After the body executes, the program returns to the line with the `while` and reevaluates the condition. If the condition evaluates to `False`, the program skips the loop body and the program continues with the line after the loop body.
  - When a `continue` statement is executed, Python skips the rest of the loop body and returns to the line with the `while` and reevaluates the condition.
  - When a `break` statement is executed, the `while` loop is terminated and the program continues with the line of code after the loop.

- Making multiple subplots

  - Matplotlib allows us to place more than one plot panel on a single figure (or window).
  - Matplotlib's pyplot module has a command, `subplot`, that creates a panel in a grid of panels. Subsequent plotting and formatting commands are conducted on the panel defined by `subplot` until another `subplot` call is made.

- More on nested loops

  - Nested loops are loops that are inside other loop(s).
  - In general, for each iteration of the outer loop, the inner loop (can) go through all of its iterations. The specifics depend on what conditions are used in the loops and whether there are commands like `continue` or `break` which alter the flow of the program.
  - The initial lines of outer and inner loops, whether they are `for` loops, `while` loops, or some of each, do not have to be back-to-back. These loops can have more complicated structures.
  - We can have more than two levels of nesting. Deeper levels of nesting are created by adding another level of indentation. Nested statements can incorporate mixtures of other nested structures (such as nested branching structures).

- Conditionals using floating-point numbers:
  - Avoid comparing two floating-point numbers to each other through the equality operator (==).
  - Safe comparison involves seeing whether the absolute value of the difference of two floating-point numbers is a "small enough" value.
  - The NumPy function `isclose` conducts safe comparisons between two floating-point values.

As powerful as basic branching and looping is, we can do even more with the introduction of the boolean data type, boolean expressions, nested branching, indefinite looping, and more complex nested looping. These concepts and structures enable us to write more powerful models and data analysis routines.

8.5.3 Self-Test Answers

Self-Test Answer 8-1

The seed value controls the exact sequence of random numbers that is generated. Changing the seed value will change the values of the random numbers that are generated. Changing the seed does not change the probability distribution any given random number generating function uses to obtain random numbers. By changing the seed, we can generate different sequences of random numbers, given calls to the same random number generating function. By setting the seed to the same value in different places of a program, we ensure that the same sequence of random numbers is created in the lines of code after the line that sets or resets the seed.

Self-Test Answer 8-2

The `uniform` function returns a decimal floating-point number between zero (inclusive) and one (exclusive). The probability distribution function is flat in that interval, so values in every equally sized range from zero (inclusive) to one (exclusive) have the same probability of occurring. The `randint` function returns an integer in a given range (inclusive on the lower bound and exclusive on the upper bound). The integers in that range all have the same probability of being returned.

Self-Test Answer 8-3

Both `is_large` and `is_tall` are boolean values. The variable `is_short` is a string. The variable `is_small` is set to whatever is given by the variable `true`. Line 2 will not execute correctly unless the variable `true` is previously defined. Note the difference the capitalized "T" makes between `True` in line 1 and `true` in line 2. In the former case, the token is a boolean value whereas in the latter the token is a variable identifier (a.k.a., variable name).

Self-Test Answer 8-4

1. True
2. True
3. True
4. True
5. False

Self-Test Answer 8-5

In the `if-elif` statement, the program moves from one condition to the other (the conditions that come after the `if` and `elif` tokens), testing to see whether the condition is `True`. If one of the conditions is `True`, the body under that condition is executed and control in the program is transferred to the line after the end of the entire `if-elif` statement structure.

In a two-level nested branching statement, the inner `if` statement is executed only if the outer `if` statement's condition permits it. That is, if the inner `if` statement is in the body of the outer `if` statement's condition line, the inner statement will only execute if the outer condition is `True`. If the inner `if` statement is in the `else` body of the outer `if` statement's condition line, the inner statement will only execute if the outer condition is `False`.

Self-Test Answer 8-6

In our solution, we split up the condition with the `and` operator:

```
1   data = 42
2   if data > -3:
3       if data < 50:
4           factor = 2
5       else:
6           factor = 34
7   else:
8       factor = 34
```

We need to put code to set `factor` to 34 in two places because that assignment needs to be made if either `data` is greater than −3 or `data` is less than 50.

Self-Test Answer 8-7

The program will execute and keep printing the message to the screen forever. The first few lines of output will be:

```
Age: 19
Age: 20
Age: 21
Age: 22
Age: 23
Age: 24
Age: 25
Age: 26
Age: 27
```

and the program will continue writing the same message, except that the age that is printed out increases by one each line. This is an example of an infinite loop.

The reason this happens is that the condition:

```
True or age < 21
```

always evaluates to `True`, since the `or` operator returns `True` if either one of the operands is `True`.

Self-Test Answer 8-8

This command:

```
plt.subplot(4, 2, 3)
```

will make the second-row, first-column plot in a four-row, two-column grid of panels the current subplot.

Self-Test Answer 8-9

Figure 8.13 shows the output of the first code: a table listing the ages corresponding to the first two decades. If we changed line 7 in Figure 8.13 as described in the second part of this Self-Test Question, the table would only list even ages:

	Decade	Age
1	Decade	Age
2	---------	
3	1	0
4	1	2
5	1	4
6	1	6
7	1	8
8	2	10
9	2	12
10	2	14
11	2	16
12	2	18

```
 1  | Decade  Age
 2  | - - - - - - - - - - -
 3  | 1        0
 4  | 1        1
 5  | 1        2
 6  | 1        3
 7  | 1        4
 8  | 1        5
 9  | 1        6
10  | 1        7
11  | 1        8
12  | 1        9
13  | 2        10
14  | 2        11
15  | 2        12
16  | 2        13
17  | 2        14
18  | 2        15
19  | 2        16
20  | 2        17
21  | 2        18
22  | 2        19
```

Figure 8.13 Output of the first code in Self-Test Question 8-9.

Self-Test Answer 8-10

Floating-point numbers should not be compared with one another using the logical equality (==) operator. Instead, we should check to see whether the absolute value of the difference between the two numbers is smaller than some tolerance. The NumPy function isclose does this for default values of absolute and relative tolerance.

9   Reading In and Writing Out Text Data

In the previous chapters, the lists and arrays of data we have looked at have been pretty small. Partly, this is because it is easier to understand small datasets, which helps us when we are learning a new tool. However, what makes using a programming language (as opposed to Excel) a better tool to analyze data is the ease with which a program written in a programming language can be scaled up to handle a large dataset. In Excel, it is not so easy to go from, say, 500 rows of data to 500 million rows of data.

The key to getting a computer to operate on large amounts of data is to store the data in a file and read the data into an array (or similar variable). After we have put the data into an array, as we have seen, we can manipulate and do calculations on those data. In this chapter, we look at a general purpose file called the **text file**. All the files we are used to – Word, Excel, image, audio, and video files – hold information, and the information they hold is related to the kind of file they are. A Word file holds a written document, an Excel file holds cells of data and the formulas to do calculations on those cells, and image, audio, and video files hold the data encoding each of those kinds of media. The data used in scientific and engineering applications are usually numbers, such as measurements, or are sentences and phrases, if the data consist of descriptions. Thus, to hold scientific and engineering data, the kind of file only needs to be able to hold numbers, sentences, and phrases. A text file does just that: it holds only numbers, letters, punctuation characters, and some special characters that signify ideas such as the end of a line, and has the added benefit of being readable both by a computer and a human being.

Many text files end in the suffix *.txt*, but the name of the file is not what makes it a text file. What makes a text file a text file is the special encoding scheme it uses to store letters, numbers, etc., in a form both computers and people can read. Python source code files are also text files. Another way of thinking about a text file is that it is a stripped-down version of a Word file. Whereas a Word file also holds information about formatting (e.g., bold, underline, font size) and pagination of the text in the file (and how the text is mixed with pictures, etc.), a text file only really holds the text of the document.

In this chapter, we will look at how to read in data that are stored in a text file and write out data to the same kind of file. Text files are handy for storing files on the order of tens to thousands of lines. Once we start hitting hundreds of thousands or millions of lines, text files will still work, but they start becoming cumbersome. The ideas behind handling data in text files, however, are similar to those used in other file formats that are designed for large datasets. In Chapter 18, we will look at some of those special formats. In addition, the way Python handles text files provides us the opportunity to make a detailed introduction to the idea of **objects** in programming. The idea of an object is foundational to the field of object-oriented programming (OOP), an incredibly useful tool for developing large programs that work. Many of the programs scientists and engineers write do not need to use objects or OOP, but as more and more packages become available for scientific and engineering work, knowledge of OOP can give us access to abilities and tools that would otherwise be closed to us.

The larger files used in this chapter are available for download. See the "Supporting Resources and Updates to This Book" section in the Preface for details.

9.1   Example of Reading In and Writing Out Text Data

Molecular dynamics is the study of the motion of the atoms that make up molecules. Such simulations can be used to study the physical properties of a molecular system. For example, this method can be used to study the folding of a protein – how will the protein move and behave in response to changes in its environment (e.g., if thermal or heat energy were added to the system)? NAMD, a molecular dynamics simulation package by Phillips et al. (2005), calculates the movements of the atoms in molecules using Newton's equations of motion. This motion is dependent on the interaction between atoms which is described in chemistry as force fields. These force fields represent a summation of interactions represented by the structure within the molecule (chemical bonds, bond angles, etc.), electrostatic interaction due to charge differences, and nonbonded forces. The NAMD package can be run on a personal computer or a parallel supercomputer with over half a million processing cores.[1]

As output, NAMD produces text files. These files contain information about the simulation run and the results of the calculations. Figure 9.1 shows the beginning lines of the 670-line text file *smallproteinB_ws_eq.log*. The NAMD run that generated this file is an energy minimization simulation of a small protein in a spherical box. The purpose of the energy minimization is to find the coordinates (the positions of the atoms of the molecule in three-dimensional space) that minimize the interaction between the atoms in the molecular system, while temperature and pressure are held constant. This gives a benchmark for the scientist to compare the results of later model runs where the environment of the molecule is changed. In that way, the scientist can see how the molecule responds to a perturbation of its environment. Without first minimizing the interactions of the molecule, a scientist would not be able to determine whether the motion of the molecule observed under the perturbation is the result of random interactions between the atoms or is a result of the change to the physical environment of the molecule. Although the exact meaning of many of the terms, phrases, and abbreviations in Figure 9.1 are special to the program and the field of computational chemistry itself, we get the sense that these lines of output describe the parameters of the model run, including information about the computer the model is being run on as well as parameters of this specific simulation run.

---

[1] www.ks.uiuc.edu/Research/namd/ (accessed August 27, 2019).

```
Charm++: standalone mode (not using charmrun)
Charm++> Running in Multicore mode:  1 threads
Charm++> Using recursive bisection (scheme 3) for topology aware
partitions
Charm++ warning> fences and atomic operations not available in native
assembly
Converse/Charm++ Commit ID:
v6.8.2-0-g26d4bd8-namd-charm-6.8.2-build-2018-Jan-11-30463
[0] isomalloc.c> Disabling isomalloc because mmap() does not work
CharmLB> Load balancer assumes all CPUs are same.
Charm++> Running on 1 unique compute nodes (4-way SMP).
Charm++> cpu topology info is gathered in 0.017 seconds.
Info: NAMD 2.13 for Win64-multicore
Info:
Info: Please visit http://www.ks.uiuc.edu/Research/namd/
Info: for updates, documentation, and support information.
Info:
Info: Please cite Phillips et al., J. Comp. Chem. 26:1781-1802 (2005)
Info: in all publications reporting results obtained with NAMD.
Info:
Info: Based on Charm++/Converse 60800 for multicore-win64
Info: Built Fri, Nov 09, 2018 2:32:31 PM by jim on europa
Info: Running on 1 processors, 1 nodes, 1 physical nodes.
Info: CPU topology information available.
Info: Charm++/Converse parallel runtime startup completed at 0.0219998 s
CkLoopLib is used in SMP with a simple dynamic scheduling (converse-level
notification) but not using node-level queue
Info: 3.25781 MB of memory in use based on GetProcessMemoryInfo
Info: Configuration file is C:\common\smallproteinB_ws_eq.conf
Info: Changed directory to C:\common
TCL: Suspending until startup complete.
Info: SIMULATION PARAMETERS:
Info: TIMESTEP              2
Info: NUMBER OF STEPS       0
Info: STEPS PER CYCLE      10
Info: LOAD BALANCER   Centralized
Info: LOAD BALANCING STRATEGY  New Load Balancers -- DEFAULT
```

Figure 9.1 Lines at the beginning of the NAMD output file, *smallproteinB_ws_eq.log*. Some line breaks and spaces were added for readability.

Later on in the file, we see lines that contain numbers that are output from the simulation. Figure 9.2 shows lines from the first five timestep numbers of the simulation. Each line block that begins with "ENERGY" contains 15 numbers. The meanings of these numbers are given in the 15 labels of the "ETITLE" line block. In addition to the ENERGY and ETITLE lines, there are a few lines of messages about the run.

In Figure 9.2, each block of three lines beginning with "ETITLE" or "ENERGY" are actually a single line in the file; they are displayed as three lines to fit on the page. In the real

```
ETITLE:        TS          BOND         ANGLE          DIHED          IMPRP    ⎫ single
             ELECT          VDW       BOUNDARY          MISC        KINETIC    ⎬ line
             TOTAL          TEMP      POTENTIAL         TOTAL3       TEMPAVG    ⎭

ENERGY:         0       116.4585       94.8698         64.8673        0.2787   ⎫ single
          -9743.7939     984.5003       0.0000          0.0000        0.0000   ⎬ line
          -8482.8193       0.0000    -8482.8193      -8482.8193        0.0000  ⎭

MINIMIZER STARTING CONJUGATE GRADIENT ALGORITHM
LINE MINIMIZER REDUCING GRADIENT FROM 2.04101e+06 TO 2041.01
ENERGY:         1        51.4445       70.9667         64.3814        0.2151   ⎫ single
          -9782.4969     967.6876       0.0000          0.0000        0.0000   ⎬ line
          -8627.8016       0.0000    -8627.8016      -8627.8016        0.0000  ⎭

ENERGY:         2        28.3905       65.0812         64.0858        0.5232   ⎫ single
          -9820.4551     952.1354       0.0000          0.0000        0.0000   ⎬ line
          -8710.2390       0.0000    -8710.2390      -8710.2390        0.0000  ⎭

ENERGY:         3        54.0500       72.9659         63.9732        1.2054   ⎫ single
          -9857.6574     937.8279       0.0000          0.0000        0.0000   ⎬ line
          -8727.6349       0.0000    -8727.6349      -8727.6349        0.0000  ⎭

ENERGY:         4       133.0166       92.2749         64.0363        2.2640   ⎫ single
          -9894.0898     924.7729       0.0000          0.0000        0.0000   ⎬ line
          -8677.7252       0.0000    -8677.7252      -8677.7252        0.0000  ⎭
```

Figure 9.2 Lines 210–222 from the NAMD output file, *smallproteinB_ws_eq.log*. Some line breaks and spaces were added for readability, which is why the number of lines shown above does not equal 13. Each block of three lines beginning with "ETITLE" or "ENERGY" are actually a single line in the file, as noted by the braces. That is, all 16 items in each of those three-line blocks above are found in a single line in the file.

world, lines of text have to end when the edge of the paper is reached and we have to start on the next line. In a file, a line continues until it reaches a newline character. So, a single line in a file can be more or less as many characters as we want. Single long lines, however, are hard to show on a piece of paper. In the "ETITLE" and "ENERGY" blocks of values, even though they are each a single line, we display that single line as three lines. When we read that line using the code in Figure 9.3, all the characters in the 16 fields are read in at once, as a single entity, and are represented as an element in the list `inputstr` in Figure 9.3.

The first label, "TS," corresponds to the **timestep** number of the simulation. A timestep (often abbreviated Δt) is the amount of time between the current state of the model and the next future state of the model, which we calculate based upon the current state of the model. The timestep number is the number of timesteps the model has run through.[2] The timestep

---

[2] Unfortunately, the term "timestep" is used in different ways by different people. Sometimes, it is used to mean the timestep number rather than Δt. We can often figure out the meaning of "timestep" from the context. If someone talks about "the timestep," they usually mean Δt. If someone talks about "at timestep x," they usually mean timestep number x.

```
1    #- Module imports:
2    import numpy as np
3    import matplotlib.pyplot as plt
4
5    #- Read in NAMD log file:
6    filein = open('smallproteinB_ws_eq.log', 'r')
7    inputstr = filein.readlines()
8    filein.close()
9
10   #- Extract and convert TS and POTENTIAL values:
11   ts_list = []
12   pot_list = []
13   for i in range(len(inputstr)):
14       line = inputstr[i]
15       if line[0:7] == 'ENERGY:':
16           line_values = line.split()
17           ts_list.append(float(line_values[1]))
18           pot_list.append(float(line_values[13]))
19
20   #- Convert lists to arrays, calculate time, and plot:
21   deltat = 2
22   time = np.array(ts_list) * deltat
23   pot = np.array(pot_list)
24
25   plt.figure()
26   plt.plot(time, pot)
27   plt.xlabel("Time [sec]")
28   plt.ylabel("Potential Energy [kcal/mol]")
29   plt.title("Potential Energy for a Small Protein")
30   plt.tight_layout()
31   plt.savefig("namd_pe_vs_t.png")
```

Figure 9.3 Code to read the file *smallproteinB_ws_eq.log* and create the plot in Figure 9.4.

number starts at zero and increments by one. Thus, at timestep number seven, the model is at time $7 \times \Delta t$. This is similar to what we saw with our random walk example of a simple prognostic model in Section 8.1, where the walker walked through the domain a step at a time. The location of the walker one step in the future depended upon the current location of the walker, and a step of the walker took a timestep amount of time (though, in that model, we did not give an actual amount of time for the timestep). In this NAMD simulation, the calculations of various kinds of energy are being calculated every timestep and each timestep lasts 2 fs (i.e., 2×10^{-15} s; the "f" means "femto" and corresponds to 10^{-15}). The ENERGY lines in Figure 9.2 show TS values 0, 1, 2, 3, and 4, telling us that these lines give the energy of the molecule at those timestep numbers, corresponding to times of 0, 2, 4, 6, and 8 fs, respectively.

Because the purpose of this NAMD run is to find the stable minimal energy of the protein, we want to see how the potential energy value corresponding to the label "POTENTIAL" (which represents the summation of all of the energies that the simulation calculated based on the force fields provided) in the output file changes with time.[3] When the value of POTENTIAL has leveled off and is more or less constant at time, that will be the stable minimal energy of the protein. The value of POTENTIAL is in units kcal/mol. Thus, our task is to read the TS and POTENTIAL values – i.e., the first and thirteenth numerical values – in the ENERGY lines, for each timestep number, and plot the potential energy versus the time. Based on that, we can evaluate whether our settings for the run (e.g., the positions of the atoms) result in a stable minimal energy of the protein and thus whether this configuration of atoms can be used as a benchmark we can compare future experimental runs (that have perturbations to the enviroment) to.

How will we go about this? We first read in each line of the file, keeping only the ones that begin with "ENERGY:". From those lines, we extract the first and thirteenth numerical values and store those in an array of timestep numbers and an array of potential energy. With those values, we create an x–y plot.

Figure 9.3 shows the code to do the tasks described above. Lines 2–3 import the needed modules. In lines 6–7, we create a file object connected to the file *smallproteinB_ws_eq.log* in read-only mode and read in each line of the file, storing each line as a string and each string as an element in the list `inputstr`. After reading the file, we close the file in line 8.

In lines 11–18, we extract the values in the ENERGY lines corresponding to the TS and POTENTIAL labels. The TS values are stored in the list `ts_list` while the POTENTIAL values are stored in the list `pot_list`. We use lists to store these values because we do not know ahead of time how many of these points there will be. Line 15 tests to see whether the line of the file is an ENERGY line. If so, the values in that line are broken-up in line 16 into a list where each value is an element in that list. Remember that in Figure 9.2, each block of three lines beginning with "ETITLE" or "ENERGY" are actually a single line in the file (it shows as three lines to fit on the paper page). Thus, a single element in the list `inputstr` – which is what `line` is – is a single string that contains all the characters in the block of three lines beginning with "ETITLE" or "ENERGY".

When `line` is broken-up into its pieces and those pieces are stored as elements in `line_values`, the TS value is the second (index 1) element in `line_values` because "ENERGY:" is the first element in `line_values`. Thus, the first numerical value is the second element in `line_values`. Likewise, the POTENTIAL value is the fourteenth (index 13) element in in `line_values`. In lines 17–18, the TS and POTENTIAL string values are converted to floating-point values and appended to the `ts_list` and `pot_list` lists. In lines

---

[3] The electrostatic interaction energy U_e is given by:

$$U_e = k\frac{q_1 q_2}{r}$$

where k is a constant, r is the distance between two charged particles, and q_1 and q_2 are the charge of the particles. An attractive electrostatic interaction energy will thus be negative, because the signs of q_1 and q_2 are different. Thus, the total potential energy of the molecule will be negative because the dominant electrostatic interaction is an attractive force and the electrostatic interaction typically has the greatest impact.

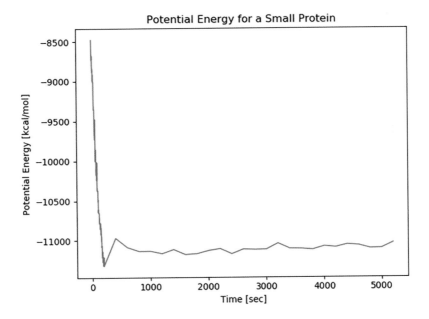

Figure 9.4 Plot of potential energy versus time values from the NAMD model run output found in *smallproteinB_ws_eq.log*. The code to generate the plot is given in Figure 9.3.

21–23, the `ts_list` and `pot_list` lists are used to calculate the arrays `time` and `pot`, which give the time of simulation in fs and the total potential energy in kcal/mol.

Lines 25–31 plot `pot` versus `time` and add appropriate annotation to the graph. We call pyplot's `tight_layout` function in line 30 to make sure the y-axis label does not get cut off.[4] Matplotlib will sometimes do that if the y-axis tick labels are very long, as is the case for the large negative potential energy values of Figure 9.4 and in similar situations with the x-axis tick labels. In line 31, we write out the plot to the image file *namd_pe_vs_t.png*.

9.2 Python Programming Essentials

The example in Section 9.1 illustrates how to use Python to read in data from a file and write out data to a file. The code in the example, however, is also our first detailed example of creating and using objects in Python programming. Thus, in this section, we first introduce what is an object and what does it mean that arrays, lists, and strings are objects. Next, we discuss how to copy variables, data, and objects. With that as background, we address the syntax of reading and writing files and how to catch file opening and other errors in Python.

[4] https://matplotlib.org/users/tight_layout_guide.html and https://matplotlib.org/api/pyplot_api.html#matplotlib.pyplot.tight_layout (both accessed September 5, 2019).

In the rest of the code examples in this section, assume that NumPy has already been imported as np.

9.2.1 Introduction to Objects

So far, nearly all the programming examples we have looked at have been written using what is called **procedural programming**. Procedural programs look at the world in terms of two entities, "data" and "functions." Data consist of numbers or other pieces of information. Functions act on the data. For instance, in this code snippet:

```
1  angle = 2.7
2  sine_of_angle = np.sin(angle)
```

the variable angle holds data or information (the value 2.7) and the function sin takes angle as input, does calculations, and returns the sine of 2.7 radians. That value is then saved in the variable sine_of_angle.

In procedural programming, these two entities, data and functions, are separate from each other. A function takes data as input and returns data as output, just like the sin function above. Additionally, there is nothing customizable about a function with respect to data. The function acts on whatever is passed into it. The function has no way of knowing whether the data that is passed into it is "correct." For instance, let us pretend we want to write code to calculate the sine of the angles in a right triangle:

```
1  angle1 = 2.7
2  angle2 = 0.5
3  sine_of_angle1 = np.sin(angle1)
4  sine_of_angle2 = np.sin(angle2)
```

This code snippet will run without an issue, and the values of the sines will be stored in the variables sine_of_angle1 and sine_of_angle2. But the program does not realize that angle1 is actually incorrect; an angle of 2.7 radians is impossible in a right triangle, as 2.7 radians is greater than $\pi/2$ radians ($= 90°$). The sin function has no way of knowing that angle1 is incorrect in this case. The sin function works with any real number that is passed into it.

In the real world, however, we do not think of things or objects as having these two features (data and functions) as separate entities. That is, real-world objects are not (usually) merely data nor merely functions. Real-world objects instead have both "states" and "behaviors." For instance, people have states (e.g., tall, short) and behaviors (e.g., playing basketball, running), in the same person and often both at the same time. These states and behaviors are also intimately connected to each other. It is because a person has certain states (e.g., their height, particular musculature) that they have certain behaviors (e.g., shooting a basketball), and a person's behaviors make use of their states. And, when a person's behaviors make use of their states, *exactly the right state* is utilized. For instance, in contrast to the sin function example described earlier, when a person shoots a basketball, the person makes use of *their own* height and musculature, not someone else's. This automatically happens in real-world objects because states and behaviors are bound together in the same object.

The aim of object-oriented programming is to imitate this in terms of software, so that "objects" in software have two entities attached to them, states and behavior, just like real-world objects. This makes the conceptual leap from the real world to programs (hopefully) less of a leap and more of a step. As a result, we can more easily implement ideas about and descriptions of the real world into instructions a computer can understand.

An object in programming is an entity or "variable" that has two entities attached to it: data and things that act on the data. In OOP lingo, instead of calling these states and behaviors, the data are called **attributes** (or **instance variables**) of the object, and the functions attached to the object that can act on the object's data are called **methods** of the object. Methods can produce a return value (like a function), act on attributes of the object in place (that is, change the values of the attributes stored in the object), or both. Methods can accept input from outside the object via input parameters or directly from the attributes stored in the object. Importantly, we *design* these methods to act on the attributes. They are not random functions someone has attached to the object. In contrast, in procedural programming, variables have only the value of the variable, with no functions (or other data) attached to the variable.

In the real world, objects are usually examples or specific realizations of some class or type. For instance, individual people are specific realizations of the class of human beings. The specific realizations, or instances, differ from one another in details but have the same pattern. For people, we all have the same general shape, organ structure, etc. In OOP, the specific realizations are called objects or **instances**, while the common pattern is called a **class**. In Section 17.2.1, we go into more detail about how this common pattern – class – is created.

So, in summary, objects are made up of attributes and methods, the structure of a common pattern for a set of objects is called its class, and specific realizations of that pattern are called "instances" or "objects" of that class.

We have referred to entities that hold data as variables and entities that act on the data as functions. This is not incorrect. But, on a deeper level, basically everything in Python, both variables and functions, are actually objects. Let us look at a number of different Python objects to get a sense of how objects work. In the next three subsections, we examine arrays, lists, and strings as objects. In Chapter 17, we will discuss how to create our own classes and further explore the object nature of the language.

9.2.2 Arrays as Objects

Thus far, we have understood array variables as referring to a bank of bins of a given number of dimensions. This is still the case. Arrays, however, are not merely variables that hold data but are objects. As such, array variables (i.e., objects) contain both attributes (that is, information or data) and methods (functions or routines that act on the attributes and information that is passed into the method call).

Attributes   What are some attributes that are attached to array objects? One attribute is called `shape` and stores the shape tuple for the array. Thus, for the following array `data`:

```
In [1]:  data = np.reshape(np.arange(12), (4,3))
         print(data)

         [[ 0  1  2]
          [ 3  4  5]
          [ 6  7  8]
          [ 9 10 11]]

In [2]:  print(data.shape)

         (4, 3)
```

the `shape` attribute is the tuple `(4, 3)`. The syntax for referring to the `shape` attribute is to specify the array object (`data`), put a period, then to specify the attribute name (`shape`). The relationship between an object and its attribute is shown by their placement relative to the period: the object is to the left and the attribute is to the right of the period.

The `shape` attribute for an array is its own separate kind of variable (i.e., object). We said above that the `shape` attribute for an array is a tuple. This means that any reference to `data.shape` is a reference to a tuple. So, if we wanted to print out the first element of the `shape` attribute of the array `data`, this would work:

```
In [1]:  print(data.shape[0])

         4
```

We can also assign an attribute to a separate variable (i.e., object):

```
In [1]:  shape_tuple = data.shape
         print(shape_tuple[0])

         4
```

Attributes can also be set to objects that themselves have attributes. When that happens, refer to the final attribute through a "chain" of attribute references. In Section 6.2.1, we were introduced to typecodes and the `dtype` attribute. That attribute is set to an object that has an attribute `char` which is a string giving the typecode of the array. For the array `data` above in the current section:

```
In [1]:  print(data.dtype.char)

         l
```

The letter "l" means the elements of `data` are "long" integers (at the end of this subsection, we will explain why the word "long" is in quotes). But what is `data.dtype`? The `dtype` attribute of the object `data` is itself an object. The `char` attribute is an attribute of the `dtype` object. At the end of the current section, we discuss array element types and typecodes in more detail. Also, when dealing with objects, there are some nuances regarding what happens if we assign an attribute or object to a separate variable and then change that separate variable. We will address these nuances in Section 9.2.5.

Methods What are some methods that are attached to array objects? Here are a few examples, applied to the array object `data` defined earlier:

```
In [1]:  print(data.sum())

         66

In [2]:  print(data.max())

         11
```

The `sum` method adds up all the elements of `data` and returns that sum. The `max` method returns the value of the largest element in `data`.

Just as we saw with attributes, we refer to the method of an object by giving the object name, a period, then the method name. Because methods are "functions," to use a method we have to call it, just as with a function. Thus, in the two cases above, there are parentheses after the method names, which contain any arguments needed when calling the method. In both cases, the argument lists are empty, because all the information needed to calculate the sum or maximum value of the array elements is already found in the object `data`. This also shows that the giveaway as to whether we are accessing attributes or calling methods of an object is whether there are parentheses after the name. If there are no parentheses, it is an attribute. If there are, we are calling a method.

We could type the name of the method without parentheses at the end, but then the interpreter would just say we specified the method itself, rather than *calling* the method. For instance, in a Jupyter notebook session, we would obtain:

```
In [1]:  print(data.max)

         <built-in method max of numpy.ndarray object at 0x7f7d7fcfc5d0>
```

The above syntax prints the identity and location in the computer's memory of the method itself. It does not actually run the method and execute its contents.

Many of the methods attached to array objects duplicate functions found in NumPy functions, as is the case for the `sum` and `max` methods. For NumPy's `shape` functions, the return value is the same as the input array object's `shape` attribute. It is something of personal preference whether to use the array method (or attribute) version or NumPy function version.

One useful array method that does not have a functional equivalent is the `astype` method. An `astype` call produces a version of the array object where the values of that array are converted into another base type (i.e., the type of each element in the array). The `astype` method requires a single input parameter which specifies the base type of the new array returned by the method. For instance, calling `astype` with the `'d'` typecode as input will create a copy of the array with the values of the elements converted to 8-byte floating-point numbers (each element occupies 8 bytes of memory in the computer):

```
In [1]:   print(data)

          [[ 0   1   2]
           [ 3   4   5]
           [ 6   7   8]
           [ 9 10 11]]

In [2]:   print(data.astype('d'))

          [[  0.   1.   2.]
           [  3.   4.   5.]
           [  6.   7.   8.]
           [  9.  10.  11.]]
```

Typically, we save the return value of `astype` as a separate array:

```
In [1]:   data_float = data.astype('d')
          print(data_float)

          [[  0.   1.   2.]
           [  3.   4.   5.]
           [  6.   7.   8.]
           [  9.  10.  11.]]
```

Listing and Naming Attributes and Methods of an Object   Some objects have a few attributes and methods while other objects have many. How can we list all the attributes and methods of an object? Perhaps the most useful way is to examine the documentation for the object's class, either online or in the Python interpreter. The online documentation is readily accessible via a web search. In the Python interpreter, we can use the `help` command, which works not only for functions (see Section 3.2.2) but also objects. For the `data` array defined earlier in this section, the command:

```
help(data)
```

typed in the Python interpreter will return documentation for the class of object. The first few lines given by `help(data)` look like:

```
Help on ndarray object:

class ndarray(builtins.object)
 |   ndarray(shape, dtype=float, buffer=None, offset=0,
 |          strides=None, order=None)
 |
 |   An array object represents a multidimensional, homogeneous array
 |   of fixed-size items.  An associated data-type object describes the
```

NumPy arrays are of the class `ndarray`. Later on in the documentation we see a section listing the attributes of `data` (here are a few lines from that listing):

```
 |   Attributes
 |   ----------
 |   T  : ndarray
 |        Transpose of the array.
 |   data : buffer
 |        The array's elements, in memory.
 |   dtype : dtype object
 |        Describes the format of the elements in the array.
 |
```

and methods (here are a few lines from that listing):

```
 |   Methods defined here:
 |
 |   __abs__(self, /)
 |        abs(self)
 |
 |   __add__(self, value, /)
 |        Return self+value.
 |
 |   __and__(self, value, /)
 |        Return self&value.
```

Often we know an object has a certain attribute or method, and we have a sense of what its name is, but we cannot remember the name exactly. To obtain a list of just the names of the attributes and methods, use the `dir` command.[5] For the `data` array defined earlier in this section, the command:

```
dir(data)
```

typed in the Python interpreter produces this list of strings (here are the first few and last few lines from that listing, with spaces and line breaks added for readability and ellipses representing the lines in between):

---

[5] https://docs.python.org/3/library/functions.html#dir (accessed January 6, 2021).

```
['T', '__abs__', '__add__', '__and__', '__array__',
 '__array_finalize__', '__array_interface__', '__array_prepare__',
 '__array_priority__', '__array_struct__',

 ...

 'shape', 'size', 'sort', 'squeeze', 'std', 'strides', 'sum',
 'swapaxes', 'take', 'tobytes', 'tofile', 'tolist', 'tostring',
 'trace', 'transpose', 'var', 'view']
```

We can save the output of `dir` to a list. For instance, `contents = dir(data)` will save the above to the list `contents`. In the discussion below, we make reference to some of the lines output by `dir` that are not shown above.

Some attribute and method names above have underscores in front of and/or behind the names whereas others do not. A few of the double-underscores in front of and behind names sound like information of some sort. The `__doc__` variable is one such attribute and refers to documentation of the object. Most of the double-underscores in front of and behind names suggest operations on or with arrays (e.g., `__add__`, `__div__`), which is what they are: those names are methods of the array object that *define* what Python will do to our data when the interpreter sees a $+$, $/$, etc., operator. In Python, an attribute or method with two underscores in front and two underscores behind are "special" and implement operator features, naming, under-the-hood components, etc.[6] The names that have no double-underscores in front and behind are names of "**public**" attributes and methods, i.e., attributes and methods normal users are expected to access or call.

Typecodes   Every element of an array has the same type. When we construct a NumPy array, Python will try to figure out what the type of the array elements should be, in order to properly hold the data, or we can use the `dtype` keyword input parameter (as part of the call to `array`, `zeros`, or another array creation function) to explicitly set what type to use to hold the elements. In scientific and engineering applications, most of the time arrays are used to hold numbers or boolean values, so the integer, floating-point, and boolean types are the most commonly used. We list the NumPy array typecodes for these types in Table 9.1.

In previous discussions on the differences between the integer and floating-point types, we focused on how the former is used to represent whole numbers in a unique way while the latter is used to represent decimal numbers in an inherently inexact way. Let us add some details to that description.

We first address how numbers are represented in memory in the computer. There are two general ways this is done. The first way is to set aside a fixed number of bytes of memory in the computer to represent a single number. The second way is to design an algorithm to represent a number that uses as much memory as needed to represent the number fully. This first way of representing numbers works well with arrays of many numbers, because we know the total number of bytes of memory that will be used by the array equals the number of bytes

---

[6] https://docs.python.org/3/reference/datamodel.html (accessed July 3, 2020).

Table 9.1 **Common NumPy array typecodes.**
Reference: https://docs.scipy.org/doc/numpy/
reference/arrays.dtypes.html (accessed
August 22, 2019).

Base type	Typecode
Boolean	`'?'`
Floating-point (8-byte)	`'d'`
Floating-point (4-byte)	`'f'`
Integer (8-byte)	`'l'`
Integer (4-byte)	`'i'`
String (Unicode)	`'U'`

per element multiplied by the number of elements. The drawback of using a fixed number of bytes to represent a number is that we limit the number of digits – either integer or decimal number – that can be stored in that fixed chunk of memory. The second way of representing numbers enables us to store all the digits of our number of interest, up to the total amount of memory available on the computer. The drawback of this method, however, is that in principle we could use up all the memory on our computer to represent a single value, something that does not work so well if we have a collection of many numbers ☺.

In the Table 9.1 listing of NumPy array data types, all the types listed are examples of using a fixed number of bytes to represent a number. In each type, the number of bytes given is how many bytes of memory are used to store an integer or floating-point element (as applicable). Thus, a 100-element array of 4-byte floating-point type elements will take up 400 bytes of memory in the computer. For boolean arrays – arrays whose elements are boolean values – each element is also fixed in size and occupies 1 byte of memory. Historically, 4-byte floating-point numbers have also been called "**single-precision**" floating-point numbers whereas 8-byte floating-point numbers have been called "**double-precision**" numbers.

The Table 9.1 listing of data types does not include any examples of the second way of representing numbers, where as many bytes of memory as needed are used to represent the number. The built-in integer representation in Python 3.x is such an example. When we assign an integer value to a scalar variable, we are using such a representation:

```
In [1]:  big_number = 8973892759874398740932874398215
         big_number.bit_length()
```

```
Out[1]:  103
```

The `big_number` object has a method `bit_length` that returns the number of **bits** that would be needed to represent the number in binary (a code made up of zeros and ones). In a computer, a single binary digit of a binary number is a bit, and can have the value of zero or one. There are 8 bits to 1 byte of memory. Thus, `big_number` requires over 12 bytes of memory to be represented in binary. The value of big_number, then, cannot be represented in a NumPy array using the 8-byte integer data type.

Because different numerical data types use different amounts of memory, we choose the data type for our arrays depending upon how much accuracy we want to retain and how much memory we have available for the array we are creating. The choice between 4- and 8-byte data types mattered more in the past, when memory was relatively limited. Today, for most applications, it makes more sense to save as many digits as possible by choosing an 8-byte integer or 8-byte floating-point data type for the arrays we make than to conserve memory. Conserving memory usually becomes important only when we have very large arrays.

How can we find out what maximum and minimum values are possible for a given array base type? NumPy provides a function, iinfo, to return this (and other) information for a given integer typecode and finfo for a given floating-point typecode. Here are the values for the 8-byte integer and floating-point types on the computer being used to write this text (line breaks and spaces are added to make the output more readable):[7]

```
In [1]:  np.iinfo('l')

Out[1]:  iinfo(min=-9223372036854775808, max=9223372036854775807,
               dtype=int64)

In [2]:  np.finfo('d')

Out[2]:  finfo(resolution=1e-15, min=-1.7976931348623157e+308,
               max=1.7976931348623157e+308, dtype=float64)
```

Python's built-in floating-point data type (that is, the type that is used when we assign a scalar value to a variable, not one of the floating-point types listed in Table 9.1) also is limited in memory size. Information about the value that can be held in this data type is found in the float_info attribute of the sys module (line breaks and spaces are added to make the output more readable):[8]

```
In [1]:  import sys
         sys.float_info

Out[1]:  sys.float_info(max=1.7976931348623157e+308, max_exp=1024,
                 max_10_exp=308, min=2.2250738585072014e-308,
                 min_exp=-1021, min_10_exp=-307, dig=15,
                 mant_dig=53, epsilon=2.220446049250313e-16,
                 radix=2, rounds=1)
```

---

[7] https://docs.scipy.org/doc/numpy/reference/generated/numpy.iinfo.html and https://docs.scipy.org/doc/numpy/reference/generated/numpy.finfo.html (both accessed July 30, 2019). Technically what is returned is a callable object, but for our purposes we can think of this as behaving as a function.

[8] https://docs.python.org/3/library/sys.html (accessed July 30, 2019).

Finally, sometimes we may find the modifier "long" used to describe either NumPy's 8-byte integer type or the built-in integer type in Python 2.7.x. Why? In Python 2.7.x, there are two built-in integer types. One is often called "int" and allocates 4 bytes of memory to store the integer value. As we have seen, this places a ceiling on the value that can be stored in a variable of that type, and the value of the maximum integer that can be stored as an "int" in Python 2.7.x is given in the sys module's `maxint` attribute. The second integer type in Python 2.7.x is called "long" and can store integers of any size, subject to available computer memory.[9] In Python 3.x, there is only one built-in integer type, often just called `int`.[10] Confusingly, this type is the same as the built-in "long" type in Python 2.7.x, and, as we saw above, can store arbitrarily large or small integers. Even more confusing, the NumPy 8-byte integer type is represented by the C `long` data type, and so in some contexts the Python 8-byte integer type might be referred to by the term "long." The bottom line: If we use Python 3.x, and refer to data types by their byte size (or by their number of bits, which is eight times the number of bytes), we will avoid most of the confusion.

More advanced information, with discussion of other kinds of element data types, is found in the NumPy documentation.[11]

9.2.3   Lists as Objects

In most of the examples we have seen so far using lists, we have used lists like arrays, to hold numbers. Lists, however, are more flexible than arrays. In particular, we can use lists to hold collections of any kind of object, and the elements in a list do not have to be the same type. Here is an example of such a list `record1`:

```
record1 = ['Seattle', 67.1, 43.2]
```

that stores daily high and low temperature (in °F) for a day at a given location and the name of the city where the measurements were taken (the values above are fictitious). The first element is a string while the last two elements are decimal floating-point numbers. When we use elements from this list, the elements will have the properties of their respective type. Thus, in these lines of code using `record1`:

```
In [1]:  print('City: ' + record1[0])

         City: Seattle

In [2]:  print('High minus low: ' + str(record1[1] - record1[2]))

         High minus low: 23.89999999999999
```

[9] https://docs.python.org/2.7/library/stdtypes.html (accessed July 27, 2019).
[10] https://docs.python.org/3.7/library/stdtypes.html (accessed July 27, 2019). See also www.geeksforgeeks.org/what-is-maximum-possible-value-of-an-integer-in-python/ (accessed July 30, 2019).
[11] https://docs.scipy.org/doc/numpy/user/basics.types.html and https://docs.scipy.org/doc/numpy-1.13.0/reference/arrays.dtypes.html (both accessed July 30, 2019).

Table 9.2 When (in general) to use lists versus arrays.

Situation	List or NumPy array?
Change the number of elements in the collection	List
Store only numbers	Array
Store only boolean values	Array
Store different kinds of objects	List
Store strings	List
Multidimensional array-like structure	Array

The use of `record1[0]` in cell `In [1]` is the use of a string. The use of `record1[1]` and `record1[2]` in cell `In [2]` are uses of floating-point numbers, which is why we can use the subtraction operator in its arithmetic sense between the two values. In the above example, the list `record1` holds values such as numbers or strings. But lists are not limited to what appear to us as "values." Lists can hold any kind of object, including other lists.

As an aside, what we see above for what Python prints out for `67.1` minus `43.2` is not a typo. As we discussed in Section 8.2.7 and above in the typecode discussion of Section 9.2.2, floating-point representations are inherently inaccurate because a limited amount of memory is used to store the value. Even if a real number "looks simple," with no repeating decimals, etc., the floating-point type cannot be assumed to represent the number exactly. Thus, the difference between floating-point numbers cannot be assumed to be exact. Python's representation of integers, via the `int` type, are exact and operations between `int` type variables are exact.

Lists also differ from arrays in that the number of elements in a list can increase or decrease, whereas the number of elements in an array is fixed when we create the array. We will see in a moment how to increase or decrease the number of elements in a list.

As a general rule, if we have collections of values that are all one type (e.g., integers), NumPy arrays are the better way to store those values. Python processes arrays faster than lists. When we have a collection of values of different types, lists are the better way to go. Lists are also usually better if we want to store objects that are neither numbers nor booleans, even if the objects all have the same type. Arrays can be used with strings and other objects, but lists are often more straightforward for this use case. Finally, if we want to create a multidimensional array-like structure, arrays are generally better to use than lists. The syntax of referencing elements and subarrays in arrays is clearer than in lists. Table 9.2 summarizes when to use lists or arrays for some common cases.

Lists, like arrays, are objects. Lists also have attributes and methods, but the list attributes are not all that useful for most users. There are a number of useful list methods, however. The `append` method can be used to add elements to the end of a list. This is a common way of making the list "grow." Thus, if we wanted to add the day's average relative humidity measurement (as a percent) to our `record1` list above, we would type:

```
record1.append(55.4)
```

Here is the Jupyter notebook session defining `record1` and appending to it:

```
In [1]:  record1 = ['Seattle', 67.1, 43.2]
         record1.append(55.4)
         print(record1)


         ['Seattle', 67.1, 43.2, 55.4]
```

Note that the call to `append` does not return anything. The addition of the 55.4 value is done "in-place." That is, the value is added to the end of `record1` without requiring the use of an assignment statement.

To remove and return the last element of the list, we can use `pop`:

```
In [1]:  record1 = ['Seattle', 67.1, 43.2]
         tmp = record1.pop()
         print(tmp)


         43.2


In [2]:  print(record1)


         ['Seattle', 67.1]
```

Note a call to `pop` does two things. First, it returns the value of whatever is at the end of the list. Second, it changes the list in-place to remove whatever was the last element.

The `insert` method places a value wherever we would like and moves the elements to the right of that location over one place to make room for the new value. The following puts the relative humidity before the high-temperature value:

```
In [1]:  record1 = ['Seattle', 67.1, 43.2]
         record1.insert(1, 55.4)
         print(record1)


         ['Seattle', 55.4, 67.1, 43.2]
```

The first argument in the call to `insert` gives the index of the location to put the value given in the second argument in the call to `insert`. Note that the use of `insert` is not the same as using element index and assignment:

```
In [1]:  record1 = ['Seattle', 67.1, 43.2]
         record1[1] = 55.4
         print(record1)


         ['Seattle', 55.4, 43.2]
```

In this example, the index 1 element is replaced by 55.4. The `insert` method does not replace any existing values but moves the existing values over to make room for the new value.

9.2.4   Strings as Objects

Although strings, like numbers, look like "values," they are also objects. Strings have attributes and methods, but the string attributes are not that useful for most users. There are a lot of useful string methods. Here we look at a few, but because the manipulation of characters from text files is the manipulation of strings, we will see a number of these methods throughout this chapter.

Methods   There are a number of methods that work on the case of characters in the string object. The upper method returns a copy of the characters in the string object where all the letters are cast into uppercase:

```
In [1]:  plot_title = "Austin's Plant Experiment 23"
         title_upper = plot_title.upper()
         print(title_upper)


         AUSTIN'S PLANT EXPERIMENT 23
```

There is no effect on nonalphabetic characters. The lower method returns a copy of the characters in the string object where all the letters are cast into lowercase. The title method returns a copy where the first letter of every word is cast into uppercase. Given the value of plot_title above:

```
In [1]:  title_lower = plot_title.lower()
         print(title_lower)


         austin's plant experiment 23


In [2]:  renew_title = title_lower.title()
         print(renew_title)


         Austin'S Plant Experiment 23
```

The title method considers the character after the apostrophe to be a new "word," and so it capitalizes the 's'. The argument lists for each of the method calls are empty. There is nothing between the parentheses for the method calls. All the information the methods need to know about to work properly – the characters stored in the string object – are already in the string object, so there is no need to pass anything into the method call.

The strip method removes **whitespace** before and after a string. In this case, whitespace includes newline characters, tabs, and similar characters:

```
In [1]:  plot_title = "   Austin's Plant Experiment 23\n   "
         print(plot_title)


            Austin's Plant Experiment 23
```

```
In [2]:  title_stripped = plot_title.strip()
         print(title_stripped)
```

```
Austin's Plant Experiment 23
```

There is an extra blank line in the printed output from In [1] that is not in the printed output from In [2]. The `strip` method does not remove *all* whitespace from a string, just the leading and trailing whitespace. Often, the fields of a text data file will contain spurious blank spaces, tabs, and newline characters. The `strip` method is a handy way of getting rid of these characters, without having to know ahead of time how many extra leading and trailing blank spaces, etc., there are.

The `split` method takes a string and returns a copy of it as a list. The method assumes that the elements of the list are the substrings in between a specified separator character. By default, if the argument list is empty, the separator character is considered to be any whitespace character:

```
In [1]:  plot_title = "Austin's Plant Experiment 23"
         title_as_list = plot_title.split()
         print(title_as_list)
```

```
["Austin's", 'Plant', 'Experiment', '23']
```

Whitespace characters include the blank space but also characters such as the tab character and newline character (those characters were discussed in Section 4.2.3).

We can also use `split` to split on a nonwhitespace character by passing it in. Given `plot_title` as defined above, the following shows the list obtained by splitting on the letter 't':

```
In [1]:  title_as_list = plot_title.split('t')
         print(title_as_list)
```

```
['Aus', "in's Plan", ' Experimen', ' 23']
```

The character we are splitting on does not end up in the substrings that become the elements of the list, because we are assuming the only role of those characters is to separate the substrings that become the elements of the list.

The `join` method can be thought of as the "undoing" of split. This method takes a list as input and joins the elements of the list together using the characters of the string object calling the method. For instance, given the list `title_as_list`:

```
In [1]:  title_as_list = ["Austin's", 'Plant', 'Experiment', '23']
         renew_title = ' '.join(title_as_list)
         print(renew_title)
```

```
Austin's Plant Experiment 23
```

The `join` method being called in the second line of `In [1]` is *not* a method of the list `title_as_list`. Instead, it is a method of the string `' '`, that is to say, a method of the blank space character (there is a space between the apostrophes; this is not the empty string). When we specify a string value, we are automatically creating a string object, and that object has attributes and methods like any other object. To see this more clearly, if we want to create a title using `title_as_list` as defined above, but to have the words separated by the word "GREAT", we would do the following:

```
In [1]:  renew_title = 'GREAT'.join(title_as_list)
         print(renew_title)
```

```
Austin'sGREATPlantGREATExperimentGREAT23
```

Whatever string the `join` method being called is attached to will be the string that connects the substrings in the list being passed into the `join` call. As a final example, to put each substring in the `title_as_list` list as defined above on separate lines, we can do the `join` call using the newline character:

```
In [1]:  renew_title = '\n'.join(title_as_list)
         print(renew_title)
```

```
Austin's
Plant
Experiment
23
```

The `print` command renders the newline character as a carriage return on the screen.

Formatting Values in a String   String objects also have the `format` method that enables us to create complex strings that include the substitution of values, nicely formatted.[12] For instance, consider again this definition of `record1`:

```
record1 = ['Seattle', 67.1, 43.2]
```

listing the fictitious daily high and low temperature (in °F) for a day in Seattle. Let us say we want to write the city name and the high minus low temperature, as we saw earlier in Section 9.2.3, but instead of writing out the difference between the high and low temperatures using the default `print` settings that prints out a lot of decimal places, we want to display the digits to the tenths place. That is, we want the string to encode characters that would provide this output once `printed`:

```
1  City: Seattle
2  High minus low: 23.9
```

---

[12] The information in this section is based on https://docs.python.org/3.7/library/string.html (accessed August 3, 2019).

This code will create the string we want (saved as the object `text`):

```
1  text = "City: {0}\nHigh minus low: {1:4.1f}".format('Seattle',
2                                                       67.1 - 43.2)
```

The key idea behind what is happening is that the string we create and save in `text` is the string specified by the characters in the double quotes, that is, the string to the left of the period before the `format` call, *except* that the parts of this "source string" with curly braces are replaced by the values in the argument list of the `format` call. Put another way, when we construct `text`, every character in the double quotes is put into `text` except the `{0}` and `{1:4.1f}` portions. For those portions of the string (which are called "**replacement fields**"),[13] we substitute the values in the argument list of the `format` call. We substitute the string `Seattle` for `{0}` and the string `23.9` for `{1:4.1f}`.

When `format` is called, the curly braces are not considered part of `text`. The only purpose of the curly braces then is to tell `format` where in the source string to make the substitution of values. Also, note that the source string contains a newline character. That character becomes part of `text` as-is and creates the return in the display when we execute `print(text)`.

How does the substitution occur? The characters inside the curly braces follow a code that tells `format` how to do the substitution:

- `{0}` means "replace me with the first (index 0) argument in the `format` calling list and format using default rules."
- `{1:4.1f}` means "replace me with the second (index 1) argument in the `format` calling list, and when doing so, format the number being passed in from the calling list so that the field is 4 characters wide (counting the decimal point) and has 1 decimal to the right of the decimal point."

The above codes follow these rules:

- The first number after the left curly brace tells us which value in the `format` argument list to use. The first argument in the argument list is addressed as 0, the second as 1, and so on.
- If only a number is given in between the curly braces, we use default conversion rules to format the value being used from the `format` argument list. In the case of strings being passed in, the default behavior is to include that string as-is into the source string.
- If we want to control the formatting of a numerical value being passed in, we put a colon after the first number after the left curly brace, and after the colon we put a value-formatting control code.
- The value-formatting control code for floating-point numbers is made up of: a number specifying the width of the field, a period, a number specifying the number of digits to

---

[13] https://docs.python.org/3.7/library/string.html#format-string-syntax (accessed August 23, 2019).

show after the decimal point, and a letter that tells `format` to write the number as a regular decimal number. Some format code letters include:

- d Format number as an integer.
- e Format number in scientific notation.
- f Format as a floating-point number.

Given these above rules for format codes, let us look at some other examples to see what happens with different values in the following code snippets:

- Example 1

```
In [1]:  text = "Mass: {0:8.2f} kg".format(6.9)
         print(text)


         Mass:      6.90 kg
```

The width of the floating-point number is set to eight characters. There are now five blank spaces between the `Mass:` and the `6.90`. Four of those blank spaces are due to the width of the floating-point number being set to eight characters (`6.90` is four characters so a total width of eight characters requires four blank spaces). The fifth blank space is the blank space between the colon after "Mass" and the left curly brace.

- Example 2

```
In [1]:  text = "Mass: {0:.2e} kg".format(6.9)
         print(text)


         Mass: 6.90e+00 kg
```

No width is explicitly set for the floating-point number (there is no number between the colon and period in the code between the curly braces). The `'2'` after the period in between the curly braces results in there being two decimal places in the factor in front of the scientific notation exponent. The letter `'e'` at the end of the code (to the left of the right curly brace) tells `format` to express the value in scientific notation.

- Example 3

```
In [1]:  text = "Mass: {0:d} kg".format(7)
         print(text)


         Mass: 7 kg
```

The letter `'d'` at the end of the code (to the left of the right curly brace) tells `format` to express the value as an integer. This requires the input argument to the `format` call is of integer type.

- Example 4

```
In [1]:  text = "Mass: {0:.0f} kg".format(6.9)
         print(text)
```

```
Mass: 7 kg
```

No digits after the decimal point are retained, so the value 6.9 is rounded up and displayed as an integer. Note that using `'d'` for a `format` input argument that is floating-point type will produce an error. So, although this example and the one above produce similar results, they operate on different types of input values.

In the above examples, we have provided explicit values or expressions using values as the arguments to `format`. This does not mean we cannot use variables instead. This code:

```
In  [1]:  record1 = ['Seattle', 67.1, 43.2]
          diff = record1[1] - record1[2]
          text = "City: {0}\nHigh minus low: {1:4.1f}".format(record1[0], diff)
          print(text)
```

```
City: Seattle
High minus low: 23.9
```

works just as before.

Finally, if the `format` argument list index is empty, the replacement fields and the `format` arguments are mapped based upon position in the string and argument list, respectively. If the formatting code is also absent, default formatting is used. Thus, we have the following:

```
In [1]:  record1 = ['Seattle', 67.1, 43.2]
         diff = record1[1] - record1[2]
         text = "City: {}\nHigh minus low: {}".format(record1[0], diff)
         print(text)
```

```
City: Seattle
High minus low: 23.89999999999999
```

when empty replacement fields ({}) are used. If we want to use a `format` argument in multiple places in the string with the replacement fields, we have to make use of the argument list index to specify which argument we want to use.

The `format` method has many more ways of customizing the formatting of replacement values in a string, including automatically dealing with percentages and left, right, or center justifying values in a field. Please see the Python documentation on the string module for details. We provide an up-to-date link to this documentation at www.cambridge.org/core/resources/pythonforscientists/refs/, ref. 18.

9.2.5 Copying Variables, Data, and Objects

In the physical world, the act of naming is usually unique and the act of copying produces a unique copy. This is to say, if we had two teddy bears and we named one Fluffy and the other Furry, even if the two bears were the same brand and model, each name would apply to only one bear and not the other. The names are unique identifiers to a specific bear. Likewise, if we were to take a book and make a copy of it, the copy would be unique and independent. If we tore out a page from the first book, the second book would not automatically lose the same page.

In the world of the computer, the acts of naming and copying are not necessarily unique. In some cases they are while in others they are not. Consider this example:

```
In [1]:  x = 24
         y = 24
         x = -17
         print(x)


         -17
```

```
In [2]:  print(y)


         24
```

We first assign the variable x and y to the same value, 24. Just as in the teddy bear example, these names are unique to the specific "24" they are assigned to, though they refer to the same value. Thus, when we reassign x to a different value, -17, the value of x changes but the value of y does not.

The same thing happens if we change the assignment of y to be x:

```
In [1]:  x = 24
         y = x
         x = -17
         print(x)


         -17
```

```
In [2]:  print(y)


         24
```

In the second line of In [1], when y is assigned to x, y is assigned to the value of x, i.e., 24. Thus, when we change x to -17, this does not affect the value of y.

Let us try something similar with lists:

```
In [1]:  x = [24, 25, 26]
         y = [24, 25, 26]
         x[0] = -17
         print(x)


         [-17, 25, 26]

In [2]:  print(y)


         [24, 25, 26]
```

In the third line of `In [1]`, we change the value of the first element of x. The result is x is changed but y is not. This matches our teddy bear intuition.

Now we change how y above is assigned:

```
In [1]:  x = [24, 25, 26]
         y = x
         x[0] = -17
         print(x)


         [-17, 25, 26]

In [2]:  print(y)


         [-17, 25, 26]
```

Whoa! What happened? By assigning y to x (second line, `In [1]` immediately above), we created a situation where any change to the elements in x also changes the same elements in y. There is something different about using y = x with integer values than with lists.

That something is the distinction between **assignment by value** versus **assignment by reference**. When we assign a variable by value, we are putting a copy of that value into a new memory location and saying that the variable refers to that copy. If another variable is the source of that copy, that variable's value will be separate from the variable that refers to the copied value. In contrast, when we assign a variable by reference, we are creating an "alias" to whatever the variable is being assigned to. That alias is merely another name to that original entity, so if we use *either the alias or the original entity's name* to change the entity, the entity is changed, whether we refer to it by the original's name or the alias.

Functionally speaking, as we have seen above, Python generally decides whether to use assignment by value versus reference in the following ways:

- When we assign a variable by *typing in the data* (e.g., an integer, a decimal number, a string, the actual values of a list), the variable is assigned by value (i.e., to a unique copy in memory of that data).
- When we assign a variable by *making a simple reference to another variable* (i.e., the right-hand side of the equal sign is only another variable).
  - *If the right-hand-side variable is a plain value*, such as an integer, a floating-point number, or a string, the left-hand-side variable is assigned by value (i.e., is a separate copy of the right-hand-side variable).
  - *If the right-hand-side variable in an object is more "complicated" than a plain value*, it is assigned by reference (i.e., the left-hand-side variable is an alias to the right-hand-side variable). The example above where we assigned a variable to a list and changes to the original affected the new variable is an example of this kind of assignment.
- When we assign a variable to the *result of an expression* on the right-hand side of the equal sign, the variable is assigned by value in the sense that it is set to a unique instance in memory of that result. Thus, later changes to the variables that make up the expression on the right-hand side do not affect the values of the variable assigned earlier.

Here is an example of the last bullet point above: When we create result arrays using array syntax operations, those result arrays are separate from the arrays that went into creating the result. Thus, a change to the input arrays will not alter the result array:

```
In [1]:  x = np.array([-1, 4, 2])
         y = np.array([ 3, 6, 9])
         z = x + y
         print(z)


         [ 2 10 11]

In [2]:  x[0] = -20
         print(x)


         [-20   4   2]

In [3]:  print(y)


         [3 6 9]

In [4]:  print(z)


         [ 2 10 11]
```

In the above example, the array z is the sum of the arrays x and y, done element-wise. In the first line of In [2], we change the x array's first element, but because that change happened *after* the creation of z, the change to x has no affect on z.

Because assignment by reference occurs when we assign a list, NumPy array, or other more complex objects to a variable, we need another means of making a memory-independent copy of such objects. Often in science and engineering programming problems, we will want to make this kind of copy in order to manipulate one version of the data without affecting the other version. Here are two ways to do so.

The first way is to use the `copy` method, which is part of list, array, and many other kinds of objects:

```
In [1]:  x = np.array([-1, 4, 2])
         y = x.copy()
         x[0] = -20
         print(x)


         [-20   4   2]

In [2]:  print(y)


         [-1   4   2]
```

The call to the `copy` method in x produces a memory-independent copy of x which we save as y. Changes in x will not affect y (and vice versa).

The second way is to use the `copy` or `deepcopy` functions that are part of the built-in copy module:

```
In [1]:  import copy
         x = np.array([-1, 4, 2])
         y = copy.copy(x)
         x[0] = -20
         print(x)


         [-20   4   2]

In [2]:  print(y)


         [-1   4   2]
```

The `deepcopy` function works the same way as above, but does a more "complete" job of copying very complex objects. The `deepcopy` function takes longer to make the copy, so the `copy` function works better for "normally" complex objects like lists and arrays. For objects that are instances of user-written classes, the `deepcopy` function is the more reliable way of making sure the entire copy is memory-independent from the original and is preferred even though it takes longer to make the copy. See Section 17.2.1 for more on user-written classes.

In our Figure 9.3 code from the example in Section 9.1, we did not have to worry about making memory-independent copies because all of the variables in our code used in one stage are set in a previous stage and are not changed afterwards. For instance, although in line 11 we create an empty list `ts_list` and append to it in line 17, after `ts_list` is complete, we only use the list (e.g., in line 11 to create the `time` array). We make no further changes to `ts_list`. In that sense, once a variable is "finished," it acts as a "read-only" entity. We read its values and do not change them. As an aside, if we had changed `ts_list` after line 22, when we created `time`, the value of `time` would not also change. Using the `array` function creates an array that is memory-independent from the source list for the array.

Finally, the rules we described in this section governing variable assignment also apply when the "assignment" occurs when objects are passed into functions and methods as part of the calling argument list. In those cases, the parameter inside the function/method will either be assigned to the argument by value or reference, following the rules of variable assignment. Thus, if we have a function:

```
def calculate(data):
    data[0] = 34
```

and we pass in a list for the parameter `data`, the change the function makes on the first element of `data` will be reflected in the list *outside the function* that was passed in:

```
In [1]:  values = [1, 5, -4]
         calculate(values)
         print(values)

         [34, 5, -4]
```

The function parameter `data` is assigned to the argument `values` by reference, not by value, so the change to `data` also changes `values`.

9.2.6   Reading and Writing Files

With the above background regarding objects and strings, we are ready to describe how to use file objects to read in and write out data to and from text files. In this section, we look at different Python constructs for doing so. With the file object, we read in the data from the file as strings of text (letters, numerals, and other characters) and use string methods to slice-and-dice the text into a useful form for scientific and engineering calculations. Another construct, the `with/as` construct, also uses file object methods, but has a simpler syntax for accessing and closing the file. Many real-world computational scenarios write out data into a specific format, so for those formats there are prewritten routines to read in the data from the text file into an array using a single line of code. In the last subsection of this section, we look at two modules/routines that are useful for text files of scientific and engineering data.

Using File Objects to Read and Write Files

A file object is a "variable" that represents the file to Python. File objects are created like any other object in Python, that is, by assignment. For text files, we **instantiate** (that is, create an instance of) a file object with the built-in `open` statement. For instance, the following line of code creates a file object `fileobj` that is connected to the file *data.txt* for reading-only (that is, to obtain data from the file but not write data out to the file):

```
fileobj = open('data.txt', 'r')
```

The first argument in the `open` statement gives the filename. The second argument sets the mode for the file:

- `'r'` for reading-only from the file,
- `'w'` for writing to a file, and
- `'a'` for appending to the file.

Once we have created the file object, we use various methods attached to the file object to interact with the file (described below). When we are done reading from or writing to a file, we close the file object using the `close` method. Thus, to close the file object `fileobj`, we execute:

```
fileobj.close()
```

To read one line from the file, we can use the file object's `readline` method. For instance, for the `fileobj` text file object, this code would read a single line of the file and save its contents as the string `aline`:

```
aline = fileobj.readline()
```

The variable `aline` contains the newline character, because each line in a file is terminated by the newline character. As we saw in our example in Section 9.1, a line in a file consists of all characters up to and including a newline character. That is, lines in a file are separated by newline characters. It does not matter how many characters are in a line. So, the visual representation of a file on a piece of paper may not accurately represent where lines are breaking in the file. Python cares about what is in the file, not how it might be displayed on a page.

To read the rest of a file that we have already started reading, or to read an entire file we have not started reading, and then put the read contents into a list, we use the `readlines` method:

```
contents = fileobj.readlines()
```

Here, `contents` is a list of strings, and each element in `contents` is a line in the `fileobj` file. Each element also contains the newline character, from the end of each line in the file.

The variable names `aline` and `contents`, as is the case most of the time with variable names in Python, are not special. We can use whatever valid variable name we would like to hold the strings we are reading in from the text file. It is, however, better to choose a variable name that is descriptive of what the variable refers to.

To write a string to the file that is defined by the file object `fileobj`, we use the `write` method attached to the file object:

```
fileobj.write(astr)
```

Here, `astr` is the string we want to write to the file. The `write` method does *not* automatically write a newline character to the file after the string is written. If we want a newline character to be added, we have to append it to the string prior to writing (e.g., `astr + '\n'`). Also, the `write` method does not have a return value. The action of the `write` method is to write data into the text file `fileobj` refers to, not to return a value to the Python interpreter for the interpreter to do calculations with.

To write a list of strings to the file, we use the `writelines` method:

```
fileobj.writelines(contents)
```

Here, `contents` is a list of strings, and, again, a newline character is *not* automatically written after the string. Thus, we have to explicitly add the newline character to each element of the list of strings if we want that character written to the file. The `writelines` method also does not have a return value, since the action of `writelines` is to act on the file, not to provide a return value to manipulate in the Python interpreter.

As a small example, we write out the contents of the following table:

Masses
5.4
9.2
3.7

to the file *masses.txt*, read the values back in, and then check the values are the same. This code will do what we want:

```
In [1]:  data = ['Masses\n', '5.4\n', '9.2\n', '3.7\n']
         fout = open('masses.txt', 'w')
         fout.writelines(data)
         fout.close()

         fin = open('masses.txt', 'r')
         data_in = fin.readlines()
         fin.close()

         print(data)
         print(data_in)


         ['Masses\n', '5.4\n', '9.2\n', '3.7\n']
         ['Masses\n', '5.4\n', '9.2\n', '3.7\n']
```

Each element of `data` has to be a string and has to terminate with a newline character, if we want to have each element on a separate line in the file *masses.txt*. When we read in the contents of the file, the newline characters also come in, as the contents of `data_in` show.

Each of the "numerical" elements of `data` and `data_in` are not numbers but strings consisting of numerals, a decimal point, and a newline character. If we want to do mathematical calculations with these elements, we need to first convert them to floating point using the

```
-1.4     4.2    19.2
 1.3    -2.1     9.4
18.0    11.3    -6.3
11.1    -3.9     4.8
```

Figure 9.5 Contents of a three-column file of numbers, *three-cols.txt.*

float function. If they expressed integers, we could use int to do the conversion. Also, when using float, we do not have to worry about removing the newline character; it is automatically removed during the conversion. Likewise, before writing out a floating point or integer number to a file, the number needs to be converted to a string (e.g., by str). We discuss more examples of this conversion in the next sections.

In the Section 9.1 example, we read in all the input of the file in lines 6–8 of Figure 9.3, almost identical to our *masses.txt* example above.

Using String Methods to Process File Content

Let us say we have read in the contents of a file from the file and now have the file contents as a list of strings. How do we manipulate them? In particular, how do we turn them into numbers (or arrays of numbers) that we can analyze? We saw earlier that Python has a host of string manipulation methods, built into string variables (a.k.a., objects). Here are a few examples of using these methods for dealing with the contents of scientific and engineering text data files.

Breaking Up Columns from a Line of Text The readline and readlines methods read in a line of text from a file as a single string. This means a line from a file consisting of multiple columns of numbers will consist of the number of columns worth of numbers separated by blank spaces (or commas, tabs, or some similar character or set of characters separating the columns). For instance, let us say we have the file given in Figure 9.5. The three columns are separated by blank spaces, but the number of blank spaces between each column differs from line to line because of the presence or absence of negative signs as well as the number of digits to the left of the decimal point.

If we read in the first line of the file using readline and save the line as the variable aline, we can use the split method to obtain a list with each column's value as a separate element in the list, with syntax like the following:

```
In [1]:   aline_list = aline.split()
          print(aline_list)

          ['-1.4', '4.2', '19.2']
```

When `split` is called with no arguments, the method removes all occurrences of all whitespace of any kind between the nonwhitespace characters of the string. Note that `aline` is a string and `aline_list` is a list of strings. At this point, none of the values are converted into numbers we can use in mathematical calculations.

Once we have this list of strings, we can convert the values to arrays of numbers, which we can then use to make calculations. Here are two ways of doing so:

- Looping. If we loop through a list of strings, we can use the `float` or `int` functions on each string to obtain a floating point or integer number, respectively. For instance:

```
aline_list = ['-1.4', '4.2', '19.2']
aline_num = np.zeros(len(aline_list), 'd')
for i in range(len(aline_list)):
    aline_num[i] = float(aline_list[i])
```

takes a list of strings `aline_list` and turns it into a NumPy array of 8-byte floating-point numbers `aline_num`. We can specify the array datatype without actually writing the `dtype` keyword; NumPy array constructors like `zeros` will understand a typecode given as the second positional input parameter.

- The `astype` method. If we convert the list of strings into a NumPy array of strings, we can use the `astype` method for type conversion to floating point or integer. For instance:

```
aline_list = ['-1.4', '4.2', '19.2']
aline_num = np.array(aline_list).astype('d')
```

takes a list of strings `aline_list`, converts it from a list to an array of strings using the `array` function, and turns that array of strings into an array of 8-byte floating-point numbers `aline_num` using the `astype` method of the array of strings.

Earlier, in Section 9.2.3, we had said that, in general, strings are better stored in lists than arrays. The temporary creation of an array from `aline_list` in line 2 above (of the second bullet) is an exception to this rule.

In the Section 9.1 example, in lines 13–18 of Figure 9.3, we loop through all the elements of `inputstr`, extract the TS and POTENTIAL portions of the lines beginning with ENERGY, and convert those string numerals into floating-point values by calling `float` (lines 17 and 18).

Joining Columns for a Line of Text   We encounter the opposite situation when we have multiple values – say three numbers – that we want to concatenate together as a single string, that we will then write out as a single line of the data file. For instance, say we had the above list `aline_list` assigned to the above string values by:

```
aline_list = ['-1.4', '4.2', '19.2']
```

To create the string `aline2` that contained the values in `aline_list` but separated by the tab character, we can use the `join` method:

```
In [1]:  aline_list = ['-1.4', '4.2', '19.2']
         aline2 = '\t'.join(aline_list)
         print(aline2)
```

```
-1.4     4.2      19.2
```

The "\t" character is the tab character and is a string like any other. As a result, it also has the `join` method. In the second line of `In [1]`, we call the `join` method of the tab character, which takes the elements of the string list `aline_list` and concatenates the elements together using the tab character.

In Try This! 9-9 and 9-11, we look at the solutions for reading and writing single and multiple-column files of numbers using the above string methods. In our Figure 9.3 chapter example, we used the `split` method to break up the fields from a line of ENERGY values in line 16. We restricted the `split` operation only to ENERGY values by putting the call in an `if` statement defined by line 15. Our Figure 9.3 does not have an example of using `join` to create a line of values for an output file.

Using the `with`/`as` Construct to Read and Write Files

In nearly all the examples in this chapter, we explicitly created a file object and explicitly closed a file object. The `with`/`as` construct provides another way of doing so, more concisely. Earlier in this section, we looked at a simple example of reading in the file *masses.txt* (reproduced here for convenience):

```
fin = open('masses.txt', 'r')
data_in = fin.readlines()
fin.close()
```

Using the `with`/`as`, we can write these lines as:

```
with open('masses.txt', 'r') as fin:
    data_in = fin.readlines()
```

The token after the `as` is the file object name and is used just as if we had created the file object using assignment. After the `with`/`as` body ends, the file is automatically closed. Thus, with the `with`/`as` construct, we do not have to explicitly call the `close` method of a file object. We can use the `with`/`as` construct for writing out to a file too.

In this example, it appears the savings is only a single line of code, but in more complicated scenarios involving potential file manipulation errors (which are introduced in Section 9.2.7), the `with`/`as` construct can make the code substantially more concise.[14] The `with`/`as`

---

[14] https://docs.python.org/3/tutorial/inputoutput.html (accessed July 13, 2020).

construct is comparatively new to Python, however. As scientific code tends to progress relatively slowly, it is useful to learn how to explicitly open and close file objects. In the present work, we will usually use file objects the traditional way.

Better Ways of Reading a File

The output file in the example of Section 9.1 required us to write a custom program to read in that file. Often, however, we can use a prewritten function to read in a text file because many files are in a common format. In this section, we mention the NumPy function `genfromtxt` and the csv module. The first works well for text files made up only (mostly) of numbers, and the second works well for text files made up of comma-separated values.[15]

Files with Only (Mostly) Numbers   The `genfromtxt` function works well for files that are mostly filled with numbers. How would we use `genfromtxt` to read the data file from Figure 9.5? The code below will produce the two-dimensional array `data` from that file:

```
from numpy import genfromtxt
data = genfromtxt('three-cols.txt', dtype='d')
```

If we run the above lines of code and print out the contents `data`, we obtain:

```
In [1]:  print(data)

         [[-1.4   4.2 19.2]
          [ 1.3  -2.1  9.4]
          [18.   11.3 -6.3]
          [11.1  -3.9  4.8]]
```

The `dtype` keyword input parameter tells the function to convert the file's values to 8-byte floating-point numbers. If left undefined, the `genfromtxt` function will still convert the numerical values to 8-byte floating-point type by default, so setting this keyword input parameter is not necessary, but if we wanted the result array `data` to be a specific type, setting `dtype` is the way to do it. Try This! 9-11 and Section 12.1 discuss some additional keyword input parameter settings for a `genfromtxt` call.

Files with Comma-Separated Values   Comma-separated values (CSV) files refer to text data files where the columns are separated by commas. CSV files are found not only in science and engineering but also business, finance, and other fields. We might think, reasonably so, that to read in a CSV file, all we need to do is apply the string method `split` using the comma (`','`) as the delimiter. This would work fine if CSV files contained only numbers. Actually, CSV files often have fields that have newline characters or commas inside the fields. For instance,

---

[15] For a discussion of many other prewritten ways of reading in input, see the SciPy Cookbook's "Input and output" page. We provide an up-to-date link to this page at www.cambridge.org/core/resources/pythonforscientists/refs/, ref. 57.

```
1   Date Bought,Item and Details,Purchase Price,Quantity
2   01/01/2015,"Test tubes: large, medium, and small",1.95,75
3   02/04/2016,"Mass spectrometer",8344.99,1
4   01/22/2013,"Beakers\n\n    Very breakable!",3.99,43
```

Figure 9.6 Contents of a CSV file containing a fictitious lab equipment inventory.

consider Figure 9.6, which shows what might be the first four lines from a CSV file containing a lab equipment inventory. The first line is a header that gives the names for each column. Lines 2–4 are data lines. The quotation marks tell us that the second element in that list should be considered a single string (e.g., 'Test tubes: large, medium, and small') and not separate items (e.g., 'Test tubes: large', ' medium', ' and small'). Unfortunately, a split(',') call cannot figure this out. Note that in a CSV file, there is usually no whitespace between commas. If there were, the whitespace would be considered part of the values for the column after the comma.

We could, using string variables and Python's programming constructs, write a parser that would properly deal with these and similar special cases. It would, however, be a lot of work. Fortunately, there is a module called csv that handles these special cases for us. Below is code using the csv module that will read the contents of a CSV text file *mydata.csv* and put the contents into the list data:

```
1   import csv
2   fileobj = open('mydata.csv', 'r')
3   readerobj = csv.reader(fileobj)
4   data = []
5   for row in readerobj:
6       data.append(row)
7   fileobj.close()
```

In line 2, we create a file object connected to *mydata.csv*, ready for reading. In line 3, we create readerobj, an iterable created by the reader function. Iterables are objects that you can loop through using a for loop, as in line 5 in the code above. The syntax of looping through an iterable is the same as looping through a list (which itself is an iterable), and each "element" in readerobj (which is set to row in the line 5 loop) represents a line of the CSV file and is appended to the data list. In line 7, we close the file object. The list data is not a list of strings but a list of lists. Each element in data is a single row in *mydata.csv*, parsed so that each column is a separate element.

If the CSV text file *mydata.csv* contained the lines shown in Figure 9.6, printing those rows using

```
print(data[:4])
```

would give (reformatted for better display):

```
[['Date Bought', 'Item and Details', 'Purchase Price', 'Quantity'],
 ['01/01/2015', 'Test tubes: large, medium, and small', '1.95', '75'],
 ['02/04/2016', 'Mass spectrometer', '8344.99', '1'],
 ['01/22/2013', 'Beakers\\n\\n    Very breakable!', '3.99', '43']]
```

The elements of the list `data[0]` are strings, as is the case for the elements of `data[1]`, `data[2]`, etc. If we want to do arithmetic on the prices or quantities, we have to convert the prices into floating-point values. The module, though, enables us to handle nicely the string elements in the file that have commas and newline characters in them. See the csv module's documentation for details.[16] An up-to-date link to the module's documentation is at www.cambridge.org/core/resources/pythonforscientists/refs/, ref. 53.

For CSV files that contain only numbers whose columns are separated by commas, using the csv module is overkill. The `genfromtxt` function works fine on those files, if we set the `delimiter` keyword input parameter to the string `', '`. Section 12.1 provides an example of such a call.

9.2.7 Catching File Opening and Other Errors

What happens if we try to open a file and that file does not exist? Python sends us a message telling us what is wrong and stops the normal execution of the program. This is an example of "throwing an **exception**." For instance, if we tried to execute this line of code:

```
fileobj = open('not_a_file.txt', 'r')
```

and the file *not_a_file.txt* does not exist, the following will be returned by the interpreter:

```
Traceback (most recent call last):
  File "<stdin>", line 1, in <module>
OSError: [Errno 2] No such file or directory: 'not_a_file.txt'
```

and the program will stop.

For many purposes, the above behavior works fine: This probably means we made a typo in our program in the line where we typed the filename, so if we run our program and get this message, we will go into the code, fix the typo, and rerun the program. But what if we *expected* this error to occur and had a plan to deal with the error? For instance, what if we knew that a data file we wanted to open was named either *data.txt* or *DATA.TXT*, but we did not know beforehand which it would be? This might occur because the program that generated the data named the file differently depending on what computer it was run on (historically, not all programming languages and operating systems recognized the difference between uppercase and lowercase letters). In such a situation, in Python we can use the `try`/`except` structure to gracefully handle the exception. This code:

---

[16] The discussion in this section also owes credit to http://stackoverflow.com/a/13472940 (accessed October 13, 2016).

```
1    try:
2        fileobj = open('data.txt', 'r')
3    except OSError:
4        fileobj = open('DATA.TXT', 'r')
```

tries to open *data.txt*. If it fails with an OSError, the code in the except block is executed (that code tries to open the file *DATA.TXT*). The OSError is known as an "exception class" (that is, a class that is used in generating exceptions). The OSError exception class, and a host of other exception classes, are built in to Python. We can also define our own exception classes. See Try This! 17-5 for details.

The try/except structure is a useful way of gracefully handling errors of many kinds, not just file errors. Pretend we are writing a program and want the execution of the program to stop when a certain condition occurs. To manually stop the program in Python we use a raise statement. Consider a function to calculate the area of a circle given a user-inputted radius. We can ensure the user will not pass in a negative radius by writing an if statement that tests for a negative radius, and if true, executes a raise statement:

```
def area(radius, pi=3.14):
    if radius < 0:
        raise ValueError('radius negative')
    area = pi * (radius**2)
    return area
```

The syntax for raise is the command raise followed by an exception class. In this case, we used the built-in exception class ValueError, which is commonly used to denote errors that have to do with bad variable values. In between the parentheses, we provide a string message that will be output by the interpreter when the raise is thrown. The act of causing an exception to happen is often called "raising" or "throwing" an exception.

So far, we have loosely been saying that an exception "stops" the execution of a program. Actually, raising an exception is not exactly the same as stopping the execution of a program. In a traditional "stop," program execution terminates and control returns to the operating system. The program cannot do anything more. When an exception in Python is raised, execution stops and sends the interpreter up one level to see if there is some code that will properly handle the error. This means that by using raise, we can gracefully handle expected errors without terminating the entire program.

Earlier, we saw how to handle a file opening exception using the try/except construct: We execute the block under the try then execute the except if an exception is raised. This same method works for exceptions raised in nonfile-opening situations. For instance, assume we have the function area as defined above (i.e., with the test for a negative radius). The code below calls the function area using a try/except construct that will gracefully recognize a negative radius and call area again with the absolute value of the radius instead as input:

```
1    rad = -2.5
2    try:
3        a = area(rad)
4    except ValueError:
5        a = area(abs(rad))
```

How does this work? When the interpreter enters the `try` block, it executes all the statements in the block one by one. If one of the statements returns an exception (as the first `area` call in line 3 will because `rad = -2.5` is negative), the interpreter looks for an `except` statement at the calling level that recognizes the exception class (in this case `ValueError`). If the interpreter finds such an `except` statement, the interpreter executes the block under that `except`. In this example, that block repeats the `area` call but with the absolute value of `rad` instead of `rad` itself. If the interpreter does not find such an `except` statement, it looks another level up for a statement that will handle the exception. This occurs all the way up to the main level, and if no handler is found there, execution of the entire program stops and control returns to the interpreter.

The examples we have seen all specify the exception class the `except` statement will handle. If we provide an `except` statement without specifying an exception class:

```
try:
    <... try statement body ...>
except:
    <... except statement body ...>
```

any exception class is handled by the `except` statement.

In the above examples, we used the exception classes `OSError` and `ValueError`. These and other built-in exception classes (e.g., `TypeError`, `ZeroDivisionError`, etc.) are listed in a good Python reference.[17] For many scientific and engineering applications, this suite of exception classes is enough for us to address the specific type of error we are protecting against. The better and more advanced approach, however, is to define special exception classes to customize handling. Try This! 17-5 introduces the topic.

9.3   Try This!

The examples in this section address the main topics of Section 9.2. We practice using methods of various kinds of objects, including the `format` method of strings. We explore how different kinds of "copies" behave. We also read and write files and write code to catch errors.

For all the exercises in this section, assume NumPy has already been imported as np.

---

[17] See http://docs.python.org/library/exceptions.html for a listing of built-in exception classes (accessed August 17, 2012).

Try This! 9-1 Integer and Floating-Point Objects: Weight and Count

Consider the following variables:

```
1    weight = 22.4
2    count = 17
```

Are `weight` or `count` objects? Do these variables have attributes and/or methods? If so, what is one of the public methods and what does it do? How can we find out what these attributes and/or methods are?

Try This! Answer 9-1

Both integer and floating-point variables are objects and have both attributes and methods. Applying the `dir` command to each variable – i.e., `dir(weight)` and `dir(count)` – in the Python interpreter will give us a listing of both the attributes and methods in the object. For the `weight` floating-point object, one method is the method `is_integer`, which returns `True` or `False` depending on whether the number is a whole number:[18]

```
In [1]:   weight = 22.4
          weight.is_integer()

Out[1]:   False

In [2]:   weight = 22.0
          weight.is_integer()

Out[2]:   True
```

We know `is_integer` is a public method (i.e., a method normal users are expected to be able to use) because its name does not begin with a single- or double-underscore. The `count` integer object also has methods. The bottom line is that nearly everything in Python is an object, even what appear to be "mere" numbers.

Try This! 9-2 Using an Array Method: Masses

Pretend we have the following array of masses (in kg):

```
masses = np.array([3.5, 2.1, 9.4, 6.7])
```

Using these values, how could we generate the following as a single string?:

```
3.5 -> 2.1 -> 9.4 -> 6.7
```

---

[18] https://docs.python.org/3.7/library/stdtypes.html#additional-methods-on-float (accessed August 22, 2019).

The above string, for instance, could be used as an illustration of some process that leads to the mass values changing.

Try This! Answer 9-2

Here is a solution, assuming `masses` is already defined:

```
1   concat_str = ''
2   masses_str = masses.astype('U')
3   for imass in masses_str:
4       concat_str += imass + ' -> '
5   concat_str = concat_str[:-4]
```

Recall that:

```
concat_str += imass + ' -> '
```

does the same thing as:

```
concat_str = concat_str + imass + ' -> '
```

The `+=` operator "adds on" (which is concatenation for strings) the result of the expression to the right of the operator to the value of the variable to the left of the operator, then reassigns the combined result to the variable to the left of the operator. Details are found in the solution for Try This! 6-5.

Although, as we said in Section 9.2.3 and Table 9.2, in many cases lists are better than arrays to use for holding strings, the conversion of an array of numerical values to an array of strings is an exception. In this case, the `astype` method of the array is really useful. If we want to convert `masses_str` to a list, we can use the `list` function:

```
masses_str = list(masses_str)
```

Also, in line 2 of the above solution to this Try This!, we create a temporary array `masses_str` to hold the string version of `masses`. We could have dispensed with the temporary array by putting the `astype` call directly in the `for` statement in line 3:

```
for imass in masses.astype('U'):
```

Also, in line 5, we slice the last four characters from `concat_str` because the way our `for` loop body is constructed, an extra `' -> '` string is added after the last value in `masses_str`.

Try This! 9-3 Using List Methods: Masses

Given the `masses_str` array from the Try This! 9-2 solution, convert it into a list `new_masses` and then, in this order, make the following changes to `new_masses`:

1. Reverse the order of the values

2. Append the string 'NaN' to the end of the list
3. Insert the string 'NaN' to the third position in the list

The strings 'NaN' is an abbreviation for the phrase, "not a number," and is a way of referring to a bad data value in a collection of good data values. In Section 15.2.2, we describe how in the pandas package the idea of "not a number" is encoded as a special floating-point value, but, for now, we let this string represent a bad data value.

Try This! Answer 9-3

Here is a solution:

```
1   new_masses = list(masses_str)
2   new_masses.reverse()
3   new_masses.append('NaN')
4   new_masses.insert(2, 'NaN')
```

The `reverse` method reverses the contents of a list. There are no arguments passed into the method call (the parentheses are empty) because all the information needed to successfully reverse the order of the elements of `new_masses` is already present in `new_masses`. The `append` method adds in the item passed in the argument list, after the last element of `new_masses`. And the `insert` call places the item that is the second argument in the call at the location given by the index that is the first argument in the call. Index 2 is the third position in a list.

The result is:

```
In [1]:   print(new_masses)

          ['6.7', '9.4', 'NaN', '2.1', '3.5', 'NaN']
```

All three methods in the solution act on the `new_masses` list in place. That is, they alter the contents of the object they are attached to, without the results of the method calls being in a return value.

Try This! 9-4 Using String Methods: Chemical Elements

Consider the following list of information on the first two chemical elements, where each chemical element itself is a list giving the name, number of protons, and number of neutrons for the chemical element:

```
atoms = [['hydrogen', 1, 0], ['helium', 2, 2]]
```

Create a string that lists the names of the chemical elements, all in capitals, in a single column, with each chemical element on its own single line. Write this solution in a general way that does not require `atoms` to be two elements long (i.e., where the solution would work even if `atoms` were to grow or shrink in size).

Try This! Answer 9-4

Because there are only two items in `atoms`, we could write this without using a loop:

```
output_str = atoms[0][0].upper() + '\n' + atoms[1][0].upper()
```

The list element specified by `atoms[0]` is this list:

```
['hydrogen', 1, 0]
```

Thus, `atoms[0][0]` refers to the first element in that list, namely the string `'hydrogen'`. That string is an object, which has the method `upper`, so `atoms[0][0].upper()` calls the `upper` method of the string `'hydrogen'`. That string all in uppercase is returned. A similar logic works with the `atoms[1][0].upper()`. Printing out the above code results in:

```
In [1]:  print(output_str)

         HYDROGEN
         HELIUM
```

If we loop through the indices of the `atoms` list, we can create `output_str` irrespective of how many items are in `atoms`:

```
1   output_str = ''
2   for i in range(len(atoms)):
3       output_str += atoms[i][0].upper() + '\n'
4   output_str = output_str[:-1]
```

We initialize `output_str` to the empty string in line 1, so we can concatenate the right-hand side of line 3 to `output_str`, and thus grow `output_str`. The iterator `i` runs through the indices of `atoms`, not the indices of the sublists that are the elements of `atoms`. In line 4, we strip off the last character, the extra newline character added by line 3 in the last iteration of the `for` loop.

Try This! 9-5 Using the String `format` Method: Precipitation

Consider the definitions of the Seattle-area precipitation data-related arrays from Figure 6.3, reproduced here for convenience:

```
1   months = np.array([1, 2, 3, 4, 5, 6, 7, 8, 9, 10, 11, 12])
2   precip = np.array([5.58, 3.54, 4.54, 2.93, 2.10, 1.47,
3                      0.49, 1.03, 1.72, 4.31, 6.50, 5.41])
```

The `precip` values are in inches. Making use of the string `format` method, write code that creates a single string that, if printed, displays the month, precipitation, and then (in curly braces), the precipitation rounded to the nearest tenths of an inch:

```
1    Month: 1,  Precip: 5.58 {5.6}
2    Month: 2,  Precip: 3.54 {3.5}
3    Month: 3,  Precip: 4.54 {4.5}
4    Month: 4,  Precip: 2.93 {2.9}
5    Month: 5,  Precip: 2.10 {2.1}
6    Month: 6,  Precip: 1.47 {1.5}
7    Month: 7,  Precip: 0.49 {0.5}
8    Month: 8,  Precip: 1.03 {1.0}
9    Month: 9,  Precip: 1.72 {1.7}
10   Month: 10, Precip: 4.31 {4.3}
11   Month: 11, Precip: 6.50 {6.5}
12   Month: 12, Precip: 5.41 {5.4}
```

Hint: The `format` method interprets two curly braces right after each other in the method's string object as meaning, "put a literal curly brace here," because the curly brace is used in the string object to delimit a field to format.[19]

Try This! Answer 9-5

Assuming `months` and `precip` are defined as above, this code will create the desired string (here called `output_str`):

```
1    output_str = ''
2    line = "Month: {0:d}, Precip: {1:.2f} {{{1:.1f}}}"
3    for i in range(np.size(months)):
4        output_str += line.format(months[i], precip[i]) + '\n'
5    output_str = output_str[:-1]
```

We saved the string whose `format` method we will use as the variable `line`, to keep the length of our line 4 code limited. The `line` string does not change each iteration of the `for` loop. Only the values of the arguments in the `format` call change each iteration.

In line 2, the first replacement field `{0:d}` means "put the value of the first argument to `format` (i.e., `months[i]`) here and format it as an integer using default rules." The portion of line 2 that says `{1:.2f}` means "put the value of the second argument to `format` (i.e., `precip[i]`) here and format it as a floating-point number using default rules for width and showing two places after the decimal." The portion of line 2 that says `{1:.1f}` means "put the value of the second argument to `format` (i.e., `precip[i]`) here and format it as a floating-point number using default rules for width and showing one place after the decimal." We are not limited to using an argument to `format` only once. Here we used the second argument to `format` (denoted in the replacement field by the index value 1, to the left of the colon) twice in the string `line`. The double left curly brace to the left of `{1:.1f}` means "put a literal left curly brace here." The double right curly brace to the right of `{1:.1f}` means "put a literal right curly brace here."

---

[19] https://docs.python.org/3.7/library/string.html#format-string-syntax (accessed August 23, 2019).

Try This! 9-6 Effect of Changing an Array Value on a Copy: Precipitation Month Numbers

Consider again the `months` array definition from Figure 6.3 that we revisited in Try This! 9-5 above. Describe what happens to `months`, `months1`, and `months2` if we execute the following code:

```
1   months1 = months
2   months2 = months.copy()
3   months[5] = -99
4   months2[3] = -999
```

Please explain *why* what happened, happened.

Try This! Answer 9-6

This is the result when we print the three arrays, after the code above executes:

```
In [1]:  print(months)

         [ 1    2    3    4    5  -99    7    8    9   10   11   12]

In [2]:  print(months1)

         [ 1    2    3    4    5  -99    7    8    9   10   11   12]

In [3]:  print(months2)

         [ 1    2    3  -999    5    6    7    8    9   10   11   12]
```

The `months[5]  =  -99` line altered both `months` and `months1`. Changing one changed the other. This is because `months1` is a reference to `months`, not a memory-independent copy of `months`. The `months2[3]  =  -999` line altered `months2`, but because `months2` was created by a call to the `copy` method of `months`, `months` was not altered. In addition, as `months2` occupies an entirely separate area of memory from the linked `months` and `months1` arrays, the change we made to `months` had no effect on `months2`.

Try This! 9-7 Relationship of a Slice to the Original Array: Precipitation Month Numbers

Consider again the `months` array definition from Figure 6.3 that we revisited in Try This! 9-5. Describe what happens to `months`, `sub_months1`, and `one_month2` if we execute the following code:

```
1   sub_months1 = months[:7]
2   one_month2 = months[8]
3   months[5] = -9
4   sub_months1[3] = -99
5   months[8] = -999
```

Please explain *why* what happened, happened.

Try This! Answer 9-7

This is the result when we print the three arrays, after the code above executes:

```
In [1]:  print(months)

         [   1    2    3  -99    5   -9    7    8 -999   10   11   12]

In [2]:  print(sub_months1)

         [  1   2   3 -99   5  -9   7]

In [3]:  print(one_month2)

         9
```

For arrays, a slice is not a memory-independent copy of those elements in the array. Thus, changes in the original array will propagate to the slice. When the "slice," however, is only a single value, the variable assigned to that element refers to a memory-independent copy of that element. Generally, if we want an array slice to be memory-independent from the original array, we need to use the copy method of the slice, as in:

```
sub_months1 = months[:7].copy()
```

For a list, slices *are* memory-independent copies. This code:

```
1   months = [1, 2, 3, 4, 5, 6, 7, 8, 9, 10, 11, 12]
2   sub_months1 = months[:7]
3   sub_months1[3] = -99
```

produces these values:

```
In [1]:  print(months)

         [1, 2, 3, 4, 5, 6, 7, 8, 9, 10, 11, 12]
```

```
In [2]:  print(sub_months1)

         [1, 2, 3, -99, 5, 6, 7]
```

A change to the slice does not affect the original list.

Try This! 9-8 Relationship of Array Syntax Expression Results to the Original Array: Precipitation Month Numbers

Consider again the `months` array definition from Figure 6.3 that we revisited in Try This! 9-5. Describe what happens to `months` and `add_yr_months` if we execute the following code:

```
1    add_yr_months = months + 12
2    months[5] = -99
3    add_yr_months[3] = -999
```

Please explain *why* what happened, happened.

Try This! Answer 9-8

This is the result when we print the two arrays, after the code above executes:

```
In [1]:  print(months)

         [ 1   2   3   4   5 -99   7   8   9  10  11  12]
```

```
In [2]:  print(add_yr_months)

         [ 13   14   15 -999   17   18   19   20   21   22   23   24]
```

The array `add_yr_months` is memory-independent from `months`, so changes to one do not affect the other. When we change the original array `months` or the derived array `add_yr_months`, the `add_yr_months = months + 12` line is not re-executed to adjust to the changes.

Try This! 9-9 Writing and Reading a Single-Column File: Volumes

Pretend we have the following array of volume measurements `volumes` (in liters):

```
volumes = np.array([1.9, 24.5, 4.4, 17.8, 8.3])
```

Write these values to a file *one-col_volumes.txt*, then read the file back in into a new array, `volumes_new`.

Try This! Answer 9-9

Figure 9.7 shows code to accomplish the task. Comment lines are included in that code to help guide the reader and to illustrate a way of using inline comments to comment on a small block of code on the side. Figure 9.8 shows the contents of the output file, *one-col_volumes.txt*, generated by the code.

In lines 1–4 above, we create a list that is a version of volumes but where the floating-point values are converted to strings and a newline character is appended to the end of each element. That appending happens by first creating a list of just newline characters that is the length of volumes (in line 1). The syntax in line 1 (specifying a list of a single value multiplied by a number) can be used to create lists of any single value, duplicated that number of times. The following shows how we can use that syntax to create a seven-element list of ones:

```
In [1]:  ones = [1]*7
         print(ones)


         [1, 1, 1, 1, 1, 1, 1]
```

```
1   outputstr = ['\n']*len(volumes)    #- Convert to string
2   for i in range(len(volumes)):      #  and add newlines
3       outputstr[i] = \
4           str(volumes[i]) + outputstr[i]
5
6   fileout = open('one-col_volumes.txt', 'w')   #- Write out
7   fileout.writelines(outputstr)                #  to the
8   fileout.close()                              #  file
9
10  filein = open('one-col_volumes.txt', 'r')    #- Read in
11  inputstr = filein.readlines()                #  from the
12  filein.close()                               #  file
13
14  volumes_new = np.zeros(len(inputstr), 'f')   #- Convert
15  for i in range(len(inputstr)):               #  string to
16      volumes_new[i] = float(inputstr[i])      #  numbers
```

Figure 9.7 Code to solve Try This! 9-9.

```
1.9
24.5
4.4
17.8
8.3
```

Figure 9.8 Contents of a one-column file of volumes, *one-col_volumes.txt*, generated by the code in Figure 9.7.

Lines 6–8 write out the list that is the string version of `volumes`. We open the file for writing in line 6, write out the list `outputstr` in line 7, and close the file object in line 8. We repeat the same operations in lines 10–12, except we open the file and read in the values into the list `inputstr`.

In lines 14–16, we convert the values of `inputstr` to floating point and fill the array `volumes_new` with those converted values. We do not have to strip off the newline character before converting the number to floating point using the built-in `float` function.

We could also create the array `volumes_new` from the list `inputstr` by turning `inputstr` into an array and then using the `astype` method to convert the values to 8-byte floating-point type:

```
volumes_new = np.array(inputstr).astype('d')
```

The above line would replace lines 14–16 and produce the same result.

Finally, we can check whether `volumes` and `volumes_new` are the same using the NumPy function `allclose`. This function tests to see whether every corresponding element in two arrays are "close" to each other, where closeness is defined using the algorithm of the `isclose` function in Section 8.2.7. If all corresponding elements are close, `True` is returned. If any corresponding elements are not close, `False` is returned. For `volumes` and `volumes_new`:

```
In [1]:   np.allclose(volumes, volumes_new)

Out[1]:   True
```

which tells us the values written out to *one-col_volumes.txt* are the "same" as the values read in from the file.

Try This! 9-10   Writing a Multicolumn File: NAMD Output

For the NAMD output file *smallproteinB_ws_eq.log* described in Section 9.1, write a program to read in the contents of the file and write out the first four numerical values in the ENERGY lines as separate columns, separated by tabs. Include a header in the file that labels each columns with the applicable ETITLE label. The first lines of the output file (please call it *namd_subset.txt*) should look like Figure 9.10.

Try This! Answer 9-10

Figure 9.9 shows a solution. In lines 2–4, we read in the contents of the *smallproteinB_ws_eq.log* file into a list of strings `inputstr`, just as we did in Figure 9.3. In lines 7 and 8, we create two empty lists, `labels_list` and `four_cols_list`, which will be used to hold the column labels and the lines of ENERGY-related values, respectively. By using a list for the column label, we can detect whether or not we have already extracted the column labels by testing to see whether the length of `labels_list` is zero or not. Because we use a list to hold the lines

```
1   #- Read in NAMD log file and first four columns:
2   filein = open('smallproteinB_ws_eq.log', 'r')
3   inputstr = filein.readlines()
4   filein.close()
5
6   #- Extract values, create list of tab-separated columns:
7   labels_list = []
8   four_cols_list = []
9   for i in range(len(inputstr)):
10      line = inputstr[i]
11
12      if (len(labels_list) == 0) and (line[0:7] == 'ETITLE:'):
13          labels_list = line.split()
14          labels_str = '\t'.join(labels_list[1:5]) + '\n'
15          four_cols_list.append(labels_str)
16
17      if line[0:7] == 'ENERGY:':
18          line_values = line.split()
19          line_str = '\t'.join(line_values[1:5]) + '\n'
20          four_cols_list.append(line_str)
21
22  #- Output list of tab-separated columns to new file:
23  fileout = open('namd_subset.txt', 'w')
24  fileout.writelines(four_cols_list)
25  fileout.close()
```

Figure 9.9 Code to solve Try This! 9-10.

of ENERGY-related values, we do not have to know ahead of time how many ENERGY lines will be read from the *smallproteinB_ws_eq.log* file.

In line 9, we loop through each element of inputstr. The first if statement block, lines 12–15, looks for the first ETITLE line of labels. If the labels have not been read in yet (i.e., len(labels_list) == 0), and line is an ETITLE line, we split the line based on the presence of whitespace, construct a string where each of the labels is separated by the tab character, and add that string to the list of times we will eventually write out to the *namd_subset.txt* file. Recall that even though the label (and ENERGY value) columns are separated by variable amounts of whitespace, this is not a problem because the split method called without input arguments intelligently eliminates the whitespace between the columns. When we select the first four values of ETITLE and ENERGY, we slice [1:5] (lines 14 and 19), not [0:4], because the first element when splitting the ETITLE or ENERGY line is the string ETITLE: or ENERGY:, respectively.

The second if statement block, lines 17–20, works like lines 12–15 except for the ENERGY values. In lines 23–25, we write out the values in four_cols_list to the *namd_subset.txt* file. Each element of four_cols_list has the newline character at the end of the string stored in that element, because of lines 14 and 19.

TS	BOND	ANGLE	DIHED
0	116.4585	94.8698	64.8673
1	51.4445	70.9667	64.3814
2	28.3905	65.0812	64.0858
3	54.0500	72.9659	63.9732
4	133.0166	92.2749	64.0363
5	43.2134	69.9780	63.9838
6	21.5603	60.1603	64.2993
7	34.2403	59.4258	64.7425
8	77.4038	70.0697	65.3177
9	146.3527	95.1719	66.0296
10	57.7400	64.5197	65.0837
11	23.1257	58.0980	64.8538
12	60.1606	70.6345	64.7660
13	175.2678	98.6394	64.8140
14	37.9562	64.1370	64.7819

Figure 9.10 Lines at the beginning of *namd_subset.txt*, a multicolumn file that contains selected ENERGY values from *smallproteinB_ws_eq.log*. The file is displayed with tab stops set every 12 characters.

Try This! 9-11 Reading a Multicolumn File: NAMD Output

Write a program that will read in the file *namd_subset.txt* that was created by the solution to Try This! 9-10, and place its contents (neglecting the header of labels) into a two-dimensional NumPy floating-point array. Do this first using a file object and string methods and a second time using the NumPy function genfromtxt. Hint: Regarding genfromtxt, the skip_header keyword input parameter sets how many lines at the beginning of the file to ignore.

Try This! Answer 9-11

Here is a solution using a file object and string methods:

```
filein = open('namd_subset.txt', 'r')
inputstr = filein.readlines()
filein.close()

multi_col_list = []
for i in range(len(inputstr)):
    line = inputstr[i]
    line_values = line.split()
    multi_col_list.append(line_values)

str_array = np.array(multi_col_list[1:])
float_array = str_array.astype('d')
```

The first 10 lines of this code are similar to code we have seen previously. We should note that the `split` method called in line 8 considers the tab character as whitespace, the string `line` is split properly so each column is a separate element in `line_values`.

The key to creating the two-dimensional array in the above code is that `multi_col_list` has to be a list of lists where each element of `multi_col_list` is the same length (in this case, a length of 4). In line 11, we leave off the column labels by slicing the list using `[1:]`. The remaining elements of `multi_col_list` are converted by `array` into a two-dimensional array of strings. Then, when we convert those strings into floating-point values in line 12, using `astype`, the result will be a two-dimensional array of floating-point values.

Here is a solution using `genfromtxt`:

```
float_array_gft = np.genfromtxt('namd_subset.txt', skip_header=1)
```

A lot simpler! By specifying the keyword input parameter `skip_header`, we skip reading in the first line in the file.

After running these two solution code snippets back-to-back, we can compare the arrays `float_array` and `float_array_gft`:

```
In [1]:  print(np.allclose(float_array, float_array_gft))

         True
```

The solutions produce the same result, so both methods work fine to read in a multicolumn file of numbers. The NumPy `genfromtxt` function also has other keyword input parameters that allow us to further customize our read-in experience, depending on the characteristics of the file.[20]

Try This! 9-12 Reading a Multicolumn CSV File: NAMD Output

Pretend we have a file *namd_subset.csv* that is the same as the file *namd_subset.txt* from Try This! 9-11, except instead of having tab-separated colums, it is a CSV file, where commas separate the columns. Repeat the task from Try This! 9-11 by writing a program that reads the file and places its contents (neglecting the header of labels) into a two-dimensional NumPy floating-point array. Do this using the csv module.

---

[20] Details are found in the NumPy User Guide's entry on using `genfromtxt` and the function's reference manual entry. We provide up-to-date links to both at www.cambridge.org/core/resources/pythonforscientists/refs/, refs. 56 and 42 respectively.

Try This! Answer 9-12

This code solves the problem:

```
1    import csv
2    import numpy as np
3
4    fileobj = open('namd_subset.csv', 'r')
5    readerobj = csv.reader(fileobj)
6    multi_col_list = []
7    for row in readerobj:
8        multi_col_list.append(row)
9    fileobj.close()
10
11   str_array = np.array(multi_col_list[1:])
12   float_array = str_array.astype('d')
```

In line 11, we select only the `[1:]` rows for converting into a two-dimensional string array `str_array` because the first element of `multi_col_list` is a list giving the file's header line values. Because of the way `readerobj` works, `multi_col_list` is a list of lists, not a list of strings.

Try This! 9-13 Reading a File Using with/as: NAMD Output

Redo Try This! 9-12 using the with/as construct.

Try This! Answer 9-13

This code solves the problem:

```
1    import csv
2    import numpy as np
3
4    with open('namd_subset.csv', 'r') as fileobj:
5        readerobj = csv.reader(fileobj)
6        multi_col_list = []
7        for row in readerobj:
8            multi_col_list.append(row)
9
10   str_array = np.array(multi_col_list[1:])
11   float_array = str_array.astype('d')
```

Basically, everything is the same except the file handling code is inside the with/as block. The looping over `readerobj` has to be in the with/as block too because it ultimately refers back to the `fileobj` file.

Try This! 9-14 Catching File Opening and Other Errors: Volumes

Consider the code in the solution to Try This! 9-9 (Figure 9.7) that reads in the file *one-col_volumes.txt* and places floating-point versions of the values into the `volumes_new` array. Place that code into a `try/except` code structure that will handle these situations:

- There is an error in opening the file, e.g., the file is not found, which produces an `OSError`.
- The file has one or more letters in it (i.e., the column does not only consist of numbers), which will cause the `float` function to return a `ValueError`.

In both cases, have the program print a pertinent message and set the value of `volumes_new` to the value `None`.

Hints: Handling of multiple exceptions in the `try` block is done with multiple `except` clauses, one following the other. The value `None` is a specially defined quantity, just as `True` and `False` are specially defined quantities. In all three cases, they are not strings and thus are not delineated by apostrophes or quotation marks. We discuss `None` in more detail in the answer to this Try This!.

Try This! Answer 9-14

Here is a solution:

```
1    try:
2        filein = open('one-col_volumes.txt', 'r')   #- Read in
3        inputstr = filein.readlines()               #  from the
4        filein.close()                              #  file
5
6        volumes_new = np.zeros(len(inputstr), 'f')  #- Convert
7        for i in range(len(inputstr)):              #  string to
8            volumes_new[i] = float(inputstr[i])     #  numbers
9    except OSError:
10       print("File opening error")
11       volumes_new = None
12   except ValueError:
13       print("Cannot convert to float:  Non-numeric values in file")
14       volumes_new = None
```

The Python interpreter goes through each line in the `try` block and tries to execute it. If, when doing so, the line of code generates an exception, the program stops executing that line, leaves the `try` block, and looks to see whether any of the `except` blocks address the exception class that was raised. We can handle as many different kinds of exception classes as we wish by adding additional `except` blocks. After an `except` block executes, control is transferred to the line after the entire series of `except` blocks. Control does not go back to the `try` block nor does the interpreter continue through the remaining `except` blocks looking for another match.

To use a `try/except` block, we only need to know what exception classes are raised by the functions and methods we are using in the `try` block (here the `open` and `float` functions). We do not need to know where in those functions a `raise` statement is made or any details about how the functions test for exception conditions. All we need to do is put the calls to those functions in the `try` block and tell Python what to do once the particular exception is thrown that we want to handle.

The `None` value is a special value that the Python interpreter treats as meaning "nothing." This means that any operations or functions we try to apply to a variable set to `None` will result in an exception being raised:[21]

```
In [1]:  volumes_new = None
         volumes_triple = volumes_new * 3


         TypeError     Traceback (most recent call last)
         <ipython-input-1-c90f03ec200e> in <module>()
               1 volumes_new = None
         ----> 2 volumes_triple = volumes_new * 3

         TypeError: unsupported operand type(s) for *: 'NoneType' and 'int'

In [2]:  volumes_sin = np.sin(volumes_new)


         AttributeError    Traceback (most recent call last)
         <ipython-input-5-7b237d8cfeb2> in <module>()
         ----> 1 volumes_sin = np.sin(volumes_new)

         AttributeError: 'NoneType' object has no attribute 'sin'
```

Essentially, the only operation we can apply to `None` is to test whether a variable is set to `None`. Assuming `volumes_new` is set to `None`, as above:

```
In [1]:  print(volumes_new is None)


         True

In [2]:  print(volumes_new is not None)


         False
```

The proper syntax for testing whether a variable is `None` is to use the operator `is`, not the logical equality operator `==`. Likewise, to test whether a variable is not `None`, we use `is not` rather than `!=`.

As a result of these characteristics of `None`, setting a variable to `None` more or less "protects" it from being inadvertently (and thus improperly) used later on in the program. In our solution above, setting `volumes_new` to `None` means any calculations we do later on using `volumes_new`, that assume the variable is set to a NumPy array, will raise an exception. This feature of "failing loudly" is a good thing, as it helps us to catch errors. The worst errors are those that occur silently.

---

[21] The error messages output are edited for clarity.

9.4   More Discipline-Specific Practice

The Discipline-Specific Jupyter notebooks for this chapter cover the following topics:

- Arrays, lists, and strings as objects.
- Copying variables, data, and objects.
- Reading and writing files.
- Catching file opening and other errors.

9.5   Chapter Review

9.5.1   Self-Test Questions

Try to do these without looking at the book or any other resources or using the Python interpreter. Answers to these Self-Test Questions are found at the end of the chapter. For all questions in this section, assume NumPy has already been imported as np.

Self-Test Question 9-1

Describe what two kinds of basic entities make up an object.

Self-Test Question 9-2

What does it mean that the two kinds of entities described in Self-Test Question 9-1 "make up" an object?

Self-Test Question 9-3

Describe how objects in object-oriented programming differ from variables in procedural programming.

Self-Test Question 9-4

Describe a few attributes and methods of NumPy array objects. Provide a few code snippets showing the use of these attributes and methods.

Self-Test Question 9-5

Describe a few attributes and methods of list objects. How do these differ from NumPy array objects? Which is better to use if all elements are numbers? Provide a few code snippets showing the use of these attributes and methods.

Self-Test Question 9-6

Describe a few attributes and methods of string objects. Provide a few code snippets showing the use of these attributes and methods.

Self-Test Question 9-7

Pretend we have a variable `length` equal to 23.7 (in m). Using the string object's `format` method, print the string:

```
The beam is 23.7 m long.
```

without typing in the number `23.7`.

Self-Test Question 9-8

What will be the values of `data1`, `data2`, `data3`, and `data4` after this code snippet is executed?

```
1   data1 = np.array([2, 7, -3, -9])
2   data2 = np.array([4, -5, 2, 16]) + data1
3   data3 = data2
4   data4 = data3
5   data1[1] = -37
6   data2[0] = -18
```

Self-Test Question 9-9

How would we create a memory-independent copy of a list or an array?

Self-Test Question 9-10

Provide code to read in a file *values.txt* made up of three tab-separated columns of numerical values. The file has no header lines. Store the values in a NumPy floating-point array called `values` that is three columns wide and has the number of rows equal to the number of lines in *values.txt*. Solve this problem once using file objects and once using the NumPy function `genfromtxt`.

Self-Test Question 9-11

Consider the following code:

```
1   fobj = open('velocities.txt', 'r')
2   velocities = fobj.readlines()
3   fobj.close()
```

Rewrite the code using the `with`/`as` construct.

Self-Test Question 9-12

Assume the *values.txt* file in Self-Test Question 9-10, rather than having tab-separated columns, is a CSV file. Use the csv module to read in the values and save them in a NumPy array as described in Self-Test Question 9-10.

Self-Test Question 9-13

Assume we have a try/except block where there are five lines of code in the body of the try block. What happens if in line 2 of that body an exception is raised? Assume the exception class is the class listed in the except statement. What if the exception class raised is *not* the class listed in the except statement?

9.5.2 Chapter Summary

In this chapter, we examined different ways of using Python to read and write text files. Along the way, we also learned more about OOP and the nature and use of objects, as Python conceptualizes files as objects, with attributes and methods attached to them. To manipulate and access text files, we use the file object's methods, with explicitly opened/closed file objects or file objects accessed within the with/as construct. Once the contents of text files are read in, we found string methods to be useful for manipulating the contents. For most daily work, however, we want to use special functions designed for reading in data, such as found in NumPy and the csv module. Specific concepts and structures we looked at include:

- Introduction to objects and typing

  - Objects have attributes (data) and methods (actions on data) bound together in the same entity. In contrast, procedural programming sees data and actions on data as separate entities.
  - Integers can be stored either in a set number of bytes of memory or by using an algorithm that enables Python to represent an integer without any limit to size, save the total memory available on the computer. Floating-point numbers are stored in a set number of bytes of memory.

- Array, list, and string objects

  - Array objects have attributes and methods. The typecode is an attribute that tells us the base type of an array.
  - List methods enable us to add on elements to the list (e.g., append) as well as remove elements from the list (e.g., pop). Lists are mutable and thus the size of lists can change. Lists can grow or shrink in size as needed.
  - Because lists and NumPy arrays have different characteristics, in some cases lists are better to use whereas arrays are better to use in other cases. In general, arrays work better for large, possibly multidimensional collections of numbers, whereas lists work better for strings and collections of objects of different types.
  - String methods provide many ways of manipulating the contents of a string that go beyond mere concatenation. The format method provides a way of automatically formatting numerical data so it looks the way we want it to (e.g., number of digits in

width, the number of places after the decimal point, etc.) when we put that numerical data into a string.

- Copying variables, data, and objects

  - Assignment by value refers to assigning a variable to the actual value of a quantity.
  - Assignment by reference refers to assigning a variable by making an alias to another variable. Variables assigned by reference are not memory-independent from the original variable the assigned variable refers to, so a change in one will change the other.
  - When we assign a variable to a plain value, such as an integer or floating-point number, this is done by value.
  - When we assign a variable to a list or other more complex entity, this is done by reference.
  - To obtain a memory-independent copy of a complex entity, we can use the `copy` method that is attached to many of these objects or the `copy` and `deepcopy` functions of the copy module.
  - When objects are passed into functions and methods as part of the calling argument list, the parameter inside the function/method will be assigned to the argument either by value or reference, following the rules of variable assignment.

- Reading and writing files

  - The `open` statement creates a file object and readies it for reading or writing.
  - The `readline` and `readlines` methods of a file object read in one line or all remaining lines of the file the object connects to, respectively. The `readlines` method returns a list of the contents of all the lines of the file, where each line of the file is a string that is stored as an element in the `readlines` generated list.
  - The `write` and `writelines` methods of a file object write one string or a list of strings to the file the object connects to, respectively. Newline characters must be part of the strings that are being written out, if desired. The methods do not automatically add newline characters to their output to the file.
  - The string method `split` is useful for splitting up input read in from a file. The string method `join` is useful for preparing values for a single line of output to a file.
  - The `with`/`as` construct provides a means to more concisely access a file object than explicitly creating a file object and closing it when finished.
  - The NumPy function `genfromtxt` reads in tables of numbers and puts them into a NumPy array. The csv module handles CSV files, including ones with complex formatting.

- Catching file opening and other errors

  - When an error occurs, an exception is thrown, causing Python to look for code that will properly handle the exception class that is raised.
  - The `try`/`except` construct provides an environment in which to run lines of code (the `try` block), and if an exception is raised, control of the program is transferred to an `except` block that can handle the particular exception class that is raised.

9.5.3   Self-Test Answers

Self-Test Answer 9-1

Objects are made up of attributes and methods. Attributes are pieces of information or data. Methods are functions or actions that act on attributes and/or information passed into the method from outside the object.

Self-Test Answer 9-2

Attributes and methods "make up" an object in the sense that an object is a container that holds both kinds of entities. Thus, when we move the object around by passing it to functions or the methods of other objects, assign the object to another variable, or copy the object to another variable, all the attributes and methods go along with the object.

Self-Test Answer 9-3

Objects in object-oriented programming are entities consisting of attributes and methods, that is, information and functions that can act on that information. Variables in procedural programming only hold information. Functions that act on those variables are entirely separate from variables and their information. Variables have to be passed into the functions in order for them to have access to the information found in the variables.

Self-Test Answer 9-4

Here are a few attributes:

- `dtype` An object describing the datatype of the array. The `dtype.char` attribute gives the typecode string for the array.
- `shape` The shape tuple for the array.
- `size` The total number of elements in the array.

Here are a few methods:

- `astype` Returns a copy of the array converted into another base type.
- `reshape` Returns the array reshaped to the shape specified by another shape tuple.
- `sum`: Returns the sum of all elements in the array.

And here are examples of using some of these attributes and methods:

```
In [1]:  data = np.array([[3, 5, -7], [1, 9, -2]])
         print(data.shape)

         (2, 3)
```

```
In [2]:   print(data.dtype)

          int64

In [3]:   print(data.dtype.char)

          l

In [4]:   print(data.reshape((3,2)))

          [[ 3   5]
           [-7   1]
           [ 9  -2]]

In [5]:   print(data.astype('d'))

          [[ 3.   5.  -7.]
           [ 1.   9.  -2.]]
```

Self-Test Answer 9-5

Lists do not really have public attributes. Here are a few methods:

- append Add an element to the end of the list.
- count Returns the number of times the value passed into the argument list occurs in the list.
- pop Remove the last element of the list and return the element.

Lists differ from NumPy arrays in a number of ways. First, lists are not fixed in size – we can add and subtract elements from lists – whereas arrays are fixed in size as long as they exist. Second, the elements of a list do not have to be the same type whereas for an array, the elements of a list have to have the same base type. Finally, NumPy arrays support a cleaner indexing syntax for dealing with multidimensional arrays, support array syntax operations, and generally run faster than lists. For a collection of items that are all numbers, arrays are generally better than lists to use.

And here are examples of using some of these methods:

```
In [1]:   data = [3, 5, -7, 1, 9, -2]
          data.append(19)
          print(data)

          [3, 5, -7, 1, 9, -2, 19]
```

```
In [2]:   item = data.pop()
          print(data)


          [3, 5, -7, 1, 9, -2]

In [3]:   print(item)


          19
```

Self-Test Answer 9-6

String objects do not really have public attributes. Here are a few methods:

- `find` Return the lowest index where the string passed in as an argument is found in the base object's contents. If the argument is not found, `-1` is returned.
- `isupper` Returns `True` if the string object is all uppercase.
- `join` Concatenate elements of a list using the string object's contents as the "glue" between the elements and return the result.
- `split` Break-up a string into a list of substrings that are set apart by a delimiter character(s).
- `strip` Return a copy of the string with leading and trailing whitespace removed.

And here are examples of using some of these methods:

```
In [1]:   label = "Sample of ash from Mt. St. Helens"
          print(label.isupper())


          False

In [2]:   print(label.split())


          ['Sample', 'of', 'ash', 'from', 'Mt.', 'St.', 'Helens']

In [3]:   single_col = '\n'.join(label.split())
          print(single_col)


          Sample
          of
          ash
          from
          Mt.
          St.
          Helens
```

Self-Test Answer 9-7

Here is a solution:

```
In [1]:  output = "The beam is {0:.1f} m long.".format(length)
         print(output)

         The beam is 23.7 m long.
```

Self-Test Answer 9-8

Here are the values of the arrays:

```
In [1]:  print(data1)

         [  2 -37  -3  -9]
```

```
In [2]:  print(data2)

         [-18   2  -1   7]
```

```
In [3]:  print(data3)

         [-18   2  -1   7]
```

```
In [4]:  print(data4)

         [-18   2  -1   7]
```

The change to `data1` in line 5 changes only `data1`. Line 2 is not reevaluated after line 5 is executed. The change to `data2` in line 6 also affects `data3` and `data4` because `data3` is essentially an alias of `data2` and `data4` is essentially an alias of `data3`.

Self-Test Answer 9-9

Use the `copy` method attached to the list or array object, such as in line 2 below:

```
In [1]:  data1 = np.array([2, 7, -3, -9])
         data2 = data1.copy()
         data1[1] = -37
         print(data1)

         [  2 -37  -3  -9]
```

```
In [2]:   print(data2)

          [ 2   7 -3 -9]
```

Changes to data1 do not affect data2, because the latter is a memory-independent copy of the former.

Self-Test Answer 9-10

Here is the answer using file objects:

```
1    filein = open('values.txt', 'r')
2    inputstr = filein.readlines()
3    filein.close()
4
5    multi_col_list = []
6    for i in range(len(inputstr)):
7        line = inputstr[i]
8        line_values = line.split()
9        multi_col_list.append(line_values)
10
11   str_array = np.array(multi_col_list)
12   values = str_array.astype('d')
```

The solution is nearly the same as for Try This! 9-11, except the filename is different in the open statement of line 1, all rows of multi_col_list are used in line 11 when converting the list to an array, and the name of the array is changed to values in line 12. This code works for files of any number of columns because the call to split to obtain the column values automatically "knows" how many columns there will be, based upon the column-separation string.

Here is the answer using the NumPy function genfromtxt:

```
values = np.genfromtxt('values.txt')
```

Even simpler!

Self-Test Answer 9-11

```
1    with open('velocities.txt', 'r') as fobj:
2        velocities = fobj.readlines()
```

Self-Test Answer 9-12

This code will solve the problem:

```
1    import csv
2    import numpy as np
3
4    fileobj = open('values.txt', 'r')
5    readerobj = csv.reader(fileobj)
6    multi_col_list = []
7    for row in readerobj:
8        multi_col_list.append(row)
9    fileobj.close()
10
11   str_array = np.array(multi_col_list)
12   values = str_array.astype('d')
```

While it is nice to name CSV files with the suffix *.csv*, CSV files are just text files that follow certain conventions. Remember, because of the way `readerobj` works, `multi_col_list` is a list of lists, not a list of strings.

Self-Test Answer 9-13

Line 2 in the `try` body stops executing (it does not successfully complete executing because it raises an exception) and program execution is transferred to the `except` block (assuming the exception class is the class listed in the `except` statement). None of the other lines in the `try` body after line 2 of the body is executed.

If the exception class raised is not the class listed in the `except` statement, the exception that was raised is not successfully "caught," and control of the program goes up one level from the level of the `try/except` statement. If there still exists no `except` statement that can handle this exception, control of the program goes up another level, and so on. If we reach the uppermost interpreter level and there is still no `except` statement that can handle this exception, the program ends.

10   Managing Files, Directories, and Programs

One of the neat features of Python is that it works very well as a "glue" language. That is, it enables us to manage many, if not all, parts of a scientific or engineering workflow, within a single computing environment. In earlier chapters, we saw how Python can be used for reading in data, doing calculations on the data, and making plots of the results. We have also seen that Python can be used to generate "data," in the form of model or simulation results, which can also be analyzed and graphed. In the past, scientists and engineers used separate programs to work on each of these steps in their workflow, and these separate programs did not communicate with each other except through files. In Python, all these workflow tasks can occur in a single interpreter session, and the data or information can be passed from one task to another by passing or accessing variables (or objects).

Python, though, can also manage the parts of a scientific or engineering workflow that are the responsibility of the operating system, such as creating, deleting, and moving files, directories, and programs. For most of us, we accomplish these tasks by manipulating icons (representing files, folders, or programs) on our computer. If the number of files we need to manage is small, the select-and-drag approach is quick and effective. But what if the number of files is large, say in the hundreds or thousands? The select-and-drag approach quickly becomes a nightmare.

Not infrequently, in scientific or engineering applications, we encounter situations where we have hundreds or thousands of files to manage. Any kind of remote sensing device – e.g., a satellite instrument, an in-situ sensor – will generate data files, and over the course of a long period of measurement, this will result in a large number of files. In this chapter, we will find that with very few lines of code, Python can tackle the management of five or five million files, and can do so in an operating system-independent way. The Python code we write to manage files on a Windows system can also work on a Mac OS X system, with no change in the code. Very cool!

A Note of Warning   Because Python's file and directory manipulation tools are so powerful (we can easily delete basically every file on our computer if we are not careful), we recommend that beginners learning to use these tools create a scratch directory or folder and do all their work in there, until they have gotten the hang of using these functions and commands. Change the **current working directory** to this scratch directory. This will decrease the potential of inadvertently messing up the file system. See Section 10.2.1 for details regarding how to change the current working directory.

```
1    #- Module imports:
2
3    import numpy as np
4    import os
5    import shutil
6
7
8    #- Function to run model and write output file:
9
10   def model_run(N_crit=0, max_eat_rate=0.1, use_energy=0.05,
11                 filename=None):
12       #+ Set random seed, number of positions, etc.:
13       np.random.seed(31493293)
14       n_positions = 5001
15       domain_shape = (200, 200)
16       food = np.zeros(domain_shape) + 5.0
17       N = np.zeros(domain_shape)
18
19       #+ Set arrays for walker:
20       energy = np.zeros(n_positions, dtype='d')
21       x = np.zeros(n_positions, dtype='l')
22       y = np.zeros(n_positions, dtype='l')
23
24       energy[0] = 0.3
25       x[0] = domain_shape[1] // 2
26       y[0] = domain_shape[0] // 2
27
28       step_x = np.array([ 0,  1, 1, 1, 0, -1, -1, -1])
29       step_y = np.array([-1, -1, 0, 1, 1,  1,  0, -1])
```

Figure 10.1 Code to create multiple model runs of a random walker under different constraints, following the algorithm of Figure 8.4. The output from the model runs are written to separate files and grouped in directories by N_{crit}.

10.1 Example of Managing Files, Directories, and Programs

In Section 8.1, we looked at a random walk model that simulates the propagation of a bacterial colony under a set of given conditions. In the current section, we create code that runs that model for different values of N_{crit}, the maximum rate at which the walker consumes food, and the amount of energy the walker uses each step. These model runs are run automatically through combinations of the above parameters, save the results of each model run to a unique file, and place output files with the same value of N_{crit} in a separate subdirectory. The code to do these tasks is found in Figure 10.1.

We import the modules in lines 3–5. In lines 10–64, we place all the code to run the model (i.e., the code from Figure 8.4) into a function. That way, each time we want to make another model run, we just call the function rather than retyping all the lines of code that make

```
30    #+ Take steps:
31    i = 1
32    while i < n_positions:
33        step_idx = np.random.randint(0, np.size(step_x))
34        xtrial = x[i-1] + step_x[step_idx]
35        ytrial = y[i-1] + step_y[step_idx]
36
37        if (xtrial >= domain_shape[1]) or (xtrial < 0):
38            continue
39        if (ytrial >= domain_shape[0]) or (ytrial < 0):
40            continue
41
42        if N[ytrial,xtrial] < N_crit:
43            N[ytrial,xtrial] += 1
44            x[i] = x[i-1]
45            y[i] = y[i-1]
46        else:
47            if energy[i-1] > 0.0:
48                x[i] = xtrial
49                y[i] = ytrial
50            else:
51                x[i] = x[i-1]
52                y[i] = y[i-1]
53
54        energy_from_eat = np.min([ max_eat_rate, food[y[i-1],x[i-1]] ])
55        energy[i] = energy[i-1] + energy_from_eat - use_energy
56        food[y[i-1],x[i-1]] -= energy_from_eat
57
58        if energy[i] < 0.0:
59            energy[i] = 0.0
60
61        if food[y[i-1],x[i-1]] < 0.0:
62            food[y[i-1],x[i-1]] = 0.0
63
64        i += 1
```

Figure 10.1   (part 2)

the calculations for the model. In this function, as is seen in line 10, the values of N_crit, max_eat_rate, and use_energy are passed in as keyword input parameters, rather than set in the body of the program as in Figure 8.4. Thus, lines 14, 19, and 20 in Figure 8.4 are not reproduced in Figure 10.1. The function model_run also takes the output filename as a keyword input parameter (line 11), so we can customize where we want the output from the model run to go.

Because most of the model code in model_run is duplicated from Figure 8.4, we do not discuss lines 10–64 in Figure 10.1. Lines 66–70 in model_run are new, however, and write out the x- and y-direction index values at each step, for the model run, as two tab-separated columns (with the first line as a header), to the output file given by filename.

```
65      #+ Write out files of positions in current working directory:
66      fileobj = open(filename, 'w')
67      fileobj.write('x' + '\t' + 'y\n')
68      for i in range(n_positions):
69          fileobj.write(str(x[i]) + '\t' + str(y[i]) + '\n')
70      fileobj.close()
71
72
73  #- Run models for various parameters and group files:
74
75  os.chdir('manage_files_progs_model')
76
77  for iN_crit in [0, 1]:
78      iN_crit_dir = 'N_crit' + str(iN_crit)
79      if not os.path.exists(iN_crit_dir):
80          os.mkdir(iN_crit_dir)
81
82      for imax_eat_rate in [0.1, 0.2, 0.3]:
83          for iuse_energy in [0.05, 0.1, 0.2]:
84              ifilename = 'N_crit' + str(iN_crit) + '-' \
85                          + 'max_eat_rate' + str(imax_eat_rate) + '-' \
86                          + 'use_energy' + str(iuse_energy) \
87                          + '.txt'
88              print( 'Calculate output: ' + ifilename)
89              model_run(N_crit = iN_crit,
90                        max_eat_rate = imax_eat_rate,
91                        use_energy = iuse_energy,
92                        filename = ifilename)
93              shutil.move(ifilename, iN_crit_dir)
94
95  os.chdir('..')
```

Figure 10.1   (part 3)

Lines 75–95 are the bulk of the main program and call the `model_run` function for each of the combinations of three model parameters. Line 75 changes the current working directory to a directory *manage_files_progs_model*, which is assumed to exist. The loop begun by line 77 iterates through two values of N_{crit}, zero and one. For each value of N_{crit}, a subdirectory is created beginning with *N_crit* and ending with either *0* or *1* (lines 78–80). In lines 82–93, we loop through all possible combinations of `max_eat_rate` and `use_energy` given possible values of `max_eat_rate` of 0.1, 0.2, and 0.3, and possible values of `use_energy` of 0.05, 0.1, and 0.2. Using those values, we construct a filename that describes the specific combination (lines 84–87), prints a message to the screen to let us know how things are progressing (line 88), calls `model_run` (lines 89–92), and moves the output file to the applicable N_{crit} subdirectory (line 93).

The message that is printed to the screen as the program is progressing includes the filename that is being generated. The first few lines printed out look like:

```
1   Calculate output: N_crit0-max_eat_rate0.1-use_energy0.05.txt
2   Calculate output: N_crit0-max_eat_rate0.1-use_energy0.1.txt
3   Calculate output: N_crit0-max_eat_rate0.1-use_energy0.2.txt
4   Calculate output: N_crit0-max_eat_rate0.2-use_energy0.05.txt
```

Each filename output is uniquely named using the values of N_{crit}, max_eat_rate, and use_energy for that particular run. The first few lines of *N_crit0-max_eat_rate0.1-use_energy0.05.txt* are:

```
1   x        y
2   100      100
3   101      100
4   102      99
5   102      98
6   103      99
7   102      100
8   102      99
```

All other output files have a similar look and each file contains 5002 lines.

We can use an operating system command, via Python, to obtain the line count of a file. On a Linux or Mac OS X system, to obtain the number of lines in the file *N_crit0-max_eat_rate0.1-use_energy0.05.txt*, the following code will work:

```
1   import os
2   filename = 'N_crit0-max_eat_rate0.1-use_energy0.05.txt'
3   os.system('wc -l ' + filename)
```

The Linux or Mac OS X wc -l command counts the number of lines in whatever file is specified after it. Windows has equivalent commands.[1] The system command of the os module executes the operating system command that is given by the string passed in to the system call.

10.2 Python Programming Essentials

As the example in Section 10.1 illustrates, Python enables us to create files and directories and move files into directories by providing tools that interface with the tools the computer's operating system provides to manage files and directories. Some of these tools directly call the operating system's tools (e.g., system). Others are Python "versions" of operating system or **shell** commands (e.g., mkdir). A shell is a text-based interface to an operating system. A common shell for Windows is PowerShell. For Linux, bash (or the Bourne Again SHell) is widely used. Mac OS X, as a Unix-based operating system, also has access to the common

---

[1] https://superuser.com/a/959037 (accessed September 3, 2019).

shells available to Linux. When combined with the other capabilities the Python programming language gives us – to do a task over and over again using loops, to execute code conditionally using branching statements – we are able to automate much of our scientific and engineering modeling and data analysis workflow, even in cases where we are managing many files or analysis cases.

In this section, we examine the tools Python gives us to manage files, directories, and programs. In Sections 10.2.1–10.2.4, we focus on operations with files and directories: what is a filename, path, and working directory, how to create and remove directories, and how to move, rename, copy, and delete files and directories. In Sections 10.2.5 and 10.2.6, we discuss how to list the contents of a directory and test individual files to see what kind of "file" something is. Because files and directories are described by strings, the task of looping through lists of strings is effectively the same as looping through files and directories. Lastly, in Section 10.2.7, we discuss how to run programs from within Python.

10.2.1 Filenames, Paths, and the Working Directory

All files have a filename: we keep track of files by their name. But, files are not sprinkled in random places on a computer. Instead, they are organized inside collections of files called folders or directories. Directories can also hold other directories, called subdirectories, which in turn can hold other files and subdirectories, and so on. The **path** to a directory or file is the sequence of directories (plus filename, if applicable) we go through to reach the directory or file at the end of that sequence. Figure 10.2 shows a **directory tree** – a hierarchical list or "tree" of files starting from the directory *MyFiles*. To reach the file *Wind.txt*, the path would start with *MyFiles*, go into the *StationData* directory, then into the *June* directory, and end with the file of interest, *Wind.txt*. On a Linux or Mac OS X system, this path would be written as:

MyFiles/StationData/June/Wind.txt

whereas on a Windows system, this path would be written as:

MyFiles\StationData\June\Wind.txt

Depending on the operating system we are using, the character that separates different directories in the path may differ. In Linux and Mac OS X, the character is a forward slash ("/"). In Windows, the character is a backslash ("\"). The backslash character in a Python string is entered in by putting a backslash after a backslash, i.e., `'\\'`. When that string is `printed` out, only one backslash displays.

Python's os.path module (the path submodule of the os module) provides tools to help us construct and manipulate paths, accounting for the difference between the directory separation characters used in different operating systems.[2] The `join` function of os.path takes a set of string arguments and joins them together into a single string with the correct

---

[2] See the Python documentation for os.path for more details. We provide an up-to-date link to the documentation at www.cambridge.org/core/resources/pythonforscientists/refs/, ref. 60.

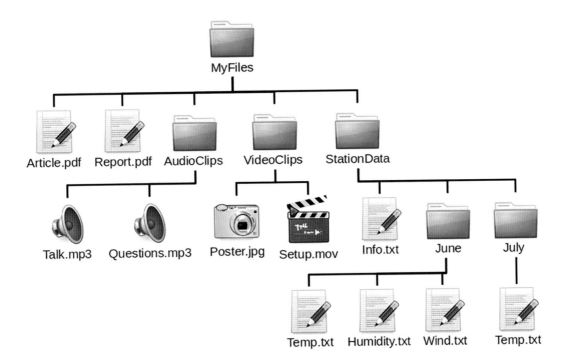

Figure 10.2 A directory tree.

character separating directories for the operating system being used. For instance, for our *Wind.txt* path example above, join would give:

```
In [1]:  import os.path
         mypath = os.path.join("MyFiles", "StationData", "June", "Wind.txt")
         print(mypath)
```

```
MyFiles/StationData/June/Wind.txt
```

The above code is executed on a Linux system, so the path separation character used is a forward slash. The result is a string, saved as mypath, that lists the directories we need to traverse to reach the final *Wind.txt* file. Thus, it describes the exact location of that specific *Wind.txt* file, relative to the present location. This is an example of a **relative path**, a path specification relative to the present location. The present location is called the current working directory. By default, files are read from and written out to the current working directory. By default, when we start Python or Jupyter using a command-line interface, the current working directory is the location we are in when we started the Python interpreter. In other environments, a different directory may be set as the current working directory by default.

In addition to functions that create a path, the os.path module also has functions to manipulate paths. The abspath function gives the **absolute path** (a.k.a., "**full path**") for a given path. The absolute path is the path that is anchored (i.e., starts from) the highest-level location of the computer. On a Windows machine, the highest-level location is a drive letter. On a Linux or Mac OS X machine, the highest-level location location is the root directory. Thus, the absolute path for every file and directory on the computer is unique. Here is an example of what `abspath` gives us:

```
In [1]:  import os.path
         mypath = os.path.join("MyFiles", "StationData", "June", "Wind.txt")
         print(os.path.abspath(mypath))
```

```
/home/jlin/MyFiles/StationData/June/Wind.txt
```

In this example, we started Python while we were in the */home/jlin* directory, so the full path to `mypath` is */home/jlin/MyFiles/StationData/June/Wind.txt*. That is, it is the full path to the current working directory plus the relative path described by `mypath`, and this entire specification is the string `abspath` returns.

The function `basename` returns the file or directory that is at the end of the path. For the above example:

```
In [1]:  import os.path
         mypath = os.path.join("MyFiles", "StationData", "June", "Wind.txt")
         print(os.path.basename(mypath))
```

```
Wind.txt
```

If we want a straightforward way to figure out what kind of file is specified by path, via the file's extension, we can use the `splitext` function to split off the file extension from the rest of the path. For example:

```
In [1]:  import os.path
         mypath = os.path.join("MyFiles", "StationData", "June", "Wind.txt")
         print(os.path.splitext(mypath))
```

```
('MyFiles/StationData/June/Wind', '.txt')
```

The `splitext` function returns a two-element tuple with the extension in the second element and the rest of the path in the first element. If the path does not have a file extension, the second element of the tuple is an empty string. We can run tests on what file extension is returned and have our program respond accordingly.

Importing the os module automatically gives us access to the os.path module. We do not also have to import the os.path module separately, if we already imported the os module. Thus, to access both os and os.path functions, a single `import os` line is adequate. We still have to specify os or os.path in front of the function we are calling, however, if we use a standard `import` statement.

In addition to managing paths, Python also allows us to manage the current working directory. There are a number of ways of obtaining and changing the current working directory. Many Interactive Development Environments (IDEs) use a graphical user-interface (GUI) to enable us to manage the current working directory. Spyder's GUI method is described at www.cambridge.org/core/resources/pythonforscientists/spyder/.

We can also use Python commands to manage the current working directory, and this method works regardless if we are running Python in a terminal, Jupyter, Spyder, or another IDE. We obtain and change the current working directory from within the interpreter session or a Python code file using the `getcwd` and `chdir` methods, respectively, of the os module. For instance:

```
In [1]:   import os
          print(os.getcwd())


          /home/jlin
```

```
In [2]:   os.chdir('/home/jlin/Desktop')
          print(os.getcwd())


          /home/jlin/Desktop
```

In this code snippet, we display the original current working directory in cell In [1], using the `getcwd` method. The return value of the method is a string. In the first line of In [2], we use the `chdir` command to set the current working directory to a different directory (this example is on a Linux system). In the final line of In [2], we display the current working directory again, to show the current working directory was successfully changed. Using `getcwd` and `chdir` to manipulate the current working directory is less intuitive, because it uses typed-in commands rather than a GUI, but these commands enable us to manipulate the current working directory through a program or script, which means we can automate that task.

In Figure 10.1, we use `chdir` in line 75 to change the current working directory to *manage_files_progs_model*. All output files generated by calls to `model_run` will then be written into that directory (after which we can move the file as we wish). In line 95, we use `chdir` again to change the current working directory to the parent directory of *manage_files_progs_model*, that is, the directory we were in before we executed line 75. For Linux, Mac OS X, and Windows, the string meaning "parent directory" is two periods, one after the other with no space between them. The os module also stores the parent directory string as the attribute `pardir`, so we could also rewrite line 95 as:

```
os.chdir(os.pardir)
```

10.2.2 Making and Removing Empty Directories

To create a new, empty directory, we can use the `mkdir` command from the os module. In this example, we create the *August* directory in the *StationData* directory in the directory tree of Figure 10.2 (assuming *MyFiles* is relative to the current working directory):[3]

---

[3] Reference for parts of this paragraph: https://docs.python.org/2/library/os.html (accessed November 9, 2016).

```
1   import os
2   mypath = os.path.join("MyFiles", "StationData", "August")
3   os.mkdir(mypath)
```

The mkdir command, however, assumes that all intermediate directories to the directory we are creating already exist. In the above example, if the *MyFiles* and *StationData* directories do not already exist, the call to mkdir will fail and return an error.

If we want to create all the intermediate directories in addition to the final directory (also called the **leaf directory**), we should use makedirs:[4]

```
1   import os
2   mypath = os.path.join("MyFiles", "SatelliteData", "Landsat")
3   os.makedirs(mypath)
```

In this example above, the makedirs command creates not only the leaf directory, *Landsat*, but also the intermediate directory *SatelliteData*, which does not exist in the Figure 10.2 directory tree. The makedirs function does all this in one fell swoop.

For both the mkdir and makedirs calls above, when they are executed in the interpreter, no message is printed by the interpreter. The commands create the directories on the computer and that is all they do. No news means the command executed "successfully." We should check, though, that we typed everything in correctly. If we made a typo, "successful" will not be the same as "the desired outcome" ☺. For both mkdir and makedirs, we are creating empty directories that have nothing inside them (except parent directories that are part of the path). We address the case of copying and deleting directory trees with files in them in Section 10.2.4.

Given makedirs, why would we ever want to use mkdir? Admittedly, makedirs is powerful, but with such power comes the potential to mess things up. Unless we know that we want to create all the intermediate directories in a path, it would be wiser to use mkdir to limit any inadvertent mistakes.

The mkdir and makedirs functions only work if the directory they are trying to make does not already exist. If the directory already exists, a FileExistsError exception is thrown. This is why in line 79 of Figure 10.1, before we use mkdir to create the directory given by iN_crit_dir, we test whether that directory already exists. The os.path function exists returns True if the path specified in the input argument already exists and False otherwise.

To remove an empty directory, we can use the os module's rmdir function. For instance, to remove the *Landsat* directory we created above, assuming the current working directory is the parent of *MyFiles* and *Landsat* is empty:

---

[4] Reference for parts of this paragraph: https://docs.python.org/2/library/os.html (accessed November 9, 2016).

```
1   import os
2   mypath = os.path.join("MyFiles", "SatelliteData", "Landsat")
3   os.rmdir(mypath)
```

Only *Landsat* is removed by the above code. The *MyFiles* and *SatelliteData* directories are unaffected. If the directory passed into `rmdir` is not empty, an `OSError` exception is thrown.

Removing a directory that is not empty is a different kind of task, as it involves going through each subdirectory under that directory (and any subdirectories under each subdirectory, and so on) and deleting all files found and subdirectories found. We address how to remove directory trees filled with files in Section 10.2.4.

10.2.3 Moving and Renaming Files and Directories

The shutil module has a number of functions that simulate shell commands. The `move` function can be used to move a file to another directory or to rename a file. The function takes two arguments, the source and its destination (both strings). If the destination is a directory, the source file is moved into the destination directory. If the destination is not a directory, the file is renamed.[5] An example of using `move` to move a file into a directory is given in line 93 of Figure 10.1: `move` places the file given by the string variable `ifilename` into the directory given by the string variable `iN_crit_dir`.

The `move` function also works for renaming directories. This example renames the Figure 10.2 *June* directory in *StationData* to *June2019* (assuming our current working directory is the parent to *MyFiles*):[6]

```
1   import os.path
2   import shutil
3   june_path = os.path.join("MyFiles", "StationData", "June")
4   new_june_path = os.path.join("MyFiles", "StationData", "June2019")
5   shutil.move(june_path, new_june_path)
```

If the current working directory is the *StationData* directory, the renaming is even easier. With the imports removed, it is:

```
shutil.move("June", "June2019")
```

There is no need to bother with `os.path.join` because we are only referring to paths made up of a single directory name and level.

We can also use shutil's `move` function to move directories (and, with it, everything in that directory) into another directory. Thus, if we wanted to create a directory called *Media* inside *MyFiles* (in the Figure 10.2 tree) and move *AudioClips* and *VideoClips* into *Media*, we would execute the following code (we are assuming the current working directory is *MyFiles*):

---

[5] Reference for parts of this paragraph: https://docs.python.org/2/library/shutil.html (accessed November 7, 2016).
[6] Reference for parts of this paragraph: https://docs.python.org/2/library/shutil.html (accessed November 9, 2016).

```
1    import os
2    import shutil
3    os.mkdir("Media")
4    shutil.move("AudioClips", "Media")
5    shutil.move("VideoClips", "Media")
```

10.2.4 Copying and Deleting Files and Directories

To copy a file, we can use shutil's `copy2` function. Like the `move` function, `copy2` also takes two arguments, the name of the source and then the name of the destination. For instance, to make a copy of all the files in the directory *June* in Figure 10.2, with the prefix *copy-* appended to the front of each filename, we can use the following code (assuming the current working directory is *June*):

```
1    import shutil
2
3    filenames = ['Temp.txt', 'Humidity.txt', 'Wind.txt']
4
5    for ifile in filenames:
6        shutil.copy2(ifile, 'copy-' + ifile)
```

For the code snippet above, the copies will be in the same directory as the original files. However, as powerful as `copy2` is, some **metadata** – information about the file – attached to the file will not transfer. This can include, for instance, access control lists (ACLs) that are attached to files in Mac OS X.[7]

The example above is the first use we have seen of a `for` loop to iterate through a list of non-numerical items. In previous chapters, and in the example in Section 10.1, we have more or less confined our use of `for` loops to looping through lists and arrays of numbers. However, `for` loops can be used to loop through any iterable structure, in this case strings. At each iteration, of the loop in line 5, `ifile` is set to each of the strings in `filenames`, one in turn. Thus, in the first iteration, `ifile` is set to `'Temp.txt'`, in the second iteration, `'Humidity.txt'`, and in the third iteration, `'Wind.txt'`.

Because `ifile` is a string at each iteration, `ifile` has all the attributes and methods of a string available to it. For instance, let us modify the problem in the above example to say we want to make a copy of all the files in the directory *June* in Figure 10.2, with the prefix *copy-* appended to the front of each filename, *but with the filename portion in the new filename in all caps.* This code would solve this problem:

---

[7] Reference for parts of this paragraph: https://docs.python.org/2/library/shutil.html (accessed November 7, 2016).

```
1  import shutil
2
3  filenames = ['Temp.txt', 'Humidity.txt', 'Wind.txt']
4
5  for ifile in filenames:
6      shutil.copy2(ifile, 'copy-' + ifile.upper())
```

The new copies would be named: *copy-TEMP.TXT*, *copy-HUMIDITY.TXT*, and *copy-WIND.TXT*. We accomplish this by calling `ifile.upper()` in line 6. This produces an uppercase version of the string `ifile`. The object nature of variables in Python, coupled with the fact that the iterator variable in a `for` loop *becomes* the object it is assigned to for that iteration of the loop, gives us a powerful tool to automate method calls, by looping through the objects that have those methods.

The os module contains a `remove` function that will remove files. In the example below, we take the files in the directory *AudioClips* in Figure 10.2 and delete them. Again, the script assumes the current working directory is *AudioClips*:

```
1  import os
2
3  filenames = ['Talk.mp3', 'Questions.mp3']
4
5  for ifile in filenames:
6      os.remove(ifile)
```

Copying and deleting directories takes more work than copying and deleting files, as the copy and deletion has to extend to all the files and subdirectories that are part of the directory we are copying or deleting. For instance, to copy the `MyFiles` directory in Figure 10.2, we need to copy *Article.pdf*, *Report.pdf*, and the *AudioClips*, *VideoClips*, and *StationData* subdirectories, and all the files and subdirectories of those subdirectories, and so on.

To make a copy of a directory, we use the shutil function `copytree`. Like the move function, it takes two arguments, strings that specify the source and the destination of the copy. The following makes a copy of *StationData* called *StationDataCopy*. Both the original and copy are in the same directory (the current working directory):[8]

```
1  import shutil
2  shutil.copytree("StationData", "StationDataCopy")
```

The `copytree` function does not copy over all the metadata. If we care about copying *all* the metadata, we probably have to write a shell script, and use the tools provided natively by the operating system.

---

[8] Reference for parts of this paragraph: https://docs.python.org/2/library/shutil.html (accessed November 9, 2016).

To remove a directory and all its subdirectories and files, we can use the shutil function rmtree. The following will remove *StationData*, and everything in *StationData*, from the current working directory:[9]

```
1  import shutil
2  shutil.rmtree("StationData")
```

If the file path passed into rmtree does not exist, Python will raise a FileNotFoundError exception.

In the examples of copytree, rmtree, and move operating on directories that we have looked at, these functions have done all the hard work for us. We have not discussed how the functions know to operate on each file and subdirectory at each level of the directory tree. The "how" we cover in Chapter 20, when we consider how to search through a directory tree. The methods discussed there, principally the idea of **recursion** (that is, of following a pattern over and over down levels), also work to copy, delete, and move directory trees. Thus, the kinds of copying, removing, and moving that we discussed in this section are also called **recursive** operations.

10.2.5 Listing the Contents of a Directory

In our earlier examples of copying, removing, and moving files, we created a list of files through which to loop. But oftentimes, the list of files we are interested in operating on is the contents of a directory. If there is a way of obtaining that, it would save us from having to type in the list of filenames. The os module's listdir function will do what we want. Consider the following examples, assuming we are in the *MyFiles* directory of the directory tree in Figure 10.2:[10]

```
In [1]:  import os
         print(os.listdir('.'))


         ['AudioClips', 'StationData', 'Report.pdf', 'Article.pdf', 'VideoClips']

In [2]:  print(os.listdir(os.getcwd()))


         ['AudioClips', 'StationData', 'Report.pdf', 'Article.pdf', 'VideoClips']
```

In the In [1] example, we pass in the current directory symbol (which in Linux, Mac OS X, and Windows is a period) to the call to listdir and the function returns the list of the contents of the current working directory (which is *MyFiles*). In the In [2] example, we pass in the return value of a call to the getcwd function, which returns the full path of the current working directory. The current directory symbol is also stored as the curdir attribute of the

---

[9] Reference for parts of this paragraph: https://docs.python.org/2/library/shutil.html (accessed November 9, 2016).
[10] Reference for parts of this paragraph: https://docs.python.org/2/library/os.html (accessed November 10, 2016).

os module. If `listdir` is called without an argument, a listing is generated of the current working directory's contents. Note, the "current directory symbol" is an operating system-specific abbreviation for the current working directory. It is not the path of that directory.

Sometimes, we are interested in listing only a subset of the contents of a directory. The glob module has a function called `glob` that will do basic pattern matching when doing a directory listing. If we are in the *MyFiles* directory of the directory tree in Figure 10.2, we obtain the following using `glob`:[11]

```
In [1]: import glob
        print(glob.glob('*'))


        ['AudioClips', 'StationData', 'Report.pdf', 'Article.pdf', 'VideoClips']

In [2]: print(glob.glob('*.pdf'))


        ['Report.pdf', 'Article.pdf']
```

In both calls to `glob`, the asterisk is a wildcard character meaning "any character or set of characters." Thus, in the In [1] `glob` call, the function looks for all contents of *MyFiles*, i.e., items of any name. In the In [2] call, the function looks for all contents whose name ends in *.pdf*. More complex pattern matching is possible. See the glob module documentation for details. We provide an up-to-date link to the documentation at www.cambridge.org/core/resources/pythonforscientists/refs/, ref. 55.

10.2.6 Testing to See What Kind of "File" Something Is

In Section 10.2.2, we learned about the os.path module's `exists` function, which tests to see whether a file or directory currently exists. Sometimes, we want to know not only if a file or directory exists but whether the item in question is a file or a directory, before attempting to do something with the item. If the item turns out to be a directory, we cannot open the item and use `readlines` on it, but we can list its contents. The functions we discuss below are in the os.path submodule. All of these functions take the filename as a string as input:[12]

- `isdir` Returns True if the file is a directory.
- `isfile` Returns True if the file is a regular file.
- `islink` Returns True if the file is a symbolic link (i.e., an alias).

For both `isfile` and `isdir`, if the file is a link, the link is followed until its termination. We can, for instance, make an alias of an alias of an alias, etc. Following a link means following that chain until we reach the original file or directory. If there is a file or directory at the end of the chain, the function will return True or False as appropriate.

---

[11] Reference for parts of this paragraph: https://docs.python.org/2/library/glob.html (accessed November 10, 2016).
[12] Reference for parts of this paragraph: https://docs.python.org/2/library/os.path.html (accessed November 7, 2016).

Here are the results from using these functions in a Jupyter notebook on files in Figure 10.2, assuming the current working directory is *MyFiles*:

```
In [1]:  import os.path
         print(os.path.isfile('Article.pdf'))

         True

In [2]:  print(os.path.isfile('AudioClips'))

         False

In [3]:  print(os.path.isdir('AudioClips'))

         True

In [4]:  print(os.path.islink('AudioClips'))

         False

In [5]:  print(os.path.islink('Article.pdf'))

         False
```

Both the file and directory above are regular items, not links (i.e., not aliases), and so the `islink` call returns `False`.

10.2.7 Running Non-Python Programs in Python

In Section 10.1, we created a function `model_run` to make a run of our random walk model. This is a common way of running a Python program, as putting the model code into a function allows us to make repeated model runs by calling the function. We are, however, not limited in Python to running only other Python programs. We can, from a Python `script` or program, run nearly any program that is available for the operating system to run. We will also see in Chapter 25 how to connect Python to functions and variables from programs written in other programming languages.

In Section 10.1, we looked at an example of using the os module's `system` function. This function takes a string as input and executes that string as a command in the operating system. Thus, the call to `system` from that example (partly reproduced here):

```
filename = 'N_crit0-max_eat_rate0.1-use_energy0.05.txt'
os.system('wc -l ' + filename)
```

is the same as if we had opened a terminal window on a Linux/Mac OS X computer and typed in:

```
wc -l N_crit0-max_eat_rate0.1-use_energy0.05.txt
```

at the prompt in the command-line interface to the operating system. Put another way, the system call runs the Linux/Mac OS X wc program that counts the number of lines in *N_crit0-max_eat_rate0.1-use_energy0.05.txt*.

Why bother with using system? Why not type in a command directly on a computer's terminal window instead (or use the operating system's GUI)? Why go to the trouble of writing this in Python? Because system processes a string, we can use all the methods, programming structures, and operators Python provides to manipulate strings to construct the command we want system to execute. This includes the myriad packages written for Python that can be used to analyze data, not just to manipulate strings. This means we can write a script to automate whatever task we want the operating system program to do and that this script can be much more sophisticated than either typing in commands directly in the terminal, using a GUI, or through a shell script.

10.3   Try This!

The examples in this section address the main topics of Section 10.2. We practice manipulating paths, conducting various file and directory management operations, and running non-Python programs from within Python.

Try This! 10-1 Paths and Current Working Directory

Consider the Figure 10.2 directory tree. If our current working directory is *MyFiles*, how would we change the current working directory to *July*?

Try This! Answer 10-1

This code solves the problem:

```
1   import os.path
2   july_path = os.path.join("StationData", "July")
3   os.chdir(july_path)
```

Try This! 10-2 Making an Empty Directory

Consider the Figure 10.2 directory tree. If our current working directory is *MyFiles*, how would we create a new directory under *MyFiles* called *Papers*?

Try This! Answer 10-2

This code solves the problem:

```
import os
os.mkdir("Papers")
```

We do not need to use `join` to create a path because the *Papers* directory will be in the current working directory.

Try This! 10-3 Removing an Empty Directory

Consider the directory *Papers* created in Try This! 10-2. If our current working directory is *MyFiles*, how would we delete *Papers* if *Papers* is empty and print out a message saying "Papers not deleted" if *Papers* is not empty?

Try This! Answer 10-3

This code solves the problem:

```
import os
try:
    os.rmdir("Papers")
except OSError:
    print("Papers not deleted")
```

The `OSError` is thrown if *Papers* is not empty. By putting the `rmdir` call into a `try`/`except` block, we can appropriately handle the case if *Papers* is not empty.

Try This! 10-4 Moving and Testing Files: Random Walk Model Runs

Consider the Figure 10.1 case of creating multiple random walk runs for various values of N_{crit}, `max_eat_rate`, and `use_energy`, and placing the output files into a *N_crit0* or *N_crit1* directory, as appropriate. In each of the *N_crit0* and *N_crit1* directories, create subdirectories named *max_eat_rate0.1*, *max_eat_rate0.2*, and *max_eat_rate0.3* and move the *.txt* files into those subdirectories, as appropriate.

Try This! Answer 10-4

The code in Figure 10.3 solves the problem. It assumes the current working directory is the directory where we have the subdirectories *N_crit0* and *N_crit1*. A few assorted items to note:

- In line 18, we customize the error message printed out in the case of an unanticipated file by including the name of the unanticipated file in the message. This makes it easier for us to diagnose what the problem might be.

```
 1   import os
 2   import shutil
 3
 4   for idir in ['N_crit0', 'N_crit1']:
 5       os.chdir(idir)
 6       os.mkdir('max_eat_rate0.1')
 7       os.mkdir('max_eat_rate0.2')
 8       os.mkdir('max_eat_rate0.3')
 9       for item in os.listdir('.'):
10           if os.path.isfile(item):
11               if item[8:23] == 'max_eat_rate0.1':
12                   shutil.move(item, 'max_eat_rate0.1')
13               elif item[8:23] == 'max_eat_rate0.2':
14                   shutil.move(item, 'max_eat_rate0.2')
15               elif item[8:23] == 'max_eat_rate0.3':
16                   shutil.move(item, 'max_eat_rate0.3')
17               else:
18                   raise ValueError("incorrect file: " + item)
19       os.chdir('..')
```

Figure 10.3 Code to solve Try This! 10-4.

- Although lines 11, 13, and 15 work as-is, we would prefer to search for whether the string `'max_eat_rate0.1'`, `'max_eat_rate0.2'`, etc., occurs *anywhere* in item rather than in a specific character range. In Section 19.2.1, we will learn about the in operator, which does exactly this.
- If our program changes the current working directory, it is a good habit at the end of the program to restore the current working directory to the location it was at at the beginning of the program. This is not a hard-and-fast rule, but it is a practice that prevents us from finding ourselves in a directory we did not expect to be in.

Try This! 10-5 Copying Directories: Random Walk Model Runs

Consider the Figure 10.1 case of creating multiple random walk runs for various values of N_{crit}, max_eat_rate, and use_energy, and placing the output files into a *N_crit0* or *N_crit1* directory, as appropriate. Make copies of *N_crit0* and *N_crit1* and call them *N_crit0_copy* and *N_crit1_copy*. All four directories can be in the same directory.

Try This! Answer 10-5

This code solves the problem assuming the current working directory is the directory that contains *N_crit0* and *N_crit1*:

```
1    import shutil
2    shutil.copytree("N_crit0", "N_crit0_copy")
3    shutil.copytree("N_crit1", "N_crit1_copy")
```

This solution will work even if we have changed *N_crit0* and *N_crit1* as specified in Try This! 10-4, because `copytree` copies all files and subdirectories in *N_crit0* and *N_crit1*, no matter how many levels of subdirectories there are.

Try This! 10-6 Deleting Directories: Random Walk Model Runs

Assume we have successfully created the *N_crit0_copy* and *N_crit1_copy* directories of Try This! 10-5. Also assume that the current working directory is *not* the directory that holds *N_crit0*, *N_crit1*, *N_crit0_copy*, and *N_crit1_copy* but *N_crit0*. How would we delete *N_crit1_copy*?

Try This! Answer 10-6

There are two main strategies for solving this problem. First, we can change the current working directory and then delete *N_crit1_copy*:

```
1    import os
2    import shutil
3    os.chdir('..')
4    shutil.rmtree('N_crit1_copy')
```

A second approach is to create a path that references *N_crit1_copy* from the current working directory:

```
1    import os
2    import shutil
3    file_path = os.path.join(os.pardir, 'N_crit1_copy')
4    shutil.rmtree(file_path)
```

Recall that `os.pardir` is the string `'..'` in Linux, Mac OS X, and Windows systems.

Both approaches work fine. The choice between approaches might boil down to whether we find it easier to think in terms of "change the current working directory to wherever the files are that we want to operate on" or "stay put and act on the files from our current location."

Try This! 10-7 Running Non-Python Programs in Python: Random Walk Model Runs

Consider the Figure 10.1 case of creating multiple random walk runs for various values of N_{crit}, `max_eat_rate`, and `use_energy`, and placing the output files into a *N_crit0* or *N_crit1* directory, as appropriate. Change the current working directory to *N_crit1* and print a list of the files in that directory. Do this once using `listdir` and once using the operating system's command for listing the contents of a directory. Assume the current working directory is the directory that holds *N_crit1*.

Try This! Answer 10-7

The code below works on a Linux or Mac OS X computer:

```
import os
os.chdir('N_crit1')
print(os.listdir())
print(os.system('ls'))
```

Recall that calling `listdir` without arguments gives the listing of the contents of the current working directory.

On a Windows computer, instead of passing in `'ls'` to the call to `system`, pass in the string `'dir'`. The other lines do not have to change if we move from one computer to another. By using the Python "versions" of operating system commands to change the current working directory (`chdir`) and list the contents of the current directory (`listdir`), we automatically have written code that can be used on any operating system Python runs on.

10.4 More Discipline-Specific Practice

The Discipline-Specific Jupyter notebooks for this chapter cover the following topics:

- Filenames, paths, and the working directory.
- Making and removing empty directories.
- Moving and renaming files and directories.
- Copying and deleting files and directories.
- Listing the contents of a directory.
- Testing to see what kind of "file" something is.
- Running non-Python programs in Python.

10.5 Chapter Review

10.5.1 Self-Test Questions

Try to do these without looking at the book or any other resources or using the Python interpreter. Answers to these Self-Test Questions are found at the end of the chapter.

Self-Test Question 10-1

Describe what is the current working directory.

Self-Test Question 10-2

Describe what the os.path module `join` does? Create a path describing the directory *data* that is located in the parent directory to the current working directory.

Self-Test Question 10-3

What is the difference between an absolute and relative path?

Self-Test Question 10-4

What do the os module's `getcwd` and `chdir` functions do?

Self-Test Question 10-5

Write code that creates a directory *data* in the parent directory *experiment*. Assume that *experiment* is in the current working directory. Before creating *data*, make sure the directory does not already exist.

Self-Test Question 10-6

Write code to remove the directory *calculations*, which is within the directory *results*. Assume the current working directory is the directory that *results* is in and that *calculations* is empty.

Self-Test Question 10-7

Write code to rename all files in the current working directory that end in *.txt* to end in *.dat*. Do *not* assume that the current working directory only contains files.

Self-Test Question 10-8

What is the difference between the shutil module's `copy2` and `copytree` functions?

Self-Test Question 10-9

What is the difference between the os module's `remove` and `rmdir` functions and shutil module's `rmtree` function?

Self-Test Question 10-10

Describe what the os module's `system` function does. What does the function accept as input?

10.5.2 Chapter Summary

Python enables us not only to create files but also to manage files and directories on our computers. We can also execute operating system commands from within the Python interpreter. As a result, we are able to write programs that can manipulate any number of files. Whether we are managing a few files or a few million files, the number of lines of code needed to accomplish the task are the same. And, by making use of Python's versions of file system utilities, we can write our programs so that they run on multiple kinds of computers. Cool! Specific topics we covered in this chapter include:

- Filenames, paths, and the working directory

  - Paths describe the sequence of directories (and the filename, if applicable) we need to traverse to reach a subdirectory or file of interest.
  - A relative path is the path to reach the desired file or subdirectory assuming we start at the current working directory.
  - An absolute path is the path to reach the desired file or subdirectory assuming we start at the topmost directory. In Linux or Mac OS X, the topmost directory is the root directory. In Windows, the topmost directory is a drive letter.
  - The os.path module's `join` function creates a path, for the computer's specific operating system, given a sequence of directories (and filename, if applicable) passed into the function as strings.
  - The current working directory is the place where the Python interpreter reads, writes, and manipulates files by default. It is the current "location," as far as the interpreter is concerned.
  - Various Interactive Development Environments (IDEs) have mechanisms to display and change the current working directory. In a Python program or interpreter session, the os module's `getcwd` returns the absolute path of the current working directory, and the os module's `chdir` function changes the current working directory to another directory of our choice.
  - Because paths are stored as strings, we can use string methods to slice-and-dice paths in whatever ways we want!

- Making and removing empty directories and moving and renaming files and directories

  - The os module's `mkdir` command creates an empty directory. Any intermediate directories specified by the path to the directory being created must already exist for the command to work successfully.
  - The os module's `makedirs` command creates an empty directory, specified by a path, and all intermediate directories needed to create that path, if those directories do not yet exist.
  - The os module's `rmdir` command removes an empty directory.
  - The shutil module's `move` command moves files or directories to new locations or renames files or directories.

- Copying and deleting files and directories

  - The shutil module's `copy2` command copies files.
  - The os module's `remove` command removes files.

- The shutil module's `copytree` command copies a directory tree, that is, everything inside the directory, including any and all files and subdirectories.
- The shutil module's `rmtree` command removes a directory tree, that is, everything inside the directory, including any and all files and subdirectories.

• Listing the contents of a directory and testing to see what kind of "file" something is

- The os module's `listdir` command lists the contents of the directory whose path is passed in as an argument.
- The glob module's `glob` command lists the contents of the directory following a pattern. The pattern supports basic wildcard matching.
- The os.path module's function `isdir` tests to see whether the item specified by a path is a directory.
- The os.path module's function `isfile` tests to see whether the item specified by a path is a file.
- The os.path module's function `islink` tests to see whether the item specified by a path is a link or alias to another file or directory.
- By testing to see what kind of "file" an item in a directory is, we can choose actions appropriate to that item. For instance, we would not try to open a directory by creating a file object.

• Running non-Python programs in Python

- The os module's `system` command will execute operating system commands passed to it in the form of a string.
- Because `system` takes a string as input, we can construct the operating system command we wish to execute by using string methods and operations. This gives us a means of automating the execution of those commands. The computer ends up "writing" its own instructions.

10.5.3 Self-Test Answers

Self-Test Answer 10-1

The current working directory is the location where file read and writes occur by default, and the location against which relative path specifications to other locations are defined.

Self-Test Answer 10-2

The `join` method creates a path to a directory or file by taking the strings that describe the elements of the path to that directory or file and places the operating system's directory separation character between the elements. For instance, the solution to the problem is:

```
import os
mypath = os.path.join('..', 'data')
```

The variable `mypath` will be the string `'../data'` on a Linux or Mac OS X system and the string `'..\\data'` on a Windows system.

Self-Test Answer 10-3

A relative path describes the path we would take to get to a file or directory assuming we were starting from the current location (or current working directory). An absolute path describes the path to a file or directory assuming we started at the topmost location on a computer's file system. On a Windows machine, the topmost location is a drive letter while on a Linux or Mac OS X system the topmost location is the root directory.

Self-Test Answer 10-4

The `getcwd` function returns the absolute path to the current working directory. The `chdir` function changes the current working directory to a specified path.

Self-Test Answer 10-5

We can solve this by using `mkdir`:

```
1   import os
2   data_path = os.path.join('experiment', 'data')
3   if not os.path.exists(data_path):
4       os.mkdir(data_path)
```

or using `makedirs`:

```
1   import os
2   data_path = os.path.join('experiment', 'data')
3   if not os.path.exists(data_path):
4       os.makedirs(data_path)
```

Both functions are part of the os module. The `mkdir` function assumes all intermediate-level directories to the final directory or file in the path already exist. The `makedirs` function creates any intermediate-level directories to the final directory or file in the path that do not already exist.

Self-Test Answer 10-6

```
1   import os
2   mypath = os.path.join("results", "calculations")
3   os.rmdir(mypath)
```

Self-Test Answer 10-7

We can obtain a listing of all contents of the directory by using `listdir`:

```
import os
import shutil
list_files = os.listdir()
for ifile in list_files:
    if os.path.isfile(ifile):
        if os.path.splitext(ifile)[-1] == '.txt':
            shutil.move(ifile, ifile[:-4] + '.dat')
```

Because the problem says we cannot assume the current working directory only contains files, we add a check in line 5 to make sure we only move files. We use `splitext` to separate the extension from `ifile` and in line 6 check for only the files that end in *.txt*.

We can also solve this using `glob` to do pattern matching on the filename suffix *.txt*:

```
import glob
import shutil
list_files = glob.glob('*.txt')
for ifile in list_files:
    shutil.move(ifile, ifile[:-4] + '.dat')
```

This second solution, however, assumes there are no directories in the current working directory that end in *.txt*.

Self-Test Answer 10-8

The shutil module's `copy2` function copies a file. The shutil module's `copytree` function copies a directory tree, including all levels of files and subdirectories inside the tree.

Self-Test Answer 10-9

The os module's `remove` function removes a file. The os module's `rmdir` function removes an empty directory. The shutil module's `rmtree` function removes a nonempty directory that can have many levels of files and subdirectories inside it, all of which are removed.

Self-Test Answer 10-10

The os module's `system` function accepts a string as input. The string is an operating system command-line interface command, which the `system` function executes as if it were the operating system.

Part II

Doing More Complex Tasks

11   Segue: How to Write Programs

As mentioned in the Preface, we have saved this chapter on how to write programs (except for a brief look in Section 6.2.7) until now because we wanted to focus our teaching of how to program on "doing" rather than on "talking about doing." So, our problems so far have been generally on the short side. As we move into longer and more complex programs, we are in a place to consider more detailed advice about how to write programs. In this chapter, we discuss three topics regarding how to write programs: the process of writing programs, the importance of testing, and the importance of style conventions.

11.1   From Blank Screen to Program: A Process to Follow

Because computers enable us to do amazing things, many people think computers are smart. But computers are not smart: computer programmers are. Computers can only store information, do calculations, make choices, and repeat a task. Whatever complex behavior a computer exhibits requires a human being to translate that behavior into these simple steps. This is not easy and requires creativity, trying things out, making mistakes, and perseverance.

Even though programming is hard work, we can break up the task of writing a program into chunks that can help us organize the task. There are many ways of doing so. Ultimately, everyone has to find a system that works for them. That being said, the four-step process we outline below works. More importantly, it has a silly acronym, so it is memorable ☺. Here are the four steps:

1. Catalog
2. Analyze
3. Outline
4. Write.

Here is one way to remember these steps: "To write a complex program, it helps to have a CAOW."

As we explain these four steps, we will describe how these steps could have been applied to create the program that solved the Section 7.1 example. In that example, we calculated the nearest-neighbor (NN) average of surface/near-surface air temperature (in °C) over the US Midwest at November 11, 2017, 18:00Z. We also converted those average values to kelvin and expressed the kelvin values as a fraction of the maximum value in the domain. The code for the solution is found in Figure 7.3.

Step 1 Catalog

In the first step, we make lists (not Python lists but lists on a piece of paper). We want to get a global sense of what we have to work with and what our goal is. Many people try to circumvent this step. They say to themselves, "I already know what data I have and what I am trying to do." That may work for a simple problem, but for complex programs, we really benefit from taking the time to list these things out, explicitly:

- Input. What input and information do we have to work with? List files, data, variables, parameters, etc.
- Output. What is the final result of the program? This could be a calculated quantity, the program itself (such as if we are designing the user-interface for a computational tool), or a file of data.
- In-between or intermediate. What are the intermediate values we need in order to create the output based on the input?

In these lists, we want to be as specific as possible. For instance, in describing the input, we do not want to say, "measurements," but "a text file of floating-point numbers in two tab-separated columns, where the first is the time of the measurement and the second is the measurement of velocity in m/s." The more detail we give, the better is our understanding of what we have to work with.

For our Figure 7.3 surface/near-surface air temperature NN averaging program, the Step 1 catalog might look like:

- Input

  - Data array of floating-point air temperature values (five rows, seven columns), in °C.
  - The offset to convert temperatures from °C to K.

- Output

  - Array of NN averaged temperature values in °C.
  - Array of NN averaged temperature values in K.
  - Array of NN averaged temperature values as a fraction of the maximum value in the domain.

- In-between or intermediate. When calculating the NN average at locations in the US Midwest, we need to store the sum of the nearest-neighbor temperatures as well as the number of nearest-neighbor points, to calculate the NN average at that location.

Step 2 Analyze

In the second step, we analyze the problem by taking the catalog from Step 1 and asking: (a) what calculations are needed to obtain the in-between or intermediate values (and from those, the final values), and (b) what computational patterns might be used to make the needed calcuations. Regarding the calculations that are needed, here are some common sources of equations and relationships for these calculations:

- Scientific laws or theories that connect variables together or describe how a quantity evolves with time (e.g., Newton's Second Law, $F = ma$).
- Statistical and mathematical measures and quantities (e.g., median, variance).
- Quantities that are derived from the input. For instance, if the input values are mass (m) and velocity (v) of an object, a quantity derived from those values would be kinetic energy $\left(E_k = \frac{1}{2}mv^2\right)$.
- Geometric relationships. For instance, say we are writing a program that makes a plot following a 4:3 ratio of width to height. Then, the 4:3 ratio is a geometric relationship that enables us to calculate the height of the plot given the width.
- Relationships connected to a computational activity. For instance, a program to manage files may be interested in adding a string showing the date of the file to the filename. The joining of that date string and the original filename is a relationship created by our concatenation activity.

Not all scientific and engineering programming tasks directly involve the analysis of data or other measurements. Scientists and engineers also write programs that enable the user to conduct visualizations (e.g., plots), create reports of benefit to stakeholders (e.g., a weather forecast), and manage the storage and distribution of data (e.g., from satellite observations). Nearly all of those tasks involve values that are not scientific data or measurements and relationships that are not scientific laws or mathematical theorems. Values of this nature include quantities related to the visualization window, the report, or the data files.

Once we have examined our catalog of input, output, and in-between or intermediate values and figured out what relationships take us from the input values to the intermediate values and finally to the output values, we need to ask whether the calculation of those relationships involves any computational "patterns." These patterns are also known as computational algorithms and implement the steps we want the computer to take to obtain the results we are after. Here are a few patterns we have already seen:

- Looping. If we need to do something over and over again, we probably need to use a loop.
- Branching. If the solution requires us to make choices, we probably need a branching statement.
- Functions. If there are any subtasks, particularly if the subtask is repeated in the program, we probably want to write that subtask as a function, or look to find a function in a pre-written package that does what we want.
- Collections of items. If we have a collection of related items, we probably want to hold them in a list or array. For calculations on numbers that apply to all (or a well-defined subset of) elements in an array, we can use array syntax. In other cases, we may have to loop over elements (or indices) in the list or array.

As we look at the calculations we need to do to turn the input into the output, and find that the calculations follow a computational pattern, at that step in our solution, we will need to use that pattern.

For our Figure 7.3 surface/near-surface air temperature nearest-neighbor (NN) averaging program, the Step 2 analysis might look like:

- Calculations

    - NN averaging. Sum up neighboring values to a location and divide by the number of those neighboring values.
    - Conversion from °C to K. Add 273.15 to the temperature in degrees Celsius.
    - Maximum temperature. For a collection of temperatures, find the largest value.
    - Temperature as a fraction of the maximum. Divide the temperature by the maximum of a collection of temperatures. Both values are in K.

- Patterns

    - Nested looping. To go through each element in a two-dimensional array, we construct a nested loop where the outer loop iterates through the row indices while the inner loop iterates through the column indices.
    - Incrementing. To increase a variable by a quantity, we add that quantity to the variable's current value and reassign the variable to the result. The syntax is `value = value + quantity`, which can be expressed more compactly as `value += quantity`. The variable `value` is an accumulator variable.
    - Defining subarrays. Slicing syntax enables us to extract a subarray from an array.
    - Array syntax. Calculations on an array variable will make the calculations on each element of the array.
    - Maximum of an array function. The NumPy function `max` will return the largest value from the input array passed into the function.

Step 3 Outline

In the third step, we take our catalog and analysis and sketch out a solution, in outline form, using comment lines. Do not write executable code.

Why go to this trouble? Many beginning programmers want to dive directly into writing executable code and go directly to Step 4. Even those who completed Steps 1 and 2 may feel outlining the code via comments is a waste of effort and time. But there are good reasons for this intermediate step between cataloging/analysis and coding.

First, just as in the writing of essays and papers, outlining helps us organize our thoughts and see how different portions of our essay or code interact with each other. We can more easily see what the calculations depend upon and what subtasks are part of a bigger task when the tasks are listed in an outline form. The outline form, with its indentations and brief descriptions, visually displays those dependencies and connections.

Second, we write comments rather than code because when we write code, we often focus on making the syntax correct. Such a focus, however, is the programming version of trying to run before we have learned to walk. At this stage, we need to sort out the order of the calculations and computational tasks. A focus on syntax distracts us from this endeavor. In fact, writing code too early can be a waste of time. Because we have not ironed out the sequence of the program, the code we write is likely incorrect. We will have to rewrite it after we have hammered out the correct sequence through trial and error.

Third, although it is true that Step 3 is more for people learning to program, and that experienced programmers can often skip this step, even experienced programmers can find it useful to outline using comments for particularly difficult sections of code. New programmers skip this step at their peril.

What, then, do we put into these comments? The style of the comments is not as important as (a) using an outline structure, with indentation (and numbering, if the code is particularly complex), and (b) being descriptive. As long as we hit these goals, it is not that important whether we use plain-English descriptions or **pseudocode** (code-like words and phrases that are similar to a programming language but do not follow the syntax of an actual programming language).

While we can describe each block of activity in the outline however we wish – using full sentences, a few phrases, or pseudocode – in many cases, we want to describe:

- What values are going into the block of activity.
- What values are coming out of the block of activity.
- What has been changed or done in the block of activity.

If we are writing a function, creating comments that describe these three points is particularly useful, because these three questions are precisely the ones users of a function need answered in order to properly use the function. The answers to these three questions are a strong basis for a docstring for the function.

For our Figure 7.3 surface/near-surface air temperature NN averaging program, an example of a Step 3 outline is given in Figure 11.1. Note how some of this outline finds its way into the final code in the form of comments describing what different sections of code do. This is another benefit of writing the outline using comments: We automatically create some in-code documentation!

Step 4 Write

In the fourth and last step, we take our outline and turn its descriptions into executable code. This last step will be easy or hard depending on how detailed is our Step 3 outline. If the outline is very detailed, each subpoint or task described in the outline will result in just a few lines of code. If the outline is not detailed enough, each subpoint or task described in the outline will require many lines of code, which will be more difficult to write. The bottom line: Programming is easiest when we break down a problem into small, discrete tasks. If we did a good job with Step 3, Step 4 will be relatively straightforward.

```
 1    #- Create data array:  Input values for the program.
 2    #
 3    #- Create array of zeros to hold averages:  Use NumPy zeros function.
 4    #
 5    #- Calculate average using nearest-neighbor algorithm.  Average is
 6    #  calculated at each location of the input array:
 7    #  * Create empty array to hold the average values calculated by this
 8    #    block of code.
 9    #  * Loop through all row indices:
10    #    - Loop through all column indices:
11    #      * Initialize the accumulator variable for the sum of the
12    #        temperatures by setting it to the value at the current
13    #        location.
14    #      * Initialize to 1 the accumulator variable for the count of
15    #        the number of neighbors to 1.
16    #      * Examine each point above, below, left, and right of the
17    #        current location:  If the point index is within the
18    #        dimensions of the input array:
19    #        - Increase the accumulator variable for the sum of the
20    #          temperatures by the temperature at that point.
21    #        - Increment the accumulator variable for the count of the
22    #          number of neighbors by 1.
23    #      * Calculate the average at that location by dividing the
24    #        sum of the temperatures by the number of neighbors.
25    #
26    #- Obtain subarray and Kelvin values as fraction of maximum value:
27    #  * Extract the subarray of values using slicing syntax.
28    #  * Convert Celsius to Kelvin.
29    #  * Calculate fraction of maximum values by dividing Kelvin
30    #    temperatures by the maximum of the temperatures.
```

Figure 11.1 A Step 3 outline for the program given in Figure 7.3.

11.2 The Importance of Testing

We have seen that computer programs are powerful: Very few lines of code can manipulate massive volumes of data and make huge numbers of calculations in a short amount of time. But, how do we know the answers programs give us are right? How do we know an inadvertent mistake in our program did not make our results garbage? We cannot, unless we test our program.

There are many kinds of tests we can subject our program to, and in Chapter 23, we examine in more detail various tools to help with software testing. For now, we mention two types of tests.

The first kind of test compares the final result of our program to a case where the correct answer is already known. Such an answer may be computed by hand, or it may be the result of previous analysis that has been conducted so many times (possibly by other computer programs) that the conventional wisdom is that the result is correct. As an example of the first kind of correct answer, consider a program to calculate the kinetic energy ($E_k = \frac{1}{2}mv^2$) of an object, given the mass (m) and velocity (v) of the object. We can calculate the value of E_k using specific values of mass and velocity, say 4 kg and 5 m/s, by hand, and compare whether the program produces that value. As an example of the second kind of correct answer, consider the spatial distribution of the seasonal average (that is, the average over a season) of surface temperature. In the field of climatology, such an analysis has been done many times over many years, and the pattern is well known. Even without seeing the actual numbers, we know that during northern Summer (June–August), the temperatures will be warmer in the Northern Hemisphere than in the Southern Hemisphere (where it is Winter), and vice versa during northern Winter (December–February). One way to check whether a climate model is "correct" is to compare the spatial distribution of the seasonal average of surface temperature with what has been observed.

In the second kind of test, we look inside the program and examine whether the values of the in-between or intermediate variables are "correct." Ideally, correct is defined by hand calculations using simpler "cases" of input. For instance, if we are doing calculations on an array of thousands of elements, we might examine just a few of the elements and check whether the values match what we would calculate by hand. Sometimes, the calculations are too complex to do by hand, and "correct" means whether the values are reasonable given what we know about the problem. To use the example of kinetic energy, if E_k was an intermediate value and too difficult to do by hand (it is not, but let us pretend), we might examine the variable and check that its value is close to a back-of-the-envelope estimate, or at the very least that it is positive (as E_k cannot be negative). Regardless of what we compare the in-between or intermediate variable's values with, the key in this second kind of test is to examine such variables as the program is executing, and ask whether they make sense.

The easiest way to do this second kind of test, to get a sense of what the values of the intermediate variables are, is to put in `print` statements throughout the program and print out the values of intermediate variables. Oftentimes, adding a note as to the correct value will help us to see whether the test passes:

```
In [1]:  m = 4.0
         v = 5.0
         E_k = 0.5 * m * (v**2)
         print("Kinetic energy should be 50: " + str(E_k))


         Kinetic energy should be 50: 50.0
```

In addition to printing out the variables, if we are writing our program using an Interactive Development Environment (IDE), we can place **breakpoints** in the program which will pause the program and allow us to examine the variables at that point. For Jupyter notebooks, there are a number of ways to simulate the setting of breakpoints, but the methods at the time of this writing are not as smooth as in an IDE such as Spyder.[1] We describe a method of running a debugger in a Jupyter notebook at www.cambridge.org/core/resources/pythonforscientists/jupyterdb/. The method described on that web page also works when running Python in a terminal window. We describe running a debugger in Spyder in detail at www.cambridge.org/core/resources/pythonforscientists/spyderdb/. Below, we give highlights of using a debugger in Spyder to illustrate how breakpointing works.

In the Spyder IDE, breakpoints are set by clicking on the line in the source code (in the Editor pane) where we want the breakpoint, then selecting the menu command Debug ⟩ ⟩ Set/Clear breakpoint. A small, red, filled circle appears in front of that line in the Editor pane, indicating a breakpoint is set there. We can set multiple breakpoints in one source code file. As we execute the program from breakpoint to breakpoint, we can observe what variables are defined and what their values are via the Variable Explorer pane. This pane is enabled by selecting the menu command View ⟩⟩ Panes ⟩⟩ Variable explorer.

We control execution via the block of toolbar buttons circled in red on Figure 11.2. These control Spyder's debugger, the tool that enables us to run the program slowly, to help us find the bugs in our program. By executing the program using the debugger (rather than the normal Python interpreter), we can run the program from breakpoint to breakpoint as well as one line at a time. The leftmost button in the circled tool bar portion starts running the program using the debugger, from the beginning of the code file. Execution continues until a breakpoint is reached or the program ends. The fifth button from the left continues execution until the next breakpoint. The other buttons provide other ways of stepping through the program and are also available as options or suboptions in the Debug menu.

Figure 11.2 shows a code file, *kinetic_energy.py*, in the Editor pane that has two breakpoints set in it. The figure shows the state of the debugger after running the code from the beginning until the first breakpoint. When the debugger reaches the breakpoint, execution stops and the line the breakpoint is defined at is not executed. As the Variable Explorer pane in the figure shows, at this point, the only variable that exists is m and it has the value 4.0. To continue executing until the next breakpoint, we can select Debug ⟩⟩ Continue or the corresponding button on the debugger button palette. The judicious use of breakpoints (and other debugger tools) enables us to see exactly what is the state of all the variables at key points in our program, so we can see whether our program is correctly calculating the in-between or intermediate values.

Thus far, we have looked at how tests let us know whether our program is doing what it should. But testing also plays an important role in the *way* we program. The temptation for most beginning (and too many advanced) programmers is to sit down and try and write out an entire program at once. This strategy, however, acts as a breeding ground for bugs. It is so easy for errors to creep in when we write many lines of code at once. The better strategy is to

---

[1] See, for instance, www.blog.pythonlibrary.org/2018/10/17/jupyter-notebook-debugging/, for details (accessed September 26, 2019).

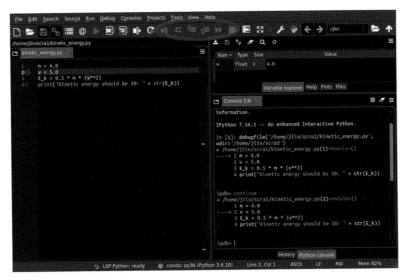

Figure 11.2 *kinetic_energy.py* in the Spyder IDE, showing breakpoints and the contents of the Variable Explorer pane after running the program to the first breakpoint. The red ellipse shows the toolbar buttons used for managing execution in the debugger.

write out *just a few lines* of code, then print out the values of the variables we calculated (or examine the values by setting breakpoints) to see whether those few lines of code give us the correct results. If so, we can go on and write another chunk of code. If not, we only have a few lines of code we have to poke through to find the source of the bug. This is *much* easier to do than trying to comb through tens or hundreds of lines of code, searching for the source of a bug.

11.3 The Importance of Style Conventions

Although the Great American Novel it is not, as far as programming languages go, Python code is pretty easy to read. Still, we can imagine that if left to their own devices, each person would use a different way of choosing variable names and other entities in a program. Some people might go for anarchy, following one naming convention in one part of their program and another in another part of their program. If Python programmers followed a style convention – rules regarding how the code should look – it would go a long way towards making code written by one programmer readable to another.

Python has such a style convention, which is described in the Python Enhancement Proposal (PEP) 8 document.[2] We have already seen one of these conventions. The convention

---

[2] We provide an up-to-date link to the document at www.cambridge.org/core/resources/pythonforscientists/refs/, ref. 64.

of indenting four spaces for the body of a function, `if` statement, etc., is from PEP 8. PEP 8 is quite detailed, but following just a few rules will make our code much more readable:

- Case for names. Nearly everything follows lowercase convention except class names which follow the CapWords convention. In the Python lowercase convention, words are all in lowercase and are separated by an underscore. In the CapWords convention, words all run together and the first letter of every word is capitalized. Thus, classes will be named, for instance, `KeplerData`, but variables, functions, and modules will be named, for instance, `kepler_data`.
- Filenames. These are also lowercase with underscores separating words, e.g., *kepler_data.py*. Recall that Python files can be imported and become modules, so the name should follow the case conventions that modules follow, which is the lowercase convention.
- Variable names. These should be as descriptive as possible but also not too long.
- Token spacing. Tokens (variables, operators, etc., in an expression) should be separated by blank spaces. The exception is multidimensional array index lists, which by convention can have spaces after the commas or not (e.g., `data[3,2]` or `data[3, 2]`).

Although these rules are useful, sometimes not following these rules will result in more readable code. Style conventions are ultimately about making code more readable. If following one particular convention makes a portion of code much less readable, strike another path for that portion.

In Chapter 21, we go into more detail regarding how to make code more usable to others.

Writing programs is not easy. It takes practice and making lots of mistakes. Attention to the topics we have addressed in this chapter – the process of writing programs, the importance of testing, and the importance of style conventions – can make the task of programming easier and the results of higher quality. The strategies and tools we described in this chapter take work, but they really can help!

12 *n*-Dimensional Diagnostic Data Analysis

In previous chapters, we looked at one-dimensional and two-dimensional arrays and used them in a variety of tasks, including data analysis, prognostic modeling, and the management of the output from multiple model runs. Arrays are, however, not limited to one or two dimensions. If we have a three-dimensional physical space – made up of length, width, and height – the measurements of some quantity every meter in length, width, and height in that space (at the same instant in time) would be nicely stored as elements in a three-dimensional array. If all such measurements are taken at regular intervals in time – say every hour – those values could be stored as elements in a four-dimensional array. The NumPy module in Python has structures that enable us to create and manipulate such *n*-dimensional (or multidimensional) arrays with an ease similar to creating and manipulating one- and two-dimensional arrays. In this chapter, we discuss the tools for such work and their application to the analysis of *n*-dimensional arrays.

The larger datasets used in this chapter are available for download. See the "Supporting Resources and Updates to This Textbook" section in the Preface for details.

12.1 Example of *n*-Dimensional Diagnostic Data Analysis

In Section 7.1, we examined the surface/near-surface air temperatures over the US Midwest at November 11, 2017, 18:00Z. These temperatures were stored in a two-dimensional array, whose values were given in Table 7.1. The longitude values of the columns in Table 7.1 were given in Table 7.2, and the latitude values of the rows in Table 7.1 were given in in Table 7.3.

These temperature values were extracted from a larger data file of surface/near-surface air temperatures at four times a day – at 0:00Z, 6:00Z, 12:00Z, and 18:00Z – every day, in the year 2017.[1] Thus, the full set of values in the data file is three-dimensional in nature: temperature values are given at different latitudes, longitudes, and times.

For our example in the current section, instead of looking at US Midwest surface/near-surface air temperatures at a single time, we consider temperatures at those locations at four times: November 11, 2017, 0:00Z, 6:00Z, 12:00Z, and 18:00Z.[2] Table 12.1 shows the data at these locations and times. The first two-dimensional block of data gives the temperatures at

---

[1] The data file for 2017 is itself part of a set of data files extending back to January 1, 1948, and going to the present. The full dataset is described in Kalnay et al. (1996).

[2] November 11, 2017, 0:00Z, 6:00Z, 12:00Z, and 18:00Z are, respectively, in Central Standard Time: November 10, 2017, 6:00 p.m., midnight between November 10 and 11, 2017, November 11, 2017, 6:00 a.m., and November 11, 2017, noon.

```
1    #- Import modules:
2    import numpy as np
3    from scipy import genfromtxt
4
5    #- Read in datafile and reshape into 3-D array:
6    air_temp_2d = genfromtxt('surf_temp_n-d_midwest.csv', delimiter=',')
7    air_temp = np.reshape(air_temp_2d, (4, 5, 7))
8
9    #- Create array of zeros to hold 24-hr average:
10   air_temp_24h_avg = np.zeros(np.shape(air_temp)[-2:],
11                               dtype=air_temp.dtype.char)
12
13   #- Calculate 24-hr average at each location:
14   avg_shape = np.shape(air_temp_24h_avg)
15   for i in range(avg_shape[0]):
16       for j in range(avg_shape[1]):
17           air_temp_24h_avg[i,j] = np.mean(air_temp[:,i,j])
18
19   #- Convert air_temp to K and print maximum value:
20   air_temp_K = air_temp + 273.15
21   print( "Maximum temperature in K: " + str(np.max(air_temp_K)) )
```

Figure 12.1 Code to read the file *surf_temp_n-d_midwest.csv*, calculate the 24-hour average surface/near-surface air temperature over the US Midwest shown in Table 12.2, convert the input temperatures to kelvin, and print the maximum input temperature in kelvin.

November 11, 2017, 0:00Z. The second two-dimensional block of data gives the temperatures at November 11, 2017, 6:00Z, and so on.

Figure 12.1 shows code that analyzes the surface/near-surface air temperature over the US Midwest from Table 12.1. The program calculates the 24-hour average temperature at each location in the US Midwest given in the data, converts the temperatures in the original data to kelvin, and prints the maximum temperature in the original data in kelvin.

The data values given in Table 12.1 are stored in a file *surf_temp_n-d_midwest.csv*, which looks exactly as Table 12.1 except there are no blank lines in *surf_temp_n-d_midwest.csv* separating the two-dimensional blocks of data. This enables us to read in the entire grid of values using `genfromtxt`, in line 6 of Figure 12.1. In line 7, we reshape the data read in from *surf_temp_n-d_midwest.csv* so that it is in the form of a three-dimensional array of four **sheets** (a sheet is a two-dimension array of rows and columns),[3] five rows, and seven columns.

In lines 10–11, we create a two-dimensional array of five rows and seven columns in which to store the 24-hour average of temperatures at each location. Those values are calculated in

---

[3] What we are calling "sheets" are also called "slices" by some. In prior chapters, we used "slice" to refer generically to subarrays and sublists (and the creation thereof). A sheet, however, is a particular kind of subarray, a "slice" that covers all rows and columns for a fixed value in the other dimension that is not a row or column. To avoid confusion between generic slicing and the slice (sheet) dimension of a three-dimensional array, we refer to that other dimension as the sheet dimension.

Table 12.1 Surface/near-surface air temperature over the US Midwest on November 11, 2017, at 0:00Z, 6:00Z, 12:00Z, and 18:00Z (in °C). The first two-dimensional grid block of values are at 0:00Z, the second grid block of values are at 6:00Z, and so on. The file *surf_temp_n-d_midwest.csv* consists of the above values except without the blank lines separating the four blocks of two-dimensional grids. Only a reasonable number of decimal places are shown in this table. More digits are available in the dataset than are shown.

```
-2.4,   -4.0,   -4.9,   -5.5,   -4.8,   -4.6,   -5.2
 2.5,    1.9,    0.0,   -2.3,   -2.3,   -2.1,   -3.1
 9.9,    8.4,    5.3,    3.1,    3.6,    3.8,    2.7
13.7,   12.7,   10.0,    9.1,    9.4,    8.5,    8.1
13.4,   14.4,   13.0,   13.4,   14.4,   13.6,   13.9

 0.2,   -1.0,   -2.4,   -3.5,   -3.6,   -3.9,   -5.0
 5.8,    3.2,    0.2,   -2.2,   -2.6,   -2.7,   -4.4
 9.9,    6.7,    3.0,    0.8,    1.2,    1.4,   -1.1
10.2,    8.4,    6.0,    4.5,    4.0,    3.2,    2.9
 9.3,    9.8,    8.0,    7.5,    7.7,    7.2,   10.3

 1.4,   -0.4,   -2.4,   -4.0,   -4.4,   -3.9,   -3.9
 5.9,    3.6,   -1.1,   -4.5,   -4.5,   -3.8,   -6.0
 8.4,    4.6,    0.4,   -1.6,   -1.0,   -1.4,   -6.4
 9.5,    5.6,    3.1,    2.6,    1.6,   -0.5,   -2.9
 9.0,    7.5,    6.1,    6.6,    6.1,    4.6,    6.6

 7.9,    7.1,    5.7,    3.6,    2.3,    1.7,    1.0
 9.9,    8.5,    7.4,    5.7,    4.9,    4.5,    2.1
12.8,   10.6,    9.1,    8.1,    8.8,    8.3,    4.0
15.4,   13.5,   11.1,   10.5,   11.3,   10.1,    6.8
17.6,   17.0,   14.7,   15.3,   16.3,   15.1,   14.4
```

lines 14–17, where we use a nested loop to go through each of the row and column indices for each of our locations in the US Midwest. At each location, we extract the temperatures at the four times and calculate the mean of those values. The 24-hour average temperature values (stored in `air_temp_24h_avg`) are given in Table 12.2.

In line 20, we convert the values of `air_temp` from degrees Celsius to kelvin. In line 21, we calculate the maximum temperature amongst all locations and times in `air_temp` and print that value out. That value is 290.75 K.

12.2 Python Programming Essentials

In Chapter 7, we examined two-dimensional arrays. In the example in Section 12.1, we extend our understanding of arrays and looping to accommodate three-dimensional arrays, as a prototype for understanding *n*-dimensional arrays. A three-dimensional array can be thought

Table 12.2 The 24-hour average of surface/near-surface air temperature over the US Midwest using values from November 11, 2017, 0:00Z, 6:00Z, 12:00Z, and 18:00Z (in °C). Values are calculated using the code in Figure 12.1 (i.e., these values are the contents of the array `air_temp_24h_avg`**).**

```
 1.78,   0.42,  -1.00,  -2.35,  -2.62,  -2.68,  -3.27
 6.03,   4.30,   1.62,  -0.82,  -1.12,  -1.03,  -2.85
10.25,   7.58,   4.45,   2.60,   3.15,   3.02,  -0.20
12.20,  10.05,   7.55,   6.67,   6.58,   5.32,   3.72
12.33,  12.18,  10.45,  10.70,  11.12,  10.12,  11.30
```

of as a stack or sequence of two-dimensional arrays. By analogy, if a two-dimensional array is a single sheet of paper, a three-dimensional array is a stack or pile of papers. A three-dimensional array would look something like Figure 12.2 and has height, width, and depth.

In this section, we explore the characteristics of a three-dimensional array as a prototype for the characteristics of *n*-dimensional arrays. We begin by examining the shape of, how to index, and how to slice (or **select**) subarrays in *n*-dimensional arrays. We then consider array syntax and functions in an *n*-dimensional array. *N*-dimensional arrays can be reshaped, and we examine both how this is done and what this tells us about the memory locations of an array's elements. For two-dimensional arrays, we found doubly nested loops could be used to access the elements in the array. For three-dimensional arrays, triple nested loops accomplish the same task, and higher nested loops are used for *n*-dimensional arrays. In this section we also explore combining loops with array syntax. Finally, we provide a summary table of some array functions.

In the rest of the code examples in this section, assume that NumPy has already been imported as `np`.

12.2.1 The Shape of and Indexing *n*-Dimensional Arrays

For a one-dimensional NumPy array, the shape is a one-element tuple. For a two-dimensional array, the shape is a two-element tuple, listing the number of rows and columns, in that order. For a three-dimensional array, the shape is a three-element tuple, listing the number of sheets, rows, and columns, in that order. Thus, for the `air_temp` array shown in Figure 12.2, the shape of the array is `(4, 5, 7)`.

As in the case of one- and two-dimensional arrays, the shape of an array is obtained using the NumPy `shape` function (e.g., line 10 of Figure 12.1) or by accessing the `shape` attribute of the array object. Also, note that the NumPy `size` function, as before, gives the total number of elements in the array, which is the product of all the elements in the shape tuple.

What is the shape of a higher-dimensional array? Shape tuples are constructed following these two rules: (1) the number of columns is always the rightmost element in the shape tuple, and (2) each dimension building off (or a collection of entities of) the previous dimensions is

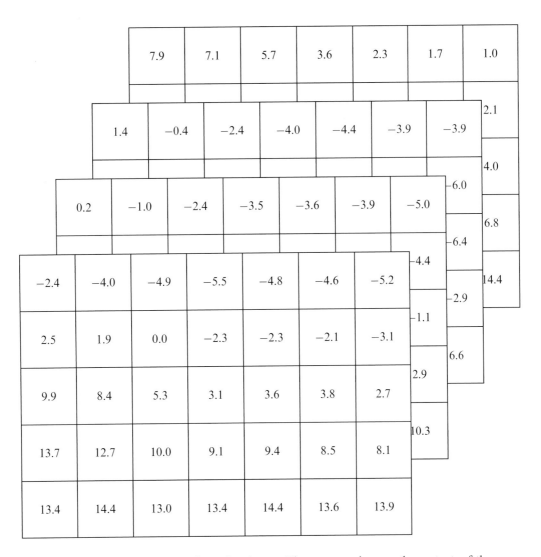

Figure 12.2 A schematic of a three-dimensional array. These array values are the contents of the array `air_temp` in the code in Figure 12.1, as shown in Table 12.1. The first (index 0) sheet is the frontmost two-dimensional array. The last (index 3) sheet is the backmost two-dimensional array.

described to the left of the description of the previous dimensions, in the shape tuple. Thus, for the three-dimensional array `air_temp` above, because there are seven columns, the last element of the shape tuple is 7. Rows are collections of columns, so the number to the left of the 7 in the shape tuple is the number of rows, 5. Sheets are grids of rows and columns, so the number of those sheets (four) is placed to the left of the 5 in the shape tuple. The shape of a four-dimensional array works the same way. If we have an array that is a four-dimensional "hypercube" of elements, that is, multiple three-dimensional arrays (a.k.a., "cubes"), the number of such three-dimensional arrays is given in the shape tuple as the element to the left of the number of sheets, which is left of the number of rows, which is left of the number of columns. This rule is followed for the shape of *n*-dimensional arrays.

Indexing an element in an *n*-dimensional array follows the same rules as in a two-dimensional array, except we reference all *n* dimensions in the sequence between the square brackets ([]), separated by commas. The first element in a dimension has index 0, the second element in that dimension has index 1, etc. For instance, if we wanted to access the first sheet, second row, and fourth column of the `air_temp` array in Figure 12.2, we would type in:

```
air_temp[0,1,3]
```

The order of the indices in the indexing specification follows the order of the dimensions in the shape tuple: sheet, row, then column. The indices for each dimension are separated by commas. Sometimes, for clarity, we put blank spaces after each comma, such as:

```
air_temp[0, 1, 3]
```

The code works the same way either way.

12.2.2 Selecting Subarrays from *n*-Dimensional Arrays

Selection (or slicing)[4] of subarrays in *n*-dimensional NumPy arrays works the same as in two-dimensional arrays, except the number of elements between the square brackets ([]) is more than two. The index ranges for each dimension are still separated by a colon, and the lower limit is inclusive while the upper limit is exclusive. Within the square brackets, the selection range for each dimension is given in the same order as in the array's shape tuple.

For instance, to extract the subarray from the three-dimensional `air_temp` array in Figure 12.2 that extracts from the first and second sheets, the second through third rows, and the last three columns, we would type in the following:

```
air_temp[0:2,1:3,4:7]
```

Figure 12.3 shows a schematic of the elements (shaded) that are selected by the code above. Because the act of selecting the first two sheets extends to the beginning of the sheet dimension (where each two-dimensional grid is considered a sheet), and the act of selecting the last three columns extends to the end of the column dimension, we can leave off the respective index values and the selection operation will extend to the end of the dimension. That is, this code will work just as well:

```
air_temp[:2,1:3,4:]
```

---

[4] In this subsection and later on in this chapter, we use "selection" and "selecting" to describe extracting a subarray from an array. In prior chapters, we called this operation "slicing," and that is the more common term. We avoid the latter term in this and related chapters to prevent confusing selection with the "sheet" dimension of a three-dimensional array, which in other works on Python is sometimes called the "slice" dimension. "Slicing slices" makes it sound as if the array selection operation only works on the sheet/slice dimension, which is not true.

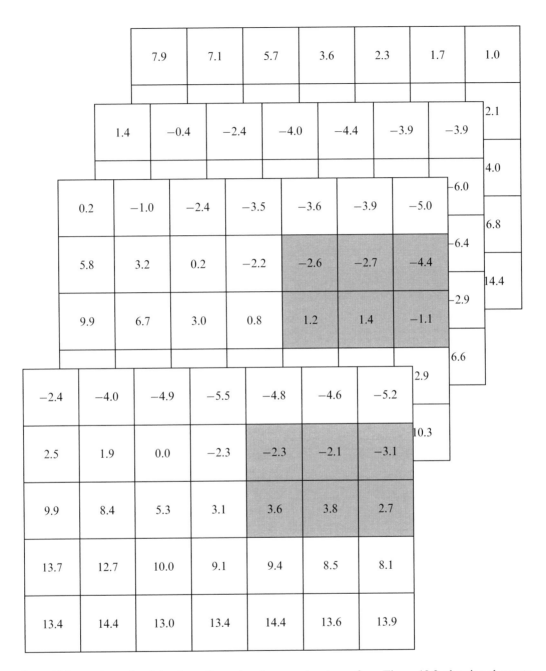

Figure 12.3 A schematic of the three-dimensional array `air_temp` from Figure 12.2, showing elements of the subarray sliced by `air_temp[0:2,1:3,4:7]` (shaded). The first sheet is shifted down relative to its position in Figure 12.2 in order to show the shaded regions of the second sheet.

To select all elements along an entire dimension, a colon without lower or upper bounds is used to specify that dimension. For instance, in line 17 of Figure 12.1, the code `air_temp[:,i,j]` selects the one-dimensional array that consists of all times in `air_temp` (the first dimension in the shape tuple) at the row and column locations given by the values of the integers `i` and `j`, respectively.

12.2.3 Array Syntax and Functions in *n*-Dimensional Arrays

Just as with one- and two-dimensional arrays, array syntax works in *n*-dimensional arrays to execute calculations on each element, if the calculation involves a scalar, or on corresponding elements between arrays, if the calculation involves two arrays of the same shape. Thus, line 20 of Figure 12.1 (reproduced here for convenience):

```
air_temp_K = air_temp + 273.15
```

adds 273.15 to each element of the three-dimensional array `air_temp` and places the result at the corresponding location in the array `air_temp_K` (which is also created by the operation). The array `air_temp_K` is the same shape as `air_temp`.

When we operate on two *n*-dimensional arrays of the same shape, the operation is applied to corresponding elements and the result is placed in an output array at the same corresponding location. Thus, if we have two three-dimensional arrays `data1` and `data2`, this operation:

```
data_sum = data1 + data2
```

adds the values at corresponding locations of `data1` and `data2` and places that sum into the same location in `data_sum`. All three arrays – `data1`, `data2`, and `data_sum` – have the same shape.

Functions that act on arrays in an element-wise fashion behave the same way for *n*-dimensional arrays. For instance, NumPy's `sin` function calculates the sine of each element of the input array and places the result in the corresponding element in the output array. Thus, for a three-dimensional array `data1`:

```
data_sin = np.sin(data1)
```

the resulting array `data_sin` is also three-dimensional and is the same size and shape as `data1`. The arrays `data1` and `data_sin` would have the same shape regardless of how many dimensions `data1` has.

Not all functions that act on arrays act solely (or primarily) element-wise on arrays. Some functions assume the arrays have a certain structure – for instance, that the array has a certain number of dimensions, or that each dimension represents a particular dimension in space or time. As those functions are specialized, we will not discuss them further. Quite a number of NumPy functions, however, allow the user to control the axis(es) the function acts on. One such function is the NumPy `sum` function. Whereas other functions may not follow the `sum` function's implementation, `sum` shows us how a function can use axis control to create additional forms of functionality without departing from the central idea of the function (which for `sum` is "adding up").

```
counts = np.array([[[3, 1, 2],
                    [1, 0, 3],
                    [1, 1, 2]],
                   [[1, 2, 2],
                    [0, 1, 2],
                    [3, 1, 2]]])
```

Figure 12.4 Fictitious counts of a species of insect in different locations of a country field. The two sheets of the array represent different times and the three-row, three-column elements of a sheet represent different locations in the field.

Pretend we have a three-dimensional array of integers `counts` as shown in Figure 12.4. This array represents counts of a species of insect at different times and locations in a country field. Each location may be a quadrat,[5] or "sampling square," used in ecology to define a standard area in which observations of organisms are conducted.

A call of `np.sum(counts)` would return 28, which is the sum of all elements in the array `counts`. If we set keyword input parameter `axis` to 0, 1, or 2, however, the `sum` function yields quite different behavior:

```
In [1]:  np.sum(counts, axis=0)

Out[1]:  array([[4, 3, 4],
                [1, 1, 5],
                [4, 2, 4]])

In [2]:  np.sum(counts, axis=1)

Out[2]:  array([[5, 2, 7],
                [4, 4, 6]])

In [3]:  np.sum(counts, axis=2)

Out[3]:  array([[6, 4, 4],
                [5, 3, 6]])
```

The `axis` keyword sets the dimension along which the summation occurs, where the dimension index is the same as the index that specifies the dimension in the array's shape tuple. If `axis` is set to 0, the summation is through the sheets. That is, we add up the values in the different sheets for each fixed row and column index. The result is a three-row, three-column two-dimensional array: For the array returned by `np.sum(counts, axis=0)`, the `[0,0]` element (which equals four) is `counts[0,0,0]` (which equals three) plus `counts[1,0,0]` (which equals one). The `[0,1]` element of the result array is `counts[0,0,1]` plus

---

[5] See https://en.wikipedia.org/w/index.php?title=Quadrat&oldid=919633941 (accessed October 25, 2019), but all data are fictitious.

counts[1,0,1], and so on. In Section 12.2.6, we will look at a way of emulating what the sum function does for axis=0 using a mixture of for loops and array syntax.

In the case where axis is set to 1, the summation occurs through the rows. That is, we add up the values in the different rows for each fixed sheet and column index. The result is a two-row, three-column two-dimensional array: For the array returned by np.sum(counts, axis=1), the [0,0] element (which equals five) is counts[0,0,0] (which equals three) plus counts[0,1,0] (which equals one) plus counts[0,2,0] (which equals one). The [0,1] element of the result array is counts[0,0,1] plus counts[0,1,1] plus counts[0,2,1], and so on. With the axis parameter, we can make the summation occur through a single dimension rather than through all elements in the array.

12.2.4 Reshaping *n*-Dimensional Arrays and Memory Locations of Array Elements

At first glance, arrays of different dimensions may seem very different from one another, as different as a line is from a plane is from a solid. But, because of how array elements are stored in the computer's memory, arrays of different dimensions are merely different "views" of the same object.

Consider the 12-element array xloc, which could describe the *x*-direction locations (in m), taken every 1 s, of a cart moving in a straight line in the *x*-direction at a constant speed of 1 m/s:

```
xloc = np.array([0, 1, 2, 3, 4, 5, 6, 7, 8, 9, 10, 11])
```

The cart begins at an initial location of 0 m. In the computer's memory, adjacent elements are stored next to one another. Thus, the element whose value above is zero is next to the element whose value above is one. The element whose value is two is stored next to the element whose value is one and the element whose value is three. The block of memory addresses of each of these elements is contiguous. That is, these elements are stored in order, one after the other, in the computer's memory.

The NumPy function reshape will transform an array into another array of a different shape, as long as the new array has the same number of elements. The first argument of reshape is the source array and the second argument of reshape is the shape tuple of the new array. Let us transform xloc into a two-dimensional array xloc_2d that has three rows and four columns:

```
In [1]:  xloc_2d = np.reshape(xloc, (3, 4))
         print(xloc_2d)

         [[ 0  1  2  3]
          [ 4  5  6  7]
          [ 8  9 10 11]]
```

Within each row, the order of the values of xloc_2d is the same as the order of values of xloc (i.e., 0, 1, 2, 3, etc.). Between rows, the value of the first element in a row is always the one that, in `xloc`, came after the value that is the last element in the previous row in `xloc_2d` (i.e., 3, 4 and 7, 8).

When we reshape a one-dimensional array, we do not move any of the values in memory. The values that were next to each other in memory remain next to each other in memory in the new, reshaped array, even if the number of dimensions of the reshaped array are different. We can see this if we transform `xloc` into a three-dimensional array `xloc_3d` that has two sheets, two rows, and three columns:

```
In [1]:  xloc_3d = np.reshape(xloc, (2, 2, 3))
         print(xloc_3d)

         [[[ 0  1  2]
           [ 3  4  5]]

          [[ 6  7  8]
           [ 9 10 11]]]
```

Within each row, the order of values is unchanged. The last element of one row is next to the first element of the next row. And the last element of one sheet is next to the first element of the next sheet. If we had another dimension, the same principle applies: each element in that dimension is a subarray that is placed next to the subarray that is the next element in that dimension. This principle continues with higher and higher dimensions.

We could use `reshape` to do the reverse: to turn a three-dimensional array into a one-dimensional array. There is a special function, however, that turns an array of any dimension into its one-dimensional equivalent, `ravel`:

```
In [1]:  print(np.ravel(xloc_3d))

         [ 0  1  2  3  4  5  6  7  8  9 10 11]
```

The `ravel` function "unravels" a multidimensional array into its one-dimensional version.

How are reshaping and raveling useful? One use is to transform an array into a form that makes it easier for us to select subranges on which to do analysis. This is the purpose of line 7 of Figure 12.1. The numbers in the *surf_temp_n-d_midwest.csv* file are stored as uninterrupted rows of regular columns, that is, as a two-dimensional grid. The values come into the program as a two-dimensional array `air_temp_2d`. By transforming that array into `air_temp` in line 7, we make it straightforward to extract the values at a single location for all times, as the `[:,i,j]` syntax does in line 17 of Figure 12.1.

One use of raveling is to simplify looping through the elements of the array. We saw in Section 7.2.5 that in order to loop through a two-dimensional array, we need two loops, one nested within the other. By tranforming a two- or *n*-dimensional array into its one-dimensional version, we can always use a single loop to go through all elements of the array. We lose the information of the dimensional structure of the *n*-dimensional array, but we now can loop through the elements of the array regardless of the number of dimensions it originally had and without changing our code (e.g., by adding another nested loop). In cases where we want to loop manually through an *n*-dimensional array, we will need to use nested loops. In Section 12.2.6, we examine triple and higher nested loops.

12.2.5 Subarrays and Index Offset Operations

In previous sections, we have looked at how the array slicing or selection syntax (using ranges specified by bounds separated by a colon) enables us to select subarrays. We have also seen how to add or subtract values from an array element's index value (e.g., i+1) to select an element in an array some elements away from the current element. Once selected, we can use those elements to make calculations, such as the running mean operation in line 14 in Figure 6.3. In this section, we use array slicing/selection with array syntax to conduct index offset operations that enable us to make calculations like the running mean operation without using loops. We briefly saw an example of this use in Try This! 5-5.

Consider again the one-dimensional `xloc` array from Section 12.2.4. Let us say we are interested in calculating how far each location of the cart is from the location 1 s previous. By inspection, we can see the difference between each location of the cart and its previous location is 1 m, but let us write code to calculate this. If we wanted to store the differences in a separate array, the code to calculate this using a `for` loop is as follows:

```
1  diffs = np.zeros(np.size(xloc)-1)
2  for i in range(np.size(xloc)):
3      diffs[i] = xloc[i+1] - xloc[i]
```

The array `diffs` is one element less in size than `xloc` because there is no location prior to the first location to calculate a difference with.

With array slicing (i.e., selection) and array syntax, however, we can solve the above problem in a single line of code:

```
diffs = xloc[1:] - xloc[0:-1]
```

Figure 12.5 shows how the above line of code works. The `xloc[1:]` slicing operation extracts the portion of `xloc` starting from the second (index 1) element going through the end of the `xloc` array. On Figure 12.5, this is shown by the thick lower brace and arrow. The `xloc[0:-1]` operation extracts the portion of `xloc` starting from the first (index 0) element going through the next to last element of the `xloc` array. On Figure 12.5, this is shown by the thick upper brace and arrow. Recall that slicing/selection ranges are inclusive at the lower bound and exclusive at the upper bound. When we make calculations using array syntax, the two subarrays are aligned with one other and the operation (here subtraction) is done element-wise between the corresponding elements in the subarrays. On Figure 12.5, this is shown by the thin arrows that connect the subarrays. The result of the subtraction operation is the bottom array in Figure 12.5. Thus, through judicious use of array slicing and array syntax, we are able to conduct index offset operations with very little code. Avoiding loops also, in general, makes the code run faster.

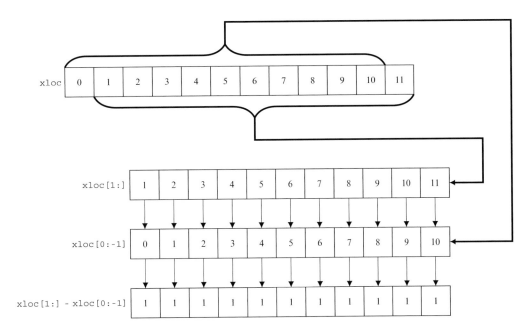

Figure 12.5 A schematic of `xloc`, `xloc[1:]`, `xloc[0:-1]`, and `xloc[1:] - xloc[0:-1]`. The thick braces and arrows show the slicing/selection ranges creating the subarrays. The thin arrows show the element-wise operation of array syntax between the subarrays to create the `xloc[1:] - xloc[0:-1]` array.

Index offset operations using array slicing and syntax also work for *n*-dimensional arrays. Consider the three-dimensional array `counts` from Figure 12.4. Let us say we want to calculate the difference in counts (at each time) between each point in the field and the point directly below it. Assuming `counts` is already defined, the following Jupyter notebook session makes the calculation, saves it as the three-dimensional array `diff`, and prints that array out:

```
In [1]:  diff = counts[:, 0:-1, :] - counts[:, 1:, :]
         print(diff)

         [[[ 2  1 -1]
          [ 0 -1  1]]

          [[ 1  1  0]
          [-3  0  0]]]
```

There are only two rows in each time sheet because the bottom-most locations in the country field have no point below them with which to calculate the difference in counts. The first `counts` subarray in the first line in cell `In [1]` extracts the points from all the rows of each sheet, except the last row (in this case, the first two rows of each sheet). In that selection, we want all time sheets and all columns, so the first and last dimension index ranges are given by lone colons. The row dimension index range is `0:-1` to specify the first row (inclusive) to the last row (exclusive). The second `counts` subarray in the first line in cell `In [1]` extracts the points from all the rows of each sheet, except the first row (in this case, the last two rows of each sheet). The row dimension index range is `1:` to specify the second row (inclusive) through all the rest of the rows.

12.2.6 Triple Nested Loops and Mixing Array Syntax/Selection and Looping

Although array syntax works fine with *n*-dimensional arrays, there are times we want to use looping with higher dimensional arrays. In this section, we examine two situations: when we want to explicitly loop through each element of an *n*-dimensional array and when we want to loop through only a subset of the dimensions of an *n*-dimensional array.

In Section 7.2.5, we found that to loop through all elements in a two-dimensional array, we use a two-level or doubly nested loop. To accomplish the same task with a three-dimensional array, we use a triple nested loop. Consider again the three-dimensional array `counts` from Figure 12.4. Let us pretend the scientist who typed in the data from the notebooks of the field observer accidentally typed 1 when the actual observation was 7. So, we want to go through each element of the array and change each occurrence of 1 to 7. Here is a triple nested loop through `counts` that will accomplish this task:

```
1  for i in range(2):
2      for j in range(3):
3          for k in range(3):
4              if counts[i,j,k] == 1:
5                  counts[i,j,k] = 7
```

The first iterator `i` goes through the indices for each sheet in `counts`, the second iterator `j` goes through each row index, and the third iterator `k` goes through each column index. Thus, if we did not know ahead of time how large is each dimension, the following would accomplish the same task as the code above for `counts` of any shape:

```
1  for i in range(np.shape(counts)[0]):
2      for j in range(np.shape(counts)[1]):
3          for k in range(np.shape(counts)[2]):
4              if counts[i,j,k] == 1:
5                  counts[i,j,k] = 7
```

In general, to go through each element of an *n*-dimensional array, we nest as many loops as there are dimensions in the array.

With higher-dimension arrays, we often find ourselves interested in looping through only one or a few dimensions of the array to access a subarray with which to do calculations. Consider again the three-dimensional array `counts` from Figure 12.4. Say we want to calculate the total insect counts at each location in the country field, over all time sheets. In this code below, we use a `for` loop to go through the indices of the time sheets (the first dimension of the array), use the subarray selection notation to obtain all rows and columns of a single time sheet, and use array syntax to add up corresponding row and column elements with the array `total` that stores the running total:

```
1   for i in range(np.shape(counts)[0]):
2       if i == 0:
3           total = counts[i,:,:]
4       else:
5           total += counts[i,:,:]
```

This produces the same result as calling the `sum` function using `axis=0`, which we saw in Section 12.2.3. In the code above, we add the `if` statement to check whether the loop is in its first iteration. If so, it creates the `total` array from the first sheet of `counts` (line 3). In subsequent iterations of the loop, we add the values of the index `i` time sheets to the corresponding values in `total` and store the result at the corresponding locations in `total` (line 5).

The `+=` and similar operators, when applied to arrays, also work according to the rules of array syntax. That is, if we have two arrays, x and y, both of the same shape, this:

```
x += y
```

is the same as:

```
x = x + y
```

where corresponding elements of x and y are added together and that result is stored in the corresponding element of x.

Our Figure 12.1 example provides another example of combining loops with array syntax. In lines 15–17, we use a doubly nested `for` loop to go through each row and column element of the three-dimensional array `air_temp`. At each row and column location, we extract all time sheets. That one-dimensional subarray is fed into the mean function (line 17) and the resulting average value is stored in a corresponding row and column location in `air_temp_24h_avg`.

With double, triple, and higher nested loops, we might wonder whether it matters which order we nest the loops. In the triple nested loop examples in this section, we looped first through the indices for sheets, then rows, and then columns. Could we have changed this order, say to go through sheets, then columns, then rows? It turns out, the order we used in our triple nested looping examples – sheets, rows, then columns – is the most efficient order.

To understand this, we need to recall: (1) how elements of an array are stored in memory, and (2) the order of activity in nested loops. Regarding the first topic, in Section 12.2.4, we saw that in NumPy arrays, elements in adjacent columns within the same row (and on the same sheet) are contiguous to (or next to) each other in the computer's memory. After that, the last element in one row is next to the first element in the following row, and the last element in the last row of a sheet is next to the first element of the first row of the next sheet.

Regarding the second topic, in a nested loop, the "fastest-varying" or "most-active" loop is the innermost loop in a nested loop. By this we mean that because the innermost loop goes through all its iterations for one iteration of the outer loop one level up, the innermost loop is more active than the outer loops. For instance, if we have a two-dimensional array `data` with six rows and four columns, this code:

```
for i in range(6):
    <... some outer calculations ...>
        for j in range(4):
            <... some inner calculations ...>
```

will run the <... *some inner calculations* ...> four times (the number of inner loop iterations) for every one time the outer loop iterates (and executes the <... *some outer calculations* ...>). The body of the innermost loop is the most active.[6] For a triple nested loop, the third nesting level or innermost loop is the fastest-varying loop, the second nesting level is the next fastest-varying loop, and the outermost loop is the slowest-varying loop.

In our triple nested loop examples above, we used these loops to go through the element indices for each array dimension, in order to access the elements in the *n*-dimensional array. Thus, because columns within a row (on the same sheet) are contiguous to each other in memory, if the fastest-varying looping index (the innermost loop) is the index for columns, each of those fastest-varying iterations will move only a short distance to find the next array value. Moving from one column index to the next is moving from one memory location to the adjacent memory location. The computer will not have to "go very far" to get the value at the next iteration. If, however, the innermost loop runs through another dimension such as the index for rows, for each iteration of that loop, the computer has to "jump over" an entire row's worth of values to get the value for the next iteration, and at the next column iteration, the computer goes back to the first row and does the row jumping for each iteration all over again. This is extra work. For a very large array, this extra work can add up. The bottom line: If we nest our loops so that the order of accessing array elements matches the order the elements are stored in memory – with the fastest-varying loop going through column indices, the row indices the next level up, and so on – our code will run faster.

12.2.7 Summary Table of Some Array Functions

The full listing of NumPy array functions is found in the NumPy Reference.[7] In this section, we summarize some array functions. We do not describe the various ways of customizing the behaviors of the functions, such as through the use of keyword input parameters.

---

[6] This is not entirely accurate. The innermost loop is part of the body of the loop a level up from the innermost loop, so technically all the activity of the innermost loop is counted as the activity of the loop one level up. That being said, the statement that the innermost loop is the most active is correct if we do not "double count" the innermost loop's activity, that is, if we separate the <... *some inner calculations* ...> from the <... *some outer calculations* ...> in our counting.

[7] The present table references the NumPy Reference. We provide an up-to-date link to the NumPy Reference at www.cambridge.org/core/resources/pythonforscientists/refs/, ref. 39.

Customization is described in other sections throughout this text (e.g., Section 12.2.3) and in the NumPy Reference.

Creating Arrays

Name	Description
arange	Create an array whose values are as if they were generated by a call to range.
array	Turn an object like a list into an array.
ones	Create an array of a given shape where each element equals one.
zeros	Create an array of a given shape where each element equals zero.

Array Inquiry and Manipulation

Name	Description
ravel	Turn an array into its one-dimensional equivalent.
reshape	Reshape the shape of an array into a new shape, given by an input shape tuple. The total number of elements in the new shape has to equal the total number of elements in the original array.
shape	Returns the shape tuple for the array, which lists the number of elements in each dimension of the array.
size	Returns the total number of elements in the array.

Mathematics

Name	Description
arccos	Return the inverse cosine (or arc cosine) of an input.
arcsin	Return the inverse sine (or arc sine) of an input.
arctan	Return the inverse tangent (or arc tangent) of an input.
cos	Return the cosine of an input (assumed to be in radians).
cumsum	Return an array whose elements are the cumulative sum of all previous values, up to and including the element in question.
exp	Calculate e^x where x is the input.
log	Return the natural logarithm of an input.
log10	Return the base-10 logarithm of an input.
sin	Return the sine of an input (assumed to be in radians).
sum	Return the sum of all elements in the array.
tan	Return the tangent of an input (assumed to be in radians).

Statistics

Name	Description
mean	Calculates the average (arithmetic mean) of the values in an input array.
median	Calculates the median of the values in an input array.
std	Calculates the standard deviation of the values in an input array.

12.3 Try This!

The examples in this section address the main topics of Section 12.2. We practice manipulating and using *n*-dimensional arrays to conduct sophisticated data analysis with very few lines of code. We also practice higher nested looping beyond a doubly nested loop.

Try This! 12-1 Shape and Indexing: Insect Counts

Consider the `counts` array from Figure 12.4. What is the shape of the array? How would we refer to the two elements in the array that equal 0?

Try This! Answer 12-1

The shape of `counts` is (2, 3, 3). The two elements in the array that equal 0 are `counts[0,1,1]` and `counts[1,1,0]`.

Try This! 12-2 Selecting Subarrays: Surface Temperature

Consider the array `air_temp_K` in Figure 12.1. What would be the syntax to select the following subarrays? What is the number of dimensions and shape for each of the subarrays?

1. All elements in the second sheet.
2. All elements in all sheets except the northernmost and southernmost locations.
3. The elements in the last sheet not including the first two latitude values and the last two longitude values.
4. All elements with the first longitude value.

Try This! Answer 12-2

Recall that `air_temp_K` is the same shape as `air_temp`. So, even though Table 12.1 shows the values of `air_temp`, not `air_temp_K`, we can still obtain the selection bounds for the desired subarrays and the resulting shape. We do not need to have the array values to answer this question:

1. `air_temp_K[1,:,:]`, which has two dimensions and shape (5, 7). A colon without an upper or lower bound selects all elements in that dimension.
2. `air_temp_K[:,1:-1,:]`, which has three dimensions and shape (4, 3, 7).
3. `air_temp_K[-1,2:,:-2]`, which has two dimensions and shape (3, 5). A range of indices lacking an upper or lower bound to the right or left of the colon selects through and including the end or beginning of the dimension, respectively. The array selection syntax `air_temp_K[3,2:,:-2]` and `air_temp_K[3,2:,0:4]` also work, as we can index a dimension from the front or back.
4. `air_temp_K[:,:,0]`, which has two dimensions and shape (4, 5).

Try This! 12-3 Array Syntax and Functions: Insect Counts

Consider the counts array from Figure 12.4. Calculate the total counts over all time sheets for each location and then express that count as a fraction of the maximum number of total counts at a location. Thus, the largest value in the result should be one, at the location of the maximum total count.

Try This! Answer 12-3

This code solves the problem:

```
1   total = np.sum(counts, axis=0)
2   total = total / np.max(total)
```

The axis=0 keyword in the call to the sum function sums through the sheets. The call to max finds the maximum value in total. In the right-hand side of the equal sign in line 2, we use array syntax to divide each value of total by the maximum value in total, and the result replaces the old value of total.

Try This! 12-4 Reshaping Arrays: Insect Counts

Consider the counts array from Figure 12.4. What are the first five elements of the array if it is transformed into a one-dimensional array? The last five elements?

Try This! Answer 12-4

The first five elements of the one-dimensional version of counts are: 3, 1, 2, 1, and 0. The last five elements of the one-dimensional version of counts are: 1, 2, 3, 1, and 2. We can answer this by looking at the array and remembering that: (1) when an *n*-dimensional array is converted to a one-dimensional array, all elements are kept in order in memory; (2) column elements are next to one another in memory, and the last element in a row is next to the first element in the next row.

We can also solve this by using NumPy's ravel function:

```
In [1]:  np.ravel(counts)

Out[1]:  array([3, 1, 2, 1, 0, 3, 1, 1, 2, 1, 2, 2, 0, 1, 2, 3, 1, 2])
```

and reading off the appropriate values.

Try This! 12-5 Index Offset Operations: Insect Counts

Consider the counts array from Figure 12.4. How would we calculate the difference in counts (at each time sheet) between each point in the field and the point directly to the right of it?

Try This! Answer 12-5

In the notebook solution below, we save the result in the array `diff` and print out the array:

```
In [1]:  diff = counts[:,:,0:-1] - counts[:,:,1:]
         print(diff)

         [[[ 2 -1]
           [ 1 -3]
           [ 0 -1]]

          [[-1  0]
           [-1 -1]
           [ 2 -1]]]
```

The result `diff` is a three-dimensional array with two sheets, three rows, and two columns. The number of columns is less than the number of columns in `counts` because the rightmost column does not have a location in the country field to the right of it.

Try This! 12-6 More Complex Offset Operations: Insect Counts

Consider the `counts` array from Figure 12.4, but assume that instead of it having just two time sheets, it has 100 time sheets. How would we calculate the difference in counts (at each time sheet) between each point in the field and the point directly to the right of it *from the time sheet before?*

Try This! Answer 12-6

The selection and array syntax notation (whose result is saved in `diff`) for this calculation is:

```
diff = counts[1:,:,0:-1] - counts[0:-1,:,1:]
```

The `counts[1:,:,0:-1]` selection selects all columns except the last, for all time sheets except the first. The `counts[0:-1,:,1:]` selection selects all columns except the first, for all time sheets except the last. In both selection operations, all rows are selected.

By using both positive and negative indexing, we do not actually need to know there are 100 time sheets in `counts`. The above selection code works for `counts` of 2, 100, or 100 million time sheets. In all valid cases, `diff` will be a three-dimensional array. In the case for two time sheets (the actual `counts` array from Figure 12.4), however, the shape of `diff` will be `(1, 3, 2)`. A three-dimensional array that has a single time sheet is the same as a two-dimensional array, so we can use `reshape` to reshape `diff` into an array of shape `(3, 2)` without the loss of any important information. When an array has a dimension of length 1 that can be removed in this way, the dimension is called a **degenerate** dimension.

Try This! 12-7 Subarrays and Offset Operations: Surface Temperature

Consider the array `air_temp_K` in Figure 12.1. For each longitude location, calculate the difference between the three southernmost latitudes in a time sheet and the three northernmost latitudes in the previous time sheet, with the latitudes matching one-to-one between the time sheet from North to South. Put another way, in calculating the difference, the third latitude from the South in a time sheet is matched with the first latitude from the North in the previous time sheet, the second latitude from the South with the second latitude from the North in the previous time sheet, and so on.

Try This! Answer 12-7

The selection and array syntax notation (whose result is saved in `diff`) for this calculation is:

```
diff = air_temp_K[1:,-3:,:] - air_temp_K[0:-1,0:3,:]
```

The selection of offset time sheets follows the same syntax as in the solution for Try This! 12-6. The rows dimension selection `-3:` in `air_temp_K[1:,-3:,:]` selects the last three rows whereas the rows dimension selection `0:3` in `air_temp_K[0:-1,0:3,:]` selects the first three rows. Instead of `0:3`, we could also have used `:3`.

Try This! 12-8 Offset Operations and Higher Nested Looping: Insect Counts

In Try This! 12-5 we calculated the difference in counts (at each time sheet) between each point in the field and the point directly to the right of it, for the `counts` array from Figure 12.4. Do the same calculation using nested loops, without the use of array slicing/selection notation or array syntax. Place the result in an array `diff`.

Try This! Answer 12-8

Here is a solution:

```
1  diff_shape = (counts.shape[0], counts.shape[1], counts.shape[2] - 1)
2  diff = np.zeros(diff_shape, dtype='l')
3  for i in range(diff.shape[0]):
4      for j in range(diff.shape[1]):
5          for k in range(diff.shape[2]):
6              diff[i,j,k] = counts[i,j,k] - counts[i,j,k+1]
```

In line 1, we use the array attribute form of the shape of `counts` (see Section 9.2.2) in constructing the shape tuple `diff_shape` for the result array `diff`. The number of columns in `diff` will be one less than the number of columns in `counts`, so the final element in `diff_shape` is `counts.shape[2] - 1`.

In line 2, we create the array `diff` that will hold the result. Unlike with array syntax, the result array is not automatically created, so we have to explicitly do so prior to the looping through the indices of its elements. In lines 3–5, we do that looping. Again, we use the attribute form of the shape, here `diff`.

In line 6, we do the calculation. Because the number of columns in `diff` is 1 less than in `counts`, the k+1 indexing of `counts` will not exceed the number of columns of `counts`.

Try This! 12-9 Offset Operations and Higher Nested Looping: Surface Temperature

Do the same calculation as Try This! 12-7 using nested loops, without the use of array slicing/selection notation or array syntax. Place the result in an array `diff`.

Try This! Answer 12-9

Here is a solution:

```
1   diff_shape = (air_temp_K.shape[0] - 1, 3, air_temp_K.shape[2])
2   diff = np.zeros(diff_shape, dtype='d')
3   for i in range(diff.shape[0]):
4       for j in range(diff.shape[1]):
5           for k in range(diff.shape[2]):
6               diff[i,j,k] = air_temp_K[i+1,-3+j,k] - air_temp_K[i,j,k]
```

The shape for `diff` defined in line 1 is the tricky part and will govern what indexing we use in the calculation in line 6. The number of time sheets is one less than in `air_temp_K`. There are three southernmost latitudes and three northernmost latitudes, so the number of rows in `diff` is three. The calculation is conducted at all longitude values, so the number of columns in `diff` is the same as the number of columns in `air_temp_K`.

Given `diff_shape`, and our looping in lines 3–5 through the indices of `diff_shape`, in line 6 we refer to the next time sheet after the index i time sheet as i+1 and each southernmost latitude index relative to the index j (which iterates through each northernmost latitude index) as -3+j. Because j loops through 0, 1, and 2, then -3+j loops through -3, -2, and -1, respectively.

12.4 More Discipline-Specific Practice

The Discipline-Specific Jupyter notebooks for this chapter cover the following topics:

- The shape of and indexing *n*-dimensional arrays.
- Selecting subarrays from *n*-dimensional arrays.
- Array syntax and functions in *n*-dimensional arrays.
- Reshaping *n*-dimensional arrays.
- Subarrays and index offset operations.
- Triple and higher nested looping.

12.5 Chapter Review

12.5.1 Self-Test Questions

Try to do these without looking at the book or any other resources or using the Python interpreter. Answers to these Self-Test Questions are found at the end of the chapter. For all questions, assume that NumPy has already been imported as np.

Self-Test Question 12-1

How many elements are in a shape tuple for a three-dimensional array? For a ten-dimensional array?

Self-Test Question 12-2

Assume we have a four-dimensional array humidity, where the first dimension is time, the second is elevation, the third is latitude, and the fourth is longitude. All dimensions increase with increasing index. Thus, for instance, the lowest elevation's values are at index 0 for that dimension. What is the code to access the element at the third time value, the highest elevation, the fifteenth latitude, and the first longitude?

Self-Test Question 12-3

For the humidity array in Self-Test Question 12-2, select the subarray consisting of: the fifth to seventeenth time values (both inclusive), all elevations, the first four latitudes, and the last longitude. Save this subarray as the array subhumidity.

Self-Test Question 12-4

For the humidity array in Self-Test Question 12-2, if the values in the array are given as ratios, convert the values into percentages and save the result as the array humidity_percent. Use array syntax.

Self-Test Question 12-5

For the humidity array in Self-Test Question 12-2, how would we use the NumPy sum function to calculate the sum through all elevations at all times, latitudes, and longitudes? Save the result in the array sum_elev. How many dimensions is sum_elev?

Self-Test Question 12-6

Let us say we want to reshape the array data into the array new_data using the NumPy function reshape. What must be true about the shapes of each of the arrays?

Self-Test Question 12-7

For the `humidity` array in Self-Test Question 12-2, create an array `new_humidity` where all latitude and longitude values at a given time and elevation are represented by a single one-dimensional array. Do not change the memory order of the latitude/longitude values.

Self-Test Question 12-8

For the `humidity` array in Self-Test Question 12-2, what element is next to the first element in the array? That is, where is the second element of the array?

Self-Test Question 12-9

What does the NumPy `ravel` function do?

Self-Test Question 12-10

Given the array `xloc`:

```
xloc = np.array([0, 1, 2, 3, 4, 5, 6, 7, 8, 9, 10, 11])
```

calculate the array `prod_halves` whose elements are the product between the corresponding elements of the last and first halves of the array. That is, this should be the result of `prod_halves`:

```
In [1]:  print(prod_halves)

         [ 0   7 16 27 40 55]
```

Use array syntax and slicing/selection, not looping.

Self-Test Question 12-11

For the `humidity` array in Self-Test Question 12-2, calculate the array `diff` that is the current time value minus the previous time value, at every elevation, latitude, and longitude. Use array syntax and selection, not looping.

Self-Test Question 12-12

Repeat Self-Test Question 12-11 using looping instead of array syntax and selection.

Self-Test Question 12-13

Assume we have a function `process_humidity` that accepts as input a three-dimensional array dimensioned by elevation, latitude, and longitude and returns a scalar floating-point value. For the `humidity` array in Self-Test Question 12-2, write code that selects the elevation, latitude, and longitude subarray at each time value, calls `process_humidity` using that subarray as input, and saves the result as an element in a one-dimensional array `processed`. Hint: The length of `processed` should equal the number of time values.

12.5.2 Chapter Summary

In this chapter, we extended our understanding of arrays to arrays of more than two dimensions. We found that many of the behaviors and syntax of n-dimensional arrays are similar to the behaviors and syntax of one- and two-dimensional arrays. In examining n-dimensional arrays, we also extended our understanding of how array elements are stored in memory, multidimensional subarray slicing/selection, how we can use slicing/selection and array syntax to make complex index offset calculations, higher nested loops, and mixing array syntax/selection and looping. Specific topics we covered in this chapter include:

- The shape of and indexing n-dimensional arrays

  - Elements in a three-dimensional array are referenced by sheet, row, then column index inside the square brackets.
  - In the specification for an element in an n-dimensional array, the column index is always the rightmost index, the row index is to the left of the column index, the sheet index is to the left of the row index, and so on going left.
  - Along a dimension, the first element is index 0, the second element is index 1, etc.

- Selecting subarrays from n-dimensional arrays

  - Subarrays are selected by providing upper- and lower-bound indices that specify a range of elements, *for each dimension*, of the array the subarray is being selected from.
  - The upper- and lower-bound indices for a dimension are separated by a colon. The upper bound is exclusive and the lower bound is inclusive.
  - The absence of an upper and/or lower bound means the selection goes inclusively through to the end and/or beginning, respectively, *of that dimension*.

- Array syntax and functions in n-dimensional arrays

  - Operations between an array and a scalar: the operation occurs between each element of the array and the scalar and the result is placed in the corresponding location of an output array of the same shape as the original array.
  - Operations between an array and another array of the same shape happen element-wise: the operation occurs on corresponding elements and the result is placed in the corresponding location of an output array of the same shape.
  - Many NumPy functions operate on arrays in an element-wise fashion.
  - Some NumPy functions operate on arrays in more than an element-wise fashion. The `sum` function is one such function. By setting the `axis` keyword input parameter, we can control the dimension of the array through which the summation occurs.

- Reshaping *n*-dimensional arrays

  - The NumPy `reshape` function reshapes an array into a new shape, given as an input parameter.
  - When using the `reshape` function, the total number elements of the array cannot be changed.
  - The `reshape` function does not change the order of the elements in the array as they are stored in memory. Reshaping an array only changes where rows, sheets, etc., end and another begins.
  - The NumPy `ravel` function turns any array into its flattened or one-dimensional version. The `ravel` function also works without changing the order in memory in which the elements are stored.

- Memory locations of array elements

  - Consecutive column elements within a row are stored next to each other in memory.
  - The last element in a row is contiguous in memory to the first element of the next row.
  - The last element in a sheet is continuous in memory to the first element of the next sheet (which is the first element of the first row of that next sheet).
  - For higher dimensions, each element in that dimension is a subarray that is placed next to the subarray that is the next element in that dimension.

- Subarrays and index offset operations

  - By carefully selecting different ranges from a subarray, and combining them using array syntax, we can emulate index offset operations that in previous examples required a loop to conduct.
  - Using subarrays and array syntax instead of looping enables us to describe index offset operations with fewer lines of code. Avoiding loops also, in general, results in code that runs faster.

- Triple nested loops and mixing array syntax/selection and looping

  - To access all elements in an *n*-dimensional array via nested looping, there need to be *n* loops nested in each other, each iterating through all or some of the indices of a dimension.
  - The innermost loop of a nested loop structure is the "fastest-varying" or "most-active" loop. Thus, iteration through the column index, where adjacent elements are next to each other in memory, should occur in the innermost loop. Iteration through the row index should occur in the loop up one nesting level from the innermost loop, and so on.
  - We can combine looping and array selection/syntax to do multiple and recurring operations on subarrays.

12.5.3 Self-Test Answers

Self-Test Answer 12-1

The number of elements in a shape tuple equals the number of dimensions in the array. Thus, the shape tuple of a three-dimensional array has three elements. The shape tuple of a ten-dimensional array has ten elements.

Self-Test Answer 12-2

```
humidity[2, -1, 14, 0]
```

Self-Test Answer 12-3

```
subhumidity = humidity[4:17, :, 0:4, -1]
```

Self-Test Answer 12-4

```
humidity_percent = humidity * 100.0
```

Self-Test Answer 12-5

This calculates `sum_elev`:

```
sum_elev = np.sum(humidity, axis=1)
```

The `sum_elev` array has three dimensions.

Self-Test Answer 12-6

The total number of elements between `data` and `new_data` cannot be different. Thus, the product of the elements in each array's shape tuples must be identical.

Self-Test Answer 12-7

```
1   new_shape = (humidity.shape[0], humidity.shape[1],
2                 humidity.shape[2] * humidity.shape[3])
3   new_humidity = np.reshape(humidity, new_shape)
```

The `new_shape` tuple is three elements in length. The final dimension collapses the latitude/longitude values into a one-dimensional array, so the number of elements in that dimension is the product of the number of latitude and longitude points.

Self-Test Answer 12-8

The first element of the array is at the first time value, first elevation, first latitude, and first longitude. The second element is at the next longitude. Thus, that element is at the first time value, first elevation, first latitude, and second longitude.

Self-Test Answer 12-9

The `ravel` function returns a flat or one-dimensional version of the *n*-dimensional array. In doing so, the memory order of the elements in the array is preserved.

Self-Test Answer 12-10

```
prod_halves = xloc[6:] * xloc[0:6]
```

Self-Test Answer 12-11

```
diff = humidity[1:,:,:,:] - humidity[0:-1,:,:,:]
```

Self-Test Answer 12-12

Here is a solution:

```
diff_shape = (np.shape(humidity)[0] - 1,
               np.shape(humidity)[1],
               np.shape(humidity)[2],
               np.shape(humidity)[3])
diff = np.zeros(diff_shape, dtype='d')
for i in range(diff_shape[0]):
    for j in range(diff_shape[1]):
        for k in range(diff_shape[2]):
            for l in range(diff_shape[3]):
                diff[i,j,k,l] = humidity[i+1,j,k,l] - humidity[i,j,k,l]
```

We write out the elements of `diff_shape` in lines 1–4 to make it clear what goes into `diff_shape`. We could replace those lines with something more concise, such as:

```
diff_shape = list(np.shape(humidity))
diff_shape[0] -= 1
```

The `zeros` function accepts a list (or array) version of `diff_shape`; `diff_shape` does not have to be a tuple.

Self-Test Answer 12-13

```
1   processed = np.zeros(np.shape(humidity)[0], dtype='d')
2   for i in range(np.size(processed)):
3       processed[i] = process_humidity(humidity[i,:,:,:])
```

13 Basic Image Processing

In previous chapters, we have used looping and array syntax to analyze data in arrays of a number of dimensions. When we used looping constructs to examine the elements of arrays, we also found that branching statements like the `if` statement enable us to do different calculations depending upon whether certain conditions are true or not. Is it possible, however, to include branching logic into array syntax and avoid writing loops while still doing different calculations depending on whether certain conditions are true or not? In this chapter, we examine array functions and constructs that enable us to do exactly that.

We introduce these tools in the context of image processing. Images are ubiquitous in today's media-fueled world. We find them everywhere, whether snapshots from our phones, cat videos, or movies on-demand. Likewise, scientific and engineering work finds images everywhere: satellite instruments, medical scanning equipment, and digital telescopes are a few sources of imagery. When loaded into a computer, however, all images become an array of numbers. In the basic examples we look at in this chapter, we will find the array handling tools we learned to use in previous chapters also enable us to analyze the information encoded in an image file. Our discussion in this chapter will also set us up to discuss basic animation in Chapter 14.

In this chapter, because we are using image processing to motivate a more detailed examination of array syntax and related concepts, our approach to solving image processing problems utilizes lower-level tools than might be used in a real application. For real-world image processing problems, we want to take advantage of the various tools that already exist to tackle such problems. These tools include the Pillow library, the scikit-image package, and additional functions in the image submodule of Matplotlib that are not described in the present chapter. Up-to-date links to the main Pillow and scikit-image pages are at www.cambridge.org/core/resources/pythonforscientists/refs/, refs. 51 and 52 respectively.[1]

The images used in this chapter are available for download. See the "Supporting Resources and Updates to This Book" section in the Preface for details.

13.1 Example of Image Processing

Volkman et al. (2004) conducted a study to observe the formation of tuberculous granulomas (granulomas are aggregates of macrophages; macrophages are a white blood cell which fight infection) in zebrafish following infection with *Mycobacterium* (the type of bacteria

---

[1] See also van der Walt et al. (2014) for more on scikit-image.

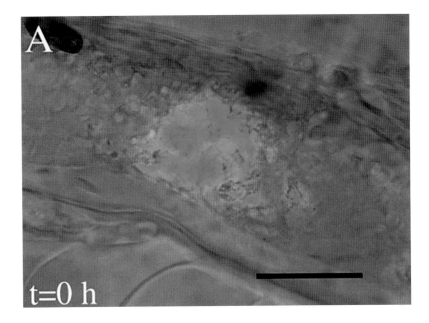

Figure 13.1 Image from Volkman et al. (2004) showing an aggregation of macrophages on the first day of granuloma formation (time $t = 0$). The horizontal black line in the lower right represents 50 μm in length.

which cause tuberculosis).[2] The bacteria were engineered to express green fluorescent protein, causing the infected macrophages to glow green. The accumulation of infected (green) macrophages into granulomas over time can be observed using a microscope, illustrated in Figure 4 of their paper. Our Figures 13.1–13.3 show three panels at different time points from Volkman et al.'s (2004) Figure 4, "Progression of Aggregates."

Our first task is to read in the first of these images, Figure 13.1, and manipulate it. We want to find all the points in the image that meet some criteria of "greenness" (which represents the accumulation of infected macrophages) to calculate the percentage of all the pixels in the image that meet that criteria, and to create a version of the image that highlights those "green" pixels in some way so they stand out even more. The code in Figure 13.4 accomplishes this task.

In previous chapters, we have given a broad description in the Example sections as to what the code does and gone into more detail in the Python Programming Essentials section. This is even more true in the current chapter, as the combination of array syntax with branching results in powerful but terse code that requires substantial context to explain fully. Please let our whirlwind description below whet your appetite for Section 13.2. Do not worry if the description only partly makes sense.

---

[2] Thanks to Cynthia Gustafson-Brown for providing this section's example and contributing to the text in this section.

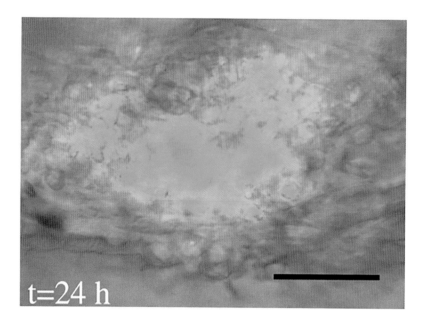

Figure 13.2 Image from Volkman et al. (2004) showing the granuloma 24 h after formation. The horizontal black line in the lower right represents 50 μm in length.

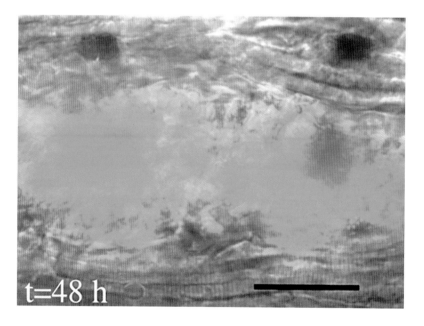

Figure 13.3 Image from Volkman et al. (2004) showing the granuloma 48 h after formation. The horizontal black line in the lower right represents 50 μm in length.

```
1    import numpy as np
2    import matplotlib.pyplot as plt
3    import matplotlib.image as mpimg
4
5    img = mpimg.imread("g004-A1.png")
6
7    thresh = 0.58
8    offset = 0.05
9    condition = np.logical_and(
10                   np.logical_and(img[:,:,1] > thresh,
11                                   img[:,:,0] < (thresh - offset)),
12                   img[:,:,2] < (thresh - offset) )
13
14   img_highlight = img.copy()
15   img_highlight[:,:,0] = np.where(condition, 1.0, img[:,:,0])
16   img_highlight[:,:,1] = np.where(condition, 0.0, img[:,:,1])
17   img_highlight[:,:,2] = np.where(condition, 1.0, img[:,:,2])
18
19   plt.imsave("g004-A1_highlight.png", img_highlight)
20
21   print( 'Percent "green": ' + \
22           str(np.sum(condition) / np.size(condition) * 100.0) )
```

Figure 13.4 Code to analyze the image in Figure 13.1. The filename of that input image is *g004-A1.png*. The output filename is *g004-A1_highlight.png*, and the image in that file is given in Figure 13.5.

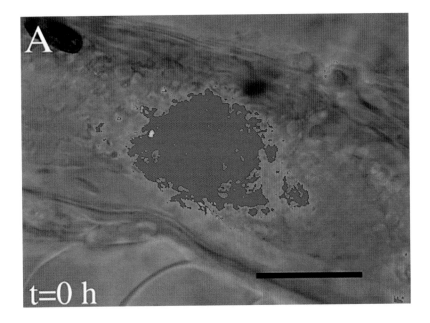

Figure 13.5 Image generated by the code in Figure 13.4 (and the first iteration of the loop in Figure 13.6). The filename of the image that is generated is *g004-A1_highlight.png*. Image generated is altered from Volkman et al. (2004).

```
1    img1 = mpimg.imread('g004-A1.png')
2    img2 = mpimg.imread('g004-A2.png')
3    img3 = mpimg.imread('g004-A3.png')
4
5    images = [img1, img2, img3]
6
7    thresh = 0.58
8    offset = 0.05
9
10   for i in range(len(images)):
11       img = images[i]
12       condition = np.logical_and(
13                   np.logical_and(img[:,:,1] > thresh,
14                                  img[:,:,0] < (thresh - offset)),
15                   img[:,:,2] < (thresh - offset) )
16
17       img_highlight = img.copy()
18       img_highlight[:,:,0] = np.where(condition, 1.0, img[:,:,0])
19       img_highlight[:,:,1] = np.where(condition, 0.0, img[:,:,1])
20       img_highlight[:,:,2] = np.where(condition, 1.0, img[:,:,2])
21
22       plt.imsave("g004-A" + str(i+1) + "_highlight.png", img_highlight)
23
24       print( 'Percent "green" frame A' + str(i+1) + ': ' + \
25               str(np.sum(condition) / np.size(condition) * 100.0) )
```

Figure 13.6 Code to analyze the images in Figures 13.1–13.3. The filenames of these input images are *g004-A1.png*, *g004-A2.png*, and *g004-A3.png*. The output filenames are *g004-A1_highlight.png*, *g004-A2_highlight.png*, and *g004-A3_highlight.png*, and the images in those files are given in Figures 13.5, 13.7, and 13.8, respectively. Assume the import lines from the code in Figure 13.4 have already been run.

Lines 1–3 of Figure 13.4 import the needed packages. Besides importing in NumPy and the pyplot module of the Matplotlib package, we also import the image module of the Matplotlib package. That last module has a function `imread` that we use in line 5 to read in the file. The return value of `imread`, it turns out, is a NumPy array, and we save that as `img`. In Section 13.2.1, we learn more about what it means to load in a Portable Network Graphics (PNG) format image as a NumPy array. For now, suffice it to say that the array stores how strong are the red, green, and blue components (or channels) of each **pixel**.

In lines 7–12, we look for all pixels that meet our criteria of "greenness." Specifically, we look for everywhere the green channel in the image is stronger than the value of `thresh` and where the red and blue channels are weaker than `thresh - offset`. The array `condition` is `True` wherever the criteria we just described is true and `False` otherwise.

In line 14, we create a copy of `img`. In this copy, `img_highlight`, we take all the pixels that meet our criteria of "greenness" and change their color to magenta. This is accomplished in lines 15–17.

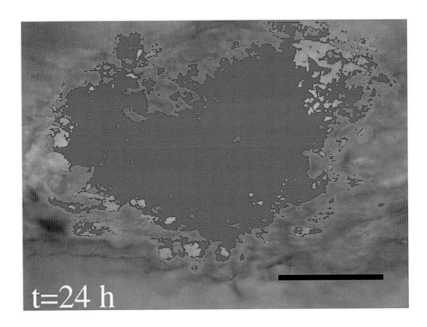

Figure 13.7 Second image generated by the code in Figure 13.6. The filename of the image that is generated is *g004-A2_highlight.png*. Image generated is altered from Volkman et al. (2004).

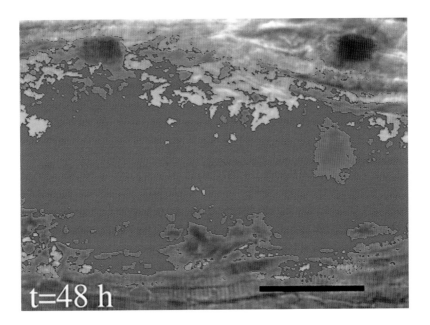

Figure 13.8 Third image generated by the code in Figure 13.6. The filename of the image that is generated is *g004-A3_highlight.png*. Image generated is altered from Volkman et al. (2004).

```
1 │ Percent "green" frame A1: 10.849044504296526
2 │ Percent "green" frame A2: 37.901115813774524
3 │ Percent "green" frame A3: 52.18737976144671
```

Figure 13.9 Screen output by the code in Figure 13.6.

In line 19, we write out `img_highlight` to a PNG file. This image is displayed in Figure 13.5. Finally, in lines 21–22, we output the percentage of pixels in the image that meets these criteria for "greenness," which is 10.85 percent. This provides us with a quantitative estimate of how large the granuloma (that is, the aggregate of these white blood cells) is, which we can use to quantitatively compare to the other images.

For our second task, we want to duplicate what we did for our first task and apply it to all three of the images from Figures 13.1–13.3. Note that all three of these images have the same number of pixels in each width and in each height. The code in Figure 13.6 accomplishes this task and produces the magenta-highlighted version in Figure 13.5 (same as the code in Figure 13.4), Figure 13.7, and Figure 13.8 as well as the output to the screen found in Figure 13.9. That output to the screen gives us a quantitative estimate of how large the granuloma is in each Figure 13.1–13.3 frame: 10.85 percent, 37.9 percent, and 52.19 percent, respectively.

The code in Figure 13.6 is nearly identical to the code in Figure 13.4. The key differences are: (1) three input images are read in in lines 1–3 of Figure 13.6, and (2) in Figure 13.6, lines 9–22 of Figure 13.4 are put inside a `for` loop that iterates through each of the three image array objects. The looping is accomplished by: (1) saving the image array objects as `img1`, `img2`, and `img3` (lines 1–3 in Figure 13.6), (2) placing those objects into a list `images` (line 5), and (3) looping through the indices of the `images` list (line 10). We use each of the iterator `i` values to assign each image array object in `images` to the variable `img` as well as to give each output filename and the values printed to screen an "A1," "A2," or "A3" label, as appropriate (lines 22–25).

13.2 Python Programming Essentials

In previous chapters, we were introduced to array syntax, looping, branching, and boolean variables and expressions. In this chapter, we use Python to do some basic image processing, but as Python represents colors in a two-dimensional image as data in a three-dimensional array, the image processing use case offers the opportunity to explore more complex manipulations of arrays using array syntax and array functions. For instance, lines 9–12 in Figure 13.4 make use of array syntax to create an array of boolean values. Lines 15–17 in Figure 13.4 make use of the array function `where` which enables the altering and manipulation of arrays based upon whether array elements meet a condition.

In this section, we first unpack how to read, display, and write images using the Matplotlib package. Next, we describe boolean arrays and how to use them. These boolean arrays are very useful when using array syntax and functions. We describe ways of using array syntax and functions to ask questions of data in arrays and describe how well these methods compare to using `for` loops and `if` statements to examine each element in an array. We end with a description of the `reduce` method attached to many NumPy functions and how to use loops to go through a list of objects.

In the rest of the code examples in this section, assume that NumPy has already been imported as `np` and Matplotlib's pyplot module is imported as `plt`.

13.2.1 Reading, Displaying, and Writing Images in Matplotlib

Generally speaking, images are made up of pixels arranged in a regular grid. Each pixel is a colored dot with a given brightness (and, in some images, a given transparency). There are a number of different ways of representing color (a number of which are also supported in Matplotlib), but in this section, we focus on the **Red–Green–Blue (RGB)** system, which decomposes the color of a pixel into red, green, and blue contributions (or "channels"). The combination of the strength of each of these channels becomes the color of the pixel.

For instance, a pixel whose red channel is 100 percent in intensity and 0 percent in green and blue intensity is a red pixel. A pixel whose red and green channels are 100 percent and whose blue channel is 0 percent is a yellow pixel. Different shades of colors are created both by altering the relative contributions of red, green, and blue as well as the actual intensity of each component. For instance, a pixel whose red and green channels are 75 percent and whose blue channel is 0 percent is an ochre yellow pixel. Note that what we perceive as "lightness" does not necessarily mean less intensity in a single channel. For instance, light blue results from red at 68 percent intensity, green at 85 percent, and blue at 90 percent, not merely from decreasing the blue channel intensity and keeping the red and green channel intensities at 0 percent.[3]

If all three red–green–blue channels are the same intensity, the pixel is a shade of gray. When all three red–green–blue channels are at 100 percent, the pixel is white. When all three red–green–blue channels are at 0 percent, the pixel is black. Thus, darker shades of gray correspond to when the three red–green–blue channels all are the same and have lower intensities. Figure 13.10 shows some colors generated by various red–green–blue intensities. These intensities are given in fractions rather than percentages, so 100 percent intensity is 1 and 0 percent intensity is 0. Percentages in between are decimal numbers between 1 and 0.[4]

When we read in, display, or write out an image, we are manipulating a grid of pixels, each of which are specified by the RGB channel intensities described above. In terms of Python, this is a three-dimensional floating-point array where the number of elements in the first dimension equals the height of the image in number of pixels, the number of elements of

---

[3] See https://www.rapidtables.com/web/color/RGB_Color.html#color-table (accessed April 25, 2019) for a list of colors and their RGB codes. In this table and many other discussions on color, 100 percent intensity is given as 255, and intensities vary from 0 to 255 rather than 0 percent to 100 percent. We discuss why later on in this section.

[4] The figure displays the intensities this way because Matplotlib expresses them in these terms for PNG image files, rather than as percentages.

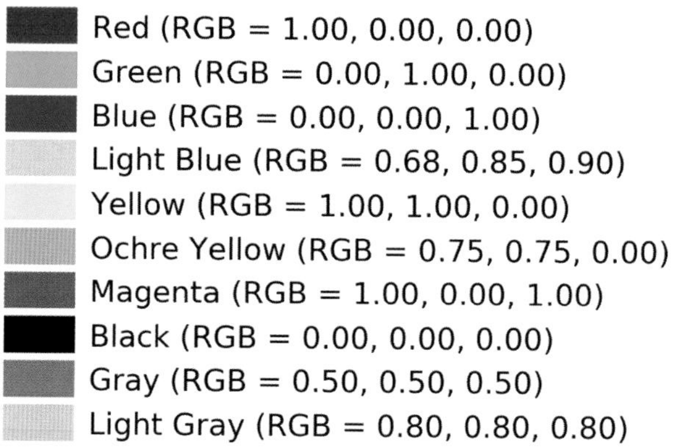

Red (RGB = 1.00, 0.00, 0.00)
Green (RGB = 0.00, 1.00, 0.00)
Blue (RGB = 0.00, 0.00, 1.00)
Light Blue (RGB = 0.68, 0.85, 0.90)
Yellow (RGB = 1.00, 1.00, 0.00)
Ochre Yellow (RGB = 0.75, 0.75, 0.00)
Magenta (RGB = 1.00, 0.00, 1.00)
Black (RGB = 0.00, 0.00, 0.00)
Gray (RGB = 0.50, 0.50, 0.50)
Light Gray (RGB = 0.80, 0.80, 0.80)

Figure 13.10 Examples of colors produced by various RGB (red–green–blue) settings. Intensities vary from 0 to 1 where 0 represents 0 percent intensity and 1 represents 100 percent intensity.

the second dimension equals the width of the image in number of pixels, and the number of elements in the third dimension is three, for the three RGB channel intensities. The values of the elements are floating-point numbers between zero and one, as the examples in Figure 13.10 show.

Because the order of dimensions (and thus the shape) for an RGB intensity array is height, width, and channels, this is different from the normal description of a three-dimensional array's shape as being sheets, rows, and columns. When selecting height and width pixels, we are selecting in the first and second dimensions of the RGB intensity array.

Let us examine a concrete example of an RGB array by printing out some slices (or selections) of the img array from Figure 13.4 (we also print out the shape of img as reference):

```
In [1]: print(np.shape(img))

        (339, 460, 3)

In [2]: print(img[169,230,:])

        [0.3647059  0.7176471  0.32941177]

In [3]: print(img[169,0,:])

        [0.50980395 0.5137255  0.5019608 ]
```

The In [1] cell shows that the img array is 339 pixels high and 460 pixels wide. The In [2] cell shows the RGB intensity triplet for a location near the center of the image: index 169 counting from the top, index 230 counting from the left. The red and blue channel values in this triplet are substantially less than the green channel values, and we suspect this is a greenish pixel. The `In [3]` shows the RGB intensity triplet for a location near the middle in height of the image (index 169 from the top) but at the left-side boundary (index 0 from the left). Because the RGB values at this location are nearly the same, we suspect this is a grayish pixel. Comparison with the coloring in the image in Figure 13.1 confirms our reasoning.

Reading an Image   The Matplotlib image module's `imread` function reads in an image (whose filename is passed in as an argument to the function call) and returns the three-dimensional NumPy array that describes the image. Line 5 in Figure 13.4 and lines 1–3 in Figure 13.6 provide examples of using `imread`.

When `imread` reads PNG files, the intensities are given as a decimal number between zero and one (inclusive on both ends of the range), representing the percentage of intensity as a fractional value. When `imread` reads files of other formats, the values in the elements of the return array may be different.[5] In general, a non-PNG image made up of pixels that each occupy 8 bits of memory (which is what most image formats use) is represented by an integer between 0 and 255 (inclusive on both ends of the range). Thus, a value of 255 represents an intensity of 1.0, a value of 254 represents an intensity of approximately 0.996, etc. The integers between 0 and 255 are used because 8 bits of memory has $2^8 = 256$ possible values. See the answer for Try This! 13-1 for more on expressing the intensities of 8-bit images.

Displaying an Image   The `imshow` function in Matplotlib's pyplot module creates a plot of the image. Given the `img` array from Figure 13.4, the following would create a plot of the image and then render the plot on the screen:

```
plt.imshow(img)
plt.show()
```

When we say `imshow` creates a plot of the image, it means that by default the image is placed in an *x*- and *y*-axis frame with ticks, tick labels, and a frame. The numbering of the *x*- and *y*-axes starts with the origin at the upper-left corner of the image. Thus, the *y*-axis tick labels count the number of pixels going down from the upper-left corner, and the *x*-axis tick labels count the number of pixels going to the right from the upper-left corner.

Writing an Image   To write out the image described by the three-dimensional NumPy array that gives the RGB intensities, we can use the `imsave` function of Matplotlib's pyplot module.[6] Line 19 in Figure 13.4 writes out the image described by `img_highlight` into the PNG file given by the string, `"g004-A1_highlight.png"`.

---

[5] https://matplotlib.org/api/_as_gen/matplotlib.pyplot.imread.html (accessed May 18, 2019).

[6] https://matplotlib.org/api/_as_gen/matplotlib.pyplot.imsave.html and https://stackoverflow.com/questions/9295026/matplotlib-plots-removing-axis-legends-and-white-spaces (both accessed May 15, 2019).

The `imsave` function technically writes out a "**Red–Green–Blue–Alpha (RGBA)**" image, which is an RGB image plus one more channel for a measure of transparency (called "alpha").[7] When the pixel is completely opaque, its alpha value is one.[8] There is no loss of information in writing out an RGB image as RGBA. However, if we write out an RGB image using `imsave` and then read it back in, the shape of the image array will be different, having four channels instead of three. The alpha channel is the last channel, i.e., index 3 in the channel dimension. The functions and routines in this section work with RGBA images as well as RGB images.[9] If we consider only the first three channels of the RGBA intensity array, this is the same as plain RGB encoding.

In some cases, we want the image we write out to be embedded in a plot frame with numbering along the *x*- and *y*-axes. To do so, we create a plot object (technically an `AxesImage` class object) using `imshow` and then save that plot object to a file using `savefig`. As an example, given the `img` array from Figure 13.4, the following creates a version of Figure 13.1 with a plot frame and numbering:

```
plt.imshow(img)
plt.savefig("g004-A1-as-plot.png")
```

The result is shown in Figure 13.11. The axis numbering is in pixels.

Finally, the Matplotlib image tutorial[10] provides a brief but nice introduction to reading in and manipulating images and inspired some ideas in this chapter. An up-to-date link to the tutorial is at www.cambridge.org/core/resources/pythonforscientists/refs/, ref. 27.

13.2.2 Boolean Arrays

In Section 8.2.2, we learned about the boolean type and logical operators to evaluate boolean algebraic expressions. With these tools, we can create more complex branching `if` statements that enable us to analyze complex situations. These tools, however, act upon scalar values, i.e., individual values rather than collections of values. When we want to use branching `if` statements to examine array values, we need to loop through the array elements and use the `if` statement and boolean expressions on individual elements.

In Section 5.2.6, we were introduced to array syntax. In the examples we have seen so far, most of the arrays have held numerical values and the array syntax operations have been arithmetic operations (e.g., addition, multiplication, etc.). As we saw with string arrays, however, arrays can hold values of other types, as long as all the elements have the same type. The boolean type is no exception.

---

[7] https://stackoverflow.com/a/45594478. This behavior is not explicitly mentioned on the `imsave` manual page (https://matplotlib.org/3.3.3/api/_as_gen/matplotlib.pyplot.imsave.html). Both pages accessed December 21, 2020.

[8] https://en.wikipedia.org/wiki/RGBA_color_model (accessed December 29, 2020).

[9] See the documentation for the `imread`, `imsave`, and `imshow` functions linked from https://matplotlib.org/api/_as_gen/matplotlib.pyplot.html (all accessed December 29, 2020).

[10] https://matplotlib.org/tutorials/introductory/images.html#sphx-glr-tutorials-introductory-images-py (accessed April 27, 2019).

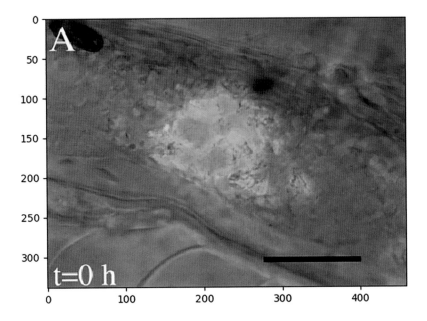

Figure 13.11 Same as Figure 13.1 except with a plot frame and numbering.
Image generated is altered from Volkman et al. (2004).

Let us revisit the atomic number examples in Section 6.2.4. We create two arrays. The first array is called `atomic_numbers` and holds the atomic numbers of three elements. The other array is called `has_more_protons_than_carbon` and holds the value `True` or `False`, depending on whether the corresponding element in `atomic_numbers` (i.e., the element with the same index) has a value greater than 6:

```
atomic_numbers = np.array([17, 2, 6, 8])
has_more_protons_than_carbon = np.array([True, False, False, True])
```

The array `has_more_protons_than_carbon` acts like any other array. We can, for instance, use a `for` loop to go through each element of the array, either by value or by index. In the following code, we access each element of `atomic_numbers` and `has_more_protons_than_carbon` (defined as above) by index and print out a message using these values:

```
for i in range(np.size(atomic_numbers)):
    print( "The element with atomic number " \
        + str(atomic_numbers[i]) \
        + " has more protons than carbon?:   " \
        + str(has_more_protons_than_carbon[i]) )
```

This loop produces this output:

```
The element with atomic number 17 has more protons than carbon?:  True
The element with atomic number 2 has more protons than carbon?:  False
The element with atomic number 6 has more protons than carbon?:  False
The element with atomic number 8 has more protons than carbon?:  True
```

In the above example, the boolean array `has_more_protons_than_carbon` was created by hand. Like any other array, however, we can have the computer calculate and assign the values to each element. We can do this by using a loop:

```
1  atomic_numbers = np.array([17, 2, 6, 8])
2  has_more_protons_than_carbon = np.zeros(np.size(atomic_numbers),
3                                          dtype='?')
4  for i in range(np.size(atomic_numbers)):
5      if atomic_numbers[i] > 6:
6          has_more_protons_than_carbon[i] = True
7      else:
8          has_more_protons_than_carbon[i] = False
```

In lines 2–3, we create a boolean array `has_more_protons_than_carbon` that is the same size as `atomic_numbers` and each element's value is initialized to `False`. The typecode `'?'` tells NumPy to create an array where each element is a boolean value. In lines 3–8, we loop through each element's index and set the value of each element of `has_more_protons_than_carbon` to `True` if the corresponding element of `atomic_numbers` is greater than 6 and `False` otherwise.

Boolean arrays, however, also support array syntax. Thus, the basic comparison operators in Table 6.2 will operate element-wise on arrays and return a boolean array with the element-wise results of the comparison. Thus, this code:

```
1  atomic_numbers = np.array([17, 2, 6, 8])
2  has_more_protons_than_carbon = atomic_numbers > 6
```

creates a boolean array `has_more_protons_than_carbon` where each element of `atomic_numbers` is checked to see if it is greater than 6, and the output array element at the corresponding location is set to `True` or `False`, accordingly. The result of the above code yields `has_more_protons_than_carbon` with the following contents:

```
In [1]:  print(has_more_protons_than_carbon)

         [ True False False  True]
```

Array syntax also works with boolean algebra operators. However, we cannot use `and`, `or`, and `not` directly, as those operators only work on scalar boolean values. Instead, we can use the NumPy functions `logical_and`, `logical_or`, and `logical_not`, respectively. Thus, the following code:

```
atomic_numbers = np.array([17, 2, 6, 8])
print( np.logical_and(atomic_numbers > 6, atomic_numbers < 10) )
print( np.logical_or(atomic_numbers < 6, atomic_numbers > 10) )
print( np.logical_not(atomic_numbers > 6) )
```

produces the following output:

```
[False False False  True]
[ True  True False False]
[False  True  True False]
```

The `logical_and` returns a boolean array whose elements are `True` where the corresponding elements in `atomic_numbers` are greater than 6 and less than 10. The `logical_or` returns a boolean array whose elements are `True` where the corresponding elements in `atomic_numbers` are greater than 10 or less than 6. The `logical_not` returns a boolean array whose elements are the opposite of what is returned by `atomic_numbers > 6`, that is, the same as asking `atomic_numbers <= 6`.

The `logical_and` and `logical_or` take two positional input arguments. If we want to chain together multiple `logical_and` or `logical_or` comparisons between arrays, we can nest the calls of these functions. Thus, this code:

```
atomic_numbers = np.array([17, 2, 6, 8])
print( np.logical_or(
          np.logical_or(atomic_numbers == 6, atomic_numbers == 10),
          atomic_numbers == 8) )
```

outputs:

```
[False False  True  True]
```

as it returns `True` only for values of `atomic_numbers` which are equal to 6 **OR** 10 **OR** 8.

In the Section 13.1 image processing example, we see this use of nesting in lines 9–12 of Figure 13.4 (and lines 12–15 of Figure 13.6). In both code blocks, we create a boolean mask `condition` that is the result of asking where:

- The green channel for all the pixels in the image (`img[:,:,1]`) has a value greater than thresh, **AND,**
- the red channel for all the pixels in the image (`img[:,:,0]`) has a value less than thresh - offset, **AND,**
- the blue channel for all the pixels in the image (`img[:,:,2]`) has a value less than thresh - offset.

Elements of `condition` are `True` if all three bullets above are `True`. The elements are `False` otherwise.

In Section 8.2.7, we were introduced to the `isclose` function, which checks to see whether two numbers are "close" to each other, returning `True` if so. Such a comparison is safer to do than strict equality (`==`) when comparing floating-point numbers. The `isclose` function also works on arrays of numbers to produce a boolean array whose values are the result of doing the comparison element-wise. Thus, given the following array `data`:

```
In [1]:  data = np.array([3.1, 3.10000000000001, 3.1001])
         print(np.isclose(data, 3.1))


         [ True   True False]
```

The first two elements of `data` are close to `3.1`, and so the first two elements of the array returned by `isclose` are `True`. The last element of `data` does not pass the `isclose` test of closeness, so the last element of the array returned by `isclose` is `False`.

13.2.3 Array Syntax and Functions and Asking Questions of Data in Arrays

In previous examples of asking questions of data in arrays, we have been limited to using `if` statements in loops. Boolean arrays enable us to write code that interrogates the data in arrays as a unit, collectively, rather than on an element-by-element basis. The result is more concise code as well as code that runs faster (in general, code using array syntax runs faster than code using loops). In this section, we examine three kinds of inquiry tasks we often are interested in when it comes to data in arrays. For our discussion of all three kinds of inquiry tasks, we consider the `months` and `precip` arrays of mean monthly total precipitation in the Seattle area defined in the code in Figure 6.7, in Section 6.2.5. For convenience, the lines from Figure 6.7 defining `months` and `precip` are repeated here:

```
months = np.array([1, 2, 3, 4, 5, 6, 7, 8, 9, 10, 11, 12])
precip = np.array([5.58, 3.54, 4.54, 2.93, 2.10, 1.47,
                   0.49, 1.03, 1.72, 4.31, 6.50, 5.41])
```

In `months`, 1 means January, 2 means February, etc. The precipitation values in `precip` are given in inches. In this section, we have added a manual line break to output from `print` statements that are too long to fit on the page, in order to improve the clarity of the output.

Conditional Calculations on Array Elements, with an Output Array of the Same Shape

In this inquiry task, we are interested in obtaining an array of the same shape as an "input" array, but with different values depending on whether a condition is true or false. For instance, pretend that for the precipitation values in `precip` defined earlier that an error was discovered in the snow/rain gauge such that all mean monthly values less than 1.1 in need to be adjusted by a factor of 0.82 (that is, the actual precipitation value is 82 percent of the value given in `precip`). All other values in `precip` do not need to be adjusted. We want a new array `adjusted_precip` containing the adjusted precipitation values, if needed, and unchanged values otherwise.

One way of creating `adjusted_precip` is to take advantage of how Python implements arithmetic using boolean values. As we saw in Section 8.2.2 for the case of scalar boolean values, an arithmetic operation on boolean values is the same as applying that operation to one (if the boolean value is `True`) or zero (if the boolean value is `False`). For boolean arrays, the arithmetic operation is conducted element-wise, using array syntax. Thus, in the following code:

```
In [1]:  mask = np.array([True, False, False, True])
         print(mask * 0.82)


         [0.82 0.   0.   0.82]
```

the result of the expression `mask * 0.82` is to return an array where every corresponding element in `mask` that is `True` will equal 0.82 while every corresponding element in `mask` that is `False` will equal 0, because where `mask` is `True`, we are multiplying 0.82 by 1, and where `mask` is `False`, we are multiplying 0.82 by 0.

For our precipitation problem, the equivalent of `mask` above is the result of a test for which points in `precip` are less than 1.1:

```
In [1]:  print(precip < 1.1)


         [False False False False False False  True  True
          False False False False]
```

If we save that boolean array to the array `is_need_adjust` and multiply that boolean by `precip * 0.82`:

```
In [1]:  is_need_adjust = precip < 1.1
         print(is_need_adjust * precip * 0.82)


         [0.    0.    0.    0.    0.    0.    0.4018 0.8446
          0.    0.    0.    0.    ]
```

we obtain an array where all values of `precip` less than 1.1 are equal to the corresponding element of `precip` times 0.82, and all other values are 0. Put another way, the expression:

```
is_need_adjust * precip * 0.82
```

returns an array containing the adjusted value for the months that need adjustment and 0 otherwise.

We have accomplished half our task. Now, we need to take all the unadjusted months and fill in those values with the unadjusted precipitation value for the appropriate month. We do this by taking advantage of how: (1) the `logical_not` function flips all `True`s into `False`s and vice versa for any boolean array passed in as an argument, and (2) the sum of any number plus zero is that number. The code below does what we want:

```
is_need_adjust = precip < 1.1
adjusted_precip = (is_need_adjust * precip * 0.82) \
                  + (np.logical_not(is_need_adjust) * precip)
```

If we print the above output array using `print(adjusted_precip)`:

```
[5.58   3.54   4.54   2.93   2.1    1.47   0.4018 0.8446
 1.72   4.31   6.5    5.41  ]
```

we see we obtain what we wanted.

We have already described lines 1–2 in the above code. What is happening in line 3? The call to `logical_not` in line 3 returns an array where each value is "opposite" the value in `is_need_adjust`. We can print out portions of line 3 to see this:

```
In [1]:  print(is_need_adjust)

         [False False False False False False  True   True
          False False False False]

In [2]:  print(np.logical_not(is_need_adjust))

         [ True   True   True   True   True   True False False
           True   True   True   True]
```

When the return value of `np.logical_not(is_need_adjust)` is multiplied by `precip`, all the places where `is_need_adjust` is `False` will be filled with the value of `precip` (because the `logical_not` turned the `False` into `True`). Thus, the parenthetical term in line 3 is an array filled with the unadjusted `precip` value where `is_need_adjust` is `False` and zero otherwise. When that array is added to the array generated by `is_need_adjust * precip * 0.82`, every element is the result of the sum of a "desired" element and a zero, which equals the "desired" element.

The arithmetic method just described works fine, but NumPy gives us another means of accomplishing the same task, with more concise syntax: the `where` function. The `where` function has two ways of being called. The way that is of relevance to conditional calculations on array elements, with an output array of the same shape, is the three-argument format. In this form, three positional arguments are passed into the call: a boolean "condition" array, a scalar or array that defines output values in the return array wherever the condition array is `True`, and a scalar or array that defines output values in the return array wherever the condition array is `False`. In the scenario above, using the array `precip`, this call to `where`:

```
1   is_need_adjust = precip < 1.1
2   adjusted_precip = np.where(is_need_adjust, precip * 0.82, precip)
```

creates the array `adjusted_precip` as before, where:

- Every element corresponding to where `is_need_adjust` is `True` is set to the corresponding element of `precip` multiplied by 0.82, and
- every element corresponding to where `is_need_adjust` is `False` is set to the corresponding element of `precip`.

Although it is often clearer to make the condition array (the boolean array `is_need_adjust` in the example above) a separate array, we can also put the boolean test directly into the `where` call as the first argument:

```
adjusted_precip = np.where(precip < 1.1, precip * 0.82, precip)
```

In the above example, all three arguments to the `where` call are arrays. If, for instance, the second argument was not an array but a single number (a scalar), then everywhere the condition is `True`, the corresponding element in the return array would have the value of that number in the second argument. A scalar third argument behaves similarly for where the condition is `False`.

One nice benefit of the `where` syntax is its readability. The result, `adjusted_precip`, is, literally, "where `precip` is less than 1.1, it is equal to `precip` times 0.82, and otherwise, it is equal to `precip`."

Conditional Calculations on Array Elements, with Only Certain Elements of the Array Retained

What if, as in the previous subsection, we want to adjust all values of `precip` that are less than 1.1 by multiplying those values by 0.82, but we only want to return those adjusted values? That is, we want a two-element array with the values:

```
[0.4018 0.8446]
```

which are the elements of `precip` (as defined in the previous section) that are less than 1.1, multiplied by 0.82.

The `where` function can also help us accomplish this task. In this case, instead of passing in three arguments to the function, we only pass in one argument, the condition array. The `where` function returns a tuple whose elements each hold an array. These arrays contain the index values (along that dimension) of the elements for which the condition is `True`. Thus:

```
In [1]: precip = np.array([5.58, 3.54, 4.54, 2.93, 2.10, 1.47,
                           0.49, 1.03, 1.72, 4.31, 6.50, 5.41])
        print(np.where(precip < 1.1))

        (array([6, 7]),)
```

The result of the call to `where` in the last line of `In [1]` is a one-element tuple. The element in that tuple is an array that lists the indices where `precip` is less than 1.1. If the array is two-dimensional:

```
In [1]:  precip_2d = np.reshape(precip, (4,3))
         print(precip_2d)


         [[5.58 3.54 4.54]
          [2.93 2.1  1.47]
          [0.49 1.03 1.72]
          [4.31 6.5  5.41]]


In [2]:  print(np.where(precip_2d < 1.1))


         (array([2, 2]), array([0, 1]))
```

The result of the call to `where` in cell `In [2]` is a tuple of two elements. The first element is the row dimension indices where the two-dimensional condition array is `True`. The second element is the column dimension indices where the two-dimensional condition array is `True`. Thus, the first element of `precip_2d` that is less than 1.1 has the (row, column) indices of (2, 0). The second element of `precip_2d` that is less than 1.1 has the (row, column) indices of (2, 1).

Once we have the tuple of the indices in each dimension for all the elements where the condition is `True`, we can use array indexing syntax to extract only the elements with these indices. For the `precip_2d` above, this code:

```
In [1]:  pts = np.where(precip_2d < 1.1)
         print(precip_2d[pts])


         [0.49 1.03]
```

extracts the points `precip_2d[2,0]` and `precip_2d[2,1]` and places them in an array. We then multiply this array by 0.82:

```
In [1]:  print(precip_2d[pts] * 0.82)


         [0.4018 0.8446]
```

Whereas `precip_2d` is two-dimensional, the result of `precip_2d[pts]` is one-dimensional.

Thus, using array syntax in making conditional calculations on array elements, with only certain elements of the array retained, is a two-step process: First, use the `where` function to obtain a tuple that provides the indices of all points where the condition is `True`. Second, use array indexing notation with the results of the `where` function call to select only those points meeting the condition. Only those points are placed into the output array.

Conditional Calculations on Array Elements, Reduced to a Scalar Value

A third common case involving arrays and conditional calculations is selecting certain elements of an array and conducting some calculations on those elements that results in a scalar value as output. For instance, we might be interested in the total rainfall in `precip` defined above for the months where the precipitation was less than 1.1 in. We can accomplish this and similar tasks by extending the methods described earlier in this section.

One method of making this kind of calculation is by using a boolean array as a mask and taking advantage of the arithmetic properties of zero and one. For instance, the following will calculate the total rainfall in `precip` for months when the precipitation was less than 1.1 in:

```
In [1]:  mask = precip < 1.1
         print(np.sum(mask * precip))
```

```
         1.52
```

As we saw earlier in this section, `mask` is a boolean array of `True`s and `False`s. When it is multiplied by `precip`, all the `True`s are treated as ones and the `False`s are treated as zeros. In the result of `mask * precip`, only the places where `mask` is `True` will have a nonzero value. When the `sum` function is applied to the result of that expression, all the elements are added up. The zero values contribute nothing to the sum so the result is only the total rainfall for the months when the precipitation was less than 1.1 in.

This method works when the function being applied to the array that has been transformed by the boolean mask handles zero values in a way different than nonzero values. The `sum` function is an example of such a function. The NumPy function `mean`, which calculates the arithmetic mean of all the elements in an array, is an example of a function that does not exhibit this characteristic. Thus, this method of using boolean masks to do a conditional calculation that reduces to a scalar will not work with `mean`.

Another way of making a conditional calculation that reduces to a scalar value that works on `mean` and other functions that the earlier method does not work on is by using the `where` function in its one-argument form, extracting only those points that meet the condition and operating on only those points. The following code uses this method to calculate the total rainfall for the months when the precipitation was less than 1.1 in:

```
In [1]:  pts = np.where(precip < 1.1)
         print(np.sum(precip[pts]))
```

```
         1.52
```

The `sum` call works on an array consisting only of those points in `precip` that meet the condition passed into the `where` function call. A call to `mean` on that subarray would also return the correct value.

The code in the Section 13.1 image processing example (Figures 13.4 and 13.6) makes use of the `where` function in its three-argument form to replace all the red, green, and blue channel values with different values if the elements of the boolean array `condition` (described in detail at the end of Section 13.2.2) are `True`. For instance, the return values of each of the `where` calls in lines 15–17 of Figure 13.4 are two-dimensional arrays giving the new red, green, and blue channel intensities for each pixel. In line 15 of Figure 13.4, the red channel in `img_highlight` is set to 1.0 (full intensity) where `condition` is `True` and is set to the existing red channel value where `condition` is `False`. The slicing or selection syntax `img[:,:,0]` selects the red channel values for each pixel in the original two-dimensional image, and `img_highlight[:,:,0]` selects the red channel values for each pixel in the to-be magenta highlighted two-dimensional image. Similarly, in line 16, the green channel values for each pixel (`img[:,:,1]` in the original image and `img_highlight[:,:,1]` in the to-be highlighted image) are set to an intensity of 0.0 when `condition` is `True`. In line 17, the blue channel values for each pixel (`img[:,:,2]` and `img_highlight[:,:,2]`) are set to an intensity of 1.0 when `condition` is `True`.

13.2.4 Performance of Looping and Array Syntax and Functions

Array syntax and array function calls are more concise than using loops to explicitly go through each element of an array and evaluate a test. In many cases, although not all cases, concise code is also more readable. Array syntax and function calls, however, have another benefit in that they often run substantially faster than using loops with `if` statements. The code in Figure 13.12 produces calculations that illustrate this difference.

The first line in Figure 13.12 imports the time module. This module has a function `time` that returns the number of seconds since the "**epoch**," which on Unix systems is January 1, 1970, 00:00:00 UTC.[11] By calculating the number of seconds since the epoch before and after a block of code, and subtracting the difference, we can use this function to obtain a rough estimate of how much time it takes to do that calculation. In Chapter 22, we discuss profiling, a more formal and detailed way of timing how long portions of a program take to run.

Lines 2–15 in Figure 13.12 are directly copied from lines 1–14 in Figure 13.4. These lines read in the original image, create the `condition` array that identifies which pixels in the image are "green," and create a copy of the image data that will be used to highlight the "green" areas with magenta.

In lines 17–21 in Figure 13.12, we calculate the amount of time it takes to replace the red channel values in `img_highlight` with the intensity value 1.0 for pixels where `condition` is `True`, using array syntax and arithmetic operations (lines 19–20). In lines 23–26, we perform the same calculation using the `where` function (line 25). In lines 28–36, we do the same calculation using `for` loops and an `if` statement (lines 32–35). In all three cases, we run the calculations 100 times (the `for` loops defined in lines 18, 24, and 29). We do this because we need the computer to take longer to do the calculations. If we do not do this, the running time estimates may be inaccurate. The calculations may be done so quickly that natural variation in computation times due to the running of normal background programs

---

[11] https://docs.python.org/3/library/time.html (accessed May 15, 2019).

```
 1   import time
 2   import numpy as np
 3   import matplotlib.pyplot as plt
 4   import matplotlib.image as mpimg
 5
 6   img = mpimg.imread("g004-A1.png")
 7
 8   thresh = 0.58
 9   offset = 0.05
10   condition = np.logical_and(
11                   np.logical_and(img[:,:,1] > thresh,
12                                   img[:,:,0] < (thresh - offset)),
13                   img[:,:,2] < (thresh - offset) )
14
15   img_highlight = img.copy()
16
17   begin_time = time.time()
18   for itrial in range(100):
19       img_highlight[:,:,0] = (condition * 1.0) + \
20                              (np.logical_not(condition) * img[:,:,0])
21   print("Array syntax time: " + str(time.time() - begin_time))
22
23   begin_time = time.time()
24   for itrial in range(100):
25       img_highlight[:,:,0] = np.where(condition, 1.0, img[:,:,0])
26   print("Array function time: " + str(time.time() - begin_time))
27
28   begin_time = time.time()
29   for itrial in range(100):
30       for i in range(np.shape(condition)[0]):
31           for j in range(np.shape(condition)[1]):
32               if condition[i,j]:
33                   img_highlight[i,j,0] = 1.0
34               else:
35                   img_highlight[i,j,0] = img[i,j,0]
36   print("Loops plus branching time: " + str(time.time() - begin_time))
```

Figure 13.12 Code that includes part of Figure 13.4, reconfigured to compare the run time between array functions and the use of explicit loops and if statements.

on the computer will be a larger influence than the running time difference due to the different ways of coding.

Figure 13.13 shows the output of running the Figure 13.12 code. These timings depend on the machine the code is run on. But, for this computer, using explicit looping plus an if statement for branching takes approximately 50 times longer than using array syntax and arithmetic operations! Compared with the where statement, explicit looping plus an if statement for branching takes approximately 134 times longer! Both array syntax and array functions are much faster than using loops with an if statement. Of the three methods, the where function is the fastest.

```
1   Array syntax time: 0.5006978511810303
2   Array function time: 0.186431884765625
3   Loops plus branching time: 25.15757703781128
```

Figure 13.13 Output from running the code in Figure 13.12. Times are in seconds.

13.2.5 The NumPy `reduce` Method

In the Figure 13.4 image processing code in Section 13.1, lines 9–12 apply a sequence of two logical "and" operations using two nested `logical_and` function calls. The code works fine but looks overly complicated. If we are interested in applying `logical_and` more than two times, the code will look even more confusing. How can we apply `logical_and` (or `logical_or`) repeatedly over a sequence of arrays?

It turns out that NumPy functions that act element-wise on arrays are based on a common "template" that defines methods common to all these functions. These functions are called **"universal functions"** or **"ufuncs."**[12] In Chapter 17, we will look more at what it means to define a "template" for a set of objects. All ufuncs have a method called `reduce` which reduces by one the array-like object being passed in and applies the ufunc in question along the first dimension (by default).[13] If we pass in a list or tuple of arrays to `reduce`, the ufunc is applied to each one of those arrays, one at a time.

The `logical_and` function is a ufunc. To apply `logical_and` to a sequence of three arrays, we can call `reduce` on a list or tuple of these arrays.[14] Thus, the code in lines 9–12 of Figure 13.4 (reproduced here for convenience):

```
1   condition = np.logical_and(
2              np.logical_and(img[:,:,1] > thresh,
3                             img[:,:,0] < (thresh - offset)),
4              img[:,:,2] < (thresh - offset) )
```

can be replaced by:

```
1   condition = np.logical_and.reduce( (img[:,:,1] > thresh,
2                                       img[:,:,0] < (thresh - offset),
3                                       img[:,:,2] < (thresh - offset)) )
```

The argument passed into the `reduce` call is a tuple whose elements are the subarrays holding the boolean results of comparisons using the intensity values for the green, red, and blue channels (respectively) for each pixel in the image. This code is clearer and easier to read.

---

[12] https://docs.scipy.org/doc/numpy/reference/ufuncs.html (accessed May 16, 2019).

[13] https://docs.scipy.org/doc/numpy/reference/generated/numpy.ufunc.reduce.html (accessed May 16, 2019).

[14] https://docs.scipy.org/doc/numpy/reference/generated/numpy.ufunc.reduce.html and https://stackoverflow.com/a/20528566 (accessed May 16, 2019).

13.2.6 Looping Through Lists of Objects

In earlier chapters, we looped through lists and arrays of numbers and lists of strings. In some cases, we looped through the collection of items directly while in other cases we looped through the collection of items by looping through the index values.

Lists (and tuples), however, can hold more than numbers or strings. They can, in fact, hold anything. We saw a little of this in Chapter 7, when we created two-dimensional arrays from lists of lists: that is, from lists whose elements are themselves lists. In line 4 in the code of Figure 13.4 (reproduced here for convenience):

```
images = [img1, img2, img3]
```

The variable `images` is a list where each element is one of the three-dimensional arrays returned by `imread` that hold the information for the pixels in each of the three images. Note that `images` is *not* a single four-dimensional array. Rather, it is a list that holds three different objects, each of which is a separate array.

As a result, in lines 10–11 of Figure 13.4 (reproduced here for convenience):

```
for i in range(len(images)):
    img = images[i]
```

as the iterator `i` goes through the values 0, 1, and 2, the value of `img` changes each iteration of the loop. The first iteration, `img` is set to the array `img1`, which is the first element of `images`. The second iteration, `img` is set to the array `img2`, which is the second element of `images`. In the final iteration, `img` is set to the array `img3`, the last element of `images`.

Later on in the body of the loop in Figure 13.4, we needed to use the value of `i` for other purposes (for filenaming in line 22 and the message that was printed out to the screen in line 24). However, if we did not need the index `i`, we could have directly looped through the values of `images` and directly made `img` the iterator. That is, we could have replaced lines 10–11 in Figure 13.4 with:

```
for img in images:
```

and all the code in lines 12–20 would work exactly the same. Iterators do not have to be numbers or strings or something with a single value. Iterators can be set to any kind of object. Whatever object is in the collection (e.g., list, tuple, etc.) given after the `in` in the `for` loop, the iterator is set to each of those objects, one in turn.

As an example of this behavior, consider the following loop:

```
1   list_objs1 = [1.2, -4, "Experiment 1", np.array([4, 7, -1])]
2   for ielement in list_objs1:
3       print(type(ielement))
```

which produces this output:

```
1   <class 'float'>
2   <class 'int'>
3   <class 'str'>
4   <class 'numpy.ndarray'>
```

The first element is a floating-point number, the second is an integer, the third is a string, and the last is a NumPy array. The variable ielement becomes each one of the elements of list_objs1, one in turn, as the loop iterates through that list.

This behavior also means that functions or methods that can act on all the kinds of objects in a list we are looping through act on each of those elements, in their turn. For instance, in the following example, we iterate through a list that holds a string, another list, and another string:

```
1   list_objs2 = ["Results", [3, -2, 11], "Experiment 2"]
2   for ielement in list_objs2:
3       print(len(ielement))
```

The len function properly works on both strings and lists, so the result of the above code is:

```
1   7
2   3
3   12
```

In this final example, we iterate through a list of strings, all of which have the method title (which returns the string with the first letter of every word capitalized):

```
1   list_objs3 = ["robert boyle", "procedures", "petri dish"]
2   for ielement in list_objs3:
3       print(ielement.title())
```

When run, the resulting output is:

```
1   Robert Boyle
2   Procedures
3   Petri Dish
```

The iterator ielement is set to each of the string objects in list_objs3 and the title method is run. As ielement changes each iteration, the result of the call to title will also change.

Loops in Python, then, are not limited to iterating only through single numbers or values but may iterate through any kind of object. When iterators are set to an object, the iterator takes on all the properties and capabilities of that object. Any methods of the object the iterator can call and any functions that can act on objects of that type can also act on the iterator.

13.3 Try This!

Although this section includes image-processing-related examples similar to those we have seen in Section 13.1, because the main topics of Section 13.2 are so widely applicable to all kinds of scientific and engineering analysis, there are Try This! exercises in this section that are unrelated to image processing.

The image-processing Try This! in this section all, in some way, connect with Figures 13.14–13.16, which show visualizations of satellite imagery of the city of Bhubaneswar in the state of Odisha, India, in nighttime, on April 30, 2019 and May 5, 2019. Between these two dates, Tropical Cyclone Fani hit the city causing substantial damage, including to the electrical

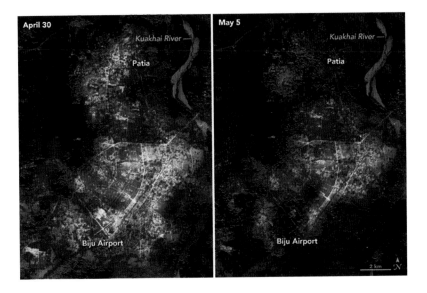

Figure 13.14 The city of Bhubaneswar in the state of Odisha, India, in nighttime, on April 30, 2019 and May 5, 2019. The images were created by Lauren Dauphin from Landsat 8 and Suomi NPP–VIIRS satellite data (Patel and Dauphin, 2019).

Figure 13.15 The base image for the April 30, 2019 image in Figure 13.14. The filename for the image is *bhubaneswar_blm_2019120_lrg.png*. See Patel and Dauphin (2019) for details.

utility system.[15] In these images, the reds and yellows express lighting intensity, and the decrease in reds and yellows from April 30 to May 5 shows the substantial drop in nighttime electricity use due to the damage from the tropical cylcone. Figure 13.14 shows annotated visualizations of the satellite imagery to facilitate side-by-side comparison. Figures 13.15 and 13.16 show the base images for Figure 13.14. That is, they are not cropped and show the underlying images without annotation.

Try This! 13-1 Reading in Images: Satellite Imagery

Write code that will read in the April 30, 2019 base image in Figure 13.15 and print out the number of pixels, number of RGB values to describe the image, dimensions of the array of RGB values, and the maximum and minimum RGB intensities.

[15] Patel and Dauphin (2019).

Figure 13.16 The base image for the May 5, 2019 image in Figure 13.14. The filename for the image is *bhubaneswar_blm_2019125_lrg.png*. See Patel and Dauphin (2019) for details.

Try This! Answer 13-1

This code will accomplish the tasks:

```
import numpy as np
import matplotlib.image as mpimg

img = mpimg.imread("bhubaneswar_blm_2019120_lrg.jpg")

print( "Number of pixels: " + str(np.size(img[:,:,0])) )
print( "Number of RGB values: " + str(np.size(img)) )
print( "Dimensions of the array of RGB values: " + str(np.shape(img)) )
print( "Maximum RGB intensity: " + str(np.max(img)) )
print( "Minimum RGB intensity: " + str(np.min(img)) )
```

and the output is the following:

```
1   Number of pixels:   10715382
2   Number of RGB values:   32146146
3   Dimensions of the array of RGB values:   (3294, 3253, 3)
4   Maximum RGB intensity:   255
5   Minimum RGB intensity:   0
```

As line 3 of the output indicates, the image is 3294 pixels high and 3253 pixels wide. The array of RGB intensities, however, are not decimal values between and including 0 and 1 but integers between and including 0 and 255. Because the image file is a Joint Photographic Experts Group (JPEG) format file, imread reads in the intensities in their 8-bit form. That is, each channel for each pixel has a possible intensity value equal to one of 256 possible values. There are 256 values because in a digital computer, values are represented by binary numbers. In a computer, a single binary digit of a binary number is a bit, and can have the value of "on" or "off" (or "one" or "zero"). Eight binary digits have a total of $2^8 = 256$ possible values, so an 8-bit image is an image whose channel intensities can have values from 0 to 255.

Matplotlib's imread function will only handle PNG files natively. For those files, the channel intensities returned are decimal values between and including 0 and 1. If we want to transform the 0 to 255 intensities into 0 to 1 values, we can divide each element in img by 255:

```
img = img / 255.0
```

In doing so, however, the array img will now occupy significantly more memory. Why?

The original intensity values, being 8-bit values, each take only one byte of memory (8 bits equals 1 byte). For NumPy arrays, we can see how many bytes an array occupies by printing out the nbytes attribute.[16] For our img array:

```
In [1]:  print(img.nbytes)

         32146146
```

This is the same value returned by the NumPy size function. Each element occupies 1 byte of memory, so the total number of elements in the array is equal to the total number of bytes of memory occupied by that array. When we convert img to decimal values by dividing by 255, the resulting array is a floating-point array. In that array (more precisely a "double-precision floating-point array"), each element occupies 8 bytes of memory, not 1 byte. There are more possible values for a decimal number than 256 possibilities. Thus, for our transformed img array:

---

[16] https://docs.scipy.org/doc/numpy/reference/generated/numpy.ndarray.nbytes.html#numpy.ndarray.nbytes and https://stackoverflow.com/a/15591157 (both accessed May 21, 2019).

```
In [1]:   print(img.nbytes)

          257169168
```

which is eight times the memory it occupied before the transformation. When an object occupies more memory, it takes longer for the computer to manipulate and do calculations with it. Thus, for larger images, we might find it better to leave the channel intensities in their 8-bit form and manipulate integers rather than transform them into decimal values.

Try This! 13-2 Boolean Arrays: Satellite Imagery

For the April 30, 2019 base image in Figure 13.15, write code that will create a boolean mask where gray pixels are `True` and nongray pixels are `False`.

Try This! Answer 13-2

This code will create the boolean mask `condition`:

```
1    import numpy as np
2    import matplotlib.image as mpimg
3
4    img = mpimg.imread("bhubaneswar_blm_2019120_lrg.png")
5
6    thresh = 20
7    diff1 = np.absolute(img[:,:,0] - img[:,:,1])
8    diff2 = np.absolute(img[:,:,1] - img[:,:,2])
9    diff3 = np.absolute(img[:,:,0] - img[:,:,2])
10
11   condition = np.logical_and(
12               np.logical_and(diff1 < thresh, diff2 < thresh),
13               diff3 < thresh )
```

To find the pixels that are gray, we take advantage of the idea that the red, green, and blue channel intensities are all nearly the same for gray pixels. We first calculate the absolute value of the differences between channel intensities for each pixel and then compare whether those differences are all within some value `thresh`. The value of `thresh` will be an integer, because the elements of `img` are all integer values between and including 0 and 255.

Try This! 13-3 Boolean Arrays and Comparing Floating Point: Temperatures

Pretend we have the following measurements from a highly accurate thermometer inside an ice water bath (a beaker filled with ice and water), saved in an array `bath_temp`. This is similar to

Try This! 8-10, except with multiple temperature values in an array. All values are in °C and are taken at various, consecutive times:

```
bath_temp = np.array( \
    [0.011394329, 0.008490942, 0.000348343, 0.00003469,
     0.000033323, 0.000020031, 0.000009932, 0.00000033,
     0.000000453, 0.000000032, 0.000000095, 0.00000054,
     0.000000002, 0.000000005, 0.000000001, 0.00000023])
```

How many measurements are "close to" the freezing point of water, $0\,^\circ\mathrm{C}$?

Try This! Answer 13-3

We use the NumPy isclose function's default definition of what constitutes "close to." Using that function to test where bath_temp is close to 0.0 yields the following boolean array:

```
In [1]:  np.isclose(bath_temp, 0.0)

Out[1]:  array([False, False, False, False, False, False, False, False, False,
                False, False, False,  True,  True,  True, False], dtype=bool)
```

The index 12–14 values of bath_temp are close to 0.0 while the rest of the values are not. To obtain the number of measurements that are close to 0.0, we can sum the boolean values returned from isclose, because all True values become 1 and False values become 0 during the summing operation:

```
In [1]:  np.sum(np.isclose(bath_temp, 0.0))

Out[1]:  3
```

We can sum up the idea behind this trick as follows: True yields an occurrence tally because it becomes a one during an arithmetic operation with the True.

Try This! 13-4 Conditional Array Calculations: Satellite Imagery

For the April 30, 2019 base image in Figure 13.15, write code that uses the boolean mask condition and pixel channel intensities array img from Try This! 13-2 to create a revised version of the base image where all areas that are *not* gray are made green, to highlight the areas that in some way represent electrically lit areas. Write that version to an image file.

Try This! Answer 13-4

The code below will solve the problem. The code assumes the import statements from the Try This! 13-2 have already been made and that condition exists as is calculated in Try This! 13-2:

```
1  import matplotlib.pyplot as plt
2
3  not_condition = np.logical_not(condition)
4  img_highlight = img.copy()
5  img_highlight[:,:,0] = np.where(not_condition, 0, img[:,:,0])
6  img_highlight[:,:,1] = np.where(not_condition, 255, img[:,:,1])
7  img_highlight[:,:,2] = np.where(not_condition, 0, img[:,:,2])
8
9  plt.imsave("bhubaneswar_blm_2019120_lrg_highlight.png", img_highlight)
```

The boolean mask from Try This! 13-2 tell us which pixels are gray. We apply the `logical_not` function on that array in line 3 to turn all occurrences of `True` to `False`, and vice versa, because we want to highlight the nongray pixels. In line 4, we make a copy of the original image's data `img` so that we do not change the values of the original image.

In lines 5–7, we use the `where` function to replace the red and blue intensities for all nongray points to 0 and the green intensities to 255. Recall that the `img` array in the Try This! 13-2 solution are integers from and including 0 to 255. Thus, to obtain 100 percent intensity in the green channel, we set the intensities for the nongray pixels to 255.

In the last line, we save the image described by the data in the `img_highlight` array as a PNG file, but we do not have to first convert the intensities to decimal values between and including 0 and 1. We can keep the values in their 8-bit integer form and the `imsave` function will write the file out correctly.

The code produces the image given in Figure 13.17. All areas that are *not* gray are shaded green.

Try This! 13-5 Subset Elements from Conditional Calculations: Summer Precipitation

In Section 6.2.5, Figure 6.7, we provided code to calculate the average Summer mean monthly precipitation in the Seattle area. For those data, without using a loop as in lines 6–14 of Figure 6.7, extract only the Summer mean monthly precipitation values and then calculate the average of those values.

Try This! Answer 13-5

This code does the calculation (the definition of `months` and `precip` are copied from Figure 6.7):

```
1  import numpy as np
2  months = np.array([1, 2, 3, 4, 5, 6, 7, 8, 9, 10, 11, 12])
3  precip = np.array([5.58, 3.54, 4.54, 2.93, 2.10, 1.47,
4                     0.49, 1.03, 1.72, 4.31, 6.50, 5.41])
5  summer_pts = np.where(np.logical_and(months >= 6, months <= 8))
6  summer_avg = np.average(precip[summer_pts])
```

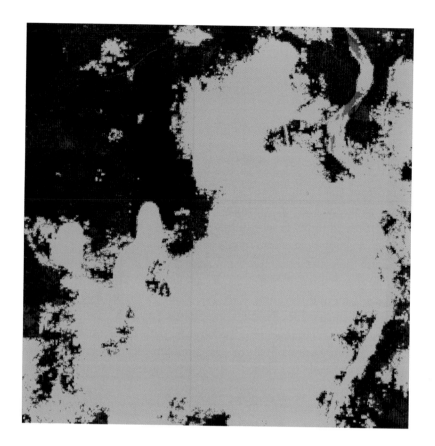

Figure 13.17 Image generated by the solution to Try This! 13-4. The filename of the image that is generated is *bhubaneswar_blm_2019120_lrg_highlight.png*. Image generated by altering an image from Patel and Dauphin (2019).

The values of `summer_pts` and `summer_avg` are:

```
In [1]:  print(summer_pts)

         (array([5, 6, 7]),)

In [2]:  print(summer_avg)

         0.9966666666666667
```

The June, July, and August months correspond to indices 5, 6, and 7, respectively, in `months` and `precip`, as the contents of `summer_pts` show. A call to the `where` function with a single boolean array as input will return a tuple that lists, for each dimension, the indices where the input array is `True`. Because the argument passed into the `where` call in line 5 of the above code is one-dimensional, the output tuple is a single-element tuple. We use `summer_pts` to extract only the elements in `precip` specified by `summer_pts`, as the output below demonstrates:

```
In [1]:   print(precip[summer_pts])

          [1.47 0.49 1.03]
```

Try This! 13-6 Scalar Result from Conditional Calculations: Satellite Imagery

For the April 30, 2019 base image in Figure 13.15, write code that will calculate what percentage of the pixels in the image are gray. Assume that the solution for Try This! 13-2 has already been run, so we have access to the `condition` and `img` variables defined in that code.

Try This! Answer 13-6

The code below, added on to the answer to Try This! 13-2, does the calculation:

```
1   percent_gray = float(np.sum(condition)) / np.size(img[:,:,0]) * 100
2   print( 'Percentage pixels are "gray":   ' + str(percent_gray) )
```

with the output:

```
Percentage pixels are "gray":   36.112982252989205
```

In line 1 of the solution above, the total number of pixels that are gray, i.e., where `condition` is `True`, equals the sum of all elements in the boolean `condition` array because during the summation operation, all `True` values are treated as one and all `False` values are treated as zero. Also, in line 1 of the solution above, we explicitly convert the result of `np.sum(condition)` into a floating-point number in order to ensure the percentage calculation uses regular division instead of integer division. This is not a problem in Python 3.x which uses regular division even if both the numerator and denominator are integers, but in Python 2.7.x, integer division is used if both the numerator and denominator are integers. By explicitly converting the numerator to a floating-point value, we ensure the code above will work the same way whether we use Python 2.7.x or 3.x.

Try This! 13-7 Scalar Result from Conditional Calculations: Summer Precipitation

In Section 6.2.5, Figure 6.7, we provided code to calculate the average Summer mean monthly precipitation in the Seattle area. Using boolean arrays, without using a loop as in lines 6–14 of Figure 6.7 or the `where` function as in Try This! 13-5, calculate the average Summer mean monthly precipitation.

Try This! Answer 13-7

This code does the calculation (the definition of `months` and `precip` are copied from Figure 6.7):

```
1  import numpy as np
2  months = np.array([1, 2, 3, 4, 5, 6, 7, 8, 9, 10, 11, 12])
3  precip = np.array([5.58, 3.54, 4.54, 2.93, 2.10, 1.47,
4                     0.49, 1.03, 1.72, 4.31, 6.50, 5.41])
5  is_summer = np.logical_and(months >= 6, months <= 8)
6  summer_avg = np.sum(is_summer * precip) / np.sum(is_summer)
```

The mask `is_summer` and the value of `summer_avg` is:[17]

```
In [1]:  print(is_summer)

         [False False False False False  True
            True  True False False False False]

In [2]:  print(summer_avg)

         0.9966666666666667
```

A carriage return and spaces are added to the above screen output so the entire line fits on a page. Arrays are still a single sequence of values.

Elements in `is_summer` that are `True` correspond to Summer months. When in line 6 we multiply `precip` by `is_summer`, every value of `precip` that has a corresponding `False` value in `is_summer` ends up contributing nothing to the `sum` of `is_summer * precip`, because `False` becomes a zero when we multiply it by a number. The `precip` elements corresponding to `True` do contribute to the sum because `True` becomes a one when we multiply it by a number. In the denominator of the expression in line 6, the total number of Summer elements is equal to the sum of all elements in `is_summer` for the same reason: `True` becomes a one and `False` becomes a zero when we add the elements in the boolean array together.

Try This! 13-8 The `reduce` Method: Satellite Imagery

Recreate the boolean mask in Try This! 13-2 by using the `reduce` method.

Try This! Answer 13-8

The lines to change in the solution to Try This! 13-2 are these (reproduced below for convenience):

---

[17] A manual line break is added to the `print` statement output to improve the clarity of the output.

```
condition = np.logical_and(
            np.logical_and(diff1 < thresh, diff2 < thresh),
            diff3 < thresh )
```

These lines are replaced by:

```
condition = np.logical_and.reduce(
            (diff1 < thresh, diff2 < thresh, diff3 < thresh) )
```

For this use, the three boolean arrays we are using `logical_and` on, in sequence, are placed into a tuple, and that tuple is passed into the call to `reduce`. There are other ways of using `reduce` which we do not describe here. See the NumPy manual entry for `reduce` for details. An up-to-date link to the manual page is at www.cambridge.org/core/resources/pythonforscientists/refs/, ref. 43.

Try This! 13-9 Looping Through Objects: Satellite Imagery

For the Figure 13.15 and 13.16 base images, write code that will calculate what percentage of the pixels in each image are gray and will print out the percentage of nongray pixels for each image. Write the program so that the reading in of the images occurs in a loop and generates a list of array objects, and the processing of those array objects occurs in a separate loop.

Try This! Answer 13-9

The code in Figure 13.18 solves the problem. Much of the code in the solution has already been described in the other Try This! exercises in this section. The two loops, however, are worth comment. In lines 4–5, we create a list of the image filenames. In line 6, we create an empty list `images` that will hold the image intensity data arrays. In line 7, we loop through the filenames. In line 8, when we read in each image file using `imread`, we immediately store the return value as an element in `images` by using the `append` method. As we saw in Section 9.2.3, the `append` list method adds what is passed in as an argument to the method call to the end of the list. In line 10, we loop through the elements of the `images` list, each of which is an image intensity data array. The iterator `img` is set to the elements of `images`, one at a time, as we go through the list.

Because we wrote this code using loops, to process more images, we only need to add the new filenames to the `files` list. None of the rest of the code needs to change to process additional images. And we can automatically create a list of files either by using loops and string conversion and concatenation (as we did in line 22 in Figure 13.6) or through the techniques from Section 10.2.5.

The output from this Try This! solution is:

```
Percentage pixels "non-gray":  63.887017747010795
Percentage pixels "non-gray":  39.6482551905289
```

```
1    import numpy as np
2    import matplotlib.image as mpimg
3
4    files = ["bhubaneswar_blm_2019120_lrg.jpg",
5             "bhubaneswar_blm_2019125_lrg.jpg"]
6    images = []
7    for ifile in files:
8        images.append(mpimg.imread(ifile))
9
10   for img in images:
11       thresh = 20
12       diff1 = np.absolute(img[:,:,0] - img[:,:,1])
13       diff2 = np.absolute(img[:,:,1] - img[:,:,2])
14       diff3 = np.absolute(img[:,:,0] - img[:,:,2])
15
16       condition = np.logical_and(
17                       np.logical_and(diff1 < thresh, diff2 < thresh),
18                       diff3 < thresh )
19
20       percent_gray = float(np.sum(condition)) / \
21                       np.size(img[:,:,0]) * 100
22       print( 'Percentage pixels "non-gray":  ' + str(100-percent_gray) )
```

Figure 13.18 Code to solve Try This! 13-9.

The first percentage is from the April 30, 2019 image and the second percentage is from the May 5, 2019 image. Thus, the tropical storm decreased the area of the city with electrical lighting by approximately $(63.89 - 39.65)/63.89 \approx 38$ percent.

13.4   More Discipline-Specific Practice

The Discipline-Specific Jupyter notebooks for this chapter cover the following topics:

- Reading, displaying, and writing images in Matplotlib.
- Boolean arrays.
- Conditional calculations on array elements, with an output array of the same shape.
- Conditional calculations on array elements, with only certain elements of the array retained.
- Conditional calculations on array elements, reduced to a scalar value.
- The NumPy reduce method.
- Looping through lists of objects.

13.5 Chapter Review

13.5.1 Self-Test Questions

Try to do these without looking at the book or any other resources or using the Python interpreter. Answers to these Self-Test Questions are found at the end of the chapter. If not otherwise stated, assume NumPy has been imported earlier as np.

Self-Test Question 13-1

What is a pixel? What function can we use to read in an image and what does that function return?

Self-Test Question 13-2

How are colors in an image represented by Matplotlib?

Self-Test Question 13-3

What are the common ways of specifying color channel intensities in an image represented in Matplotlib?

Self-Test Question 13-4

What Matplotlib functions can be used for reading, displaying, and writing images?

Self-Test Question 13-5

What is a boolean array? Create a boolean array that shows which elements of an array of the integers between and including one and five (inclusive on both ends) are less than three.

Self-Test Question 13-6

Given the following array data defined as:

```
data = np.sin(np.arange(8) / 4.0)
```

return a boolean array mask that contains all values greater than 0.5 or less than −0.5.

Self-Test Question 13-7

Consider code defining the following array:

```
measurements = np.array([[2.3,  6.3, -1.2,  3.4],
                         [1.1, -0.9,  8.2, -9.3]])
```

Show two ways of creating an array `modified` that is the same shape as `measurements`, where all the negative values are replaced by the value -999.0 and all the other values are the corresponding values of `measurements` multiplied by 100.

Self-Test Question 13-8

Given the `measurements` array from Self-Test Question 13-7, extract all elements that are positive and place them in an array `positive`. Describe the shape of `positive`.

Self-Test Question 13-9

Given the `measurements` array from Self-Test Question 13-7, take the average of all elements that are negative and save the value as the variable `avg_neg`. Do this using two different methods.

Self-Test Question 13-10

What does a call to the `time` function in the time module do?

Self-Test Question 13-11

What is a ufunc?

Self-Test Question 13-12

Describe what the `reduce` method of a NumPy array does. Given the following array:

```
1  expt1 = np.array([[7.4,   6.0, -8.7,   2.0],
2                    [3.0,   4.9, -1.5, 10.9]])
```

use the `reduce` method to create a boolean mask called `mask` that shows where the values are greater than 1.0, less than 4.0, and is an "integer" value (i.e., close to an integer). Hint: The NumPy function `floor` takes a number and returns the value without the decimal portion.[18]

Self-Test Question 13-13

Given three arrays, `data1`, `data2`, and `data3`, how can we print out the shapes of each array without more than one `print` statement?

13.5.2 Chapter Summary

In this chapter, we took a basic look at image processing. There are other ways of encoding color information besides the methods described in this chapter. And we did not address anything dealing with advanced graphics, such as found in gaming and two- and three-dimensional computer graphics. The color model we did describe, however, permits pixel-by-pixel manipulation of two-dimensional images.

---

[18]   See the `floor` manual page for details. An up-to-date link to the page is at www.cambridge.org/core/resources/pythonforscientists/refs/, ref. 41.

Although this chapter focused on image processing, the functions and methods described in this chapter to manipulate images can also be used for the general manipulation of arrays of data. As such, they are powerful tools that enable us to concisely perform array calculations that involve asking questions of the data.

Specific topics we covered in this chapter include:

- The RGB color representation model for two-dimensional images

  - Each pixel is made up of intensities in the red, green, and blue channels.
  - The intensities for each channel and pixel are given in a three-dimensional array with the shape (*height*, *width*, 3), where *height* is the number of rows of pixels in the image and *width* is the number of columns of pixels in the image.
  - Intensities can be given as a decimal number between and including 0 and 1 or an integer between and including 0 and 255. In both cases, the range expresses an intensity between 0 and 100 percent.
  - Many images are 8-bit images, where each channel for each pixel is represented by 8 bits of memory. For those images, the intensity of a channel is a value between and including 0 and 255.
  - The RGBA model includes a channel expressing a measure of transparency.

- Reading, displaying, and writing images

  - The Matplotlib image module's `imread` function reads in an image and returns the three-dimensional NumPy array describing the image.
  - The Matplotlib pyplot module's `imshow` function will take an array of channel intensities and create a plot of the image.
  - The Matplotlib pyplot module's `imsave` function will take an array of channel intensities and write out the corresponding image.

- Boolean arrays

  - Boolean arrays are arrays of boolean values.
  - When engaging in arithmetic operations using boolean arrays, `True` elements are treated as one and `False` elements are treated as zero.
  - The `logical_and`, `logical_or`, `logical_not` NumPy functions implement the logical operators `and`, `or`, and `not`, respectively, so that they work element-wise on arrays.
  - The `reduce` method of NumPy ufuncs enables us to apply a NumPy function one by one over a sequence of arrays.

- Conditional calculations on array elements

  - An output array of the same shape: Can be returned by

    * The expression:

      $$(<mask> * <true>) + (\texttt{logical\_not}(<mask>) * <false>)$$

where *<mask>* is a boolean mask, *<true>* is the value of the return array where *<mask>* is `True`, and *<false>* is the value of the return array where *mask* is `False`, or

   * the NumPy `where` function, called with three arguments.

  - An output array retaining only certain elements from the input array: Use the NumPy `where` function, called with one argument to obtain a tuple of the indices fulfilling a condition, and use array indexing to select only those elements corresponding to those indices.
  - A scalar return value: Use one of the above methods to obtain output arrays with the same shape as the input array or only certain elements from the input array and provide that array as the input to a NumPy function that uses those values and returns a scalar output.

• A timing of how fast a snippet of code runs

  - The time module's function `time` returns the number of seconds since the epoch.
  - To calculate how much a time a snippet of code takes to run, call `time` before the snippet and save its return value and call `time` after the snippet. The difference between the times before and after the snippet is a rough estimate of how long it takes to run the code snippet.

• Looping through lists of objects: Lists can contain any kind of object as an element. When we loop through lists of objects, the iterator variable is set to each object, in turn.

13.5.3 Self-Test Answers

Self-Test Answer 13-1

A pixel is a single "dot" in an image. The image submodule of Matplotlib has a function called `imread` that can read in an image file. The `imread` function returns a three-dimensional NumPy array where the size of the first dimension is the number of rows in the image (the image height), the size of the second dimension is the number of columns in the image (the image width), and the size of the third dimension is three.

Self-Test Answer 13-2

There are a few ways, but one way is to represent the color of each pixel as the combination of red, green, and blue intensities.

Self-Test Answer 13-3

Color channel intensities vary from 0 to 100 percent (inclusive at both ends). One way is to represent this range as a decimal number between zero and one (inclusive at both ends). This is the default representation for a PNG image read into Matplotlib. Commonly, color channel intensities for a pixel in 8-bit images are represented as an integer between 0 and 255 (inclusive at both ends).

Self-Test Answer 13-4

- Reading. The image submodule's `imread` function.
- Displaying. The pyplot submodule's `imshow` function, coupled with pyplot's `show` method.
- Writing. The pyplot submodule's `imsave` function.

Self-Test Answer 13-5

A boolean array is an array whose elements are boolean values, that is, equal to either `True` or `False`. We could create the boolean array requested (which is called `mask`), by inspecting (in our mind) the array of integers one through five:

```
mask = np.array([True, True, False, False, False])
```

Here is another way that first uses `arange` to construct the array of integers between one and five and the comparison operator and array syntax to create the boolean array:

```
mask = (np.arange(5) + 1) < 3
```

The result of both is:

```
In [1]: print(mask)

        [ True   True False False False]
```

Self-Test Answer 13-6

Here is a solution and output showing the contents of `data` and `mask`:[19]

```
In [1]: mask = np.logical_or(data > 0.9, data < 0.3)
        print(data)

        [0.         0.24740396 0.47942554 0.68163876 0.84147098 0.94898462
         0.99749499 0.98398595]

In [2]: print(mask)

        [ True   True False False False   True   True   True]
```

---

[19] A manual line break is added to the `print` statement output to improve the clarity of the output.

Self-Test Answer 13-7

Here is one way using arithmetic and boolean arrays:

```
is_negative = measurements < 0.0
modified = (is_negative * -999.0) \
           + (np.logical_not(is_negative) * measurements * 100.0)
```

Here is another way that is shorter using the `where` function:

```
modified = np.where(measurements < 0.0, -999.0, measurements * 100.0)
```

The values in `modified` (by both of the above methods) are:

```
In [1]:  print(modified)

         [[ 230.  630. -999.  340.]
          [ 110. -999.  820. -999.]]
```

Both `measurements` and `modified` are two-dimensional arrays. Because of array syntax and how NumPy functions behave, the shape information of the input array is automatically preserved during the calculations to obtain the return array.

Self-Test Answer 13-8

This code solves the first part of the problem:

```
pos_pts = np.where(measurements > 0.0)
positive = measurements[pos_pts]
```

with the contents and shape of `positive` being:

```
In [1]:  print(positive)

         [2.3 6.3 3.4 1.1 8.2]
```

```
In [2]:  print(np.shape(positive))

         (5,)
```

Whereas `measurements` is two-dimensional, `positive` is one-dimensional.

Self-Test Answer 13-9

Here is one way using arithmetic and boolean arrays:

```
is_negative = measurements < 0.0
avg_neg = np.sum(is_negative * measurements) / np.sum(is_negative)
```

Here is another way that is shorter using the `where` function:

```
neg_pts = np.where(measurements < 0.0)
neg_meas = measurements[neg_pts]
avg_neg = np.sum(neg_meas) / np.size(neg_meas)
```

The value of `avg_neg` is:

```
In [1]:  print(avg_neg)

         -3.8000000000000003
```

Self-Test Answer 13-10

It returns the number of seconds since the epoch. The epoch is a specific date that acts as a time "datum" for a computer.

Self-Test Answer 13-11

A ufunc or universal function is a NumPy function that acts element-wise on arrays.

Self-Test Answer 13-12

The `reduce` method applies a NumPy ufunc along an axis. When used on a sequence of arrays given in a tuple (or list), `reduce` will apply the ufunc repeatedly on each element of the tuple. Although the `reduce` method can do more than what we just described, in this chapter we focused on using `reduce` to do a repeated operation on a sequence of arrays.

To create the final mask, we first create three different masks for each subcondition and then use `reduce` as part of `logical_and`:

```
is_gt1 = expt1 > 1.0
is_lt4 = expt1 < 4.0
is_int = np.isclose(expt1, np.floor(expt1))
mask = np.logical_and.reduce((is_gt1, is_lt4, is_int))
```

The result of each mask is:

```
In [1]:  print(is_gt1)

         [[ True  True False  True]
          [ True  True False  True]]
```

```
In [2]:  print(is_lt4)

         [[False False  True  True]
          [ True False  True False]]
```

```
In [3]:  print(is_int)
```

```
[[False   True False   True]
 [ True False False False]]
```

```
In [4]:  print(mask)
```

```
[[False False False   True]
 [ True False False False]]
```

Self-Test Answer 13-13

This code solves the problem:

```
1  for iarray in [data1, data2, data3]:
2      print(iarray.shape)
```

We can also use the NumPy `shape` function on `iarray` to obtain the shape of each array as the loop goes through each element of the list in line 1.

14   Contour Plots and Animation

Line and scatter plots, which we described in Chapters 4 and 5, are the bread and butter of scientific visualization. There are, however, many other kinds of graphs used to display and analyze scientific and engineering information. In this chapter, we examine two such kinds of visualization: contour plots and animations of two-dimensional plots and images. Contour plots are a common way of presenting data whose values change based on the location in two-dimensional space. Perhaps the most common contour plot from everyday life is the topographic map – often used by hikers and backpackers, in addition to surveyors, geologists, hydrologists, and other Earth scientists – which shows ground elevation as contour lines on a map. In Section 7.1, we saw a few examples of contour plots showing both contour lines and colored shading to represent the temperature at different locations (Figures 7.1, 7.2, and 7.4). Contour plots, however, are found in all branches of the natural sciences and engineering. Animations are a logical extension of line, scatter, and contour plotting, as well as image processing. By displaying a plot or image in the same frame, but changing what is being displayed over time, we can create the impression of motion.

We can create both contour plots and animations using the Matplotlib package. In our prior use of Matplotlib for line and scatter plots, we used the plotting functions provided as part of the pyplot submodule. To make contour plots, we make use of pyplot functions as well as Matplotlib objects (see Section 9.2.1 for an introduction to objects). These ideas are demonstrated in Section 14.1 and explained in Section 14.2.

The code and ideas in this chapter do not lend themselves as well to line-by-line footnoting and referencing as in prior chapters. Many of the ideas and portions of code are from or built from those given in the cartopy manual (Met Office, 2010–2015), the Matplotlib documentation,[1] and the documentation found via `help` in the Python interpreter. Additional references are given at www.cambridge.org/core/resources/pythonforscientists/refs14/.

The animation code in this chapter runs as written when Python is executed from a terminal window. For Jupyter and Spyder, some slight adjustments need to be made. These adjustments may also be different for different operating systems. Details are provided at www.cambridge.org/core/resources/pythonforscientists/anim-mpl/.

Finally, the larger datasets used in this chapter are available for download. See the "Supporting Resources and Updates to This Textbook" section in the Preface for details.

---

[1] We provide an up-to-date link to the documentation at www.cambridge.org/core/resources/pythonforscientists/refs/, ref. 25. Also see Hunter (2007).

14.1   Example of Making Contour Plots and Animations

As we saw earlier, Figure 7.1 shows a shaded (a.k.a., filled) contour plot of the surface/near-surface air temperature over the continental United States at November 11, 2017, 18:00Z, and Figure 7.2 shows a zoom-in of that plot over the US Midwest. Both these plots have the same shading and contour values, suggesting that we can save ourselves some work by making the code to generate the temperature contour maps flexible so that it can be used for a large region as well as a small region. If we put that code into a function we can call repeatedly, we can use it for regions of any size and avoid duplicating the code for creating the contour plots, overlaying the map, etc. Figure 14.1 shows code for such a function, and Figure 14.2 shows calls to that function that can generate Figures 7.1 and 7.2.[2]

Figure 14.1 begins with import lines for the needed modules, objects, and routines from the NumPy, Matplotlib, and cartopy packages (lines 1–7). After the import lines, we define the function `plot_map` (starting with line 10), which creates a contour plot for a specific region. This region is defined in two ways. First, we specify the longitude and latitude values of the data being plotted (the parameters `lon2d` and `lat2d`, respectively). These are in two-dimensional form and the arrays are the same shape as the array of data values being contoured against. The data being plotted at the `lon2d` and `lat2d` locations are given in the parameter `air_temp`. Because `air_temp` is a regular grid in longitude and latitude, the values in each column of `lon2d` are constant, and the values in each row of `lat2d` are constant. Second, we specify the longitude and latitude values the *x*- and *y*-axes are labeled at using the parameters `plot_xticks` and `plot_yticks`. These are one-dimensional arrays. The values of the contour levels are given in the parameter `levels`. The `plot_map` function also takes two pieces of optional input, `output_file` and `objs`, which define a filename for an image file version of the plot and Matplotlib objects on which the map can be plotted. The latter are useful when we use this function for creating animations. These parameters are listed in lines 10–13 and described in the docstring provided in lines 14–46.

```
1   #- Module imports:
2
3   import numpy as np
4   import matplotlib.pyplot as plt
5   import cartopy.mpl.ticker as cmt
6   import cartopy.crs as ccrs
7   import cartopy.feature as cf
```

Figure 14.1 Code for a function `plot_map` that can be used to generate the contour plots of Figures 7.1 and 7.2, as well as the contour plots for each panel of the animation described in Figures 14.3 and 14.3.

---

[2] The code actually used to create Figures 7.1 and 7.2 is slightly different than the code in Figure 14.1. The latter code is simplified, to make our discussion in this section more straightforward. However, the images generated, excepting the title on top, look identical.

```
8    #- Define the plot_map function:
9
10   def plot_map(lon2d, lat2d, air_temp, levels,
11                plot_xticks, plot_yticks,
12                output_file = None,
13                fig = None):
14       """Make a contour plot on the continental U.S. or part thereof.
15
16       Positional Input Parameters:
17           lon2d : 2-D array
18               Longitude values of each air_temp element.
19
20           lat2d : 2-D array
21               Latitude values of each air_temp element.
22
23           air_temp : 2-D array
24               Temperatures at a single moment in time.
25
26           levels : 1-D array
27               List of contour levels.
28
29           plot_xticks : 1-D array
30               Array of x-ticks to plot, ascending, in deg E.
31
32           plot_yticks : 1-D array
33               Array of y-ticks to plot, ascending, in deg N.
34
35       Keyword Input Parameters:
36           output_file : string
37               Output filename for an image file of the plot.
38
39           fig : Figure
40               If None, the plot is created on a new Figure object which
41               is returned by the function.  If not None, assumed to be
42               the matplotlib Figure object to overlay the map onto.
43
44       Returns:
45           The matplotlib Figure object.
46       """
```

Figure 14.1 (part 2)

The function plot_map begins by calculating the longitude and latitude range (minimum and maximum values) and the central longitude of the map projection for the plot (lines 48–50). In lines 53–61, we check to see whether an existing Matplotlib Figure object has been passed in via the fig keyword input parameter. If not, we create a new figure (line 59) and plot an Axes object that includes a map following a cylindrical projection (lines 60–61). If an existing Figure object has been passed in via the fig keyword input parameter, we

```
47    #+ Set figure initialization related parameters:
48    lon_range_plot = [plot_xticks[0], plot_xticks[-1]]
49    lat_range_plot = [plot_yticks[0], plot_yticks[-1]]
50    central_lon = (lon_range_plot[1] - lon_range_plot[0]) / 2.0
51
52    #+ Create or prep figure and set up map plot:
53    if fig is not None:
54        figure = plt.figure(fig.number)
55        ax = plt.gca(projection=ccrs.PlateCarree( \
56                     central_longitude=central_lon))
57        plt.cla()
58    else:
59        figure, temp = plt.subplots(1, 1)
60        ax = plt.axes(projection=ccrs.PlateCarree( \
61                      central_longitude=central_lon))
62
63    #+ Customize map plot:
64    ax.set_extent([lon_range_plot[0], lon_range_plot[1],
65                   lat_range_plot[0], lat_range_plot[1]],
66                   crs=ccrs.PlateCarree())
67    ax.add_feature(cf.BORDERS, linestyle=':')
68    ax.add_feature(cf.COASTLINE)
69    ax.add_feature(cf.LAKES, alpha=0.5)
70    ax.add_feature(cf.RIVERS)
71
72    states_50m = \
73        cf.NaturalEarthFeature('cultural',
74                               'admin_1_states_provinces_lines',
75                               '50m',
76                               facecolor='none')
77    ax.add_feature(states_50m, edgecolor='gray')
78    ax.set_xticks(plot_xticks, crs=ccrs.PlateCarree())
79    ax.set_yticks(plot_yticks, crs=ccrs.PlateCarree())
80    lon_formatter = cmt.LongitudeFormatter(zero_direction_label=True)
81    lat_formatter = cmt.LatitudeFormatter()
82    ax.xaxis.set_major_formatter(lon_formatter)
83    ax.yaxis.set_major_formatter(lat_formatter)
```

Figure 14.1   (part 3)

"activate" that object (line 54) and a new map projection axes (lines 55–56) for later plotting, and clear the current axes on the figure (line 57) so previous plots are not overlain.

In lines 64–83, we customize the characteristics of the contour plot and the map the plot is overlain on. This includes setting the extent of the plot, adding country and state borders, the coastline, lakes, and rivers. The tick labels for the *x*- and *y*-axes are set in lines 78–79 (using the values passed into the function via plot_xticks and plot_yticks), and the values are formatted to look like latitude and longitude values in lines 80–83 (e.g., "40° N").

```
84     #+ Make a contour plot of the temperature:
85     mymapf = ax.contourf(lon2d, lat2d, air_temp, levels=levels,
86                          transform=ccrs.PlateCarree(),
87                          cmap=plt.get_cmap('plasma'))
88     mymap = ax.contour(lon2d, lat2d, air_temp, levels=levels,
89                        transform=ccrs.PlateCarree(),
90                        linewidths=1.0,
91                        colors='k')
92
93     #+ Add contour line labels and colorbar:
94     plt.clabel(mymap, fmt='%d', fontsize=10)
95     if fig is None:
96         plt.colorbar(mymapf, orientation='horizontal')
97
98     #+ Write out plot and return changed matplotlib Figure object:
99     if output_file is not None:
100        plt.savefig(output_file, dpi=300)
101
102    return figure
```

Figure 14.1 (part 4)

```
1    #--> Create map of temperature over the continental U.S.:
2    plot_map(lon2d_cont, lat2d_cont, air_temp_cont, levels,
3            np.array([230, 240, 250, 260, 270, 280, 290, 300]),
4            np.array([25, 30, 35, 40, 45, 50, 55]),
5            output_file = "surf_temp_2-d_continental_us.png")
6
7    #--> Create map of temperature over the U.S. Midwest:
8    plot_map(lon2d_mid, lat2d_mid, air_temp_mid, levels,
9            np.array([260, 270, 280]),
10           np.array([30, 35, 40, 45]),
11           output_file = "surf_temp_2-d_midwest_us.png")
```

Figure 14.2 Calls to generate the contour plots of Figures 7.1 and 7.2. Assume the `lon2d_cont`, `lat2d_cont`, `air_temp_cont` `lon2d_mid`, `lat2d_mid`, `air_temp_mid`, and `levels` arrays are already defined. The `lon2d_cont`, `lat2d_cont`, and `air_temp_cont` arrays are large, and thus their contents are not listed. The `lon2d_mid` and `lat2d_mid` are based on the values in Table 7.2 and 7.3, cast into two dimensions. The `air_temp_mid` data are the same as in Table 7.1, although using all the numerical digits available in the dataset. The `levels` array starts at −45, ends at 45, steps every 5, and is 19 elements long.

In lines 85–91, we first create the shaded or filled contour map for the temperatures in `air_temp` (using `contourf`) and then overlay the contour lines corresponding to that map (using `contour`). The contour levels used are given by the `levels` parameter. The filled contour map uses the "plasma" **color map** (a table that maps values to fill colors) whereas

```
1    plt.ion()
2    fig = plot_map(lon2d_mid, lat2d_mid,
3                   air_temp_mid_1day[0,:,:], levels,
4                   np.array([260, 270, 280]),
5                   np.array([30, 35, 40, 45]))
6    plt.pause(1.0)
7    for i in [1, 2, 3]:
8        fig = plot_map(lon2d_mid, lat2d_mid,
9                       air_temp_mid_1day[i,:,:], levels,
10                      np.array([260, 270, 280]),
11                      np.array([30, 35, 40, 45]),
12                      fig = fig)
13       plt.draw()
14       plt.pause(1.0)
```

Figure 14.3 Code to generate the animation whose panels are shown in Figure 14.4. The `lon2d_mid`, `lat2d_mid` and `levels` arrays are the same as in Figure 14.2. The `air_temp_mid_1day` array is the same as the shape (4, 5, 7) `air_temp` array in Figure 12.1. The values of `air_temp_mid_1day` are similar to those given in Table 12.1, although all digits available in the original dataset are used.

the contour lines are colored black. Contour line labels are added in line 94. A color bar legend is drawn in line 96 if an existing `Figure` object has not been passed in via the `fig` keyword input parameter. The function finishes by writing out the plot to an image file, if `output_file` is defined, and the completed `Figure` object `figure` is returned, to be reused as desired.

Figure 14.2 shows calls to the `plot_map` function defined in Figure 14.1 that can generate Figures 7.1 and 7.2. Lines 2–5 of Figure 14.2 generate the map of surface/near-surface air temperature over the continental US while lines 8–11 generate the same map over the US Midwest.

Figure 14.3 shows code to generate four frames of animation of surface/near-surface air temperature over the US Midwest, at four different times (November 11, 2017, at 0:00Z, 6:00Z, 12:00Z, and 18:00Z). Those panels are shown in Figure 14.4. When run, the animation displays on the screen. Using other tools, we can generate an *.mp4* video from these panels, which is available for view at www.cambridge.org/core/resources/pythonforscientists/anim-fig14-4/. Note that the images cannot be easily written into a *.mov* or *.mp4* video file using the techniques in this chapter. Matplotlib provides other frameworks, such as the animation submodule, to create such animations. We do not cover this submodule in the present work (see the online documentation; we provide an up-to-date link to that documentation at www.cambridge.org/core/resources/pythonforscientists/refs/, ref. 22).

In line 1 of Figure 14.3, we turn on Matplotlib's interactive mode, so each frame will update live. In lines 2–5, we create the contour map using the values at the first time (the first sheet of the `air_temp_mid_1day` array). In line 6, we pause for 0.1 s, before plotting

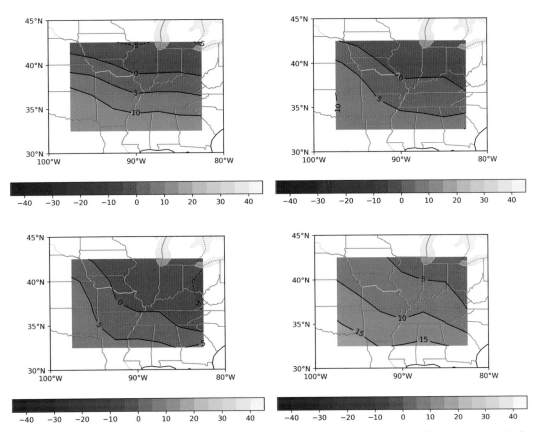

Figure 14.4 Panels that make up the animation created by the code in Figure 14.3. The panels show (left to right, then top to bottom) plots of surface/near-surface air temperature over the US Midwest on November 11, 2017, at 0:00Z, 6:00Z, 12:00Z, and 18:00Z (in °C).

the contour map for the next time. Otherwise, our eyes would not be able to easily perceive the shift between the first and second contour maps.

In lines 8–14, we plot the contour maps for the remaining three times. We loop through the indices for the remaining sheets in `air_temp_mid_1day` (line 7) and extract the values for each time in line 9, which is used to generate an updated contour map on the existing `Figure` object. In line 13, we draw the updated figure. In line 14, we pause for 0.1 s before the next iteration of the loop.

14.2 Python Programming Essentials

In this section, we apply our previous examination of the nature of objects and object-oriented programming (OOP) in order to create contour plots and basic animations of graphs. We begin this section with a description of the basics of Matplotlib's object Application

Programming Interface (API). Whether adding to contour plots, animating images, or customizing plots beyond what we saw in Chapter 5, the object API enables us to do much more than pyplot functions alone can do. Next, we describe how to use Matplotlib to create line contour plots and shaded contour plots. In applications related to Earth science, we often want to superimpose maps of political or geographical boundaries onto a contour plot, so we describe how to use the cartopy package to overlay maps onto our contour maps. We then describe basic animation using Matplotlib. Our last topic – a description of a new data structure, dictionaries, and how they can enable us to create more flexible functions – may seem out of place, but dictionaries help us write plotting functions that accommodate tweaking of a plot.

In the rest of the code examples in this section, assume NumPy has already been imported as `np` and the Matplotlib pyplot submodule has already been imported as `plt`.

14.2.1 An Introduction to Matplotlib's Object API

When we look at a graph, what kind of entities or components do we see? At the most "macro" level, a graph exists on a piece of paper or in the window of a computer screen. On that page or window, the outermost component of a graph is the frame of the graph, which usually includes horizontal (x) and vertical (y) axes. Inside the frame defined by the axes, a graph has various lines, symbols, patches of color, and text. Outside the axes frame, there are often other components that, while being outside the frame, nonetheless are connected in some way to what is plotted inside the frame: tick marks and labels, axes titles, a color bar, etc.

The key idea behind the Matplotlib object API is that each of these components of a graph is represented as an object in the program. Each of those components or objects is an instance or specific realization of a class or kind of object. Thus, in the Figure 7.1 filled contour plot of the surface/near-surface air temperature over the continental United States, there will be objects representing the page/window the graph is on, the frame and axes of the graph, each of the lines on the graph, each of the patches of color on the graph, etc. The object representing the page/window the graph is on is a `Figure` class object. The plot or graph itself (i.e., the frame, axes, and contents) of the graph is an `Axes` class object. The contour lines are each instances of the `Line2D` class, and so on. Each of these objects, as we saw in Section 9.2.1, has attributes and methods, which mirror the states and behaviors that their real-world analogues have. Thus, the Matplotlib `Figure` object representing a page/window has a method `add_axes`, which places a plot/graph onto the page/window. A `Figure` object can make one or more `add_axes` calls, just as in the real world we can place one or more sets of graph frames and axes onto a single piece of paper.

For details on all the classes that are part of Matplotlib, please see the documentation.[3] In the rest of the present section, we briefly describe a few key classes, some of their attributes and/or methods, and some of the kinds of tasks (including those shown in the Section 14.1 example) the Matplotlib object interface is particularly useful for (and when to use pyplot

---

[3] An up-to-date link to the documentation is provided at www.cambridge.org/core/resources/pythonforscientists/refs/, ref. 25.

commands instead). Attributes and methods that specifically relate to the contour plotting and the map overlay tasks in Figure 14.1 are described in more detail in Sections 14.2.2 and 14.2.3.

The Figure Class Objects of this class represent the page or window one or more plots are placed upon. Pyplot commands that create objects of this class include:

- figure This returns a new Figure object, with the name (which is generally an integer or string) passed in as a parameter in the call to figure. If the name that is passed into the call is already defined, the Figure object returned by the call to the figure command is the preexisting Figure object. This is similar to the functionality we described in Section 5.2.3. Figure 14.1, line 54 gives an example of the use of this command and of saving the object it returns to a variable, for later use.
- subplots This creates both a Figure object and an Axes object on that page/window. The command returns two values, the Figure object and Axes object(s) created by the command. Figure 14.1, line 59 gives an example of the use of this command, for the case of placing a single Axes object onto the figure. When called with two input parameters, specifying the number of rows and columns in a grid of multiple panels, the subplots command creates the Axes objects in that grid in a new figure.

  In line 59 of Figure 14.1, we save the Axes object to a temporary variable temp, because later on (lines 60–61), we reinitialize the Axes object on the figure and assign it to ax. We discuss lines 60–61 in Section 14.2.3.

  In Section 8.2.5, we described how to use the subplot command to generate multiple plots or panels in a single figure. This command is different than subplots (note the "s" as the end of the latter): subplot is used to create a single Axes object in a grid of multiple panels, whereas subplots can be used to create many such Axes objects in a grid of multiple panels. Although we can save the Axes object(s) generated by both subplot and subplots, in Chapter 8, we called the subplot command without saving its return values, unlike in Figure 14.1.

Once a Figure object is created, we can utilize the attributes and methods attached to the object. Some of these attributes include:

- axes List of Axes objects that are on the page/window.
- number An integer that represents the name of the page/window. Figure 14.1, line 54 shows an example of referencing this attribute.

One of the methods a Figure object has is the add_axes method. This method creates an Axes object and places the corresponding graph frame/axes at a specified location on the figure. The location is specified by a four-element tuple: the first element of the tuple gives the location of the left side of the plot frame, the second element the location of the bottom of the plot frame, the third element the width of the plot frame, and the last element the height of the plot frame. Each element is given in **normalized coordinates**, where the positions and distances are given in values that are fractions (i.e., between zero and one) of the width and height of the page/window (or domain of interest). Thus, for a `Figure` object called `figure`, the call `figure.add_axes((0.1, 0.2, 0.5, 0.3))` will make:

- The left edge of the plot frame at a horizontal location in the page/window that is 10 percent the width of the page/window, measured from the left side of the page/window.
- The bottom edge of the plot frame at a vertical location in the page/window that is 20 percent the height of the page/window, measured from the bottom of the page/window.
- The width of the plot frame is 50 percent the width of the page/window.
- The height of the plot frame is 30 percent the height of the page/window.

The `Axes` Class   Objects of this class represent the plot or graph itself: the frame, axes, and contents of the graph. When we call the pyplot `plot` command by itself, an `Axes` object is automatically created on a `Figure` object, although this all happens under the hood. To return an `Axes` object that we can assign to a variable, and thus use later on, we can call:

- The `add_axes` method of a `Figure` object (described earlier).
- The pyplot `gca` command. Gets the current axes on the current figure whose specifications match the input parameters given in the `gca` call. If the axes do not yet exist, they are created. Figure 14.1, lines 55–56 shows an example.
- The pyplot `axes` command. Creates axes on the current figure and makes these the current axes. Figure 14.1, lines 60–61 shows an example.

The pyplot command `cla` will clear the current axes in the current figure (e.g., Figure 14.1, line 57).

Because `Axes` objects represent the plot or graph and its contents, there are quite a few attributes and methods attached to these objects. Many of the attributes are best accessed and manipulated through methods of the `Axes` object. Here are a few attributes that we see used in Figure 14.1:

- `xaxis` Set to an `XAxis` object that represents the *x*-axis. That object itself has attributes and methods. Figure 14.1, line 82 shows how we use the `set_major_formatter` method attached to the object stored in the `xaxis` attribute to define the formatter (defined in line 80) that will be used for the longitude labels.
- `yaxis`: Set to a `YAxis` object that represents the *y*-axis, similar to the `xaxis` attribute. Figure 14.1, line 83 shows an example referencing `yaxis` that mirrors what is done in line 82 referencing `xaxis`.

`Axes` object methods include:

- `axis` Like the pyplot `axis` function (see Section 5.2.2), this method controls properties of the *x*- and *y*-axes in the plot. To set the *x*-axis minimum and maximum values, and the *y*-axis minimum and maximum values, pass in a list with those four values, in that order. If we pass in the string `'off'`, both the *x*- and *y*-axis tick marks and labeling are turned off.
- `contour` Creates a contour line plot in the given `Axes` object. We discuss this method more in Section 14.2.2. Figure 14.1, lines 88–91 shows an example.

- `contourf` Creates a filled contour plot in the given Axes object. We discuss this method more in Section 14.2.2. Figure 14.1, lines 85–87 shows an example.
- `imshow` Displays an image in the given Axes object and creates an AxesImage object. This works similarly to the pyplot function imshow, described in Section 13.2.1. We discuss this method and its return value in more detail below in the section on the AxesImage class.
- `plot` Creates a line plot. This method works the same as the pyplot plot command, except the line plot is drawn into the Axes object the method is attached to. A call to this method will return a list of the Line2D objects that are part of the line plot. Thus, assuming the one-dimensional NumPy arrays x1, y1, x2, and y2 are already defined, the following call, given an Axes object ax:

```
lines = ax.plot(x1, y1, 'o--',
                x2, y2, '*-')
```

will produce a two-element list lines, where the first element is the Line2D object from the curve plotted using x1 and y1, and the second element is the Line2D object from the curve plotted using x2 and y2.

- `scatter` Creates a scatter plot. This method is also similar to the pyplot scatter command, except the line plot is drawn into the Axes object the method is attached to.
- `set_xticks` Sets the *x*-axis tick marks that will be labeled. Accepts a list of those values as a positional input parameter. Figure 14.1, line 78 shows an example.
- `set_yticks` Sets the *y*-axis tick marks that will be labeled. Accepts a list of those values as a positional input parameter. Figure 14.1, line 79 shows an example.

The Line2D Class   Objects of this class describe a line on the plot. Similar to the Axes class, many of the attributes are best accessed and manipulated through methods of the Line2D object. A few of these methods include:

- `set_linestyle` Set the linestyle of the line.
- `set_marker` Set the marker of the line.
- `set_xdata` Set the *x*-axis data values of the line. This overwrites any currently existing values. We describe how this can be used for animation in Section 14.2.4.
- `set_ydata` Set the *y*-axis data values of the line. This overwrites any currently existing values. We describe how this can be used for animation in Section 14.2.4.

The AxesImage Class   Our chapter example from Section 14.1 creates a line and filled contour plot and does not directly manipulate an image. Matplotlib's object API, however, also has objects to enable us to do image manipulation, which we briefly describe here. An AxesImage object represents an image in Matplotlib. As described earlier, the Axes object's imshow method will create an AxesImage image object.

In Section 13.2.1, we used the pyplot command imshow with a three-dimensional array that encoded Red–Green–Blue (RGB) values. This gives us fine-grained control over the color that will be displayed when the image is rendered with show. For both the pyplot command `imshow` and the `Axes` object method `imshow`, however, we can also pass in a two-dimensional array of data, and assign colors based upon mapping the range of data values to a color map. Thus, for a given two-dimensional NumPy array `data`, and an `Axes` object `ax`, this call:

```
img = ax.imshow(data, cmap=plt.cm.rainbow, aspect='auto')
```

creates an `AxesImage` object in the `ax` plot, following the "rainbow" color map, and saves it as the object `img`. The Matplotlib module `cm` defines color maps and utilities to operate on color maps. The `plt.cm.rainbow` object is an instance of the `LinearSegmentedColormap` class.[4] The minimum value in `data` will match the minimum color in the color map, and the maximum value in `data` will match the maximum color in the color map. By setting the `aspect` keyword input parameter to `'auto'`, the image is scaled to fill the extent of the `Axes` object `ax` (i.e., the image's aspect ratio – the ratio of the *y*-axis unit to the *x*-axis unit – matches the plot's). Each element in `data`, which represents a "pixel" of color, is rendered as a "chunk" of color in the plot, scaled so that all the elements in `data` are drawn in `ax`.[5]

Here are a few `AxesImage` methods to manipulate the image:

- `set_cmap` Set the color map used for coloring the data. A color map object (e.g., `plt.cm.rainbow`) or the name of a prewritten color map is passed in as input.
- `set_data` Set the data values of the image. This overwrites any currently existing values. We describe how this can be used for animation in Section 14.2.4.

Although there are many more classes in Matplotlib that describe practically every kind of entity we might want to put on a graph, the above classes cover many of the most common use cases for scientific and engineering work. See the Matplotlib documentation for details.[6]

When to Use pyplot versus the Matplotlib Object Interface   Throughout our discussion of Matplotlib's object API, we have seen repeatedly how many methods of Matplotlib objects (especially `Axes` objects) are named the same as similar (if not identical) commands in the pyplot module. This is not accidental. In a way, we can think of the pyplot commands as operating on "objects," the current figure and plot axes. The Matplotlib object API just makes that more explicit. Thus, the choice between pyplot and the Matplotlib object interface ultimately comes down to the kind of task we want to do. We can use both interfaces on the same plot and go back and forth between them.

In general, the pyplot interface is most useful when we want to create a graph with as few lines of code as possible; pyplot commands are also easier to use when we have few figures or axes to manage or manipulate. The Matplotlib object interface is most useful when we are

---

[4] See the Matplotlib online documentation for more on color maps. An up-to-date link to the color map documentation is at www.cambridge.org/core/resources/pythonforscientists/refs/, ref. 23.

[5] The `imshow` method permits a variety of interpolation schemes to be used. An up-to-date link to the page describing interpolations for `imshow` is at www.cambridge.org/core/resources/pythonforscientists/refs/, ref. 28.

[6] An up-to-date link to the documentation is at www.cambridge.org/core/resources/pythonforscientists/refs/, ref. 25.

interested in low-level customization of plots. Some kinds of tasks that the object interface is useful for include:

- Arbitrary placement. For instance, the `add_axes` method of `Figure` objects enables us to place a plot anywhere on the page/window of a figure.
- Multiple placement. For instance, the `add_axes` method of `Figure` objects also enables us to place as many plots as we wish on a single figure.
- Change formatting. A plot consists of many entities (lines, text, data, etc.). Through the use of the appropriate methods, we can change many if not most of those entities and rerender the plot to reflect those changes.
- Provide more flexible customization than would otherwise easily be possible. We saw earlier in our discussion of the `Axes` object's `imshow` method that we can pass in the cm module's predefined color map objects (such as the `rainbow` instance of the `LinearSegmentedColormap` class) so the method knows what color map to use. We could have specified the rainbow color map by passing in the string `"rainbow"` to `imshow`'s `cmap` keyword, because it is a predefined color map. We would need to pass in a color map object for a custom color map we have defined.

    As another example, in Section 5.2.2, we learned how to use the pyplot `text` command to place text in a plot. In that example, the location of the text was specified in data coordinates. We can, however, specify the location of the text in **axis coordinates** if we set the `transform` keyword input parameter of the `text` command to the `Axes` object's `transAxes` attribute object. In axis coordinates, the location along each axis is a value from zero to one, where zero means the beginning of the axis and one is the end of the axis. Thus:

```
plt.text(0.1, 0.2, 'Provisional', transform=ax.transAxes)
```

would plot the word "Provisional" at a location 10 percent along the x-axis and 20 percent along the y-axis. The object `ax` is the `Axes` object on which the text is being placed.

14.2.2 Line and Shaded Contour Plots

As we saw in Figure 14.1, lines 85–91, we can create line contour plots by using an `Axes` object's `contour` method and shaded (or filled) contour plots by using the `contourf` method. Both are used in ways similar to the `plot` method: We pass in the data via the parameter list when we call the method. In the case of `contour` and `contourf`, the basic syntax of the method calls, if we have an `Axes` object called `ax`, is:

```
ax.contour(X, Y, Z, levels=levels)
ax.contourf(X, Y, Z, levels=levels)
```

where `X` is an array of the x-axis values for each of the data points in `Z`, `Y` is an array of the y-axis values for each of the data points in `Z`, and `Z` is a two-dimensional array of values the contour map follows. The `levels` keyword input parameter is set to a one-dimensional array which specifies the values of the contour levels (the array has to be in increasing order). Just like `plot`, `contour` and `contourf` also exist in pyplot as commands.

If X and Y are two-dimensional arrays of the same shape as Z, the x- and y-axis values for elements of Z are given in the corresponding locations of X and Y. Thus, for the top left panel of Figure 14.4, the x-axis (longitude, in degrees East) values passed in for X are:

```
[[262.5 265.   267.5 270.   272.5 275.   277.5]
 [262.5 265.   267.5 270.   272.5 275.   277.5]
 [262.5 265.   267.5 270.   272.5 275.   277.5]
 [262.5 265.   267.5 270.   272.5 275.   277.5]
 [262.5 265.   267.5 270.   272.5 275.   277.5]]
```

The x-values are the same for every row in a given column. Each row is the same as the values in Table 7.2. The y-axis (latitude, in degrees North) values passed in for Y are:

```
[[42.5 42.5 42.5 42.5 42.5 42.5 42.5]
 [40.   40.   40.   40.   40.   40.   40. ]
 [37.5 37.5 37.5 37.5 37.5 37.5 37.5]
 [35.   35.   35.   35.   35.   35.   35. ]
 [32.5 32.5 32.5 32.5 32.5 32.5 32.5]]
```

In this array, every row has a constant latitude value. Each column is the same as the values in Table 7.3. The NumPy function `meshgrid` can construct arrays such as X and Y out of one-dimensional arrays of the x- and y-values of the elements along each dimension. The answer to Try This! 14-2 describes how to use `meshgrid`.

Finally, the temperature values (in °C) passed in for Z are:

```
[[-2.4 -4.   -4.9 -5.5 -4.8 -4.6 -5.2]
 [ 2.5  1.9  0.   -2.3 -2.3 -2.1 -3.1]
 [ 9.9  8.4  5.3  3.1  3.6  3.8  2.7]
 [13.7 12.7 10.   9.1  9.4  8.5  8.1]
 [13.4 14.4 13.   13.4 14.4 13.6 13.9]]
```

In the case of a regular grid – which is the case we have here – we can pass in a one-dimensional array of the x-axis values instead for X, and a one-dimensional array of the y-axis values instead for Y, and the method will treat those values as if we passed in the two-dimensional version of the values, similar to the ones shown above.

As with other plotting methods such as `plot`, we can customize the contour plots created by `contour` and `contourf` methods using keyword input parameters. Examples include:

- cmap Set the color map. In the example of Figure 14.1, line 87, we use the pyplot `get_cmap` function to return the predefined "plasma" color map, and set that color map object to the keyword input parameter. The `cmap` keyword can also be set to the string name of the predefined color map.

- `colors` Set the color of the contour lines. When set to `'k'` (black), negative contours are, by default, rendered as dashed lines.[7]
- `linewidths` Set the thickness of the contour lines (in points). Figure 14.1, line 90 shows an example.

We will discuss the `transform` keyword input parameter (lines 86 and 89 in Figure 14.1) in Section 14.2.3.

In addition to keyword input parameters, Matplotlib also provides functions that operate on the object `contour` and `contourf` return (technically a `QuadContourSet` class object), to further customize the contour plot. This is why in lines 85 and 88 in Figure 14.1, we save the results of the calls to `contourf` and `contour` as the objects `mymapf` and `mymap`. Two such pyplot functions that act on these `QuadContourSet` objects include:

- `clabel` Add contour line labels onto the line contour map object passed into the function's call. Figure 14.1, line 94 shows an example. In that call, `fmt` sets the labels to integer values (the format code d means integer) and `fontsize` sets the font size of the label to 10 points. See the Try This! 14-2 solution discussion for more on the `fmt` format code.
- `colorbar` Add a **color bar** to the filled contour map object passed into the function's call. This bar shows what colors map to what values in the contour plot. Figure 14.1, line 96 shows an example. In that call, `orientation` sets how the color bar will be oriented.

14.2.3 Using cartopy to Overlay Maps

Because of the curvature of the Earth's surface, the translation of a regular latitude–longitude grid into coordinates on a map depends upon the kind of map projection being used and other characteristics of the map. In addition, maps often include features such as political boundaries, coastlines, lakes, and rivers. In concert with Matplotlib's plotting routines, the cartopy package accomplishes the tasks associated with overlaying maps onto plots.

Here is a summary of the steps to creating a map and then superimposing a contour plot on the map:

1. Create an instance of the Matplotlib `Axes` class, with the `projection` keyword set to an instance of a cartopy projection class that describes *the map to be displayed*.
2. Draw continents, borders, etc., and set formatting using methods of this Matplotlib `Axes` object.
3. Create the contour plot using the pyplot or `Axes` object method versions of `contour` or `contourf`, with the `transform` keyword set to the cartopy map projection object that describes the system *the data is on*.

The two-dimensional latitude and longitude coordinates of the dataset values are automatically transformed into map projection values when using the Matplotlib contouring commands.

---

[7] See the Matplotlib manual entry for `contour` and `contourf` for details.

Below, we expand on each of these three steps. Assume that the following import statements have been made, in addition to the standard NumPy and pyplot imports:

```
import cartopy.mpl.ticker as cmt
import cartopy.crs as ccrs
import cartopy.feature as cf
```

The `ticker` submodule has classes to enable longitude and latitude formatting. The `crs` submodule contains the coordinate projection classes. The `feature` submodule contains objects for drawing geographical and geopolitical features.

Step 1: Create an Instance of the Matplotlib `Axes` Class   As we saw in Section 14.2.1, there are various ways of creating a Matplotlib `Axes` object. In Figure 14.1, lines 55–56 and 60–61, we see examples of two of those ways, using the pyplot `gca` and `axes` commands, respectively.[8] These lines, however, also illustrate the first step of overlaying a map using cartopy: When we create the `Axes` object for the graph, we need to pass in a cartopy projection object describing the map that will be displayed and set it to the `projection` keyword input parameter. In both these Figure 14.1 examples, the projection object being passed in is a `PlateCarree` class object, which encodes a cylindrical map projection.[9] In both these examples, the `PlateCarree` object is customized so that its central longitude is set to the value of `central_lon`.

As an aside, the result of the `gca` or `axes` call is more precisely a `GeoAxes-Subplot` object. A `GeoAxesSubplot` object is an object that has all the capabilities of an `Axes` object with extra capabilities added on top of it. To properly understand what that means, we need to understand **inheritance**, which is addressed in Chapter 17. For now, suffice it to say that we can talk of the result of a `gca` or `axes` call, with a cartopy projection object passed in, as an `Axes` object, because although the `GeoAxesSubplot` object has more attributes and methods, it functions similarly to an `Axes` object. In this chapter, we refer to this object both ways.

Step 2: Draw Continents, Borders, etc., and Set Formatting   Once we have our `GeoAxesSub plot` (which is a kind of `Axes` object), we can utilize methods attached to it to customize the plot by adding continental boundaries, borders, and other kinds of formatting. Methods we can use include:

- `add_feature` Adds features to maps such as borders, coastlines, lakes, rivers, etc. Figure 14.1, lines 67–70 and 77 show examples. In some of those examples, the feature is stored as an attribute in the feature submodule (e.g., `BORDERS` in line 67). In another case, we have to create an object that describes that feature, such as in lines 72–76 where we create an

---

[8] There are also `Figure` method versions of the pyplot `gca` and `axes` commands. As of this writing, those versions do not work in the Figure 14.1 `plot_map` function. The pyplot versions need to be used instead.

[9] Classically, maps following this projection can be thought of as being created by wrapping a piece of paper around a globe and drawing lines from points on the globe to the piece of paper.

instance of the `NaturalEarthFeature` class, `states_50m`, and pass that into the call to `add_feature`.

- `set_extent` Sets the extent of the plot, defined by a four-element list passed in as a positional input parameter, specifying the *x*-axis minimum, *x*-axis maximum, *y*-axis minimum, and *y*-axis maximum, respectively. The values are given in data coordinates. The keyword input parameter `crs` is set to the cartopy map projection object for the coordinate system the extent data is in.

  In the Figure 14.1, line 64–66, example, the projection object describing the extent data is a regular Cartesian-like grid in longitude and latitude (i.e., a cylindrical projection grid). Thus, the `crs` keyword input parameter is set to a `PlateCarree` object, which is (essentially) the same as the map display projection in this case. In Try This! 14-3, we consider a case where the data (and extent) are on a different projection object than the map display projection.

- `set_xticks` Sets the *x*-axis tick marks to be labeled, and accepts a list of those values. As described with `set_extent` above, the keyword input parameter `crs` should be set to a cartopy map projection object that describes the tick data coordinate system. Figure 14.1, line 78, shows an example.

- `set_yticks` Sets the *y*-axis tick marks that will be labeled, and accepts a list of those values. As described with `set_extent` above, the keyword input parameter `crs` should be set to a cartopy map projection object that describes the tick data coordinate system. Figure 14.1, line 79 shows an example.

In the Figure 14.1 examples of calling `set_extent`, `set_xticks`, and `set_yticks`, the map projection object being passed in is the one that describes the extent or tick data coordinate system, not necessarily the projection object used to create the `GeoAxesSubplot` object of the map. We can thus translate data given in one projection into a plot of another projection. Although in Figure 14.1 the data projection and map display projections are the same, in Try This! 14-3 we consider a case where the data and map display projections are different.

In addition to calls of the above methods, Figure 14.1 also shows examples of more detailed formatting. For instance, in lines 80–81, we create `LongitudeFormatter` and `LatitudeFormatter` objects to enable us to add the "°N" and "°W" labels to the latitude and longitude *x*- and *y*-axis tick labels. These formatter objects are passed into the `xaxis` and `yaxis` attributes (lines 82–83, using the `set_major_formatter` method), to be used when the labels are written out.

Step 3: Create the Contour Plot   In this final step, we call the `contour` and/or `contourf` methods (or the pyplot command versions thereof) to create the contour plot, as was described in Section 14.2.2. When we do, we set the `transform` keyword input parameter in the `contour` or `contourf` call to the map projection *in which the data are given*. Thus, in the Figure 14.1, lines 86 and 89, examples, the `transform` keywords input parameter's `PlateCarree` object tells Python that the `lon2d`, `lat2d`, and `air_temp` values are in regular latitude/longitude coordinates. Python translates the values in the `transform` coordinates into the map coordinates needed to display them on the `GeoAxesSubplot` plot. In the Figure 14.1, lines 86 and 89, examples, the only difference is the `GeoAxesSubplot` projection is a `PlateCarree` object with a central longitude that is explicitly set, but in more complex scenarios, we can use this to regrid data from one projection into another for display. Again, in Try This! 14-3, we consider a case where the data and map display projections are different.

The cartopy online documentation provides a list of the map projections that are available in the package. We provide an up-to-date link to this list at www.cambridge.org/core/ resources/pythonforscientists/refs/, ref. 20

14.2.4 Basic Animation Using Matplotlib

All animation, whether hand-drawn or computer-generated, works on the same principle: draw a figure, move the figure into a new position, redraw the figure, move the figure, and repeat. The illusion of motion is created by showing a person successive snapshots of the figure as it changes over time.

Our Figure 14.3 example shows these steps. After turning on Matplotlib's interactive mode using the `ion` command (line 1), we make a contour plot of the temperature at the first time in the animation (November 11, 2017, at 0:00Z). We use the `plot_map` function (lines 2–5), just as we did in Figure 14.2 to make individual plots, except we now save the `Figure` object as the variable `fig`. This way, in later calls to `plot_map`, we can pass in that same `Figure` object, erase the old map in that window (inside `plot_map`, by calling the pyplot `cla` command), and overlay a new map in that window. That is exactly what we do in the `for` loop (Figure 14.3, lines 7–12). For each iteration of the loop, we select a new time sheet of temperature (the sheet of index i, in line 9). A call to the `draw` command (line 13) renders the new image, and between the display of each figure, we pause for a second by using the pyplot `pause` function. If we did not pause, the animation would run by faster than our eyes could process.

To create an animation of an image or line plot, we follow essentially the same steps as with the contour map animation. However, in the cases of the `AxesImage` object for an image or the `Line2D` object for a line in a line plot, there are methods that exist that enable us to replace the image or line data as needed, and then redraw the image or line. For `AxesImage` objects, the `set_data` method enables us to replace the image data for each frame of the animation. For the `Line2D` object, the `set_xdata` and `set_ydata` methods enable us to replace the x- and y values, respectively, that define the line. Below are examples of image and line animations.

Image Animation Example   Figures 13.1–13.3 show three panels of tuberculous granulomas at three different times. We can create an animation using these three panels using the code in Figure 14.5. After importing the pyplot and image submodules in lines 1–2 of Figure 14.5, we create a `Figure` and `Axes` object (line 4). We then use the `axis` method of the `Axes` object ax to turn off both the x- and y-axis tick marks and labeling (line 5). In line 7, we turn

```
1    import matplotlib.pyplot as plt
2    import matplotlib.image as mpimg
3
4    fig, ax = plt.subplots(1, 1)
5    ax.axis('off')
6
7    plt.ion()
8
9    img1 = mpimg.imread('g004-A1.png')
10   img2 = mpimg.imread('g004-A2.png')
11   img3 = mpimg.imread('g004-A3.png')
12
13   images = [img1, img2, img3]
14
15   for i in range(len(images)):
16       if i == 0:
17           img = ax.imshow(images[i])
18       else:
19           img.set_data(images[i])
20       plt.draw()
21       plt.pause(1.0)
```

Figure 14.5 Code to display an animation of the tuberculous granuloma images from Figures 13.1– 13.3.

on Matplotlib's interactive mode (so we continually update the rendering of graphic objects). Lines 9–13 in Figure 14.5 are the same as lines 1–5 in Figure 13.6: these lines read in the data from the three image files and put the data objects into a list images.

Lines 15–21 of Figure 14.5 do the actual animation. We use a for loop to go through the indices of each element of images. For the first iteration of the loop, we have to create the AxesImage object based on the first image. To do this we call the imshow method that is attached to the Axes object ax (line 17); this call occurs only when the iterator i is equal to zero (line 16). For the other two images, we use the AxesImage object's set_data method to replace the data in the AxesImage object (line 19). We render the plot using the pyplot draw function (line 20). After each iteration, we wait one second so the animation does not speed through too quickly (line 21).

Line Animation Example   For animating lines, we can independently set the x and y data that make up a single line by using the set_xdata and set_ydata methods that are part of a Line2D object. Figure 14.6 shows code that animates the first 20 steps of the random walk model described in Figure 8.4. (See also Figure 14.7.) We set that number of steps in line 4 of Figure 14.6. In lines 5–7, we create the Figure and Axes objects and set the x- and y-axis labels for the plot. Lines 8–9 set the range of the x- and y-axes to following the maximum and minimum locations for only the first 20 points (plus and minus one location index value to give a nice margin around the random walker's path). Line 11's call to ion

```
1    import numpy as np
2    import matplotlib.pyplot as plt
3
4    num_steps = 20
5    fig, ax = plt.subplots(1, 1)
6    plt.xlabel("X-Direction Index")
7    plt.ylabel("Y-Direction Index")
8    ax.axis([np.min(x[:num_steps])-1, np.max(x[:num_steps])+1,
9             np.min(y[:num_steps])-1, np.max(y[:num_steps])+1])
10
11   plt.ion()
12
13   for i in range(num_steps):
14       if i == 0:
15           lines = ax.plot(x[0], y[0], '-o')
16       else:
17           lines[0].set_xdata(x[:i])
18           lines[0].set_ydata(y[:i])
19       plt.title("Path After " + str(i+1) + " Steps")
20       plt.draw()
21       plt.pause(1.0)
```

Figure 14.6 Code to display the animation of the first 20 steps of the random walk model described by Figure 8.4. The x and y arrays are as defined in Figure 8.4. The final panel generated by this animation is given in Figure 14.7.

activates Matplotlib's interactive mode. Although not used here, the pyplot function `ioff` turns off the interactive mode.

The animation is created in the looping defined by lines 13–21. Similar to what we did in Figure 14.5, we create the `Line2D` object in the first iteration of the `for` loop. The call to the `plot` method creates a tuple of `Line2D` objects (line 15). Because only one line is defined in line 15, the tuple `lines` has only one element. That element, a `Line2D` object, is accessed in lines 17 and 18 (i.e., `lines[0]`) when the x- and y-values of the line are replaced. In lines 17 and 18, we select all elements of the x and y arrays from the first value to the value before index i (i.e., the `[:i]` array slicing). This is because we want our animation to show the random walk path growing as time progresses. Each iteration, then, needs to plot the entire path up to and including that iteration's step. Finally, in lines 19–21, we plot an updated title (which show the number of steps taken at each frame of the animation), redraw the line, and pause for one second.

There are better ways of creating animations in Python than the method we used in the above examples. In Matplotlib, use of the animation module is one such way.[10] At a certain point, scientific and engineering animations are no different than animations for gaming or

---

[10] See www.cambridge.org/core/resources/pythonforscientists/refs/, refs. 22 and 21, for up-to-date links to documentation on the animation submodule and a simple example.

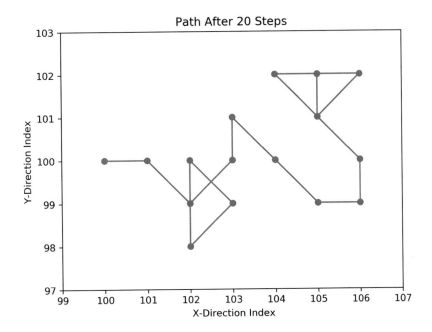

Figure 14.7 Final panel from the animation created by the code in Figure 14.6.

other applications. Pygame, for instance, is a basic package used for creating games that can also be used for visualizing simulations.[11] The richness of the Python ecosystem of packages enables us to choose the right package to do the specific scientific or engineering animation we are interested in.

14.2.5 Flexible Functions and Dictionaries

As we see with the calls to pyplot commands in our Section 14.1 example, graphs can require a lot of tweaking. Font, font size, line width, color, symbols, color bar, and map projection are just a few of the many different aspects of a graph we might want to control. In the function calls of Section 14.1, no single call had more than a few handfuls of positional and keyword input arguments. Still, we can conceive of situations where we might have several tens of parameters we would want to set in a function call. Is our only choice to type out every input parameter in the calling line each time we make the function call? Thankfully, no. Python offers the ability of using the *computer* to control the arguments that are passed into a function (or method) call. How to do so, and how to create our own functions that can take advantage of these features, is the focus of this section. But before we examine this motivating example, we need to introduce the **dictionary**.

---

[11] See www.cambridge.org/core/resources/pythonforscientists/refs/, ref. 36, for an up-to-date link to the Pygame documentation.

So far, when we have needed to store more than one piece of information, we have put them into lists/tuples or arrays. Once in those list-like structures, we pick out one or more values by specifying an index or range of indices (via slicing). But, an index does not tell us very much about what an element holds. All it says is that this is the position of the element. Many times, it is more useful for us to reference values not by a **positional index** but by something more meaningful. For instance, pretend we have a list of settings for a graph:

```
settings = ['--', 5.0, '*', 20.0]
```

where the first element describes the line style, the second element the line width (in points), the third element the marker used, and the last element the marker size (in point).[12] In a list, however, each element is referenced only by its position. Thus, to reference the marker size, we would type `settings[3]`. But, there is nothing in that notation to suggest we are referencing the marker size.

A dictionary provides the functionality to fix this problem. Like lists and tuples, dictionaries are also collections of elements. However, dictionaries, instead of being ordered, are *unordered* collections whose elements are referenced by **keys**, not by position. Keys can be anything that can be uniquely named and sorted. In practice, keys are usually integers or strings. Values can be anything. And when we say "anything," we mean *anything*, just like elements of lists and tuples. Thus, a key becomes the "label" to the value they are attached to. Note, dictionaries are unordered but are iterable over their keys. We discuss this in Section 16.2.1.

Curly braces ({ }) delimit a dictionary. The elements of a dictionary are "key:value" pairs, separated by a colon. Each key:value pair element of a dictionary is separated by commas. Thus, if we make the `settings` list into a dictionary, with the keys for each value set to a string that describes the value associated with it, we would have:

```
settings = {'linestyle':'--', 'linewidth':5.0,
            'marker':'*', 'markersize':20.0}
```

Dictionary elements are referenced like lists, except the key is given in place of the element address. With the above dictionary, the notation `settings['markersize']` would return the floating-point value 20.0. To set the value associated with the `'markersize'` key to 18.0, we combine assignment with the dictionary referencing mechanism:

```
settings['markersize'] = 18.0
```

That syntax also works to add new elements into a dictionary. Thus, to add the `'markeredgewidth'` key and set it to the floating-point value 1.0, we can use the following assignment line:

```
settings['markeredgewidth'] = 1.0
```

That line works even though the `'markeredgewidth'` key does not yet exist in the `settings` dictionary. To delete a key:value pair from a dictionary, we use the `del` command. Thus, to remove the `'linestyle'` key and its associated value from `settings`, we execute:

---

[12] These were some of the keyword input parameter settings from Figure 5.1.

```
1    import matplotlib.pyplot as plt
2    import numpy as np
3
4    def plot_maze_times(xdata, ydata, linewidth=1.0, markersize=3.0,
5                        filename='default.png'):
6        plt.figure()
7        plt.plot(xdata, ydata, linewidth=linewidth, marker='o',
8                 markersize=markersize)
9        plt.axis([0, np.max(xdata), 0, np.max(ydata)])
10       plt.xlabel('Dosage (mg)')
11       plt.ylabel('Time to Complete Maze (sec)')
12       plt.text(0.1, 10, 'Provisional')
13       plt.savefig(filename, dpi=300)
```

Figure 14.8 Solution to Try This! 5-2.

```
del(settings['linestyle'])
```

Standard dictionaries are not ordered. That is, we should not consider the key:value pairs in a dictionary to be in any kind of order. If we use strings as keys, it may look to us as if the key:value pairs are in an order (say, alphabetically), but they are not ordered. There is no "first," "second," etc., element in a dictionary.

There is a lot more we can do with a dictionary. Dictionaries may seem just like lists, with the slight difference that elements are unordered and values can be referenced by strings, but it turns out these differences make dictionaries very useful, for many different purposes. In Section 16.2.1, we discuss these features plus other aspects of dictionaries, including how to loop through all or some of the key:value pairs in the dictionary. In the rest of this section, we focus on the use of dictionaries that we suggested at the beginning of the section: to help us write functions that can better handle many positional and keyword input parameters.

Consider the `plot_maze_times` function we wrote to solve Try This! 5-2 (reproduced in Figure 14.8 for convenience). Pretend we want that solution to handle more keyword parameters than given in the current `def` statement. In fact, let us say we want our solution to handle all of the possible keyword input parameters the pyplot `plot` command can handle, that is, we want to be able to pass in *any* `plot` keyword input parameter as a keyword input parameter for `plot_maze_times`. The most direct solution – to write out all of `plot`'s keyword input parameters in `plot_maze_times`'s `def` line – would be a lot of work. And if the programmers of `plot` decided to add in more keyword input parameters, we would have to manually type those into our `plot_maze_times` solution too. What else can we do?

Our clue comes from a closer look at the parameter list in lines 4–5 of Figure 14.8. First, we see that the two positional input parameters are separated by commas and are given in an order. That is, they look like two elements in a list. Second, we see that all the keyword input parameters are in the form of an assignment. There is a name (e.g., `linewidth`) that is set equal to a value (e.g., 1.0). A name, however, is nothing more than a string without quotation marks. A keyword input parameter, then, is really a string label connected to a value (so too are all assignments: a string label connected to a value). But, if a keyword input parameter is a string connected to a value, this is exactly what a dictionary element is: a key:value pair, where the key is a string. A dictionary, then, can hold a collection of keyword input parameters.

Python provides a mechanisms by which we can refer to a collection of positional input parameters in a function as a list and a collection of keyword input parameters in a function as a dictionary. In the latter dictionary, the name of the keyword input parameter is given as a string and is the key in each dictionary element while the value that parameter is set to is the corresponding value in the dictionary element. When representing a collection of positional input parameters as a list, an asterisk is appended in front of the list name in the function's `def` statement and when we call the function. When representing a collection of keyword input parameters as a dictionary, two asterisks are appended in front of the dictionary name in the function's `def` statement and when we call the function. Thus, our Figure 14.8 `def` statement could be rewritten as:

```
def plot_maze_times(*args, **kwds):
```

where the list `args` represents the collection of positional input parameters and the dictionary `kwds` represents the collection of keyword input parameters and their default values. We have to put the asterisks in front of `args` and `kwds` to distinguishes them from two regular positional input parameters. The names "args" and "kwds" are not what are important. Instead, the asterisks in front of the variable names are what are important.

If we use the `*args/**kwds` method of specifying input parameters, we have to rewrite how we refer to parameters in the body of the function, because the dummy variables `xdata`, `ydata`, etc., no longer exist. Specifically, the positional input parameters are now referred to by position in the tuple `args` and the keyword input parameters are now referred to by the keys in the `kwds` dictionary. In lines 5–6 of Figure 14.9, we create the **local** variables `xdata` and `ydata` by referring to the appropriate items in the `args` tuple. A local variable is one whose scope – existence – is limited. In this case, `xdata` and `ydata` only exist in the function in which they are defined. For the keyword input parameters, it is a little more complicated because we need to give the local variables corresponding to those parameters the value passed in and found in `kwds`, if it is set in the calling line, and a default value otherwise. In lines 8–21, we check if the key for `'linewidth'`, `'markersize'`, and `'filename'` are in `kwds` (using the `in` syntax). If so, the local variable corresponding to that keyword input parameter is set to the keyword input parameter value. If not, that local variable is set to a default.

When we call a function that uses the `*args/**kwds` mechanism, we can call it by explicitly listing the input parameters or by passing in a list/tuple and/or dictionary of the input parameters. Thus, all the following are legitimate calls to `plot_maze_times` as defined in Figure 14.9:

```
1   import matplotlib.pyplot as plt
2   import numpy as np
3
4   def plot_maze_times(*args, **kwds):
5       xdata = args[0]
6       ydata = args[1]
7
8       if 'linewidth' in kwds:
9           linewidth = kwds['linewidth']
10      else:
11          linewidth = 1.0
12
13      if 'markersize' in kwds:
14          markersize = kwds['markersize']
15      else:
16          markersize = 3.0
17
18      if 'filename' in kwds:
19          filename = kwds['filename']
20      else:
21          filename = 'default.png'
22
23      plt.figure()
24      plt.plot(xdata, ydata, linewidth=linewidth, marker='o',
25              markersize=markersize)
26      plt.axis([0, np.max(xdata), 0, np.max(ydata)])
27      plt.xlabel('Dosage (mg)')
28      plt.ylabel('Time to Complete Maze (sec)')
29      plt.text(0.1, 10, 'Provisional')
30      plt.savefig(filename, dpi=300)
```

Figure 14.9 The Figure 14.8 solution modified to use the `*args`/`**kwds` mechanism of passing in input parameters.

```
1   xdata = [0.0, 0.08, 0.2, 0.37, 0.6, 0.84, 1.02, 1.2]
2   ydata = [105, 98, 54, 50, 65, 81, 182, 210]
3
4   plot_maze_times(xdata, ydata)
5   plot_maze_times(xdata, ydata, linewidth=5.0)
6   plot_maze_times(xdata, ydata, linewidth=5.0, filename='test.png')
7
8   myargs = [xdata, ydata]
9   mykwds = {'linewidth':5.0, 'filename':'test.png'}
10  plot_maze_times(*myargs, **mykwds)
```

The calls in line 6 and 10 do exactly the same thing.

```
1    import matplotlib.pyplot as plt
2    import numpy as np
3
4    def plot_maze_times(*args, **kwds):
5        xdata = args[0]
6        ydata = args[1]
7
8        modified_kwds = kwds.copy()
9
10       if 'linewidth' not in kwds:
11           modified_kwds['linewidth'] = 1.0
12
13       if 'markersize' not in kwds:
14           modified_kwds['markersize'] = 3.0
15
16       if 'filename' not in kwds:
17           modified_kwds['filename'] = 'default.png'
18
19       plt.figure()
20       modified_kwds['marker'] = 'o'
21       plt.plot(*args, **modified_kwds)
22       plt.axis([0, np.max(xdata), 0, np.max(ydata)])
23       plt.xlabel('Dosage (mg)')
24       plt.ylabel('Time to Complete Maze (sec)')
25       plt.text(0.1, 10, 'Provisional')
26       plt.savefig(modified_kwds['filename'], dpi=300)
```

Figure 14.10 The Figure 14.8 solution modified to use the *args/**kwds mechanism of passing in input parameters and passing all keyword input parameters to the call of plot.

In addition to more flexibility at calling time, passing in keyword input parameters using dictionaries allows us to make use of keyword input parameters we are not aware of, ahead of time, in calls in our function body. In Figure 14.10, we pass a copy of the entire keyword input parameter dictionary kwds, with modifications, to the plot call in line 21. That copied and modified dictionary is modified_kwds. The dictionary begins as a copy of kwds in line 8, is changed in lines 10–17 and 20, and is passed in in line 21. By passing in modified_kwds, any keyword input parameter that was specified in the call of plot_maze_times – even a parameter we were not aware of when we wrote plot_maze_times – will also make it to plot. Thus, this call will work fine:

```
xdata = [0.0, 0.08, 0.2, 0.37, 0.6, 0.84, 1.02, 1.2]
ydata = [105, 98, 54, 50, 65, 81, 182, 210]
plot_maze_times(xdata, ydata, linestyle='-')
```

even though plot_maze_times has no recognition of the linestyle keyword input parameter!

An aside on programming terminology: The `*args/**kwds` mechanism is the way Python accomplishes **overloading**, the creation of different versions of a method to handle different numbers of and kinds of input parameters. The interface of a method, defined by its name, the number of and kinds of input parameters, and the kind of return value, is known as the method's **signatures**. So, overloading is the practice of writing different versions of the same method to accommodate different signatures.

14.3 Try This!

The examples in this section address the main topics of Section 14.2. We practice using the Matplotlib object API and making different kinds of contour plots. We also create a basic animation and practice using dictionaries as part of writing more flexible functions.

For all the exercises in this section, assume NumPy has already been imported as `np` and the Matplotlib pyplot submodule has already been imported as `plt`.

Try This! 14-1 Matplotlib's Object API: Air Quality

Table 7.7 lists one day of hourly particulate matter ($PM_{2.5}$) concentration (in $\mu g/m^3$) measurements at four locations in Beijing, China. Using Matplotlib's object API, create a single figure consisting of four panels, where each panel is a line plot showing $PM_{2.5}$ versus hour number over that day. Use the `add_axes` method, rather than `subplots`, to lay out the panels on the figure in an irregular way (i.e., not as a regular grid). Assume the array `pm25` exists, as defined in Figure 7.7.

Try This! Answer 14-1

See Figures 14.11 and 14.12: Figure 14.12 shows a plot of the four columns of data from the array `pm25` in Table 7.7. The x-axis on all graphs is hour in the day (starting at hour 0) and the y-axis is the hourly $PM_{2.5}$ concentrations. Axes labels and titling are left off to make the code simpler to read. Figure 14.11 shows the code that generates the plot in Figure 14.12.

In line 1 of Figure 14.11, we create the `Figure` object `fig`. The figure is 5 in wide and 3 in tall. Lines 2–5 create the `Axes` objects and place them on `fig`. The tuple passed in gives the left-lower locations, then the width and height of the plot. All values are given in normalized coordinates and are expressed in fractions of the length of the appropriate axis.

In line 7, we create the array `hours`, to hold the hour in the day each measurement is given. The array begins at 0 and ends at 23. There are 24 values. In lines 8–11, we call the `plot` method attached to each `Axes` object. The plot is made at the location on `fig` the `Axes` object is placed at, using the `add_axes` call. The `ax1` object specifies a graph with red circles, the `ax2` object specifies a graph with blue stars, the `ax3` object specifies a graph with green squares, and the `ax4` object specifies a graph with black diamonds. The particulate matter concentrations for each station is a column in the `pm25` array. In line 13, we write the figure to a file.

```
1    fig = plt.figure(figsize=(5, 3))
2    ax1 = fig.add_axes((0.1, 0.7, 0.2, 0.2))
3    ax2 = fig.add_axes((0.4, 0.7, 0.2, 0.2))
4    ax3 = fig.add_axes((0.1, 0.4, 0.2, 0.2))
5    ax4 = fig.add_axes((0.4, 0.1, 0.5, 0.5))
6
7    hours = np.arange(np.shape(pm25)[0])
8    ax1.plot(hours, pm25[:,0], 'ro')
9    ax2.plot(hours, pm25[:,1], 'b*')
10   ax3.plot(hours, pm25[:,2], 'gs')
11   ax4.plot(hours, pm25[:,3], 'kD')
12
13   plt.savefig('api_example.png', dpi=300)
```

Figure 14.11 Code to solve Try This! 14-1. The plot it generates is shown in Figure 14.12.

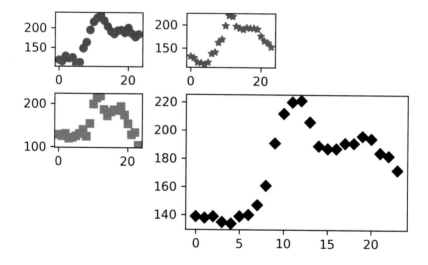

Figure 14.12 Plot created by the code in Figure 14.11.

A few notes about this solution: First, by using add_axes, we are able to place each plot anywhere on the figure we desire. In this case, we create a graph where the plot for one of the measurement stations is more prominent, while the plots for the other stations are less prominent. Second, in normalized coordinates, the width and height of each graph is the same: 0.5 (or 50 percent of the length of the *x*- and *y*-axes, respectively) for the large plot and 0.2 (or 20 percent of the length of the *x*- and *y*-axes, respectively) for the small plots. Why, then, are the plots rectangular rather than square? Because normalized coordinates specify locations and lengths in fractions of the *x*- and *y*-axes, the size of the figure (as set by the `figsize` keyword input parameter in the `figure` call) also sets the relative widths and heights of each of the plots on that figure and thus the rectangular aspect ratio.

```
1    axes = [ax1, ax2, ax3, ax4]
2    color_marker = ['ro', 'b*', 'gs', 'kD']
3    for i in range(4):
4        axes[i].plot(hours, pm25[:,i], color_marker[i])
```

Figure 14.13 Alternate code using a loop to do the same tasks as lines 8–11 of Figure 14.11.

Finally, we note that the code of Figure 14.11 contains a number of repeated tasks, such as the repeated calls to `plot`. The only difference between each `plot` call in lines 8–11 is a different color-marker specification string, a different column index in `pm25`, and a different `Axes` object. This suggests that we might put these calls into a loop that iterates through the column indices and place the color-marker strings and `Axes` objects into lists that are also referenced by the same column indices. The elements of a list can be any type or mix of types. So, there is no problem for the list to hold a collection of `Axes` objects.

Figure 14.13 shows this alternate way of using a loop to accomplish the tasks of lines 8–11 in Figure 14.11. Although the Figure 14.13 solution also requires us to manually type in `axes` and `color_marker`, we could further automate the creation of these plots by using the `append` method of the `axes` list object to add in the `Axes` objects as they are created by `add_axes` (which in turn could be put into a loop). We could also create a function that would generate color-marker pairs – based on the Table 5.2 and 5.3 values for markers and colors, respectively – to automatically fill the `color_marker` list.

Try This! 14-2 Line and Shaded Contour Plots: Elevations on a Hill

Pretend we have elevations (in meters above mean sea level) taken on a small hill. The elevations are given in the file *hill.txt*, as shown in Figure 14.14. Each location is 2 m away from each other (in North–South and East–West directions). The grid of elevations in Figure 14.14 is oriented so the northernmost row is the top row and the westernmost column is the left column. Write a program that reads in the data from *hill.txt* and stores it in a two-dimensional array, creates a line and shaded contour plot of the data using a grayscale color map, adds contour labels and a color bar for the shading, and saves the plot to the image file *hill.png*.

Try This! Answer 14-2

The code in Figure 14.15 creates the desired plot. The plot generated by the code is shown in Figure 14.16.

```
1.0,  1.8,  1.3,  1.4,  1.6,  1.5,  1.9,  1.9
1.2,  1.5,  1.5,  1.5,  1.4,  1.5,  1.7,  1.7
1.5,  1.7,  1.8,  1.8,  1.7,  1.6,  1.7,  1.5
1.2,  1.8,  2.0,  1.9,  1.9,  1.6,  1.7,  1.5
1.1,  2.0,  2.0,  2.1,  1.9,  2.0,  1.9,  1.6
1.4,  1.8,  2.0,  2.0,  1.8,  1.8,  1.6,  1.4
1.2,  1.6,  1.5,  1.4,  1.6,  1.7,  1.7,  1.3
1.2,  1.5,  1.5,  1.5,  1.6,  1.6,  1.5,  1.3
1.1,  1.4,  1.3,  1.4,  1.5,  1.6,  1.5,  1.1
```

Figure 14.14 Fictitious elevations (in m above mean sea level) on a small hill, stored in *hill.txt*. The locations of each row are 2 m from each other. The locations of each column are 2 m from each other. The top row is the northernmost row and the leftmost column is the westernmost column. See Try This! 14-2 for additional information.

After importing the relevant NumPy and Matplotlib pyplot modules (Figure 14.15, lines 1–2), we use the NumPy function genfromtxt to read the contents of *hill.txt* directly into a two-dimensional array elev (line 4). In lines 6 and 7, we create one-dimensional arrays that give the East–West (x) and North–South (y) distances from the northwest corner of the array. This follows the orientation of the data as shown in Figure 14.14.

In line 8, we use a NumPy function called meshgrid that produces "two-dimensional" versions of x and y. That is, the arrays x2d and y2d are two-dimensional arrays of the same shape as elev whose elements are the x and y values, respectively, for each corresponding location in elev. Thus, x2d looks like the following:

```
[[ 0,  2,  4,  6,  8, 10, 12, 14]
 [ 0,  2,  4,  6,  8, 10, 12, 14]
 [ 0,  2,  4,  6,  8, 10, 12, 14]
 [ 0,  2,  4,  6,  8, 10, 12, 14]
 [ 0,  2,  4,  6,  8, 10, 12, 14]
 [ 0,  2,  4,  6,  8, 10, 12, 14]
 [ 0,  2,  4,  6,  8, 10, 12, 14]
 [ 0,  2,  4,  6,  8, 10, 12, 14]
 [ 0,  2,  4,  6,  8, 10, 12, 14]]
```

and y2d looks like the following:

```
[[ 0,   0,   0,   0,   0,   0,   0,   0]
 [ 2,   2,   2,   2,   2,   2,   2,   2]
 [ 4,   4,   4,   4,   4,   4,   4,   4]
 [ 6,   6,   6,   6,   6,   6,   6,   6]
 [ 8,   8,   8,   8,   8,   8,   8,   8]
 [10,  10,  10,  10,  10,  10,  10,  10]
 [12,  12,  12,  12,  12,  12,  12,  12]
 [14,  14,  14,  14,  14,  14,  14,  14]
 [16,  16,  16,  16,  16,  16,  16,  16]]
```

The columns of x2d have all the same values because on a regular Cartesian grid, all elements in a column have the same *x*-axis value. The rows of y2d have all the same values because on a regular Cartesian grid, all elements in a row have the same *y*-axis value. This is the same as the structure of lon2d and lat2d in Figure 14.1.

When meshgrid is used with two input parameters (specifying a two-dimensional grid), two return values are produced. Similar to our use of subplots in line 59 of Figure 14.1, we can set them to variables in a single line by separating the variables being assigned by a comma (i.e., the syntax x2d, y2d, = ... in line 8).

In line 10, we create the Figure and Axes objects, fig and ax, respectively, using the subplots function. A single graph is placed on the page/window. Later on, we use methods attached to the ax object. We do not, however, explicitly use the fig object.

We create a one-dimensional array for specifying the contour levels in line 12. The result is an array of values from 0 to 3 (inclusive on both ends), incremented by 0.25. In lines 13–14, we create a filled contour plot object with the predefined "Greys" color map. That color map ranges from white to black from its minimum to maximum, transitioning through different shades of gray. In lines 15–17, we create a contour line plot object to place contour lines on top of the filled contour plot.

In line 18, we set the x-axis lower and upper limits to the minimum and maximum values in x but the y-axis limits to the *reverse* – the lower and upper limits are set to the maximum and minimum values in y. We need to do this because the default setting for making a plot in Matplotlib is to put the origin in the lower-left-hand corner. For the array elev, however, the origin is in the upper-right-hand corner (the northwest corner). By flipping around the y-axis, the plot will be oriented the same way the data are presented in *hill.txt*. This adjustment is not necessary for the example in Section 14.1 because latitude values increase as we move northwards.

We call the clabel command in line 19, operating on the lined contour plot object hillmap. We set the fmt keyword input parameter to the format code '%.2f' to create floating-point number contour labels with two decimal places. This looks like the string formatting codes that we discussed in Section 9.2.4 except instead of the code being in curly braces and after a colon,

```
1    import numpy as np
2    import matplotlib.pyplot as plt
3
4    elev = np.genfromtxt('hill.txt', delimiter=', ')
5
6    x = np.arange(np.shape(elev)[1]) * 2
7    y = np.arange(np.shape(elev)[0]) * 2
8    x2d, y2d = np.meshgrid(x, y)
9
10   fig, ax = plt.subplots(1, 1)
11
12   levels = np.arange(13) / 4.0
13   hillmapf = ax.contourf(x2d, y2d, elev, levels=levels,
14                       cmap=plt.get_cmap('Greys'))
15   hillmap = ax.contour(x2d, y2d, elev, levels=levels,
16                     linewidths=1.0,
17                     colors='k')
18   plt.axis([np.min(x), np.max(x), np.max(y), np.min(y)])
19   plt.clabel(hillmap, fmt='%.2f', fontsize=10)
20   plt.colorbar(hillmapf, orientation='horizontal')
21
22   plt.savefig('hill.png', dpi=300)
```

Figure 14.15 Code to create the contour plot of elevations described Try This! 14-2. The plot is shown in Figure 14.16.

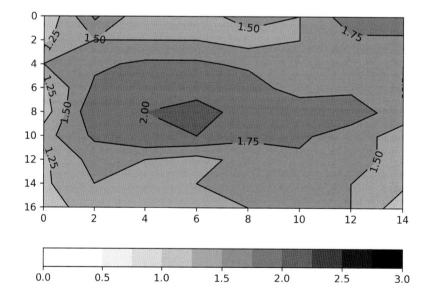

Figure 14.16 Contour plot of elevations produced by the code in Figure 14.15.

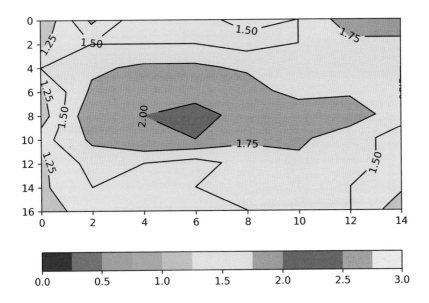

Figure 14.17 Contour plot of elevations produced by the code in Figure 14.15 except with the color map "terrain" used instead of "Greys."

the code is after a percentage sign. The use by `clabel` of a percentage sign comes from the "old" way Python used to do value formatting. For our purposes, the rationale is not important. We finish up by adding a color bar at the bottom (line 20) and writing the plot file out to *hill.png*.

We can change the color map used for the filled contour plot by changing the string we pass into the `get_cmap` call in line 14 of Figure 14.15. For instance, because we are plotting elevations, it might be nice to use the "terrain" color map instead. This color map is designed for showing surface terrain features, ranging from blues (oceans) to greens (lower elevation land) and browns (higher elevation land). To do so, we pass in `'terrain'` instead of `'Greys'` in line 14 of Figure 14.15. Figure 14.17 shows Figure 14.16 except using the "terrain" color map.

Try This! 14-3 Using cartopy to Overlay Maps: Atmospheric Quantity

Pretend the fictitious data in *hill.txt* from Try This! 14-2, instead of being elevations taken every 2 m, are measurements of some quantity over the continental United States (perhaps the concentration of some atmospheric chemical). These measurements are taken every 10 degrees in longitude, ranging from 130° W to 60° W, and every 5 degrees in latitude, ranging from 60° N to 20° N.

Write a program that reads in the data from *hill.txt* and creates a line and shaded contour plot of the data using a grayscale color map, as in Try This! 14-2, except on a Lambert conformal projection map of the region. Save the plot to the image file *meas_lambert.png*.

```
1    import numpy as np
2    import matplotlib.pyplot as plt
3    import cartopy.crs as ccrs
4    import cartopy.feature as cf
5
6    meas = np.genfromtxt('hill.txt', delimiter=', ')
7
8    x = np.array([-130, -120, -110, -100, -90, -80, -70, -60])
9    y = np.array([60, 55, 50, 45, 40, 35, 30, 25, 20])
10   x2d, y2d = np.meshgrid(x, y)
11
12   plt.figure()
13   ax = plt.axes(projection=ccrs.LambertConformal())
14   ax.set_extent([-140, -50, 15, 65], crs=ccrs.PlateCarree())
15   ax.add_feature(cf.BORDERS, linestyle=':')
16   ax.add_feature(cf.COASTLINE)
17
18   levels = np.arange(13) / 4.0
19   measmapf = ax.contourf(x2d, y2d, meas, levels=levels,
20                           transform=ccrs.PlateCarree(),
21                           cmap=plt.get_cmap('Greys'))
22   measmap = ax.contour(x2d, y2d, meas, levels=levels,
23                           transform=ccrs.PlateCarree(),
24                           linewidths=1.0,
25                           colors='k')
26   plt.clabel(measmap, fmt='%.2f', fontsize=10)
27   plt.colorbar(measmapf, orientation='horizontal')
28
29   plt.savefig('meas_lambert.png', dpi=300)
```

Figure 14.18 Code to create a line and filled contour plot on a Lambert conformal projection map of fictitious measurements over the continental United States from *hill.txt* (as described in Try This! 14-3). The plot generated is given in Figure 14.19.

Try This! Answer 14-3

Figure 14.18 shows the code to solve the problem and Figure 14.19 shows the plot that is generated. The code is very similar to the solution to Try This! 14-2 shown in Figure 14.15. The minor changes include the renaming of the `elev` array in Figure 14.15 to `meas` in Figure 14.18. The major differences are:

- In lines 8–9 of Figure 14.18, the arrays x and y are set to the longitude and latitude values of each element in `meas`. The latitude values start from 60° N and decrease to 20° N, so we do not have to manually flip around the *y*-axis as in line 18 of Figure 14.15.

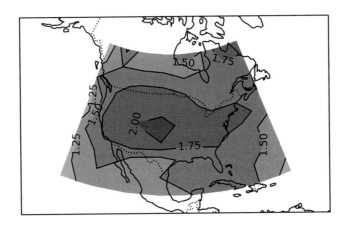

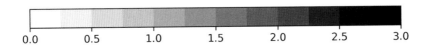

Figure 14.19 Contour plot of measurements produced by the code in Figure 14.18.

- In line 13 of Figure 14.18, the projection used to generate the `ax` object is the Lambert conformal projection, which is passed in via the `projection` keyword input parameter. This is the projection the map will display on.
- In line 14 of Figure 14.18, the extent of the display map is provided in regular, Cartesian-like latitude–longitude coordinates. Thus, the `crs` keyword input parameter is set to a `PlateCarree` object, which is the projection of the data and its extents.
- In lines 15–16 of Figure 14.18, we draw national borders and the coastline.
- Because the values of `meas` are given in regular, Cartesian-like latitude–longitude coordinates, the `transform` keyword input parameter for the calls to `contourf` and `contour` are set to a `PlateCarree` object, which is the projection of the `meas` data (lines 20 and 23 of Figure 14.18).

Because we plotted the values onto a Lambert conformal projection map, the plot in Figure 14.19 looks "warped" compared with the plot in Figure 14.16. This is the result of the distortions introduced when trying to represent a curved surface (which we assume the measurements are taken on) onto a flat map.

Try This! 14-4 Basic Animation: Trajectory of a Projectile

The trajectory for a projectile (ignoring air resistance) fired from an origin with an initial velocity of $\vec{v}_0$ at an angle θ_0 from the horizontal surface follows a curve whose horizontal (x) and vertical (y) position values vary with each other in the following way:[13]

$$y = x \tan \theta_0 - \frac{g x^2}{2 \left(v_0 \cos \theta_0 \right)^2}$$

where v_0 is the magnitude of the initial velocity and g is gravitational acceleration. The range of the projectile (R, the x-distance where the projectile hits the horizontal surface) is given by:[14]

$$R = \frac{v_0^2 \sin 2\theta_0}{g}$$

Create an animation of the trajectory of a projectile fired at 5 m/s at an angle of 30 degrees. Assume $g = 9.81$ m/s^2.

Try This! Answer 14-4

Figure 14.20 shows the code that produces the animation. Figure 14.21 shows some frames of the animation superimposed on each other. Because we know the range R of the projectile, we know the maximum value of the x-position of the trajectory. Thus, in line 11 of Figure 14.20, we create an array x that holds 51 points from $x = 0$ to $x = R$, inclusive at both ends. With that x array, we can calculate all the values of y using the equation given in the problem.

In lines 15–19 of Figure 14.20, we create the figure and plot frame and the x- and y-axis labels. We set the axes extents to show the entire trajectory and add 5 percent extra at the top of the y-axis extent to enable us to see the entire outline of the marker representing the projectile.

In lines 23–30 of Figure 14.20, we loop through each of the x and y positions, plotting each using a black circle. In the first iteration, we create the tuple of Line2D objects lines by plotting out the first point. For subsequent iterations, we update the point represented by the Line2D object by setting the x- and y-data to new values. In line 29 we render the plot, and in line 30, we pause 0.1 s between each frame of our animation.

Try This! 14-5 Flexible Functions and Dictionaries: Trajectory of a Projectile

Write a function plot_trajectory to plot the trajectory of a projectile and write the plot out to the file *trajectory.png*. The function should accept the x- and y-positions along the trajectory as positional input parameters. The figure should have an overall title that is set by keyword input parameter title. The function should use the **args/**kwds to enable the passing of additional keyword input parameters to control the plotting. Using the calculated trajectory

---

[13] Resnick and Halliday (1977, p. 57).
[14] Resnick and Halliday (1977, p. 58).

from Try This! 14-4, call `plot_trajectory` to make a plot, demonstrating the use of keyword input parameters that are not explicitly defined in `plot_trajectory`.

```
1   import numpy as np
2   import matplotlib.pyplot as plt
3
4   v_0 = 5.0          #- in m/s
5   theta_0 = 30.0  #- in degrees
6   g = 9.81            #- in m/s^2
7
8   theta_0_rad = theta_0 * np.pi / 180.0
9   R = (v_0**2) * np.sin(2.0 * theta_0_rad) / g
10
11  x = np.arange(51) / 50.0 * R
12  y = (x * np.tan(theta_0_rad)) - \
13      (g * x**2 / (2.0 * (v_0 * np.cos(theta_0_rad))**2))
14
15  fig, ax = plt.subplots(1, 1)
16  ax.axis([np.min(x), np.max(x),
17          np.min(y), np.max(y) + (0.05 * np.max(y))])
18  plt.xlabel("X-Position [m]")
19  plt.ylabel("Y-Position [m]")
20
21  plt.ion()
22
23  for i in range(np.size(x)):
24      if i == 0:
25          lines = ax.plot(x[i], y[i], 'ko', markersize=10.0)
26      else:
27          lines[0].set_xdata(x[i])
28          lines[0].set_ydata(y[i])
29      plt.draw()
30      plt.pause(0.1)
```

Figure 14.20 Code to create an animation of the trajectory of a projectile, as described in Try This! 14-4. A plot showing some frames of the animation superimposed on each other, and thus part of the trajectory, is given in Figure 14.21.

Try This! Answer 14-5

Assume that the arrays x and y are already defined as in the answer to Try This! 14-4. Given that, Figure 14.22 shows the code to define and call `plot_trajectory`, and the plot generated by that call is given in Figure 14.23. Note that in lines 9–13 of Figure 14.22, we remove the `title` keyword input parameter (if defined) from the keywords dictionary passed into the `plot` call because that keyword is not recognized by the pyplot `plot` command. If the `title` keyword input parameter is not defined in the call, the local variable `title` is set to a default value, `'Trajectory'`.

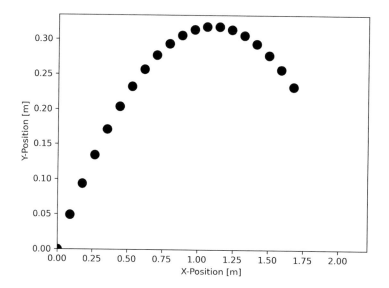

Figure 14.21 Plot showing some frames of the animation of Try This! 14-4 superimposed on each other.

```
1   import matplotlib.pyplot as plt
2
3   def plot_trajectory(*args, **kwds):
4       xdata = args[0]
5       ydata = args[1]
6
7       modified_kwds = kwds.copy()
8
9       if 'title' in modified_kwds:
10          title = modified_kwds['title']
11          del(modified_kwds['title'])
12      else:
13          title = 'Trajectory'
14
15      plt.figure()
16      plt.plot(xdata, ydata, **modified_kwds)
17      plt.xlabel('X-Position')
18      plt.ylabel('Y-Position')
19      plt.title(title)
20      plt.savefig('trajectory.png', dpi=300)
21
22  plot_trajectory(x, y, linestyle='', markersize=10.0, marker='*', color='k')
```

Figure 14.22 Code to define and call `plot_trajectory`, as described in Try This! 14-5. The plot generated is given in Figure 14.23.

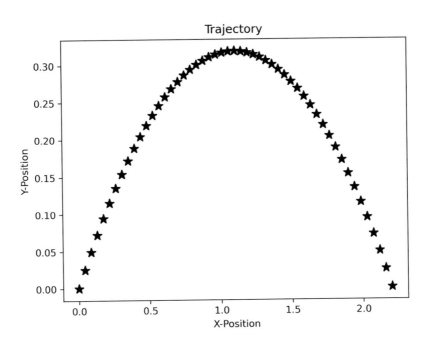

Figure 14.23 Plot generated by the code in Figure 14.22.

In the call to `plot_trajectory` in line 22, none of the keywords used has been defined in `plot_trajectory`, but all are understood by `plot` to create the graph in Figure 14.23.

14.4 More Discipline-Specific Practice

The Discipline-Specific Jupyter notebooks for this chapter cover the following topics:

- Introduction to Matplotlib's object API.
- Line and shaded contour plots.
- Using cartopy to overlay maps.
- Basic animation using Matplotlib.
- Flexible functions and dictionaries.

14.5 Chapter Review

14.5.1 Self-Test Questions

Try to do these without looking at the book or any other resources or using the Python interpreter. Answers to these Self-Test Questions are found at the end of the chapter. If not otherwise stated, assume NumPy has been imported earlier as `np` and Matplotlib's pyplot module has been imported earlier as `plt`.

Self-Test Question 14-1

Describe the differences between a `Figure`, `Axes`, and `Line2D` object.

Self-Test Question 14-2

What do pyplot functions `subplot` and `subplots` do and how do they differ from each other?

Self-Test Question 14-3

What is the difference between normalized and data coordinates?

Self-Test Question 14-4

List and describe some `Axes` object methods.

Self-Test Question 14-5

What are some circumstances to use the Matplotlib object API instead of the pyplot interface?

Self-Test Question 14-6

Define a contour line plot and a filled contour plot. What are the Matplotlib commands or methods to create these kinds of plots?

Self-Test Question 14-7

What are some ways of setting the color map for a filled contour plot?

Self-Test Question 14-8

Summarize the steps to creating a contour line plot or filled contour plot overlain on a map.

Self-Test Question 14-9

What are the steps to creating a basic animation of images? Of a line plot?

Self-Test Question 14-10

Pretend we have the following dictionary describing measurements and metadata from observations of wildlife in a meadow:

```
data = {'hour_in_day':   np.array([9, 11, 13, 15, 17]),
        'bird_count':    np.array([3,  5,  2,  2,  4]),
        'rabbit_count':  np.array([2,  1,  0,  0,  3]),
        'location':      '123 Main St.',
        'date':          'May 1, 2020'}
```

Hour 9 denotes 9:00 a.m. and hour 13 denotes 1:00 p.m. Using this dictionary, create a plot with bird and rabbit counts on the *y*-axis plotted against the hour in the day on the *x*-axis. Add titling to the plot, making use of the metadata in the dictionary.

14.5.2 Chapter Summary

In this chapter, we examined how to make contour plots and basic animation using Matplotlib. Along the way, we considered more advanced ways of using Matplotlib as well as ways of using dictionaries to write more flexible functions. Specific topics we covered in this chapter include:

- Introduction to Matplotlib's object API

  - The `Figure` object describes the page/window that holds one or more graphs. The pyplot `figure` and `subplots` commands are ways of creating `Figure` objects.
  - The `Axes` object represents the graph or plot itself.
  - A `Figure` object's `add_axes` method enables us to place an arbitrary number of `Axes` objects at arbitrary locations on the page/window.
  - The pyplot command `gca`, given certain specifications, gets an existing `Axes` object or creates a new `Axes` object.
  - The pyplot command `axes` creates an `Axes` object.
  - An `Axes` object contains methods to create the curves or shading of the graph or plot. These include `contour`, `contourf`, and `plot` for contour line plots, filled contour plots, and *x*–*y* line plots, respectively.
  - A `Line2D` object describes a line or curve on a plot.
  - An `AxesImage` object describes an image.
  - The pyplot interface to Matplotlib is used for "standard" kind of plots that do not require a lot of customization. The Matplotlib object API gives tools to create highly customized plots.

- Line and shaded contour plots

  - The pyplot `contour` and `contourf` commands create contour line plots and filled contour plots, respectively. These come as methods to `Axes` objects as well as pyplot commands.
  - The `levels` keyword input parameter specifies which levels to contour.
  - The `cmap` keyword input parameter specifes the color map to use with filled contour plots.
  - The `colors` keyword input parameter sets the color of the contour lines.
  - The pyplot `clabel` command puts labels onto contour lines.
  - The pyplot `colorbar` command places a key to the color bar onto the plot.
  - The NumPy `meshgrid` function creates two-dimensional arrays that hold the *x*- and *y*-values in a regular, Cartesian-like two-dimensional grid.

- Using cartopy to overlay maps

  - We can overlay maps by setting a map projection when we create the `Axes` object, customizing the map as desired, and then calling the plotting methods (e.g., `contour` and `contourf`) of the `Axes` object to place the contour plot onto the map.
  - When calling `contour` or `contourf` methods attached to an `Axes` object with a display map projection, we use the `transform` keyword input parameter to specify the projection the data being passed into the call are defined on.
  - Plot customization methods such as `set_extent`, `set_xticks`, etc., also need to specify, using the `crs` keyword input parameter, the projection the data being used in the customization are defined on.
  - `LongitudeFormatter` and `LatitudeFormatter` objects enable us to customize the formatting of longitude and latitude labels on the plot.

- Basic animation using Matplotlib

  - We can create basic animations in Matplotlib by iterating through plots on a figure.
  - For images and lines, we can update the image or line that is plotted by updating the data for the image or line.
  - For images described by `AxesImage` objects, the `set_data` method updates the image displayed with new data.
  - For lines described by `Line2D` objects, the `set_xdata` and `set_ydata` methods update the x- and y-values of the graph.
  - The pyplot `draw` and `pause` commands render the updated graph and pause execution for a specified amount of time, respectively.

- Flexible functions and dictionaries

  - A dictionary is an unordered collection of "values" that are indexed by a "key."
  - Keys can be anything uniquely named and sorted. Strings make good keys. Values can be any Python object or value.
  - Dictionaries are delimited by curly braces. Key:value pairs are separated by commas, and keys are separated from values by colons.
  - For a dictionary `data`, the syntax `data['title']` returns the value associated with the key `'title'`.
  - An assignment statement using a dictionary such as:

    ```
    data['title'] = 'Experiment 1'
    ```

    (where `data` is a dictionary) sets the value of the key:value pair associated with the key `'title'` to the string `'Experiment 1'`. If that key is not currently in the dictionary, the key is added to the dictionary with the value being `'Experiment 1'`.
  - The `del` command can be used to remove key:value pairs from a dictionary.
  - The `*args/**kwds` mechanism uses Python tuples and dictionaries to create and call functions using collections of positional and keyword input parameters whose composition may change with use.

14.5.3 Self-Test Answers

Self-Test Answer 14-1

A `Figure` object represents the page or window a graph or graphs are on. An `Axes` object represents a graph or plot. A `Line2D` object represents a curve on a graph.

Self-Test Answer 14-2

Both functions create an `Axes` object(s) in the context of a single or multipanel figure. With `subplot`, the `Axes` object of a single panel is created and returned. With `subplots`, multiple `Axes` objects are created and returned.

Self-Test Answer 14-3

Both coordinate systems are ways of describing a location in the plot frame. Normalized coordinates range from zero to one across each of the axes. Data coordinates range from the minimum data value of each axis to the maximum data value of each axis.

Self-Test Answer 14-4

A number of the pyplot plotting functions are methods of `Axes` objects. The `contour` method creates a contour line plot. The `contourf` method creates a filled contour plot. The `plot` method creates an x–y line plot. The `imshow` method puts an image on the axes.

Self-Test Answer 14-5

The object interface gives us more fine-grained control. When we wish to create a plot using that control, such as for arbitrary placement of plot elements, the object framework can be the better interface to use.

Self-Test Answer 14-6

A contour line plot is a contour plot that includes only the contour lines. A filled contour plot shades the regions between contour levels. The `contour` command creates a contour line plot, and the `contourf` command creates a filled contour plot.

Self-Test Answer 14-7

The ways below all set the `cmap` keyword input parameter in a call to `contourf`. The first way is to set `cmap` to an instance of a color map class, such as a `LinearSegmentedColormap` object. Predefined color map objects are in the cm module. A second way is to set `cmap` to the return from the pyplot `get_cmap` function. Finally, we can set `cmap` to a string that corresponds to the name of the predefined color map.

Self-Test Answer 14-8

First, we create an `Axes` object while setting the `projection` keyword input parameter to the display map projection. Second, we customize the map with features, grid lines, etc.

Finally, we create the contour plot using the appropriate Matplotlib commands or methods. When creating the plots, we specify the map projection the data are on via the `transform` keyword input parameter to `contour` and `contourf`.

Self-Test Answer 14-9

In both cases, we set up a loop and iterate through the images or *x*- and *y*-axis values for the line plot. In the case of images, we use the `Axes` object's `imshow` method to show the first `AxesImage` object and then use the method `set_data` attached to that object to update the image as the loop iterates. In the case of a line plot, we use the `Axes` object's `plot` method to create the first line plot to show. In subsequent iterations, we use the `set_xdata` and `set_ydata` methods of the `Line2D` object returned by the `plot` call to update the line plot. After the image or line data are updated, we call the pyplot command `draw` to render the new animation "frame" and call `pause` to wait a little time before the next animation frame.

Self-Test Answer 14-10

This code solves the problem:

```
1   plt.figure()
2   plt.plot(data['hour_in_day'], data['bird_count'],
3           'ro-', label="Birds")
4   plt.plot(data['hour_in_day'], data['rabbit_count'],
5           'bs--', label="Rabbits")
6   plt.title('Animal Counts at ' + data['location'] + \
7           ' on ' + data['date'])
8   plt.xlabel('Hour in Day')
9   plt.ylabel('Counts')
10  plt.legend()
```

15 Handling Missing Data

Although experiments and data are crucial to the fields of science and engineering, scientists and engineers also know experiments and data are messy. The number of ways data collection can fail – people forget to turn on the right switch at the right time, insects fly into critical components of an experiment, broken instruments, etc. – are myriad. Data-analysis routines then need to handle the cases of missing or bad data.

In this chapter, we examine some Python tools for handling missing or bad data (from here, for brevity, we refer to both kinds of data as "missing" data). All three of the approaches we examine use the same methodology: Within an array of values, define a "flag" that lets the user know if a given value is missing. How these approaches differ is the way that flag is defined and how we use that flag. In the first approach, we set aside a potential data value as representing a missing value and use boolean arrays or expressions to process the data. In the second approach, we define a special floating-point value – the Institute of Electrical and Electronics Engineers (IEEE) Not a Number (NaN) value – to represent missing data and define a special array-like class (`Series` for one-dimensional collections) that understands how to process data using that convention.[1] The third approach uses a special kind of an array, a **masked array**, that contains a mask in addition to the data to describe which elements contain missing values or not. We then use functions that understand masked arrays to process the contents.

15.1 Example of Handling Missing Data

Pretend we make the following observations of the number of ducks at a pond every hour over the course of a day, from 7:00 a.m. to 5:00 p.m.:

Hour in the day	Number of ducks
7	5
8	3
9	4
10	—
11	2
12	3
13	6
14	—
15	7
16	6
17	7

---

[1] The IEEE is an electrical engineering professional society that sets standards for electronic and computational devices.

```
1   import numpy as np
2   import matplotlib.pyplot as plt
3
4   #- Define data:
5   hour_in_day = np.arange(11) + 7
6   duck_count = np.array([5, 3, 4, -999, 2, 3, 6, -999, 7, 6, 7])
7
8   #- Calculate mean:
9   num_ok_values = np.sum(duck_count >= 0)
10  sum_ok_values = np.sum( np.where(duck_count >= 0, duck_count, 0) )
11  print("Mean: " + str(sum_ok_values / num_ok_values))
12
13  #- Select non-missing data for plotting:
14  ok_pts = np.where(duck_count >= 0)
15  x = hour_in_day[ok_pts]
16  y = duck_count[ok_pts]
17
18  #- Plot graph:
19  plt.figure()
20  plt.plot(x, y, 'ko-')
21  plt.title("Approach 1:  Boolean Arrays")
22  plt.xlabel("Hour in the Day")
23  plt.ylabel("Count")
24  plt.savefig("missing_data1.png")
```

Figure 15.1 Code to calculate the hourly mean count of ducks and plot the count versus hour in the day for Approach 1 in Section 15.1. The plot generated is shown in Figure 15.4.

As the table shows, there are two hours, at 10 a.m. and 2 p.m., when we were unable to make measurements. We want to calculate the hourly average number of ducks over this period in the day and plot the number of ducks versus the hour in the day. The problem is what to do with the two hours for which we have no data. We cannot assume the number of ducks during those hours was zero or some other number, because we did not make such a measurement. In the plot, we need to somehow leave those points out – preferably representing those values as blank spaces – and in the calculation of the average, we also need to ignore those times.

Below, we solve this problem using three different approaches:

Approach 1 Define a data value as missing and process the data using boolean arrays or expressions (Figure 15.1).

Approach 2 Use IEEE NaN values to represent missing values, create a `Series` object, and process the data using functions that work on such objects (Figure 15.2).

Approach 3 Use masked arrays to represent missing values and process the data using functions that work with such objects (Figure 15.3).

All three approaches produce the same result for the average, 4.7778 (if truncated to four decimal places). The graphs that result from Approaches 1 and 2 are shown in Figure 15.4.

```
1    import numpy as np
2    import matplotlib.pyplot as plt
3    import pandas as pd
4
5    #- Define data:
6    hour_in_day = np.arange(11) + 7
7    duck_count = np.array([5, 3, 4, np.nan, 2, 3, 6, np.nan, 7, 6, 7])
8    duck_count = pd.Series(duck_count)
9
10   #- Calculate mean:
11   print("Mean: " + str(duck_count.mean()))
12
13   #- Plot graph:
14   plt.figure()
15   plt.plot(hour_in_day, duck_count, 'ko-')
16   plt.title("Approach 2:  Use IEEE NaNs")
17   plt.xlabel("Hour in the Day")
18   plt.ylabel("Count")
19   plt.savefig("missing_data2.png")
```

Figure 15.2 Code to calculate the hourly mean count of ducks and plot the count versus hour in the day for Approach 2 in Section 15.1. The plot generated is shown in Figure 15.4.

```
1    import numpy as np
2    import matplotlib.pyplot as plt
3    import numpy.ma as ma
4
5    #- Define data:
6    hour_in_day = np.arange(11) + 7
7    duck_count = np.array([5, 3, 4, -999, 2, 3, 6, -999, 7, 6, 7])
8    duck_count = ma.masked_values(duck_count, -999)
9
10   #- Calculate mean:
11   print("Mean: " + str(ma.mean(duck_count)))
12
13   #- Plot graph:
14   plt.figure()
15   plt.plot(hour_in_day, duck_count, 'ko-')
16   plt.title("Approach 3:  Use Masked Arrays")
17   plt.xlabel("Hour in the Day")
18   plt.ylabel("Count")
19   plt.savefig("missing_data3.png")
```

Figure 15.3 Code to calculate the hourly mean count of ducks and plot the count versus hour in the day for Approach 3 in Section 15.1. The plot generated is the same as the Approach 2 plot shown in Figure 15.4, except with a different plot title.

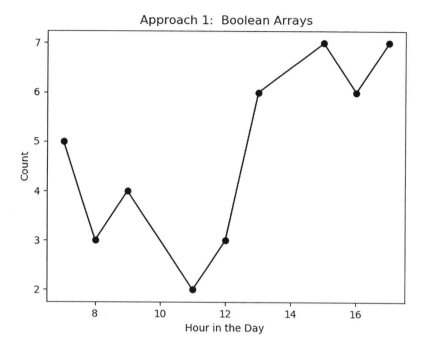

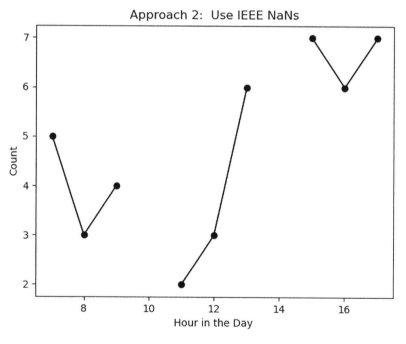

Figure 15.4 Plots generated by the code in Figure 15.1 (Approach 1, top) and Figure 15.2 (Approach 2, bottom).

The graph resulting from Approach 3 is the same as Approach 2, except with a different title, so it is not shown. As with previous chapters, we save our detailed discussion of why the code works the way it does for Section 15.2. For now, we briefly describe how the various parts of each code block functions.

In Figure 15.1, we start by importing the NumPy and Matplotlib pyplot modules (lines 1–2). In lines 5–6, we create the arrays holding the hours in the day and the counts of ducks. For the missing values, we set the count to −999, because there can be no negative counts. In lines 9–11, to calculate the mean, we separate out the calculation of the number of nonmissing values (`num_ok_values`) from the sum of all the nonmissing duck counts (`sum_ok_values`). We make use of boolean arrays and functions that act on boolean arrays to make these calculations. The mean is then the sum of all the nonmissing duck counts divided by the number of nonmissing duck counts. For plotting, we extract only those points that are nonmissing (lines 14–16). As a result, in the plot generated using those points (lines 19–24), although no marker is shown at the times where there the count is missing (hours 10 and 14), the line that connects the markers together still goes through those hours, so the graph (Figure 15.4) does not truly show the data are missing at those times.

In Figure 15.2, in addition to importing NumPy and pyplot, we also import the pandas package (line 3); pandas is a Python package used in data analysis, especially of tabular data. When we create the array holding the count of ducks (line 7), instead of using −999 as the missing value, we use the NumPy module attribute `np.nan`. This is a special floating-point number that is used to designate NaN. In line 8, we convert the array `duck_count` to a pandas `Series` object. In doing so, when in line 11 we calculate the mean of `duck_count` using the `mean` method attached to `Series` objects, the calculation automatically ignores the missing values. Likewise, when we plot `duck_count` as a `Series` object, the missing values are left blank with no connecting lines going through those times (Figure 15.4).

In Figure 15.3, in addition to importing NumPy and pyplot, we also import the ma module of NumPy (line 3). The ma module contains functions and classes for creating and manipulating masked arrays. As with Figure 15.1, we use −999 in line 7 to denote missing values. In line 8, we convert the NumPy array `duck_count` into a masked array. Thus, in line 11, when we calculate the mean of `duck_count` using the ma function `mean`, the calculation automatically ignores the missing values. Here too, when we plot `duck_count` as a masked array object, the missing values are left blank with no connecting lines going through those times.

15.2   Python Programming Essentials

In this section, we describe in more detail how the three approaches in Section 15.1 work. In the rest of the code examples in this section, assume NumPy has been imported as `np`, the Matplotlib pyplot submodule has been imported as `plt`, pandas has been imported as `pd`, and the ma submodule of NumPy has been imported as `ma`.

15.2.1 Approach 1: Define a Data Value as Missing and Process with Boolean Arrays or Expressions

In Section 13.2.2, we saw how boolean arrays coupled with array syntax and the `where` function enable us to manipulate values in an array depending on whether certain criteria are met. In Figure 15.1, we use these techniques to process only the nonmissing values. Here, we go through lines from this figure (excepting comment lines) and examine the contents of different variables, to see how the calculations are made. We also describe how we can duplicate this approach using looping rather than array syntax.

Consider lines 1–7 of Figure 15.1 (reproduced here for convenience). If we print out the contents of the condition in line 9 instead of executing line 9:

```
In [1]:  import numpy as np
         import matplotlib.pyplot as plt

         hour_in_day = np.arange(11) + 7
         duck_count = np.array([5, 3, 4, -999, 2, 3, 6, -999, 7, 6, 7])

         print(duck_count >= 0)
```

```
[ True  True  True False  True  True  True False  True  True  True]
```

we find that the condition is a boolean array where the elements corresponding to a missing value (a negative value) in `duck_count` are `False` but all other elements are `True`. As a boolean array, when used in arithmetic calculations, behaves as if `True`s equal one and `False`s equal zero, using the NumPy `sum` function on this boolean array will return the number of `True` elements, i.e., the number of nonmissing values.

If we execute lines 1–9 in Figure 15.1 and print out the argument to the `sum` call in line 10:

```
In [1]:  print(np.where(duck_count >= 0, duck_count, 0))
```

```
[5 3 4 0 2 3 6 0 7 6 7]
```

the result is the values of the `duck_count` array, with all missing values set to zero. When this array is summed, the missing values contribute nothing to the total. Thus, line 10 in Figure 15.1 gives us the total of all nonmissing values in `duck_count`. Line 11 gives us the hourly mean counts using only the nonmissing values.

If we continue on in the program, executing up to and through line 14 in Figure 15.1, we can inspect the contents of `ok_pts`:

```
In [1]:   print(ok_pts)

          (array([ 0,   1,   2,   4,   5,   6,   8,   9,  10]),)
```

which is a tuple with one element. That element is an array that lists the indices of all the elements of `duck_count` which are not negative, i.e., that are nonmissing. In lines 15 and 16, we use `ok_pts` to select those elements in `hour_in_day` and `duck_count` that are nonmissing, and save just those values to the arrays x and y. Thus, when we plot those values in line 20, only the nonmissing values are plotted (though the line connecting the points goes through the missing value hours).

In the Figure 15.1 example, the data are all integers, so if we had to do an equality test, it would work exactly. If the array elements are floating point, we would need to test for "closeness" rather than equality. For instance, if we want a boolean array where `True` corresponds to the missing values (e.g., where a floating-point version of `duck_count` is close to -999), the NumPy function `isclose` does what we want:

```
In   [1]:   duck_count = np.array([5, 3, 4, -999, 2, 3, 6, -999, 7, 6, 7], dtype='d')
            print(np.isclose(duck_count, -999))

            [False False False  True False False False  True False False False]
```

Calling the NumPy function `logical_not` on this array will turn all the `Falses` to `Trues`, and vice versa, and thus would produce a boolean array where `True` corresponds to nonmissing values.

In our Figure 15.1, we did not consider operations using the `duck_count` array, but the use of boolean arrays and array syntax can also handle that situation. For instance, say we wanted to convert our hourly counts into per-minute equivalents by dividing by 60. We can use the `where` function to do this calculation only on nonmissing values. Elements corresponding to missing value times (i.e., where `duck_count  >=  0` is `False`) would be set to -999:

```
In  [1]:   duck_count_per_min = np.where(duck_count >= 0,
                                          duck_count / 60.0, -999)
           print(duck_count_per_min)

           [ 8.33333333e-02  5.00000000e-02  6.66666667e-02 -9.99000000e+02
             3.33333333e-02  5.00000000e-02  1.00000000e-01 -9.99000000e+02
             1.16666667e-01  1.00000000e-01  1.16666667e-01]
```

For conditions that are more complex than can be expressed using array syntax, we need to use explicit looping using `for` and/or `while` loops, coupled with `if` and other branching statements, to find the missing and nonmissing values and process each accordingly. For instance, the following code will duplicate the functionality of lines 9–16 in our Figure 15.1 example (although the calculations are done in a slightly different order) as well as calculate `duck_count_per_min`:

```
1   x = []
2   y = []
3   duck_count_per_min = np.zeros(np.size(duck_count) - 999.0
4   for i in range(np.size(hour_in_day)):
5       if duck_count[i] >= 0:
6           x.append(hour_in_day[i])
7           y.append(duck_count[i])
8           duck_count_per_min[i] = duck_count[i] / 60.0
9   x = np.array(x)
10  y = np.array(y)
11
12  num_ok_values = np.size(y)
13  sum_ok_values = np.sum(y)
14  print("Mean: " + str(sum_ok_values / num_ok_values))
```

We use lists (lines 1–10) to hold the nonmissing values of `x` and `y` because we do not know before entering the loop how many missing values there will be. In line 3, we prefill `duck_count_per_min` with −999 values so only the nonmissing elements are replaced with the correct values within the `for` loop. In lines 12–14, we duplicate the same functionality as in Figure 15.1, but in our `for` loop solution, we could have used the NumPy mean function, as `x` and `y` have no missing values to skew the calculation. In this looping solution, there is only one branching statement, but the granular control a loop provides can more easily accommodate highly complex cases than using boolean arrays with array syntax.

15.2.2 Approach 2: Use `Series` and IEEE NaN Values

The pandas package has two main classes, the `Series` and `DataFrame` classes, that are "enhanced" versions of one-dimensional and two-dimensional (respectively) NumPy arrays. Each row of a `Series` object, and each row and column in a `DataFrame` object, can be assigned labels beyond the integer indices NumPy arrays have. The package provides tools operating on objects of these two classes that can accomplish very sophisticated analysis in very few lines of code. Handling missing values is only one task we can use the package for. In Chapters 18 and 19, we discuss additional features of pandas. For now, we focus on how these pandas objects handle missing values. Our discussion focuses on `Series` objects, but `DataFrame` objects handle missing objects, for the two-dimensional case, the way `Series` objects do.

By default, when converting a NumPy array (or similar collection of values) into a `Series` object, pandas assumes the IEEE NaN floating-point value represents a missing value.[2] The statistical and plotting functions written to operate on `Series` objects automatically handle missing data appropriately. Because the IEEE NaN value is stored in NumPy as the attribute `nan`, in line 7 of Figure 15.2, when we create the NumPy array holding the duck counts, every element of missing data is set to `np.nan`. After the NumPy array is converted to a pandas `Series` object in line 8, the call to the `Series` object method `mean` (line 11) calculates the mean of the data in the object, excluding all the NaN elements.

The `np.nan` is a floating-point number like any other floating-point number. Why, then, is it not enough to create a NumPy array with the missing values set to NaN (as in line 7)? Why do we have to convert the array to a `Series` object? If we use the NumPy `average` function on a NumPy array with NaN values:

```
In [1]:  duck_count = np.array([5, 3, 4, np.nan, 2, 3, 6, np.nan, 7, 6, 7])
         print(np.average(duck_count))
```

```
nan
```

To a NumPy function or operator, a NaN value is literally "not a number." Thus, any function or operator acting on a NaN also produces "not a number." The `Series` object, however, interprets the NaN value as "missing." As a result, functions that operate on `Series` objects compensate accordingly. This includes plotting using Matplotlib: NaN elements are not plotted out and connecting lines do not intersect with the *x*-locations of those elements (as the second panel of Figure 15.4 shows). Note that NaN is a floating-point number. It is not the same type as `None`.

We can also see this behavior when doing operations, such as in the calculation of duck counts per minute as we did in Section 15.2.1. Assuming `duck_count` is a `Series` object, we can calculate `duck_count_per_min` with the following:

```
In [1]:  duck_count_per_min = duck_count / 60.0
         print(duck_count_per_min)
```

```
0        0.083333
1        0.050000
2        0.066667
3             NaN
4        0.033333
5        0.050000
6        0.100000
7             NaN
8        0.116667
9        0.100000
10       0.116667
dtype: float64
```

---

2 https://pandas.pydata.org/pandas-docs/stable/user_guide/missing_data.html (accessed June 20, 2020).

Values in `duck_count` that are NaN are NaN in `duck_count_per_min`. `Series` objects are printed in a different format than NumPy array objects.

The data for a `Series` object are stored as an attribute `values`. If we want to "convert" the `Series` object into a NumPy array, we can: (1) use the `fillna` method which returns a copy of the object with the NaN values replaced by another value, then (2) access the `values` attribute of that `Series` object directly (output reformatted for better display):

```
In [1]:  filled_duck_count = duck_count.fillna(-999)
         filled_duck_count.values

Out[1]:  array([   5.,      3.,      4.,  -999.,     2.,     3.,     6.,
               -999.,     7.,      6.,      7.])
```

For more information on the pandas package, besides Chapters 18 and 19, see the online documentation. We provide an up-to-date link to the documentation at www.cambridge.org/core/resources/pythonforscientists/refs/, ref. 46.

15.2.3 Approach 3: Use Masked Arrays

The NumPy package submodule, ma, defines a masked array, which is an object that contains both array data and a mask that tells us which elements of the accompanying array are missing or not. The mask is the same shape as the underlying array and is a boolean array. When the mask element is `True`, the corresponding element in the data array is missing (i.e., that element should be "masked out"). If the mask element is `False`, the element is not missing (i.e., is valid).

As shown in line 3 of Figure 15.3, to use the masked array features we have to first import the ma submodule. To create a masked array, we can use any of a number of functions from ma. In line 8 of Figure 15.3, we use the `masked_values` function, which we use with two positional input parameters: the first is the data array and the second gives what value in the data array corresponds to missing (i.e., to `True` in the mask). The contents of the masked array object show both the data and the mask (output reformatted for better display):

```
In [1]:  duck_count = np.array([5, 3, 4, -999, 2, 3, 6, -999, 7, 6, 7])
         duck_count = ma.masked_values(duck_count, -999)
         duck_count

Out[1]:  masked_array(data=[5, 3, 4, --, 2, 3, 6, --, 7, 6, 7],
                  mask=[False, False, False,  True, False, False, False,
                         True, False, False, False],
             fill_value=-999)
```

As far as the masked array is concerned, the elements of the accompanying data that are masked out are truly missing, as is shown by the – symbol for those elements in the output above. Note, if the data array is floating point, `masked_values` will use "close to in value" to decide which elements are to be masked.[3]

---
[3] https://numpy.org/doc/1.18/reference/maskedarray.generic.html (accessed June 20, 2020).

Similar to the `Series` object discussed in Section 15.2.2, operations using masked arrays handle the masked-out values and compensate accordingly. Functions using masked arrays ignore the missing values (as appropriate) in their calculations. When plotting using Matplotlib, masked-out values are not plotted and connecting lines will not intersect with the *x*-locations of those elements. Operations with a masked-out value will yield a masked-out value, as we can see in a calculation of duck counts per minute similar to what we did in Section 15.2.1. Assuming `duck_count` is a masked array, we can calculate `duck_count_per_min` with the following (output reformatted for better display):

```
In [1]:  duck_count_per_min = duck_count / 60.0
         duck_count_per_min
```

```
Out[1]:  masked_array(data=[0.08333333333333333, 0.05, 0.06666666666666667,
                     --, 0.03333333333333333, 0.05, 0.1, --,
                     0.11666666666666667, 0.1, 0.11666666666666667],
                 mask=[False, False, False,  True, False, False, False,
                     True, False, False, False],
             fill_value=-999)
```

The result of an operation using a masked array is another masked array.

Masked arrays also have a `fill_value` attribute. If a masked array is converted into a regular NumPy array, the masked-out elements are replaced with `fill_value`. The `filled` method returns such a NumPy array:

```
In [1]:  print(duck_count.filled())
```

```
[   5    3    4 -999    2    3    6 -999    7    6    7]
```

The `get_fill_value` method returns the current `fill_value` for the masked array, and the `set_fill_value` method will set the `fill_value` of the masked array.

For more information on masked arrays, see the online documentation. We provide an up-to-date link to the documentation at www.cambridge.org/core/resources/pythonforscientists/refs/, ref. 38.

15.2.4 Which Approach Is Better?

In general, using `Series` or masked arrays is usually the better choice. For many calculations and operations, they seamlessly handle missing values. Between these, the pandas classes are probably better for most one- and two-dimensional cases, as the package has additional features that masked arrays do not support. For higher-dimensional cases, it is not as clear that pandas classes have an edge. If the problem requires manipulation of array elements one by one, by looping, although it is possible to loop through the elements in `Series` and masked array objects, it is not recommended. These objects are not meant to be used that way and thus do not perform as well in that mode. That does not mean the problem then *has* to be solved using regular NumPy arrays, but that might be the better approach to take. In the end, whatever solution we choose is a balance of code readability and extensibility, performance, and memory usage. For the most difficult cases, we may have to try different approaches on some limited test data to see which is more likely to work better.

The approaches we have discussed in this section are not the only ones for dealing with missing values, but the strategies we described are common ones. Different branches of science and engineering have different needs with regards to data, including how to handle missing data, and there are various efforts to develop packages that meet those specific needs. Although we will not discuss those efforts in detail in the present work, researchers in fields with such specialized programming tools will probably find functions and classes to handle missing data that are tailored for their particular needs.

15.3 Try This!

The examples in this section address the main topics of Section 15.2. We practice using the three different approaches of handling missing values and examine what results when using Matplotlib to plot the missing and nonmissing values.

For all the exercises in this section, assume NumPy has been imported as np, the Matplotlib pyplot submodule has been imported as plt, pandas has been imported as pd, and the ma submodule of NumPy has been imported as ma.

Try This! 15-1 Approach 1 Calculations (Array Syntax): Melting Ice

Pretend a student is doing an experiment to find out how long it takes to melt an ice cube at room temperature. The student takes a tray of eight ice cubes from the freezer, puts each cube out on separate plates, and waits for them to melt. The student records the time each ice cube melts:

Ice cube number	Time to melt
1	130 min
2	135 min
3	141 min
4	142 min
5	142 min
6	144 min
7	150 min
8	150 min

A few hours later, the student finds that the watch has not advanced at all and that the watch is stuck. The student concludes the last two data points are unreliable. As part of the lab report on this experiment, though, the student wants to calculate how much these two data points affect the average time to melt.

Using Approach 1 from Section 15.1, calculate the average time to melt with and without the bad data points. Do this using array syntax. This problem can be solved using array slicing, because the bad data points are at the end of the array, but try to do it without using array slicing.

Try This! Answer 15-1

This code solves the problem:

```
1  melt_time = np.array([130, 135, 141, 142, 142, 144, 150, 150])
2
3  good_data = np.logical_not(np.isclose(melt_time, 150))
4  avg_time_with_bad_data = np.average(melt_time)
5  avg_time_without_bad_data = np.sum(melt_time * good_data) / \
6                              np.sum(good_data)
7  print(avg_time_with_bad_data - avg_time_without_bad_data)
```

and the difference that is printed out is 2.75.

Because the `melt_time` array's values are all integers, we could have used strict equality in line 3 to check whether the value of `melt_time` equaled 150 min, but to make the code more generalizable regardless of whether the time is an integer or floating point, we use NumPy's `isclose` function (see Section 8.2.7 for details).

The `good_data` array is `True` for all good data points and `False` for all bad data points. Thus, in line 5, when we multiply `melt_time` by `good_data`, only the good data-point values are nonzero. The bad data values will become zero and will not contribute to the sum. The sum of `good_data` gives the total number of good data points, because the `False` values in that array are converted to zeros in the summing of that array (line 6).

Try This! 15-2 Approach 1 Calculations (Loop): Melting Ice

Redo Try This! 15-1 by using a loop instead of array syntax.

Try This! Answer 15-2

This code solves the problem:

```
1  melt_time = np.array([130, 135, 141, 142, 142, 144, 150, 150])
2
3  melt_time_good = []
4  for i in range(np.size(melt_time)):
5      if not np.isclose(melt_time[i], 150):
6          melt_time_good.append(melt_time[i])
7
8  avg_time_with_bad_data = np.average(melt_time)
9  avg_time_without_bad_data = np.average(melt_time_good)
10 print(avg_time_with_bad_data - avg_time_without_bad_data)
```

We use a list for `melt_time_good` because (in theory) we do not know ahead of time how many good data points there will be. The NumPy `average` function will work on lists of numbers (line 9), not just arrays of numbers (line 8), so we do not have to convert `melt_time_good` to an array before passing it into `average`.

Try This! 15-3 Approach 2 Plot: The Mouse Maze

Consider the mouse maze data that we took as part of Try This! 5-1. We now realize our timer was broken and that all times less than 60 s are suspect. Repeat Try This! 5-1 with the bad data removed from the plot and save the plot as *mazetimes_missing2.png*. Do this using Approach 2 from Section 15.1.

Try This! Answer 15-3

This code will solve the problem:

```
1   xdata = np.array([0.0, 0.08, 0.2, 0.37, 0.6, 0.84, 1.02, 1.2])
2   ydata = np.array([105, 98, 54, 50, 65, 81, 182, 210])
3   ydata = np.where(ydata < 60, np.nan, ydata)
4   ydata = pd.Series(ydata)
5
6   plt.plot(xdata, ydata, linewidth=5.0, marker='o', markersize=15.0)
7   plt.axis([0, np.max(xdata), 0, np.max(ydata)])
8   plt.xlabel('Dosage (mg)')
9   plt.ylabel('Time to Complete Maze (sec)')
10  plt.text(0.1, 10, 'Provisional')
11  plt.savefig('mazetimes_missing2.png', dpi=300)
```

We assume pandas has already been imported as pd. Unlike the answer for the original Try This! 5-1, we store our data as NumPy arrays because we use array syntax in line 3 of our solution above to test to see what data points are less than 60 s. The value of those points are set to NaN while the other data values remain unchanged. The plot generated is given in Figure 15.5. In comparison with Figure 5.6, the missing data points are absent in Figure 15.5.

Try This! 15-4 Approach 3 Plot: The Mouse Maze

Repeat Try This! 15-3 using Approach 3 from Section 15.1.

Try This! Answer 15-4

This code creates the necessary masked array:

```
1   xdata = np.array([0.0, 0.08, 0.2, 0.37, 0.6, 0.84, 1.02, 1.2])
2   ydata = np.array([105, 98, 54, 50, 65, 81, 182, 210])
3   ydata = ma.masked_less(ydata, 60)
```

We assume the ma submodule of NumPy has already been imported as ma. The ma module's `masked_less` function creates a masked array where all values in the first input parameter that are less than the value in the second input parameter are considered missing. The `masked_greater` function does the same thing except the missing values are those greater than the value in the second input parameter.

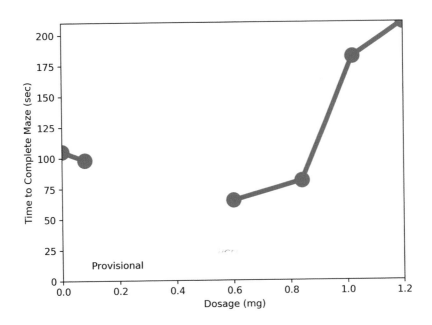

Figure 15.5 Plot generated from the code in the answer to Try This! 15-3.

The code to plot `ydata` versus `xdata` is the same as lines 6–10 of the answer to Try This! 15-3.

Try This! 15-5 Approach 3 Contour Map: Elevations on a Hill

In Try This! 14-2, we made a contour plot of fictitious elevations on a small hill. The values are found in Figure 14.14. Pretend that we now discover all elevations 2 m and greater are incorrect. Make a contour plot of the elevations showing the areas affected by the missing values as blank space. Use Approach 3 from Section 15.1.

Try This! Answer 15-5

The solution to this is nearly exactly the same as in Figure 14.15. This line:

```
elev = ma.masked_greater_equal(elev, 2.0)
```

is added right after line 4 in Figure 14.15 (and the `savefig` filename is changed). The ma module's `masked_greater_equal` function masks out data with values greater than or equal to the value of the second positional input parameter.[4] The `masked_less_equal` does the same for values less than or equal to the value of the second positional input parameter. Matplotlib's contouring routines are able to use masked arrays and turn the areas affected by the missing values into blank areas. The plot is shown in Figure 15.6.

---

[4] https://numpy.org/doc/1.18/reference/generated/numpy.ma.masked_greater_equal.html (accessed June 23, 2020).

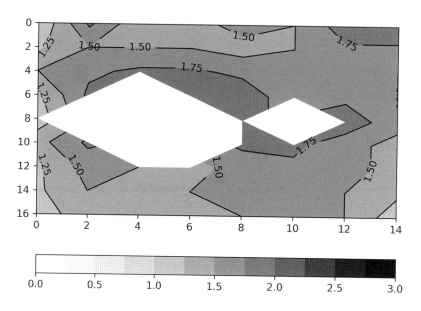

Figure 15.6 Contour plot generated by the solution to Try This! 15-5.

15.4   More Discipline-Specific Practice

The Discipline-Specific Jupyter notebooks for this chapter cover the following topics:

- Define a data value as missing and use boolean arrays and expressions.
- `Series` and IEEE NaN values.
- Masked arrays.

15.5   Chapter Review

15.5.1 Self-Test Questions

Try to do these without looking at the book or any other resources or using the Python interpreter. Answers to these Self-Test Questions are found at the end of the chapter. If not otherwise stated, assume NumPy has been imported as np, the Matplotlib pyplot submodule has been imported as plt, pandas has been imported as pd, and the ma submodule of NumPy has been imported as ma.

Self-Test Question 15-1

What is a boolean array?

Self-Test Question 15-2

Pretend we have a one-dimensional array of measurements called `values`. What is the result of the expression `values >= 0`?

Self-Test Question 15-3

Pretend we have a one-dimensional array `data` where elements that are missing have the value of 1×10^{10}. Give the code to obtain the sum of all the good values in `data`, using Approach 1 in Section 15.1 (both using boolean arrays and array syntax and using looping).

Self-Test Question 15-4

What is an IEEE NaN value?

Self-Test Question 15-5

Solve Self-Test Question 15-3 using Approach 2 in Section 15.1. We did not explicitly discuss how to take a sum of a `Series` object, but based on what we did discuss, make an inference.

Self-Test Question 15-6

What is a masked array?

Self-Test Question 15-7

Solve Self-Test Question 15-3 using Approach 3 in Section 15.1. We did not explicitly discuss how to take a sum of a masked array object, but based on what we did discuss, make an inference.

15.5.2 Chapter Summary

In this chapter, we described three approaches of handling data with missing data. All three approaches define a "flag" that lets the user and applicable functions and operators know if a given value is missing. Specific topics we covered in this chapter include:

- Defining a data value as missing and process with boolean arrays or expressions

  - Boolean arrays can be used to denote which elements of an array match the missing data value and which do not.
  - Boolean arrays can be used arithmetically to select the nonmissing values in a data array and to zero-out the missing values.
  - The NumPy `where` function (in its three-argument form) can be used to turn missing-value elements into another value.
  - The NumPy `where` function (in its one-argument form) can be used to select only those points that are nonmissing.

- The NumPy `isclose` function checks whether a floating-point number is "close to" another floating-point number, thus working around the inherent inaccuracy of the floating-point representation.
- We can apply tests to look for missing value elements through the array syntax operating on boolean arrays or through explicit looping through the elements of the arrays with branching inside the loop(s).
- More complex missing value-related conditions may be more easily expressed through explicit looping rather than array syntax operating on boolean arrays.

- Using `Series` and IEEE NaN values to handle missing values

  - The IEEE NaN is a special floating-point value that denotes "not a number."
  - The pandas `Series` object is an analogue to a one-dimensional array that has additional capabilities, including the proper handling of missing values. The pandas `DataFrame` object is an analogue to a two-dimensional array.
  - A `Series` object can be created through a call to `Series`.
  - In a `Series` object, by default, missing values are represented by the NaN value.
  - Outside of a `Series` or similar object, arithmetic operations on NaN values result in another NaN. So, for instance, it is not possible to calculate the mean of a NumPy array with some NaNs by using the NumPy `average` function.
  - The NumPy `where` function can be used to search for missing values that are defined as a specific data value and replace them with NaN values.
  - Functions written to utilize `Series` objects properly handle missing values, as appropriate. Arithmetic operations using a missing value result in a missing value. Functions that can be calculated ignoring missing values (e.g., the average of the array) are done accordingly.
  - Matplotlib plotting commands properly display the missing values of `Series` objects as blank areas with no marker or connecting line.
  - The `fillna` method of a `Series` object returns a copy of the `Series` object with the NaN values replaced by another value.
  - The `values` attribute of a `Series` object contains the values of the object as a NumPy array.

- Using masked arrays to handle missing values

  - A masked array object holds both the data values of an array and a boolean array mask of the same shape.
  - Elements of the mask in a masked array are `True` if the corresponding data value is "masked out" (i.e., missing) and `False` otherwise (i.e., the corresponding data value is a good data value).
  - Some functions that create masked arrays:

    * `masked_values` Masks out values in the input data array (first positional input parameter) that are close to a given value (specified in the second positional input parameter).

* `masked_greater` Masks out values in the input data array (first positional input parameter) that are greater than a given value (specified in the second positional input parameter).
* `masked_less` Masks out values in the input data array (first positional input parameter) that are less than a given value (specified in the second positional input parameter).
* `masked_greater_equal` Masks out values in the input data array (first positional input parameter) that are greater than or equal to a given value (specified in the second positional input parameter).
* `masked_less_equal` Masks out values in the input data array (first positional input parameter) that are less than or equal to a given value (specified in the second positional input parameter).

- Functions written to utilize masked array objects properly handle missing values, as appropriate. Arithmetic operations using a masked-out value result in a masked-out value. Functions that can be calculated ignoring masked-out values (e.g., the average of the array) are done accordingly.
- Matplotlib plotting commands properly display the missing values of masked array objects as blank areas with no marker or connecting line. For contour plots, areas affected by the missing values are left blank.
- The `filled` method of a masked array returns a NumPy array with all masked-out values replaced by the value stored in the masked array's `fill_value` attribute. The `get_fill_value` and `set_fill_value` methods get or set the value of `fill_value`.

15.5.3 Self-Test Answers

Self-Test Answer 15-1

A boolean array is an array whose elements are either `True` or `False`.

Self-Test Answer 15-2

The result is a one-dimensional boolean array of the same size as `values`. Every element in the boolean array will be `True` if the corresponding element in `values` is greater than or equal to zero. Otherwise the element in the boolean array will be `False`.

Self-Test Answer 15-3

This code obtains the sum using array syntax:

```
is_good = np.logical_not(np.isclose(data, 1e10))
sum_data = np.sum(data * is_good)
```

This code obtains the sum using looping:

```
sum_data = 0.0
for i in range(np.size(data)):
    if not np.isclose(data[i], 1e10):
        sum_data += data[i]
```

There are other ways of using boolean arrays, array syntax, and/or looping to solve this problem.

Self-Test Answer 15-4

This is a special floating-point value that is defined as representing "not a number."

Self-Test Answer 15-5

This code obtains the sum:

```
data = np.where(np.isclose(data, 1e10), np.nan, data)
sum_data = pd.Series(data).sum()
```

The NumPy sum function also can handle the Series object properly, so:

```
sum_data = np.sum(pd.Series(data))
```

also works. The pandas documentation reference at the end of Section 15.2.2 lists the functions and methods that use Series objects.

Self-Test Answer 15-6

A masked array is an array that holds both data values and a boolean mask that says which of the data values are good values and which are not. If a boolean mask element is True, this means the corresponding data value is a missing value (i.e., is "masked out").

Self-Test Answer 15-7

This code obtains the sum:

```
sum_data = ma.sum(ma.masked_values(data, 1e10))
```

The masked array documentation reference at the end of Section 15.2.3 lists the functions and methods that use masked array objects.

Part III

Advanced Programming Concepts

16 More Data and Execution Structures

One way of thinking of what a programming language does is that it provides various structures to store data and information (e.g., variables, lists, arrays, object attributes) and various structures to execute tasks (e.g., assignment, branching, looping, functions, object methods). A program, then, is some combination of these structures that accomplish the calculation, visualization, or other task the user is interested in conducting.

The data and information and execution structures we have studied and used in Parts I and II enable us to accomplish the majority of the computational tasks in science and engineering. Additional programming language structures, however, provide additional capabilities. They enable us to more easily accomplish basic tasks and write programs with capabilities that would be difficult to write using only the structures from Parts I and II.

In this chapter, we introduce additional data and execution structures. Two other execution structures are also covered in Part III: Chapter 17 extends our previous discussion on classes and objects by describing how to create our own classes, and Chapter 20 examines recursive programming. For our chapter example, we apply various combinations of data and execution structures to solve a statistical data analysis problem in four different ways. By solving it in different ways, we illustrate what capabilities different structures provide.

16.1 Example of Using More Advanced Data and Execution Structures

Pretend we have three data files named *data0001.txt*, *data0002.txt*, and *data0003.txt*.[1] Each data file contains a single column of numerical data (that is, there is one value on each line) of differing lengths. The data files have no headers. Thus, the NumPy function `genfromtxt` will read in all the data in a file and return the data in a one-dimensional array. We want to write a program that, after reading in the data in each file into its own NumPy array, calculates the mean, median, and standard deviation of the values in each data file, and saves the values to variables for possible later use. The NumPy function for calculating the mean is `average` or `mean`, the median is `median`, and the standard deviation is `std`.

We consider four different solutions of the above task and the strengths and weaknesses of these different kinds of solutions. In previous chapters, we have given a broad description in the Example sections as to what the code does and gone into more detail in the Python Programming Essentials section. In this section, the example plays a slightly different role:

---

[1] This section is adapted from Chapter 6 of Lin (2012).

```
1    import numpy as np
2
3    data1 = np.genfromtxt('data0001.txt')
4    data2 = np.genfromtxt('data0002.txt')
5    data3 = np.genfromtxt('data0003.txt')
6
7    mean1 = np.mean(data1)
8    median1 = np.median(data1)
9    stddev1 = np.std(data1)
10
11   mean2 = np.mean(data2)
12   median2 = np.median(data2)
13   stddev2 = np.std(data2)
14
15   mean3 = np.mean(data3)
16   median3 = np.median(data3)
17   stddev3 = np.std(data3)
```

Figure 16.1 Code for Solution 1, using explicit calling of functions and the storing of the result in variables.

it illustrates how the *way* we use various programming structures gives us programs of very different capabilities. Thus, in the present section we explain the example in more detail than in previous chapters. The example also does not use most of the new data and execution structures we describe in Section 16.2. But, the example shows why the structures we introduce in Section 16.2 are important, and in that way motivates the Python content of the chapter.

16.1.1 Solution 1: Explicitly Call Functions and Store Results in Variables

In this solution (Figure 16.1), the data and results from the calls to the functions are stored in a variable. Each variable has a different name, and we assign those variables explicitly (that is, by typing the variable name, the equal sign, and then the variable's value). Likewise, calls to function are done explicitly. Each time we want to call a function, we type in the name of the function with parentheses.

In lines 3–5 in Figure 16.1, we read in the three data files and save the results as three different arrays. In lines 7–9, we use the `data1` array to calculate the mean, median, and standard deviation of the values in `data1` and save the values into variables with names that describe what they hold (e.g., `mean1`). We repeat this in lines 11–13 for the `data2` values and in lines 15–17 for the `data3` values.

This program works fine and solves our task. But, we can see that the program is written so that anytime we specify a variable, whether a filename, data variable, or an analysis function, *we have to type it in*. This is fine when we have three files, but what if we have a thousand files? Very quickly, this kind of programming becomes not very fun.

```
1   import numpy as np
2
3   num_files = 3
4   means = np.zeros(num_files, dtype='d')
5   medians = np.zeros(num_files, dtype='d')
6   stddevs = np.zeros(num_files, dtype='d')
7
8   for i in range(num_files):
9       filename = 'data' + ('000'+str(i+1))[-4:] + '.txt'
10      data = np.genfromtxt(filename)
11      means[i] = np.mean(data)
12      medians[i] = np.median(data)
13      stddevs[i] = np.std(data)
```

Figure 16.2 Code for Solution 2, using explicit calling of functions and the storing of the results in arrays.

16.1.2 Solution 2: Explicitly Call Functions and Store Results in Arrays

One approach to improving on the Solution 1 code is to put the results (mean, median, and standard deviation) into arrays and have the element's position in the array correspond to the file that generated the result. Thus, the first element of the array that holds the means corresponds to the mean of the data in *data0001.txt*, etc. We can then use a for loop to read each file, make the calculations using each file's data, and store the results in the results arrays. Thus, we do not have to type in the names of the mean, median, and standard deviation variables to store the results. Figure 16.2 shows the code for such a solution.

In lines 3–6 of Figure 16.2, we set the number of files and create the arrays that will hold the results of the mean, median, and standard deviation calculations on each file's data. In line 8, we loop through the indices of the elements of these results arrays.

In line 9, we use the index for the results array (i) to construct the filename of which we will calculate the mean, median, and standard deviation. We can do this because the filenames of the data files in question are numbered, starting with one. Note how the way we construct the numerical part of the filename – by adding on three zeros to the front of i+1 and slicing the last four digits of the resulting string – enables us to handle filenames from *data0001.txt* to *data9999.txt*. In line 10, we read in the data from filename. In lines 11–13, we calculate the mean, median, and standard deviations of that data, saving the result in the index i element of the means, medians, and stddevs results arrays.

This code is slightly more compact than our Solution 1 code, which is a plus. More importantly, Solution 2 scales up to any num_files number of files. But, what if the filenames are not numbered? How then do we relate the element position of the means, medians, and stddevs results arrays to the file the calculated quantity is based on? Variable names (e.g., mean1, stddev3) *do* convey information and connect labels to values. By putting the results into arrays where element indices do not naturally connect to the file's name, we lose the ability to encode that information.

```
1    import numpy as np
2
3    means = {}      #- Initialize as empty dictionaries
4    medians = {}
5    stddevs = {}
6
7    list_of_files = ['data0001.txt', 'data0002.txt', 'data0003.txt']
8
9    for ifile in list_of_files:
10       data = np.genfromtxt(ifile)
11       means[ifile] = np.mean(data)
12       medians[ifile] = np.median(data)
13       stddevs[ifile] = np.std(data)
```

Figure 16.3 Code for Solution 3, using explicit calling of functions and the storing of the results in dictionaries.

16.1.3 Solution 3: Explicitly Call Functions and Store Results in Dictionaries

Before looking at this solution, we first need to ask how dictionaries might be useful for our problem. We previously said variable names connect labels to values, meaning that a string (the variable name) is associated with a value (scalar, array, etc.). In Python, there is a construct that can associate a string with a value: a dictionary. From that perspective, setting a value to a key that is the variable name (or something similar) – as we do with dictionaries – is effectively the same as assigning a variable with an equal sign. However, dictionaries allow us to do this *dynamically* (i.e., we do not have to type in "variable equals value") and will accommodate *any* string, not just those that have numerals in them.

Figure 16.3 shows a solution that uses dictionaries to hold the statistical results. These dictionaries are initialized as empty dictionaries in lines 3–5, to enable us to add key:value pairs into the dictionaries later in the program. The keys for the dictionary entries are the filenames, not indices (as with storing the results in a list or array). So, we can use filenames that do not necessarily have anything to do with a numerical index. To illustrate this flexibility, we loop through (line 9) a list of the filenames (line 7) instead of constructing each filename through string operations based on an array index. We type in the names of these files in line 7, but as we saw in Section 10.2.5, we can obtain a list of files in a directory through the os module's `listdir` function and obtain a list of files matching a pattern using the glob module's `glob` function. We could, then, replace line 7 in Figure 16.3 with:[2]

```
list_of_files = glob.glob("data*.txt")
```

assuming the glob module is already imported.

It would not be as easy to generate a list of files to loop through if we used arrays to store the results as in Figure 16.2. This is because dictionaries store values based on keys, not on order. The approach of storing results in arrays only works if the files have some

---

[2] See http://docs.python.org/library/glob.html for details on the glob module (accessed August 16, 2012).

```
1    import numpy as np
2    import glob
3
4    metrics = {'mean':np.mean, 'median':np.median, 'stddev':np.std}
5    list_of_files = glob.glob("data*.txt")
6
7    results = {}                        #- initialize results
8    for imetric in metrics.keys():      #  dictionary for each
9        results[imetric] = {}           #  statistical metric
10
11   for ifile in list_of_files:
12       data = np.genfromtxt(ifile)
13       for imetric in metrics.keys():
14           results[imetric][ifile] = metrics[imetric](data)
```

Figure 16.4 Code for Solution 4, storing the results and functions in dictionaries.

sort of natural order. But filenames can be named using characters that have no natural and commonly understood ordering, such as underscores and hyphens. By using dictionaries to store results in Figure 16.3, we can accommodate any unique filename.

This accommodation also enables us to refer to the statistical results more intelligently in our code. If we want to access, say, the mean of *data0001.txt*, we type in means['data0001.txt']. In contrast, in our Figure 16.2 solution, we would have referred to that result as means[0], which is less clear.

16.1.4 Solution 4: Store Results and Functions in Dictionaries

Although Solution 3 works quite well, can we make our program flexible enough to calculate more than just the mean, median, and standard deviation? What if we wanted to calculate 10 metrics? 30? 100? Here too, Python dictionaries save the day. Recall that key:value pairs enable us to put *anything* in as the value, even functions and other dictionaries. So, we refactor our solution to store the function objects themselves in a dictionary of functions, linked to the string keys 'mean', 'median', and 'stddev'. We also make a results dictionary that holds the dictionaries of the mean, median, and standard deviation results; results is a dictionary of dictionaries. Figure 16.4 shows this solution.

In line 4 of Figure 16.4, we create a dictionary whose values are the NumPy mean, median, and stddev functions themselves. Thus, a reference to a key in this dictionary references the related function, which can be called like any function:

```
In [1]:   import numpy as np
          metrics = {'mean':np.mean, 'median':np.median, 'stddev':np.std}
          data = np.array([1, 2, 3])
          metrics['mean'](data)
```

```
Out [1]:   2.0
```

In the above example, `metrics['mean']` extracts the NumPy `mean` function, which we call with values of the array `data`. The value of `metrics['mean']` is the NumPy `mean` function itself, not merely the name of the function.

In line 5, we construct the list of files using a call to the glob module's `glob` function, as described in Section 16.1.3. In line 7, we initialize the `results` dictionary. In lines 8–9, we create empty dictionaries that will hold the results for each statistical metric. That is, if after line 9 we execute:

```
In [1]:  print(results)

         {'mean': {}, 'median': {}, 'stddev': {}}
```

we see that the contents of `results` is a set of key:value pairs where the key are the keys in `metrics` and the values associated with those keys are empty dictionaries. These empty dictionaries are then ready to be filled by calling the metrics on the data from each file. So, if we run the Figure 16.4 code and then examine `results`, we would obtain something like the following (output reformatted for better display):

```
In [1]:  print(results)

         {'mean': {'data0001.txt': 2.0, 'data0003.txt': 7.5,
                   'data0002.txt': 3.5},
          'median': {'data0001.txt': 2.0, 'data0003.txt': 7.5,
                     'data0002.txt': 3.5},
          'stddev': {'data0001.txt': 0.816496580927726,
                     'data0003.txt': 1.118033988749895,
                     'data0002.txt': 0.816496580927726}}
```

The contents of each file in the above example consist of three or four values: *data0001.txt* has 1.0, 2.0, and 3.0, *data0002.txt* has 2.5, 3.5, and 4.5, and *data0003.txt* has 6.0, 7.0, 8.0, and 9.0. The order in which the key:value pairs are listed is not necessarily in a natural order, either of the keys or the values. This is as we would expect if dictionaries are unordered collections.

In lines 11–14 of Figure 16.4, we loop through all the files in `list_of_files`. For each file, we loop through each of the statistical metrics in the `metrics` dictionary. In line 14, we call the metric function given by the `imetric` key, passing in the `data` values read in from `ifile`, and save the result in the `results[imetric]` dictionary, attached to the `ifile` key. Remember, both iterators `ifile` and `imetric` are strings.

This program is now more flexible: it calculates mean, median, and standard deviation *for as many files as there are* in the working directory that match `"data*.txt"` and can be extended to calculate *as many statistical metrics as desired*. If we want to access files with a different naming pattern, we just change the search pattern in line 5. If we want to add another statistical metric, we just add another entry in the `metrics` dictionary in line 4. In Section 17.2.5, we will see how to add a metric automatically. Only two lines need to change: *nothing else in the program needs to change.* This kind of flexible program is really only possible because of the features of the dictionary data structure. The structures a programming language offers, then, dictate the kind of programs we can write, and the presence of additional data and execution structures with additional capabilities enables us to write programs that otherwise would be difficult to write.

16.2   Python Programming Essentials

In Section 16.1, we saw how the data and execution structures available to us impact the kind of programs we can write. With the right set of structures, we can write more flexible, concise, and powerful programs. In this section, we introduce some new data and execution structures and describe how and why they might be used. We also fill in our description of the dictionary data structure, to help us understand the syntax in some of the solutions in our Section 16.1 example.

16.2.1   More Data Structures

A data structure is an object that stores a collection of items. Depending on the purpose of the structure, the object is designed to store and access the items in a particular way. Some of the data structures we have seen so far include: lists, tuples, NumPy arrays, Series objects, and masked arrays. In this section, we introduce the following data structures: **stacks, queues,** and **sets**. Although these data structures are not used that much in science and engineering applications, they illustrate data structure characteristics that are not present in the data structures we have already covered. For instance, lists, tuples, NumPy arrays, Series objects, and masked arrays are all, in some way, **random-access** data structures. That is, if we know the "address" of an element (e.g., its index or key), we can obtain the value of that element anytime we want. In contrast, in a **limited-access** data structure, there are conditions that need to be met regarding when we can access an element. In general, we cannot obtain the value of any element anytime we want.

The first two new data structures we examine – the stack and the queue – are limited-access data structures. The third new data structure – the set – is not a limited-access data structure but does differ from the data structures we have already covered in that duplicate values are not allowed. We end this section by extending the description of dictionaries we started in Section 14.2.5.

Stacks

A stack is one of the most basic data structures. It consists of a collection of items that is in an order but cannot be arbitrarily accessed. All access to the stack occurs at the "top" of the stack, the place where we can add (i.e., "push") an item or remove (i.e., "pop") an item. The best way to explain a stack is to think of a stack of something physical, say a stack of books. When we add a book to the stack, it comes in at the top of the stack. When we remove a book from the stack, it comes off from the top of the stack. If we want to remove just the fourth book down from the top of the stack, we cannot just remove that book alone. We first have to remove the first, then the second, and then the third books, remove the fourth book, then replace the third, then second, then first books back to their original positions. Removing and adding always happen one at a time, through the top. In that way, the stack is limited-access.

A stack is known as a "**first in, last out (FILO)**" data structure. The first item we put into a stack is the last item that is removed from the stack. Put another way, a stack is also a "**last in, first out (LIFO)**" data structure. The last item we push to the stack is the first one popped off, because it is at the top.

In terms of size, a stack is unlimited. Practically, the amount of memory a computer has limits the number of items a stack can hold. But, unlike an array whose number of elements is fixed when the array is created, the abstraction of a stack can grow and shrink as needed, without limit.

Because a stack is limited-access, even though the items in the stack are in an order, we cannot use the order to access a specific element in the stack (as we can do using the element index in an array). A basic stack does not have a method to let us see what is in, say, the fifth element in the stack. A basic stack only has methods to push an item, pop an item, and tell us whether the stack is empty. Some also consider a "peek" method – one that shows us what would be popped without actually popping the item at top – a part of a basic stack.

Given the limitations of a stack, why is it useful? Because modern computers are so fast and have so much memory, the importance of conserving programming resources is not as great as in the past. But it still takes resources to support the flexibility other data structures offer. For instance, if the particular problem we want to solve only requires interaction with the data structure at one location, why do we need to use the computational resources to support random-access?

Additionally, there are certain problems for which the characteristics of a stack automatically solves the problem. For instance, consider an "undo" function on a word processor. Say we execute a sequence of commands, for instance, the following in the order listed:

1. Type the characters "Experiment 1: Dissolving NaCl".
2. Make "Experiment 1" bold.
3. Type a blank line under the words.
4. Type the characters "Supplies:".
5. Make "Supplies:" italicized.

If we store the commands in a stack, then undoing our last command is equivalent to popping that command from the top of the stack. The undoing of every command before the "current" command is also a popping operation, and the addition of a new command is a push. The FILO/LIFO nature of a stack automatically implements what an undo function does: between the adding into and removing from the stack, the order of access of the elements is reversed.

Although we can, in Python, duplicate the behavior of a stack by using other data structures – a list's pop and append methods duplicate a stack's pop and push methods, assuming the end of the list is top – Python also provides a stack class called LifoQueue as part of the queue built-in module.[3] In this class, the pop method is called get and the push method is called put. Here is an example of building and using a small stack:

```
In [1]:  import queue

         mystack = queue.LifoQueue()
         mystack.put(1.2)
         mystack.put(-4.5)
         mystack.put(8.7)

         print(mystack)

         while not mystack.empty():
             print(mystack.get())

         print(mystack.empty())

         <queue.LifoQueue object at 0x7fd5f0265ba8>
         8.7
         -4.5
         1.2
         True
```

After adding in three items to the stack, we try to print out the contents of the stack. That attempt does not show the contents but merely tells us mystack is a LifoQueue object. We do not have random-access to the items in the stack. From the output, we can see that as we pop each item using the get method, we obtain the latest element. We put in the elements in a certain order but they are taken out in reverse order. As they are taken out, the stack decreases in size until it is empty.

Queues

Like a stack, a queue is also a limited-access data structure. However, whereas in a stack the adding and removing of items all occur at one place, in a queue, adding to the queue occurs at one end of the queue (the "**back**") while removing from the queue occurs at the other end of the queue (the "**front**"). In this, a queue is exactly like a line to see a bank teller or store cashier. People get in line at the back of the line and the person at the front of the

---

[3] https://docs.python.org/3/library/queue.html (accessed June 29, 2020).

line is the next customer served. The operation of adding an item to the back of the queue is called "**enqueue**ing." The operation to remove an item from the front of the queue is called "**dequeue**ing."

A queue is a "**first in, first out (FIFO)**" or "**last in, last out (LILO)**" data structure. The first customer in line is the first customer served. Like a stack, a queue is unlimited in size. Also like a stack, a queue is limited-access, and so even though the items in the queue are in an order, we cannot use the order to access a specific element in a stack. A basic queue has methods to enqueue, dequeue, and say whether the queue is empty.

Queues are useful for applications where we want to create a line of some sort. For instance, let us say we were writing an application controlling a set of gates in a mouse maze, which are lined up one next to the other like the starting gates of a horse race. So, as a set of mice approaches these gates, each one is assigned to their own gate. This program reads in and stores the identification tag of the mice as they arrive at their gate, waits for a certain amount of time, then opens the gates for the mice to go on. The gates are opened in the order of each mouse's arrival. A queue would be the natural data structure to hold the objects describing each mouse, as it naturally enforces the FIFO/LILO behavior the gates are supposed to follow.

Python's built-in queue module also has a `Queue` class that implements the queue data structure.[4] The enqueue command is `put` and the dequeue command is `get`. The `empty` method tells whether the `Queue` object is empty. Here is an example of creating and using a small `Queue` object:

```
In [1]:  import queue

         myqueue = queue.Queue()
         myqueue.put(1.2)
         myqueue.put(-4.5)
         myqueue.put(8.7)

         print(myqueue)

         while not myqueue.empty():
             print(myqueue.get())

         print(myqueue.empty())


         <queue.Queue object at 0x7fd5f0265860>
         1.2
         -4.5
         8.7
         True
```

---

[4] https://docs.python.org/3/library/queue.html (accessed June 29, 2020).

As with the stack, we cannot see the contents of the queue. We can only interact with the queue by adding at the back and removing at the front. The order we add in the items is the order they come back out, as the output from the `while` loop shows us.

Sets

A set is a data structure whose items are all unique. With all the other data structures we have seen so far (e.g., lists, arrays, stacks), multiple items can have the same value. For instance, in an array, the first element could be a 7 but so too could the third element or the fifty-sixth element. There is no issue with having duplicate values across different elements. With a set, there is only one of any given value in the set.

As a result, when we add an item to a set, if that item already exists in the set, nothing happens. When we remove an item from a set, we are removing the item based upon its value; we are not removing the item from a certain location. In this way, the set is an unordered data structure. There is not a first, second, etc., position for the items in the set. However, although a set is unordered, it is iterable.

The operations we can do on a set are not as limited as with stacks and queues. We can add items, remove items, find out how many items there are, check to see what the items are, etc. But, because sets are unordered and are not connected to any kind of addressing system (such as an index with an array or a key with a dictionary), operations using such an addressing system do not exist.

Because a set does not store duplicates, this is a natural data structure for handling unique values. For instance, if we have a dataset listing all the species of insects that have been observed over the course of a year in a meadow, and we want to know how many species have been observed, we can add all the records of species names we have to a set, and only the unique ones will remain. Thus, the length of that set will be the number of species that have been observed.

Python implements a set using the `set` command. We can create a set by passing in a collection of items (using the list syntax):

```
In [1]:  myset = set([4, -8, 9])
         print(myset)

         {-8, 9, 4}

In [2]:  myset = set([4, -8, 4, 9])
         print(myset)

         {-8, 9, 4}
```

If there are duplicate items being passed in, the set command ignores the duplicate. When we print out the set, curly braces are used to delimit the set. This does not mean the items are in a dictionary. Note the absence of keys.

We can also create an empty set and add items to the set. The method `add` adds items to the set and the method `remove` removes items from the set. The `len` function tells us how many items are in the set:

```
In [1]:  myset = set()
         myset.add(4)
         myset.add(-8)
         myset.add(9)
         myset.add(-8)
         print(myset)
         print(len(myset))


         myset.remove(9)
         print(myset)
         print(len(myset))


         {-8, 9, 4}
         3
         {-8, 4}
         2
```

If we try to add an item to a set that already has that item, nothing happens.

More on Dictionaries

We introduced dictionaries in Section 14.2.5. In that section, we saw how to create a dictionary, add and replace key:value pairs to the dictionary, and delete key:value pairs from the dictionary. Dictionaries have additional capabilities, however, that give us a lot of flexibility in how we interact with dictionaries.

Dictionaries have the method `keys` that returns all of the keys in the dictionary and the method `values` that returns all the values in the dictionary. Using the dictionary `settings` from Section 14.2.5, these methods return:

```
In [1]:  settings = {'linestyle':'--', 'linewidth':5.0,
                      'marker':'*', 'markersize':20.0}
         print(settings.keys())
         print(settings.values())

         dict_keys(['linestyle', 'linewidth', 'marker', 'markersize'])
         dict_values(['--', 5.0, '*', 20.0])
```

Although `keys` and `values` return objects that are similar to lists, we need to use the `list` function to convert them to lists. The result of these methods are iterable and thus can be looped over. Assuming `settings` defined as above:

```
In [1]:  for ikey in settings.keys():
             print(ikey + ' : ' + str(settings[ikey]))


         linestyle : --
         linewidth : 5.0
         marker : *
         markersize : 20.0
```

```
In [2]:  for ivalue in settings.values():
             print(ivalue)


         --
         5.0
         *
         20.0
```

Although it appears the key:value pairs of `settings` are stored in alphabetical order by keys, we should not make such an assumption. Standard dictionaries should be assumed to be unordered data structures. Note, the `values` method returns a collection that has no further connection to the keys associated with those values. Thus, if knowledge of the key is important, we want to loop through the keys rather than the values.

With lists and arrays, we can loop through the indices (using the `range` function) or through the values of the list or array (by directly looping through the list or array). With dictionaries, in order to loop through the values of the dictionary, we have to use the `values` method, as described above. To loop through the keys, we can loop through the results of the `keys` method, as described above, or *through the dictionary directly*. That is, if we use a `for` loop through a dictionary directly, we will loop through the keys, not through the values:

```
In [1]:  for ikey in settings:
             print(ikey + ' : ' + str(settings[ikey]))


         linestyle : --
         linewidth : 5.0
         marker : *
         markersize : 20.0
```

This differs from the behavior with lists and arrays. Dictionaries, although unordered, are iterable over their keys.

Loops are particularly useful when we want to automate the filling, replacement, or deleting of key:value pairs in the dictionary. For instance, the following fills a dictionary where the first five letters of the alphabet are keys connected to the integers zero through four:

```
In [1]:  letters = 'abcde'
         mydict = {}
         for i in range(5):
             mydict[letters[i]] = i
         print(mydict)


         {'a': 0, 'b': 1, 'c': 2, 'd': 3, 'e': 4}
```

Lines 11–13 of Figure 16.3 work similarly, adding the mean, median, and standard deviation values for the file given by `ifile` to the dictionaries `means`, `medians`, and `stddevs`. Lines 9 and 14 in Figure 16.3 do a similar assignment in the dictionaries `results` and `results[imetric]`, respectively (recall that `results` is a dictionary of dictionaries).

As a side note, because dictionaries are unordered collections, if we add or remove key:value pairs from a dictionary, we do not have to worry about whether we are altering the order of the items in the dictionary. There is no order! In contrast, if we remove an element from the middle of a list, the list has to move everything to the right of that element (i.e., collapse the list) in order to keep everything in order. A list does this automatically when we use the appropriate object methods (e.g., the `remove` method), but with a dictionary, this is not a consideration to begin with.

If we want to see whether a specific key is in a dictionary, we can use the `in` operator to test whether a given key (to the left of `in`) is in a given dictionary (to the right of `in`). A boolean value is returned. Thus, for `mydict` above:

```
In [1]:  print('a' in mydict)
         print('z' in mydict)


         True
         False
```

because the dictionary `mydict` does have a key:value element with the key `'a'`, but it does not have a key:value element with the key `'z'`. To check whether a specific key is not in a dictionary, use the `not in` operator. These are examples of **membership testing**. In Section 19.2.1, we consider membership testing more broadly, as it is a form of searching.

Dictionaries are unordered collections, but this does not mean we cannot access or manipulate items in a dictionary in a specific order. The easiest way to do so is to obtain the keys of a dictionary using `keys` and sort the keys in the desired order. For instance, for the dictionary `mydict` above, we can use the built-in `sorted` function to make sure the keys are sorted in alphabetical order:

```
In [1]:  sorted_keys = sorted(mydict.keys())
         for ikey in sorted_keys:
             print(ikey + ' : ' + str(mydict[ikey]))


         a : 0
         b : 1
         c : 2
         d : 3
         e : 4
```

In Chapter 19, we discuss sorting in more detail. The `sorted` function has the capability to conduct highly customized sorts, including those based on multiple criteria.

16.2.2 More Execution Structures

So far, we have seen basic programming structures such as assignment, branching, and looping. Functions enable us to reuse portions of code. Objects enable us to bind attributes and methods together in a single entity, which we can then access and call (respectively). For NumPy arrays, we have seen that we can use array syntax to implicitly loop through all elements in the array. In this section, we describe two additional execution structures: **list comprehensions** and **lambda functions**. Both are associated with another style of programming called functional programming, which we do not discuss in the present work. See the Python documentation for more information on functional programming. We provide an up-to-date link to a "how to" on this topic at www.cambridge.org/core/resources/pythonforscientists/refs/, ref. 14. For our purposes, both list comprehensions and lambda functions provide a way of concisely defining lists and functions.

List Comprehensions

A list comprehension is a structure that enables us to concisely connect list creation and looping. Using the constructs we learned earlier, to create a list of the numbers 0, 2, 4, 6, 8, we would do the following:

```
In [1]:  values = []
         for i in range(5):
             values.append(i * 2)
         print(values)


         [0, 2, 4, 6, 8]
```

Using a list comprehension, looping and creating the list collapses into one line:

```
In [1]:   values = [i*2 for i in range(5)]
          print(values)


          [0, 2, 4, 6, 8]
```

What is happening? The list `values` is being constructed so that each element is the result of an expression embedded in an iteration of `for i in range(5)`. For each iteration, the expression in front of the iteration portion (i.e., `i*2`) is evaluated, and that value is saved as that element in the list.

We can also add an `if` test to each iteration in a list comprehension.[5] With the `if` statement, the list comprehension adds an element to the list only if the condition is `True`. For instance:

```
In [1]:   values = [i*2 for i in range(5) if i > 2]
          print(values)


          [6, 8]
```

only adds the elements 6 and 8 to the list because the 0, 2, and 4 values before were created for the `i` values 0, 1, and 2.

lambda Functions

Similar to how a list comprehension creates a list using a loop, all in a single line, a lambda function or expression creates a function, all in a single line.[6] The syntax to define a function of *<function name>* that optionally takes a sequence of comma-separated input parameters given by *<args>* is of the form:

> *<function name>* = `lambda` *<args>* : *<return value expression>*

Thus we can create a function called `double_it` which multiples a value by two by:

```
In [1]:   double_it = lambda x: x*2
          print(double_it(4))


          8
```

We call lambda functions the same way we would call a function defined using the `def` statement. In general, lambda functions are appropriate for short functions. For longer functions, the regular multiline syntax using the `def` statement is clearer.

---

[5] https://docs.python.org/3/howto/functional.html (accessed June 29, 2020).
[6] This entire section references ideas from https://docs.python.org/3/howto/functional.html (accessed June 29, 2020).

16.2.3 When to Use Different Data and Execution Structures

One way to answer the question, "When do we use different data and execution structures?" is by pointing to the outline of the present text. Different scientific and engineering work-flow tasks require different kinds and combinations of data and execution structures to accomplish. Another way of answering this question is to point at summaries of the data structures (Appendix B) and programming topics (Appendix C) the present text covers. In this section, we describe some questions and principles we can use to help us decide what data and execution structures to use.

We organize our discussion around these categories: storage structure, repetition structure, and clarity of code. These are not, by any means, the only considerations. Performance, which is briefly touched on here, is a major issue that warrants more attention (see Chapter 22 for more on performance). But, this list provides some of the topics we may consider as we choose between different programming structures.

Storage Structure

Data structures, by definition, store items. Our first consideration is what needs to be stored. Do we care about only the values, or does it matter what order the values are in? Do we care about only the unique values or is the presence of duplicate values something that is of interest? For instance, if we had a collection of counts of the number of birds observed and wanted to know how many times the count equaled a specific value, we would not want to use a set to store this collection, because a set does not store duplicate values.

Second, we need to consider the ease (or difficulty) with which we want to access or manipulate the data. Do we want random-access or is limited-access acceptable? How easy does it need to be to add new elements or delete elements? Although we might conclude that we should usually use the most flexible data structure – for instance, a list because it is random-access, ordered, mutable, and can include items of different types – that flexibility comes at the cost of increased memory use and decreased speed of execution. As many scientific and engineering use cases involve large amounts of numerical data, using a list is often a waste of resources. In such cases, it makes more sense to use a NumPy array. In contrast, if we are managing a collection of strings of an indeterminate number, a list is probably the better option because a list is mutable and because the elements of a list do not have to be all the same size in memory.

Third, we should consider whether the method of accessing the data in the data structure – whether by key, value, or order (index) – helps or hinders us. As we saw in our Section 16.1 example, dictionaries can be useful as there are benefits in connecting a string filename with the value generated using the data found in that file.

Finally, we should consider whether we need to store metadata with our data. For instance, if we want to keep track of whether some items are missing, we would want to use a data structure, such as a `Series` object or a masked array, that stores that additional information.

If we have little metadata to handle, there is no need to allocate the additional memory and processing resources that would be used in a data structure that had metadata-handling capabilities.

Repetition Structure

In this category, we want to consider what is being repeated in the program and how often it is being repeated to help us understand how we should program that repetition. If what is being repeated is a very complex calculation, with more than a few branching conditions, we probably want to use explicit looping and manipulation of individual elements. The calculation probably cannot be done using array syntax or list comprehensions (although parts may be). If the number of repetitions is high, it becomes more important to use data and execution structures that are efficient at scale. NumPy arrays work better than lists, in general, for large numbers of repeated calculations.

Clarity of Code

It has been said that we read code more than we write code. This can involve reading code that someone else has written that we want to make use of or reading code we wrote in the past (even the recent past) to improve or extend it. Clarity of code, then, is a priority. If we cannot make heads-or-tails of code we are reading, it becomes nearly useless to us.

Python, in general, generates clear code. Through our choices of data and execution structures, however, we can write programs that are clearer or not. Clarity is somewhat subjective, but here are two criteria we can use to help in our evaluation:

- Length. How many lines do we use to accomplish a task?
- Density. How many tasks are we accomplishing per line?

Generally, if our code has too many lines (over a few hundred), it is more difficult to read. When we read long programs or functions, we have to flip repeatedly from one part of the file to another, which is taxing. Regarding density, if we are accomplishing too many tasks per line of code, that too can be difficult to read. If we are accomplishing too few tasks per line of code, the code might be easier to read, but the code may be quite long.

For instance, array syntax can increase our code density, which can make our program shorter than it otherwise would be. As long as we do not try to do too many tasks in a single line of code, the use of array syntax when possible can make our code clearer (and faster). Still, in problems with many branching conditions, array syntax might end up creating a program that is so terse as to be confusing.

In the end, deciding what data and execution structures to use in our programs requires balancing many different costs and benefits. Storage structures, repetition structures, and clarity of code are some major areas where our choices can incur certain costs while obtaining certain benefits. We encounter a few more considerations in later chapters as we examine more advanced programming concepts.

16.3   Try This!

The examples in this section address the main topics of Section 16.2. We practice using a stack, queue, and set. We make use of dictionaries to conduct more complex data analysis tasks. We also practice writing list comprehensions and lambda functions.

Try This! 16-1 Using a Stack: Beaver Tracking

Pretend we have tagged five beavers with radio transmitters. After a beaver has traveled one mile,[7] its transmitter radios back the identification number of the beaver (a four-digit number). Assuming we store the identification numbers in a stack, adding them to the stack as they come in, what code prints out the identification numbers in reverse order from which they came in (i.e., the last beaver's identification number is printed first)?

Try This! Answer 16-1

Below is a solution that also fills the stack with five fictional identification numbers:

```
In [1]:  import queue

         beaver_ids = queue.LifoQueue()
         beaver_ids.put(1243)
         beaver_ids.put(9384)
         beaver_ids.put(2352)
         beaver_ids.put(4722)
         beaver_ids.put(1198)

         while not beaver_ids.empty():
             print(beaver_ids.get())


         1198
         4722
         2352
         9384
         1243
```

Because a stack is defined to be LIFO, when we repeatedly pop the value at the top by using the get method, we automatically obtain the identification numbers in reverse order.

Try This! 16-2 Using a Queue: Beaver Tracking

Pretend we have the same beaver tracking data as in Try This! 16-1, but instead of being stored in a stack the identification numbers are stored in a queue. Write a program that prints out only the identification number of the third beaver to arrive.

---

[7] Ideas from www.beaversolutions.shington.edu/agree/?sign=idaacom/beaver-facts-education/beaver-behavior-and-biology/ (accessed July 16, 2020), but all data are fictitious.

Try This! Answer 16-2

Below is a solution that also fills the queue with five fictional identification numbers:

```
In [1]:  import queue

         beaver_ids = queue.Queue()
         beaver_ids.put(1243)
         beaver_ids.put(9384)
         beaver_ids.put(2352)
         beaver_ids.put(4722)
         beaver_ids.put(1198)

         for i in range(3):
             id_number = beaver_ids.get()
         print(id_number)

         2352
```

Because a queue is defined to be FIFO, we obtain the third identification number received by dequeueing three times. Recall the dequeue method is called `get` in the queue module's `Queue` class.

The way this solution is written, the first and second identification numbers are thrown away. If we wanted to save them, we would need to put them into another data structure or back into `beaver_ids`. In the latter case, we would need to make sure we enqueued enough times to place the first identification number back to the front of the queue, if we desired to have `beaver_ids` back the way we found it.

Try This! 16-3 Using a Set: Wolf Tracking

Pretend we have a region where we have tagged a population of wolves with radio transmitters. Every time a wolf leaves the region, its identification number is saved in the list `leaves`. Every time a wolf enters the region, its identification number is saved in the list `enters`. Over a period of time, the same wolf can leave and enter the region multiple times. Write a program to calculate and print out how many individual wolves have left the region and how many have entered into the region.

Try This! Answer 16-3

To solve this problem, we need to eliminate the duplicates from the `leaves` and `enters` lists. The following code solves the problem:

```
1   print( "Number wolves left: " + str(len(set(leaves))) )
2   print( "Number wolves entered: " + str(len(set(enters))) )
```

Once the lists are converted into sets, all elements are unique. Thus, the length of those sets are the number of unique wolves from those lists.

Try This! 16-4 Initializing a Dictionary of Dictionaries: Beaver Tracking

Pretend we have the beaver tracking data as in Try This! 16-1, but in addition to transmitting a 4-digit integer identification number, information from the following fields (left column) is also provided:

Field	Default value
Nickname	Empty string
Date of birth	Empty string
Mother's ID number	−1
Father's ID number	−1

Write a program that creates and fills a dictionary `info` whose contents are key:value pairs where each key is the beaver's identification number and the value is a dictionary that holds the information on the beaver. That dictionary should have keys expressing the fields listed above and be set to the default value given (right column). Do this for five fictional identification numbers and print out the resulting `info` dictionary.

Try This! Answer 16-4

Below is a solution using the five fictional identification numbers from the solution to Try This! 16-1 (output reformatted for better display):

```
In [1]:  info = {}
         for iid in [1243, 9384, 2352, 4722, 1198]:
             info[iid] = {}
             for ifield in ['nickname', 'dob']:
                 info[iid][ifield] = ''
             for ifield in ['mother id', 'father id']:
                 info[iid][ifield] = -1

         print(info)

         {1243: {'nickname': '', 'dob': '', 'mother id': -1, 'father id': -1},
          9384: {'nickname': '', 'dob': '', 'mother id': -1, 'father id': -1},
          2352: {'nickname': '', 'dob': '', 'mother id': -1, 'father id': -1},
          4722: {'nickname': '', 'dob': '', 'mother id': -1, 'father id': -1},
          1198: {'nickname': '', 'dob': '', 'mother id': -1, 'father id': -1}}
```

We initialize empty dictionaries before we fill the dictionaries. Remember that the values in the key:value pairs can be basically any Python construct. In the `info` case, the values are dictionaries. Those dictionaries have values that are either strings or integers.

Note that the keys for `info` are integers (the identification numbers) while the keys for the values of `info` are strings. We can also mix strings and integers (and other types that have unique values) as our keys in the same dictionary.

We might ask why we would want to use integer keys because lists also use integer indices to reference values in the list. For the case of integer keys, the syntax to refer to elements in dictionary and lists is the same. With lists, however, we need to create all the elements in between the indices that have values. In the `info` dictionary above, even though the first identification number is 1198, if `info` were a list, that list would need to also have elements with indices 0 through 1197. When we have fewer items to store than the full range of integer keys would allow, a dictionary is probably the better route to go.

Because there are so few items in `info` and its values, it would have been around the same number of lines to type in the dictionary and its key:value pairs manually. But, by using loops, the solution above is more easily extended if there are additional beavers being tracked or additional fields that need to be stored.

Try This! 16-5 Filling a Dictionary with Calculation Results: Data Analysis

This Try This! is adapted from Exercise 18 in Lin (2012). Assume we have the following list of files:

```
list_of_files = \
    ['data0001.txt', 'data0002.txt', 'data0003.txt']
```

Create a dictionary `filenum` where the keys are the filenames and the value is the file number as an integer (i.e., *data0001.txt* has a file number of 1). Fill the dictionary automatically, assuming that we have a list `list_of_files`. Hints: To convert a string into an integer, use the `int` function on the string. The list and array subrange slicing syntax also works on strings.

Try This! Answer 16-5

Here is a program to fill `filenum`:

```
filenum = {}
list_of_files = \
    ['data0001.txt', 'data0002.txt', 'data0003.txt']
for ifile in list_of_files:
    filenum[ifile] = int(ifile[4:8])
```

We initialize an empty dictionary before we fill it. To obtain the integer version of the file number, we use the slicing syntax on the filename to obtain just the file number portion.

Try This! 16-6 Conditional Manipulation of a Dictionary: Wolf Tracking

Pretend we have the lists of identification numbers, `leaves` and `enters`, as described in Try This! 16-3. Create a dictionary `leaves_counts` that stores the number of times each wolf identification number occurs in `leaves`. Store that number as a value connected with the identification number as the key. Create a dictionary `enters_counts` the same way using the list `enters`. Using these dictionaries, print the identification number of all wolves who leave the region more times than they enter the region.

Hints: List objects have a method `count` that accepts a single input parameter and counts the number of occurrences of that input in the list. Also, because all of our tagged wolves originally started in the region, none of the wolves on the `enters` list is not also on the `leaves` list (because the wolf needs to first leave the region before entering it).

Try This! Answer 16-6

This code solves the problem:

```
1   leaves_counts = {}
2   for ikey in set(leaves):
3       leaves_counts[ikey] = leaves.count(ikey)
4
5   enters_counts = {}
6   for ikey in set(enters):
7       enters_counts[ikey] = enters.count(ikey)
8
9   for ikey in leaves_counts:
10      if ikey in enters_counts:
11          if leaves_counts[ikey] > enters_counts[ikey]:
12              print(ikey)
13      else:
14          print(ikey)
```

In lines 1–3 and 5–7, we count the number of occurrences of each identification number and save those counts in the `leaves_counts` and `enters_counts` dictionaries, as values tied to the identification numbers as keys. In both loops, we use `set` to obtain a collection of the identification numbers, with no duplicates.

In lines 9–14, we loop through all the keys in `leaves_counts`. In line 10, we check to see if each key is also a key in `enters_counts`. If it is, we check to see if the number of occurrences of the identification number (the key) in `leaves` is more than in `enters` (line 11). If so, we print out the identification number (line 12). If the key is not also a key in `enters_counts` (line 13), that identification number is also a case where the number of times the wolf left is more than the number of times the wolf entered (the latter being 0, because the identification number is absent from `enters`), and thus that identification number is also printed out (line 14).

We could have used the `keys` method for our looping in line 9 instead:

```
for ikey in leaves_counts.keys():
```

The above would work the same as line 9.

Try This! 16-7 Extending a Routine to Calculate More Metrics: Data Analysis

This Try This! is adapted from Exercise 19 in Lin (2012). Extend the Section 16.1 Solution 4 (in Section 16.1.4) so that it also calculates the skew and kurtosis (these are statistical quantities) of each file's data. Hint: The stats module of the SciPy package (`scipy.stats`) has functions `skew` and `kurtosis` that do the calculations.

Try This! Answer 16-7

This code solves the problem:

```
1   import numpy as np
2   import glob
3   import scipy.stats
4
5   metrics = {'mean':np.mean, 'median':np.median, 'stddev':np.std,
6              'skew':scipy.stats.skew, 'kurtosis':scipy.stats.kurtosis}
7   list_of_files = glob.glob("data*.txt")
8
9   results = {}                      #- Initialize results
10  for imetric in metrics.keys():    #  dictionary for each
11      results[imetric] = {}         #  statistical metric
12
13  for ifile in list_of_files:
14      data = np.genfromtxt(ifile)
15      for imetric in metrics.keys():
16          results[imetric][ifile] = metrics[imetric](data)
```

The only change we needed to make to the solution in Figure 16.4 (besides adding the `scipy.stats` import) was to add the key:value pairs for the `skew` and `kurtosis` functions to the `metrics` dictionary. The rest of the code in Figure 16.4 is unchanged!

Try This! 16-8 List Comprehensions and lambda Functions: Wolf Tracking

Pretend we have the lists of identification numbers, `leaves` and `enters`, as described in Try This! 16-3. The last digit of an identification number is sometimes a "check-sum" digit. A check-sum digit is a value that is not used as part of the identifier but is generated using an algorithm based on the values of the other digits in the identification number. Create a lambda function check_sum that returns the check-sum digit (the last digit) assuming an input identification number that is four digits long. Then, use this lambda function in a list comprehension to create lists leaves_csum and enters_csum that hold the check-sum digits for the corresponding identification numbers in leaves and enters, respectively.

Try This! Answer 16-8

This code solves the problem:

```
1   check_sum = lambda x: int(str(x)[-1])
2   leaves_csum = [check_sum(ival) for ival in leaves]
3   enters_csum = [check_sum(ival) for ival in enters]
```

In the function defined in line 1, str(x) converts x into a string, the [-1] extracts the last character of that string, and int converts that character back into an integer. In lines 2 and 3, we use the check_sum lambda function on each of the values in leaves and enters, respectively.

Try This! 16-9 Multiparameter lambda Functions: Beaver and Wolf Tracking

Pretend we have four-digit identification numbers as given in the beaver and wolf tracking scenarios of Try This! 16-1 and Try This! 16-3. Write a lambda function three_ids_same that will compare whether three identification numbers, input into the function, are the same.

Try This! Answer 16-9

This code solves the problem:

```
three_ids_same = lambda x, y, z: (x == y) and (y == z)
```

When writing a lambda function with multiple positional input parameters, we list each parameter one after the other, separated by commas, to the left of the colon.

16.4 More Discipline-Specific Practice

The Discipline-Specific Jupyter notebooks for this chapter cover the following topics:

- Stacks.
- Queues.
- Sets.
- More on dictionaries.
- List comprehensions.
- lambda functions.

16.5   Chapter Review

16.5.1 Self-Test Questions

Try to do these without looking at the book or any other resources or using the Python interpreter. Answers to these Self-Test Questions are found at the end of the chapter.

Self-Test Question 16-1

Describe what a stack is.

Self-Test Question 16-2

What does it mean to "push" an item to a stack or "pop" an item from a stack?

Self-Test Question 16-3

If we remove all the items from a stack, what order do they come out?

Self-Test Question 16-4

Describe what a queue is.

Self-Test Question 16-5

What do "enqueue" and "dequeue" mean?

Self-Test Question 16-6

What does it mean that a queue is "LILO?"

Self-Test Question 16-7

If we execute the following code:

```
import queue

myqueue = queue.Queue()
myqueue.put(7.0)
myqueue.put(8.4)
myqueue.put(3.9)
myqueue.get()
myqueue.put(-1.8)
myqueue.get()
```

at the end of the code, what is the value of the item at the front of `myqueue`?

Self-Test Question 16-8

What is a set?

Self-Test Question 16-9

How many items is in `values` if:

```
values = set([3, 6, 2, 4, 3, 6, 1])
```

What command returns the number of items in `values`?

Self-Test Question 16-10

What dictionary method returns all the keys in a dictionary? All the values?

Self-Test Question 16-11

Pretend we have the following list of field names (as strings) that are characteristics of a ball dropped from a balloon: `['mass', 'velocity', 'acceleration', 'Cdrag', 'area']`. Create a dictionary `projectile` that is populated by key:value pairs where these field names are the keys and −1.0 is the initial value. Use a loop to do this assigning of the key:value pairs.

Self-Test Question 16-12

For the `projectile` dictionary in Self-Test Question 16-11, what would be the expression to see whether `projectile` contained the key `'density'`?

Self-Test Question 16-13

Are dictionaries ordered or unordered? If unordered, is it possible to examine key:value pairs in a dictionary in an order? If so, how?

Self-Test Question 16-14

What does the following line of code do:

```
sequence = [i+2 for i in range(5)]
```

Self-Test Question 16-15

Create a lambda function `exponent` that takes two input parameters, x and y, and returns x^y.

Self-Test Question 16-16

Describe how our choices of data and execution structures can affect the clarity of our code.

16.5.2 Chapter Summary

In this chapter, we examined additional data and execution structures. These structures are interesting but, for our purposes, it is the new problem-solving capabilities these structures give us that are golden. For instance, as we have seen, if we limit ourselves to using variables, assignment, and function calls in our programs, everything in our programs has to be specified by typing. A structure like a Python dictionary, however, enables us to dynamically associate a name with a variable or function (or anything else), which is essentially what variable assignment does. Thus, dictionaries enable us to add, remove, or change a "variable" on the fly, yielding programs that are more concise and flexible. Other data and execution structures give us other capabilities. By leveraging such modern programming language features, we can create programs that do more scientific work, faster, and better. Specific topics we covered in this chapter include:

- Stacks and queues

  - Both stacks and queues are limited-access data structures. We cannot see the entire contents, all at once, of a stack or a queue.
  - In a stack, we add and remove items from the top of the stack. In a queue, we add items at the back of the queue and remove items from the front of the queue.
  - Adding an element to a stack is called pushing to the stack. Removing an element from a stack is called popping from the stack. Adding an element to a queue is called enqueueing. Removing an element from a queue is called dequeueing.
  - Stacks are LIFO/FILO structures. Queues are LILO/FIFO structures.
  - When we remove from a stack, the order is reversed compared with the order we added to the stack. When we remove from a queue, the order is the same compared with the order we added to the queue.

- Sets

  - A set is an unordered, iterable data structure with no duplicates. Each item in the set is unique.
  - We access items in a set by their values. There are no indices or keys connected to the values.
  - The `add` method adds an item to a set. If the item already exists in the set, nothing happens.

- More on dictionaries

  - Dictionaries are unordered collections of key:value pairs.
  - The `keys` method returns the keys in a dictionary. The `values` method returns the values in a dictionary.
  - When we loop through a dictionary directly, we are looping through the keys. In this sense the dictionary is iterable.
  - We can check to see whether a key is in a dictionary by membership testing. Thus, if we have a dictionary `data`, the expression `'units' in data` returns `True` if `'units'` is a key in the dictionary and `False` if not.
  - To access the key:value pairs in a dictionary in a specified order, we can obtain the keys, sort them, then loop through those sorted keys to access the values in that order. The built-in `sorted` function can do this sorting.

- List comprehensions and lambda functions

  - List comprehensions enable us to automatically create a list in a single line, with or without the application of a condition.
  - lambda functions enable us to define a function in a single line.
  - Both list comprehensions and lambda functions are good when the complexity of the calculations (either to obtain the values of the list or the return value of the function) are modest. For more complex situations, using an explicit `for` loop or `def` statement is better.

- When to use different data and execution structures

  - Storage structure, repetition structure, and clarity of code are three areas we need to consider when deciding what structures to use in our program.
  - Storage structure: Does order matter? Are duplicates okay? How easy should it be to access values? Will the method of accessing values give us additional functionality? Do we need to store metadata with our data in a single structure?
  - Repetition structure: Are the calculations so complex as to make explicit looping and manipulation a better choice than the implicit looping found in array syntax?
  - Clarity of code: Will the structures we use make the code too long to easily read and understand? Will the structures we use make the code too terse to easily read and understand?

16.5.3 Self-Test Answers

Self-Test Answer 16-1

A stack is a limited-access data structure that can add and remove items only from the top of the stack. As items are added to the stack, new items are placed on top of existing items.

Self-Test Answer 16-2

Pushing an item to a stack means to add an item to the top of the stack. Popping an item from a stack means remove the item at the top of the stack.

Self-Test Answer 16-3

Because a stack is LIFO, the items come out of a stack in reverse order from the order they entered into the stack.

Self-Test Answer 16-4

A queue is a limited-access data structure that adds at the back and removes from the front. It simulates a line for a bank teller or store cashier.

Self-Test Answer 16-5

Enqueue means to add to a queue (at the back of the queue). Dequeue means to remove from a queue (at the front of the queue).

Self-Test Answer 16-6

Because a queue is LILO (which is equivalent to it being FIFO), the first items put into the queue are also the first items removed from the queue. This means a queue preserves, during removal, the order of the items as they were inserted into the queue.

Self-Test Answer 16-7

```
3.9
```

Self-Test Answer 16-8

A set is an unordered but iterable collection of unique items. There are no duplicate items in a set.

Self-Test Answer 16-9

There are five items in `values`. The `len` function returns the number of items in `values`.

Self-Test Answer 16-10

The `keys` method returns all the keys in a dictionary. The `values` method returns all the values in a dictionary. Neither are called with input arguments.

Self-Test Answer 16-11

```
1  fields = ['mass', 'velocity', 'acceleration', 'Cdrag', 'area']
2  projectile = {}
3  for ikey in fields:
4      projectile[ikey] = -1.0
```

Self-Test Answer 16-12

```
'density' in projectile
```

Self-Test Answer 16-13

Dictionaries are unordered. We can examine key:value pairs in a dictionary in an order if we first sort the keys into an order of interest.

Self-Test Answer 16-14

The code creates the variable `sequence` which is a list with the following contents:

```
[2, 3, 4, 5, 6]
```

Self-Test Answer 16-15

```
exponent = lambda x, y: x**y
```

Self-Test Answer 16-16

The clarity of our code suffers when we have too many lines of code and/or when our code is so dense (does a lot in very few lines) that it is difficult to read. We need to choose data and execution structures that both do enough, so our code does not become too long, and are still readable.

17 Classes and Inheritance

In Chapter 9, we were introduced to objects and object-oriented programming (OOP). In the objects we considered, however, the classes those objects were instances of were all written by others. We just made use of the templates or common patterns that describe the array, list, string, and other objects discussed in that chapter to make arrays, lists, strings, etc.

In the present chapter, we examine how to create our own classes. That is, we examine how to write the template or common pattern that defines a class (or type) that we can then use to create objects of that class. Although the use of classes others have developed is useful, it is when we write our own classes that we are fully leveraging the benefits of OOP. We do not have to create our own classes to solve most science and engineering problems, but when the problems are particularly complex and involved, using classes to manage and structure our code can result in code that is much easier to understand, use, and maintain.

This chapter is unique in that instead of providing one motivating example in Section 17.1, we provide two very different kinds of examples in that section. We describe the reasons for this in that section.

17.1 Examples of Classes and Inheritance

We provide two examples of using classes and **inheritance** in this section. The first example is a scientific modeling example. The second example is a scientific bibliography example. There are four reasons for these two examples. First, we include the first example because it illustrates how to use a computer to run a basic deterministic, prognostic model, one of the key applications of scientific and engineering computing. Turbulent mixing, climate modeling, and space weather are just a few applications of such models. However, such a model, because it involves solving a differential equation, is best understood if one has taken differential calculus. Thus, the second reason for providing two examples in this chapter is to give a nonmathematical example for those who will find such an example to be more instructive. Third, just showing more examples is helpful to understanding how classes and inheritance work. Finally, the scientific bibliography example, even more than the scientific modeling example, shows why creating our own classes can be so powerful. The programming details illustrated in the examples in this section are discussed in Section 17.2. In particular, we elaborate on how the scientific bibliography example shows the power of OOP in Section 17.2.4.

17.1.1 Scientific Modeling Example

In Chapter 8, we considered a simple prognostic model – a model that is predictive, where we calculate how a variable changes with time. This is in contrast to a diagnostic calculation, where the quantity being calculated is based on other variables at the present moment in time (those other variables could, themselves, be time evolving). The model we considered in Section 8.1 was a nondeterministic, prognostic, random walk model, where we used randomness to drive the motion of the portion of a bacterial colony being modeled. In the present section, we consider a deterministic, prognostic model, one where later values of the variable being modeled are entirely dependent upon the prior values. In particular, we consider models that are solutions of first-order ordinary differential equations.

Let us break down the phrase, "first-order ordinary differential equations." But, before we do so, we need to know what a derivative is. Fundamentally, a derivative is a slope. Just as the slope of a line tells us the ratio of the rise-over-the-run for that straight line, a derivative tells us how fast one variable (the dependent variable) is changing as another variable (the independent variable) changes. If y is the dependent variable and t is the independent variable, the derivative is written mathematically as:

$$\frac{dy}{dt}$$

and the "d"s express the idea of a "differential," that is, an infinitesimally (or vanishingly) small change. Thus, dy/dt is the quantity "an infinitesimal change in y" over "an infinitesimal change in t." That is:

$$\frac{dy}{dt} \approx \frac{\Delta y}{\Delta t}$$

where Δ means "a discrete (or small but measurable) change in," and the approximation between the "d" differential and Δ "discrete change" is more exact the smaller the discrete change. For our purposes, that is all we need to know! There are more mathematical details to derivatives that are covered in calculus, but for now, the key idea is that a derivative is a ratio that expresses how much one variable changes as another variable changes.

A differential equation is an equation that has at least one derivative term somewhere. An ordinary differential equation is an equation whose derivative term(s) have only one independent variable and one dependent variable, and those variables are the same for all derivatives in the equation. This:

$$\frac{dy}{dt} = y + 7$$

is an ordinary differential equation, where y is the dependent variable and t is the independent variable. But:

$$\frac{\partial y}{\partial t} + \frac{\partial y}{\partial x} = y + 7$$

is not an ordinary differential equation, because the two derivative terms are in terms of different independent variables (i.e., t and x). In fact, because y now changes with both t and

x, the "d" differential aspect of a derivative needs to be written using a "∂" instead of a "d," and we call the derivatives with the ∂ symbol "partial derivatives."

Lastly, a first-order differential equation is one that has only a first-order derivative, that is, a derivative that expresses how the dependent variable changes with the independent variable. A second-order derivative expresses how a *first-order derivative* changes with the independent variable. In terms of calculus notation, the derivative dy/dt we saw earlier is a first-order derivative, but:

$$\frac{d}{dt}\left(\frac{dy}{dt}\right) = \frac{d^2y}{dt^2}$$

is a second-order derivative. Thus, when we take all this together, an equation like:

$$\frac{dy}{dt} = y + 7$$

is an example of a first-order ordinary differential equation. The prognostic models we study in the present section look like this equation.

How do we use a computer to solve an equation that looks like this? By "solve," we mean: as the independent variable t changes, given an equation for dy/dt, how do we calculate the value of y? Consider that earlier we said we can think of the derivative as:

$$\frac{dy}{dt} \approx \frac{\Delta y}{\Delta t}$$

If so, and if we replace the "$\approx$" by an equal sign, we can reconfigure the above equation in the following way:

$$\frac{dy}{dt} = \frac{\Delta y}{\Delta t}$$
$$= \frac{y(t) - y(t - \Delta t)}{\Delta t}$$

where $y(t)$ means "y at time t," and $y(t - \Delta t)$ means "y at time $t - \Delta t$." Note $y(t)$ does *not* mean "y times t," and $y(t - \Delta t)$ does not mean "y times $t - \Delta t$." Time "$t - \Delta t$" means the time that is a "discrete bit of time Δt" before time t.

We can multiply both sides by Δt and move terms around to obtain:

$$y(t) = y(t - \Delta t) + \frac{dy}{dt}\Delta t \tag{17.1}$$

This equation says that the value of y at the current time t equals the value of y some little bit of time before the current time t (precisely Δt amount of time before) plus the derivative dy/dt times Δt. The derivative dy/dt is also calculated at time $t - \Delta t$. That is, if dy/dt was given by this equation:

$$\frac{dy}{dt} = y + 7$$

then the dy/dt term in Equation (17.1) (which is dy/dt at time $t - \Delta t$) is calculated by taking the value of y at time $t - \Delta t$ plus 7.

```
 1    #- Module imports:
 2
 3    import numpy as np
 4    import matplotlib.pyplot as plt
 5
 6
 7    #- Class definitions:
 8
 9    class Model(object):
10        def __init__(self, dt=None, nsteps=None, yinit=None):
11            self.dt = dt
12            self.nsteps = nsteps
13            self.t = np.zeros(nsteps+1, dtype='d')
14            self.y = np.zeros(nsteps+1, dtype='d')
15            self.t[0] = 0.0
16            self.y[0] = yinit
17
18        def run(self):
19            for i in range(1, self.nsteps+1):
20                self.t[i] = i * self.dt
21                self.y[i] = self.y[i-1] + \
22                          (self.tend(self.y[i-1]) * self.dt)
```

Figure 17.1 Code defining a parent class `Model`, two child classes `LogisticEquation` and `FallingObject`, and a main program that runs two models and generates the plots in Figure 17.2.

```
23   class LogisticEquation(Model):
24       def __init__(self, dt=2.0, nsteps=400, Ninit=1.0):
25           super().__init__(dt=dt, nsteps=nsteps, yinit=Ninit)
26           self.a = 0.05        # 1/min
27           self.K = 5e9         # CFU/ml
28
29       def tend(self, N):
30           return (self.a * N) * (1.0 - (N / self.K))
31
32
33   class FallingObject(Model):
34       def __init__(self, dt=0.1, nsteps=30, vinit=0.0):
35           super().__init__(dt=dt, nsteps=nsteps, yinit=vinit)
36           self.g = 9.81        # m/s^2
37           self.Cd = 2.0        # unitless
38           self.rhoa = 1.275    # kg/m^3
39           self.A = 0.5         # m^2
40           self.m = 2.0         # kg
41
42       def tend(self, v):
43           return self.g - (self.Cd * self.rhoa * self.A / \
44                          (2.0 * self.m) * (v**2))
```

Figure 17.1 (part 2)

```
45   #- Main program:

46

47   bacteria = LogisticEquation()
48   fall = FallingObject()

49

50   bacteria.run()
51   fall.run()

52

53   plt.figure(1)
54   plt.plot(bacteria.t, bacteria.y, 'k-o')
55   plt.xlabel('Time (min)')
56   plt.ylabel('Population (CFU/ml)')
57   plt.title('Bacteria Population')
58   plt.savefig('logistic_model.png')

59

60   plt.figure(2)
61   plt.plot(fall.t, fall.y, 'k-o')
62   plt.xlabel('Time (sec)')
63   plt.ylabel('Velocity (m/s)')
64   plt.title('Falling Object')
65   plt.savefig('falling_object.png')
```

Figure 17.1   (part 3)

Why does Equation (17.1) solve the differential equation? Because the value of y at the current time is entirely given by the value of y and dy/dt *at a time in the past*. Thus, as long as we are given an initial or starting value of y at time zero (and dy/dt depends on y), we can calculate y at any later time by calculating y at times at Δt increments. That is to say, we calculate y at $t = \Delta t$ by using y at $t = 0$, then we calculate y at $t = 2\Delta t$ by using the y at $t = \Delta t$ we just calculated, and so on.

Equation (17.1) is only one way of using a computer to solve a differential equation. It is called "Euler's method," and while it is useful for instruction, it has a number of failings as an algorithm. For real-world modeling applications, other "numerical methods" are preferred, but this topic is outside the scope of the present work. See Press et al. (1989) for more on numerical methods.

Figure 17.1 shows the code defining the classes Model, LogisticEquation, and FallingObject. The Model class defines attributes and methods that apply to both the LogisticEquation and FallingObject classes. By defining them in Model, we do not have to also define them again in LogisticEquation and FallingObject.

The LogisticEquation class defines a logistic equation model that can be used to model the growth of a population, such as a population of bacteria. The logistic equation is the differential equation:

$$\frac{dN}{dt} = aN \left(1 - \frac{N}{K} \right)$$

where N is the population in CFU/ml (CFU means "colony forming unit," such as a bacterium) and is the dependent variable that is changing with time. The independent variable is time (t) in minutes. The other variables are constants. For modeling bacteria population using this equation, a good value of a is 0.05 min^{-1}, based on the growth rate of the bacterium *E. coli* ($\approx$ 20 min). A good value of K is the approximate density of this bacterium that can be achieved in liquid media (5×10^9 CFU/ml).

The `FallingObject` class defines a model for the velocity of a specific object falling under the influence of gravity and encountering air resistance. The differential equation governing this fall is:

$$\frac{dv}{dt} = g - \frac{C_d \rho_a A}{2m} v^2$$

where v is the velocity in m/s (where positive v is pointing towards the ground) and is the dependent variable that is changing with time. The independent variable is time (t) in seconds. The other variables are constants. For Earth's gravitational field near the surface, g, the gravitational acceleration, is about 9.81 m/s^2. The drag coefficient, C_d, expresses the degree to which the object goes through the air smoothly or not. For our specific object, we take $C_d = 2$ (the coefficient is unitless). The air density ρ_a is approximately 1.275 kg/m^3. For our specific object, the cross-sectional area (A) is 0.5 m^2 and the mass (m) is 2 kg.

The code in Figure 17.1 also provides a main program that creates two modeling objects, runs the models, and plots the results in Figure 17.2. This code illustrates how we can create different classes of models (`LogisticEquation`, and `FallingObject`), share common attributes and methods by making reference to a separate **parent** class (`Model`), create and run the model objects we have created, and use the results from the model runs.

Figure 17.1 opens with import statements (lines 3–4). Lines 9–22 define the `Model` class. This class is a parent class, that is, a class that will provide attributes and methods for **child** classes that **inherit** from `Model`. Besides defining attributes that apply to `LogisticEquation` and `FallingObject` (lines 11–16), including the arrays that hold the values of the independent variable (`self.t`) and the dependent variable (`self.y`), the class also defines a method `run` that calculates `self.nsteps` number of values of the dependent and independent variables. The method `run` has no input parameters (besides `self`) because all the information it needs is stored as attributes of the object. The calculation of each element of `self.y` that is done in lines 21–22 follows Equation (17.1). Thus, the method `tend` that is called in line 22 has to return the value of dy/dt evaluated using `self.y[i-1]` (that is, at the time at the `i-1` index in the array). The `tend` method is not written in `Model` but is written in `LogisticEquation` and `FallingObject`. The versions of `self.tend` are different between those two classes, because the equation for the derivatives (dN/dt and dv/dt) are different between the two models.

One final note about `Model`: In the definition of the `__init__` method, the three keyword input parameters all have a default value of `None`. Why? We have to remember that the purpose of `Model` is to define attributes and methods that apply to both `LogisticEquation` and `FallingObject`. We should not create instances of the `Model` class. By setting the

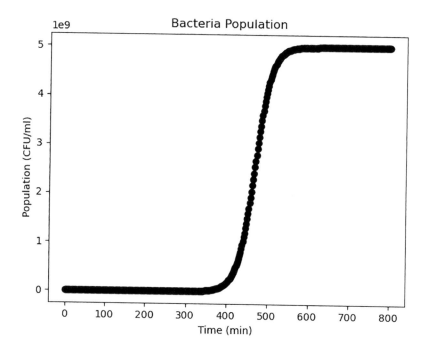

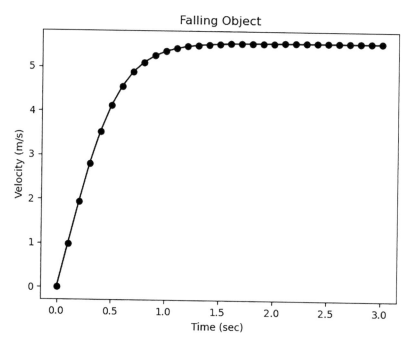

Figure 17.2 Plots generated by the code in Figure 17.1.

```
1    #- Module import:
2
3    import operator
4
5
6    #- Class definitions:
7
8    class AuthoredResource(object):
9        def __init__(self, authorlast, authorfirst, title):
10           self.authorlast = authorlast
11           self.authorfirst = authorfirst
12           self.title = title
13
14
15   class Book(AuthoredResource):
16       def __init__(self, authorlast, authorfirst,
17                       title, place, publisher, year):
18           super().__init__(authorlast, authorfirst, title)
19           self.place = place
20           self.publisher = publisher
21           self.year = year
22
23       def write_bib_entry(self):
24           return self.authorlast \
25               + ', ' + self.authorfirst \
26               + ', ' + self.title \
27               + ', ' + self.place \
28               + ': ' + self.publisher + ', ' \
29               + self.year + '.'
```

Figure 17.3 Code defining a parent class AuthoredResources, two child classes Book and Article, a Bibliography class, and a main program that creates instances of the last three classes and generates the output in Figure 17.4.

input parameters of Model's __init__ method to None, we ensure that if we try to create an instance of Model, it will not work. In our definitions of LogisticEquation and FallingObject, we override these defaults to None in the lines calling super (lines 25 and 35). How super works is described in Section 17.2.2.

Lines 23–30 define one of those child classes, the LogisticEquation class, and lines 33–44 define another one of those child classes, the FallingObject class. In each of these class definitions, we assign attributes using assignment statements (i.e., lines 11–16, 20–22, 26–27, and 36–40), and we define methods by using def statements (i.e., lines 10, 18, 24, 29, 34, and 42). Both these classes define their own version of the method tend (lines 29–30 and 42–44), as described earlier.

In the main program, we create an instance of LogisticEquation in line 47 of Figure 17.1 and an instance of FallingObject in line 48. We execute the run methods of each object in lines 47–48. In lines 53–58 and 60–65, on separate plots, we plot the results from the

```
30   class Article(AuthoredResource):
31       def __init__(self, authorlast, authorfirst,
32                    articletitle, journaltitle,
33                    volume, pages, year):
34           super().__init__(authorlast, authorfirst, articletitle)
35           self.journaltitle = journaltitle
36           self.volume = volume
37           self.pages = pages
38           self.year = year
39
40       def write_bib_entry(self):
41           return self.authorlast \
42               + ', ' + self.authorfirst \
43               + ' (' + self.year + '): ' \
44               + '"' + self.title + ',"' ' \
45               + self.journaltitle + ', ' \
46               + self.volume + ', ' \
47               + self.pages + '.'
```

Figure 17.3 (part 2)

model simulations: the bacteria populations (`bacteria.y`) versus time (`bacteria.t`) and the object velocities (`fall.y`) versus time (`fall.t`).

For a more complex example of using OOP to decompose a scientific model, see Lin (2009).

17.1.2 Scientific Bibliography Example

Publications play an important role in science and engineering in distilling, documenting, and sharing the results of studies and projects.[1] Bibliographies or list of references in these papers and reports document the resources consulted.

Figure 17.3 shows the code defining the classes `AuthoredResource`, `Book`, `Article`, and `Bibliography`. The `AuthoredResource` is a parent class that defines attributes `Book` and `Article` (and any other resource that has a single author and title) share. `Book` and `Article` class objects represent a book and article, respectively. Both make use of `AuthoredResource` and add and/or customize attributes and methods so that objects of those classes behave in the way we expect them to behave. For instance, for `Book` objects, we expect the bibliographic entries (generated by `write_bib_entry`) to include publisher information and put the year at the end. For `Article` objects, however, we expect the year to be in parentheses after the author's name and for the article's title to be in quotation marks. The `Bibliography` class stores a list of `Book` and `Article` objects which make up the bibliography.

---

[1] Parts of this section are adapted from Chapter 7 of Lin (2012).

```
48   class Bibliography(object):
49       def __init__(self, entrieslist):
50           self.entrieslist = entrieslist
51
52       def sort_entries_alpha(self):
53           tmp = sorted(self.entrieslist,
54                     key=operator.attrgetter('authorlast',
55                                             'authorfirst'))
56           self.entrieslist = tmp
57           del(tmp)
58
59       def write_bibliog_alpha(self):
60           self.sort_entries_alpha()
61           output = ''
62           for ientry in self.entrieslist:
63               output = output \
64                        + ientry.write_bib_entry() + '\n\n'
65           return output[:-2]
66
67
68   #- Main program:
69
70   balewa = Book('Balewa', 'Jessica', 'Numerical Methods',
71                 'New York', 'NumComp Press', '2010')
72   jones = Book('Jones', 'Timothy', 'Art and Visualization',
73                'Seattle', 'RenderAV', '2005')
74   wong = Article('Wong', 'Peilan', 'Protein Folding',
75                  'J. Bio. Adv. C', '19', '1-11', '2017')
76
77   bibliog = Bibliography([wong, jones, balewa])
78   print(bibliog.write_bibliog_alpha())
```

Figure 17.3   (part 3)

Figure 17.3 opens with an import statement (line 3). Lines 8–12 define the Authored Resource class. This class is a parent class, that is, a class that provides attributes and methods for child classes that inherit from AuthoredResource. AuthoredResource sets the attributes that store the author's last and first names and the title of the resource. Lines 15–29 define the Book class. Besides setting attributes, this class definition also defines the method write_bib_entry which generates the correctly formatted bibliographic entry for a book. Lines 30–47 define the Article class. Besides setting attributes, this class definition also defines the method write_bib_entry which generates the correctly formatted bibliographic entry for an article.

Lines 48–65 of Figure 17.3 define the Bibliography class. Objects of this class accept a list of resource objects (i.e., objects of the Book or Article class) which is saved as the self.entrieslist attribute. Two methods exist to manipulate this list: sort_entries_alpha ensures the self.entrieslist list is sorted by author's last

```
Balewa, Jessica, Numerical Methods, New York: NumComp Press, 2010.

Jones, Timothy, Art and Visualization, Seattle: RenderAV, 2005.

Wong, Peilan (2017): "Protein Folding," J. Bio. Adv. C, 19, 1-11.
```

Figure 17.4 Output from the main program in Figure 17.3.

name then first name, and `write_bibliog_alpha` outputs a bibliography of the resources in `self.entrieslist`, in alphabetical order by author, and formatted appropriately. In the main program (lines 70–78), we create one `Article` object and two `Book` objects (using ficticious data). We put these objects into a `Bibliography` object. Finally, we display the bibliography generated using these objects by calling the `write_bibliog_alpha` method of the `bibliog` object (line 78). Output from the main program in Figure 17.3 is shown in Figure 17.4.

In our Figure 17.3 classes, all the **constructors** use positional input parameters. In contrast, in our Figure 17.1 classes, all the constructors use keyword input parameters. Both mechanisms (or a mixture of the two) are acceptable. Using keyword input parameters gives us the ability to set default values, which in Figure 17.1 we used to make sure we could not create an instance of `Model`. But, positional input parameters are also fine to use.

17.2   Python Programming Essentials

Recall from our introduction to object-oriented programming (OOP) that attributes and methods are held together in one entity.[2] That entity is a realization of a pattern or form for an object called a `class`. That realization is called an instance or object of that `class`. We describe four topics in this section: how to define a class in Python, what is and how to use inheritance, using `sorted` for more complex sorting, and why should we create our own classes. Keep the code examples in Figures 17.1 and 17.3 handy as we discuss these topics.

17.2.1   Defining and Using a Class

Class definitions start with the `class` statement, which has syntax of this form:

> `class  <class name>(<parent class>):`

The *<class name>* follows the CapWords convention, where every word in the name is capitalized. The *<parent class>* is set to `object`, if there is no parent class. If there is a parent class, the name of that class is inside these parentheses. The idea of parent classes is

---

[2] Parts of this section are adapted from Chapter 7 of Lin (2012).

addressed in Section 17.2.2. For instance, in Figure 17.1, `Model` has no parent class, and so the class definition line is (line 9, reproduced here for convenience):

```
class Model(object):
```

However, `LogisticEquation` has a parent class (i.e., `Model`), and so its class definition line is (line 23, reproduced here for convenience):

```
class LogisticEquation(Model):
```

The block following the `class` line is the class definition. Within the definition, we refer to the instance of the class as `self`. We need some token in the `class` statement to represent the object that is the instance of the class because this statement is the template to create such objects. No such objects exist until the `class` statement executes. Once a real instance of the class is created, we "substitute in" that object's name for `self`. We can create as many instances of a class as we want, and all of those instances will have different names. For instance, in lines 70–73 of Figure 17.3, we create two instances of the `Book` class, named `balewa` and `jones`.

Attributes and Methods

References to attributes of instances of the class have to have `self` in front of the attribute name. For example, in the `FallingObject` class, the air density rhoa attribute (line 38, Figure 17.1) is referred to as `self.rhoa` in the class definition. If a variable does not have `self` in front of its name, the variable is not an attribute. Rather, it is a local variable, local to the method it is defined in. In a class definition, attributes are set by the normal assignment notation with the attribute name to the left of the equal sign and the value the attribute is being assigned to to the right of the equal sign.

When calling a method of an instance of the class, we follow a similar convention involving `self`.[3] In the call, `self` must be in front of the method name. For example, in the `Bibliography` class definition, when we call the `sort_entries_alpha` method (line 60, Figure 17.3), we type `self.sort_entries_alpha()`.

Methods are defined using the `def` statement. When we define a method, it looks just like we are defining a function except for two differences: the definition is inside the body of a `class` statement, and the first argument in any method is `self`. Even a method without any "real" parameters will have `self` as an argument in the `def` line. See, for instance, line 59 in Figure 17.3. The use of `self` is how Python tells a method, "make use of all the previously defined attributes and methods in this instance," because `self` represents the instance. It is as if `self` is being passed in as an argument. However, `self` is not really an argument when we call the method. We never type `self` or put an argument in its position when we call the method.

---

[3] There are class methods that are not attached to instances of a class, but are attached to the class itself. These are outside the scope of the present work.

Finally, we can write methods that return values (e.g., write_bib_entry in the Article class in Figure 17.3), save new values to attributes (e.g., in Figure 17.3, line 56 assigns a new value to the entrieslist attribute), do both, or do neither. Methods can also call other methods, both those defined elsewhere and those defined in the class (e.g., line 13 in Figure 17.1 and line 60 in Figure 17.3, respectively).

Constructors and Creating and Using Objects

Usually, the first method we define in a class is the __init__ method. This method is called the constructor and is automatically executed when we create an instance of the class (note its specific, special name). In fact, the arguments to __init__ are the arguments for creating an instance of the class. In lines 70–71 of Figure 17.3, we create the object balewa. Six arguments are passed in when building the object. These arguments match the parameters (excluding self) in lines 16–17 of Figure 17.3. That is, the arguments here:

```
balewa = Book('Balewa', 'Jessica', 'Numerical Methods',
              'New York', 'NumComp Press', '2010')
```

match the non-self parameters here:

```
def __init__(self, authorlast, authorfirst,
             title, place, publisher, year):
```

The __init__ method is not called by typing in __init__ but by creating the object, by typing the name of the class and then an argument list within parentheses (the list can also be empty, as in lines 47 and 48 of Figure 17.1). Because the __init__ method is called when we create (or instantiate) an instance of a class, we often put in code to initialize attributes in the object based upon the arguments that are passed into the constructor.

When we create an object of a class, the type of that object is the class. If after creating the balewa object in lines 70–71 in Figure 17.3 we printed out the object, we would see:

```
In [1]:  balewa = Book('Balewa', 'Jessica', 'Numerical Methods',
                        'New York', 'NumComp Press', '2010')
         print(balewa)

         <__main__.Book object at 0x7f4324960d68>
```

We can make as many instances of a class as we wish, just as we can make as many integer variables, list objects, or array objects as we wish. Each object is separate from one another, but all the objects of a class are constructed following the same template, the class statement for that class.

Once we create an instance of a class, we can access the attributes and methods of that object by using the same syntax as we saw in Chapter 9: we type the name of the object, a period, then the attribute name or method name. When calling a method, we also add parentheses for an argument list. Examples of using attributes include `bacteria.t` and `bacteria.y` in lines 54 of Figure 17.1. An example of calling an object's method is the call `bibliog.write_bibliog_alpha()` in line 78 of Figure 17.3.

Private Attributes and Methods

In Section 9.2.2, we saw that attributes and methods without underscores in front of their names are considered public while those with double-underscores in front and back are "special," implementing operator features, naming, under-the-hood components, etc. Sometimes, when creating our own classes, we want to create attributes and methods for our use in the class definition but that we do not want a user of our class to access. In Python, we delineate these as **private** by putting one or two underscores in front of the attribute of method name (and either zero or one underscore behind the name). One underscore in front of the name means the attribute or method has "normal privacy" whereas two underscores mean the attribute or method is "very private."

In the one-underscore (normal privacy) case, the interpreter does not do anything to prevent someone from accessing or changing the attribute or method. For instance, in this example of a `RightTriangle` class, where we provide the lengths of the legs (`self.a` and `self.b`) and calculate the length of the hypotenuse, we save the hypotenuse as `self._c`:

```
In [1]:  import numpy as np

         class RightTriangle(object):
             def __init__(self, a, b):
                 self.a = a
                 self.b = b
                 self._c = np.sqrt(a**2 + b**2)

             def write_lengths(self):
                 return str([self.a, self.b, self._c])

         triangle = RightTriangle(3, 4)
         print(triangle.write_lengths())
         triangle._c = 17.0
         print(triangle.write_lengths())

         [3, 4, 5.0]
         [3, 4, 17.0]
```

When we set `triangle._c` to 17.0 in the next-to-last line of the code, the value of the `_c` attribute is changed, even though there is a single-underscore in front of the `c`. In the normal privacy situation, the purpose of the underscore is to enable the author of the class to tell the user, "I would not mess with this if I were you." It is up to the user of the class to heed that warning or not.

With two underscores in front of the attribute or method, the Python interpreter engages in "name mangling" (i.e., temporarily "changing" the name of the attribute or method, behind-the-scenes in the interpreter session) that affords some protection.[4] For instance, consider the same code as above but with the attribute holding the length of the hypotenuse having two underscores in front:

```
In [1]:  import numpy as np

         class RightTriangle(object):
             def __init__(self, a, b):
                 self.a = a
                 self.b = b
                 self.__c = np.sqrt(a**2 + b**2)

             def write_lengths(self):
                 return str([self.a, self.b, self.__c])

         triangle = RightTriangle(3, 4)
         print(triangle.write_lengths())
         triangle.__c = 17.0
         print(triangle.write_lengths())


         [3, 4, 5.0]
         [3, 4, 5.0]
```

In this case, the next to last line of code that sets `triangle.__c` to 17.0 has no effect.

While it may seem as if the double-underscore route is preferred when we want to create a private attribute or method, it turns out that for many or most cases, the single-underscore mechanisms works better. For scientific applications, where sometimes we desire greater flexibility and customization, the use of double-underscores in front can make using the class more difficult than desired. It is often enough to just tell users not to mess with something. They usually listen to the warning.

17.2.2 Inheritance

As we define our own classes, we may notice that different classes share similarities with each other. In Figure 17.3, for instance, both the `Book` and `Article` classes share some attributes (e.g., `self.authorlast`). In Figure 17.1, both the `LogisticEquation` and `FallingObject` classes share the `run` method. The idea of inheritance is to make it easy to reuse the attributes and methods that are shared between classes.

If we have a collection of classes that all share some attributes and methods, the basic idea of inheritance is to define a separate class that contains those shared attributes and methods

---

[4] https://docs.python.org/3/tutorial/classes.html#private-variables (accessed July 3, 2020).

and then to use that class as the "**base**" on which to build each of the collection of classes with the shared attributes and methods. Thus, in Figure 17.3, we define the `AuthoredResource` class which defines three attributes and assigns them in the constructor. All the classes that use `AuthoredResource` as a base template to build from (i.e., `Book` and `Article`) automatically start with those three attributes and the constructor `AuthoredResource` defines.

The class with the shared attributes and methods is called the "base," "**super**," or "parent" class. The class that uses the class with the shared attributes and methods, adding to or changing it, is called a "**derived**," "**sub**," or "child" class. The following phrasing is all interchangeable: derived classes inherit from base classes, subclasses inherit from superclasses, and child classes inherit from parent classes. As we saw in Section 17.2.1, a child class specifies a parent class to inherit from by putting the name of the parent class in the parentheses after the name of the child class. Thus, in:

```
class LogisticEquation(Model):
```

`Model` is the parent class and `LogisticEquation` is the child class. The topmost parent class is a built-in class called `object` that defines all the low-level infrastructure to support a class. If we are defining a class that has no other parent, we put `object` in the parentheses of the `class` statement.

Inheritance is most useful when the collection of classes we are defining have a "superset/-subset" nature to them. That is to say, some group of classes is in some way a subset of a broader category. In our Section 17.1 examples, the broader category of `Model` encompasses specific kinds of models, `LogisticEquation` and `FallingObject` (and others not yet defined). The broader category of `AuthoredResource` encompasses specific kinds of resources, `Book` and `Article` (and others not yet defined).

Inheritance can cover multiple "generations" or "levels." One class can be a parent to a child, and that child in turn can be a parent to another child (or "grandchild"), and so on. For instance, in Matplotlib, the `Artist` class is the parent to the (private) `_AxesBase` class. In turn, `_AxesBase` is the parent to the `Axes` class, which in turn is the parent to the `PolarAxes` and `GeoAxes` classes.[5] We also call all parents, grandparents, etc., of a class the "**ancestors**" of that class. Because every child builds off of its parent, and thus starts with the attributes and methods of their parent, if we change or add an attribute or method to the topmost parent of an inheritance lineage, the change or addition propagates to all the children, grandchildren, etc., in that lineage! In this way, a properly constructed inheritance structure is a very powerful tool for writing code that is easier to maintain.

Checking Type When Using Inheritance   Python gives us functions/expressions for checking whether an object is of a specific class or whether it is of a specific class or any of the ancestors

---

[5] https://matplotlib.org/3.1.0/api/artist_api.html (accessed July 4, 2020).

of that class. To accomplish the former, we can use an expression using the `type` function. To accomplish the latter, we can use the `isinstance` function. Pretend the following lines of code are run after line 78 in Figure 17.3 is executed:

```
In [1]:  print(type(balewa) == Book)
         print(type(balewa) == AuthoredResource)
         print(isinstance(balewa, Book))
         print(isinstance(balewa, AuthoredResource))
         print(isinstance(balewa, Article))


         True
         False
         True
         True
         False
```

The first two lines of `In [1]` check to see whether the object `balewa` is specifically a `Book` object or `AuthoredResource` object, respectively. It is the former but not the latter. The last three lines of `In [1]` check to see if `balewa` is an instance of `Book`, `AuthoredResource`, or `Article`, in the sense that it is of that specific class *or an ancestor*. That is, the built-in function `isinstance` also returns `True` for all **descendants** of the class given as the second input parameter. Thus, the `In [1]` third and fourth line calls of `isinstance` return `True` because `balewa` is of type `Book` and a descendant of `AuthoredResource`. However, `balewa` is not of type `Article` or any of its descendants (of which there are none), and so `isinstance` returns `False`.

If we do not know the possible class name of an object, we can check the output of a `type` call on the object with the result of a `type` call on an object of the possible class we want to check against. For instance, if we have an object `values` and we want to see if that is a list object:

```
type(values) == type([])
```

does what we want. We have passed in an empty list to the second `type` call.

Overriding   A common way of using inheritance is to define a method in a parent class and then override that definition in the child class. That way, the version in the child class is customized for that class' more specific use case. In our examples in Figures 17.1 and 17.3, we override the `Model` and `AuthoredResource` constructors in the children of those classes. Once a method is overridden, it essentially replaces the parent version of that method.

When we override a method, however, we often want to add on functionality to the parent version of the method, rather than jettison the parent version's functionality entirely. That is, we often want to have our new method execute the parent's version of the method and also execute additional lines of code. In Figure 17.3, we want the `Book` constructor to assign the attributes `self.place`, `self.publisher`, and `self.year`, but also to assign the common attributes defined in the parent class `AuthoredResource` (i.e., `self.authorlast`, `self.authorfirst`, and `self.title`). We can do this by using the `super` function.[6] A call to the `super` function references the parent of that class. If we put a period after the call, then a method name, we reference the parent's version of the method. Thus, line 18 in Figure 17.3 (reproduced here for convenience, without the indentation):

```
super().__init__(authorlast, authorfirst, title)
```

calls the `AuthoredResource` version of `__init__`, passing in the values of `authorlast`, `authorfirst`, and `title` as arguments.

Python supports **multiple inheritance**, when a class has more than one parent class. In that case, in general, the child class inherits all the attributes and methods defined in the parent classes. We can add multiple parent classes by listing them in the parenthesis in the `class` statement. Each of the parent classes is separated by commas in that listing. We do not cover multiple inheritance in any further depth in the present work.

17.2.3 More Sophisticated Sorting Using `sorted`

In Section 16.2.1, we saw how we could used the built-in `sorted` function to sort a list. When we sort values such as strings or integers, there is a natural order to use: alphabetically or ascending by value, respectively. But how do we sort user-defined objects, including objects of different types?

Lines 52–57 of Figure 17.3 define the `sort_entries_alpha` method, which sorts the `self.entrieslist` attribute and replaces the old `self.entrieslist` attribute with the sorted version. The method uses the built-in `sorted` function, which takes a keyword parameter `key` that gives the key used for sorting the argument of `sorted`. The key is generated using the `attrgetter` function of the operator module. This function takes the attributes of the names listed as arguments to `attrgetter` out of the elements of the item being sorted. The attribute names passed into `attrgetter` are strings, and thus we refer to the attributes of interest by their string names, not by typing in their names. This makes the program much easier to write. In lines 54–55, `attrgetter` has two arguments: `sorted` indexes `self.entrieslist` by the `attrgetter`'s first argument attribute name first then the second. Thus, when we sort the `Book` and `Article` objects in `self.entrieslist`, we do so first on the basis of the author's last name, then on the basis of the author's first name.

Through the key mechanism, `sorted` can be used to sort any objects based on any criteria. After all, when we create our own classes, *we* decide what the attributes are! But the `sort_entries_alpha` method also shows us something about the benefits of creating our own classes, as we will see in Section 17.2.4.

---

[6] https://docs.python.org/3/library/functions.html#super (accessed July 4, 2020).

17.2.4 Why Create Our Own Classes?

Having described OOP in detail, we might reasonably ask, "why the big deal?" On one level, it seems like classes are just a different way of organizing data and functions. Instead of putting them in libraries (or modules), we put them in a class. Even the behavior of inheritance, which enables us to easily reuse attributes and methods defined in a class, seems like it can be duplicated using functions. It seems like a lot of extra trouble to define a class, create objects, and call methods. If all we need is a single class with a single instance of that class, OOP is a lot of trouble for little benefit. For relatively simple programs, it is probably better to use procedural programming instead. For more complex programs, however, by properly designing our classes so our objects are used in conjunction with each other, an object-oriented structure can make our code much simpler to write and understand, easier to debug, and less prone to error.

Our Section 17.1.2 example illustrates how this can work. Recall that in that example we created `Book` and `Article` classes to manage information related to book and article resources. Both of these classes are also children of `AuthoredResource`. We also created a `Bibliography` class that manages a bibliography, given instances of `Book` and `Article` objects. In the `Bibliography` class, we defined a method `sort_entries_alpha` that sorts the `Book` and `Article` objects and a method `write_bibliog_alpha` that generates the bibliography using those objects. Let us consider what each method shows us about OOP.

What `sort_entries_alpha` Illustrates About OOP   This method does two main tasks. First, it sorts a list of items that are totally differently structured from each other based on two shared types of data (attributes). Second, it accomplishes the sort using a sorting function that does not care about the details of the items being sorted, only that the items have these two shared types of data. In other words, the sorting function does not care about the item type, only that all item types have the attributes `authorlast` and `authorfirst`.

This does not seem that big of a deal, but think about how we would have had to do such sorting using procedural programming. First, each instance would have been a list, with a label of what kind of resource it is. For instance, the line 74–75 wong object in Figure 17.3 would be represented by this list:

```
wong_array = ['article', 'Wong', 'Peilan', 'Protein Folding',
              'J. Bio. Adv. C', '19', '1-11', '2017']
```

The procedural programming sorting function we would write would need to know which elements of this list we would want to sort with (here the second and third elements of the list). *But the locations where those data are in each list would likely be different*, depending on the resource type (e.g., a book versus an article). Thus, in our sorting function, we would need to write multiple `if` tests (based on the resource type) to extract the correct field in the list to sort by. But, if we changed the key we wanted to sort by – say, from the author's name to the date of publication – we would have to *change the element index we used to sort against*. This means *manually* changing the code of the `if` tests in our sorting function.

It is easy to make such a manual code change and test that the change works if we have only a few resource types, but what if we have tens or hundreds of resource types? We would have to change each of the tens or hundreds of if statements in our sorting function. What a nightmare! Think of the number of possible bugs we might introduce just from keystroke errors alone! But, in object-oriented programming, we can switch the sorting key at will (just the attribute names in lines 54–55 of Figure 17.3) and support as many resource types as we want *without requiring additional code changes.* There are no if tests to change. This is the power of OOP over procedural programming. Code structured using an OOP framework naturally results in programs that are much more flexible and extensible, resulting in dramatically fewer bugs.

What `write_bibliog_alpha` Illustrates About OOP Here too, let us ask how would we write a function in procedural programming that outputs an alphabetized bibliography? We might write something like the following sketch:

```
1    def write_bibliog_function(list_of_entries):
2        <... open output string or file ...>
3
4        for i in range(len(list_of_entries)):
5            ientryarray = list_of_entries[i]
6
7            if ientryarray[0] == "book":
8                <... call function for bibliography entry
9                     for a book, and save to output string
10                    or file ...>
11
12           elif ientryarray[0] = "article":
13               <... call function for bibliography entry
14                    for an article, and save to output string
15                    or file ...>
16
17           <... handle additional resource types ...>
18
19       <... close output string or file ...>
```

Again, in this solution sketch, we see how in procedural programming we have to write if tests in our function to make sure we format each resource's bibliographic entry correctly, depending on resource type (e.g., book or article), by calling the specific function that formats that resource's entry. In fact, for *every* function that deals with multiple resource types, we need this tree of if tests. If we introduce another resource type (say a web page), we would need to add another if test in *all functions* where we have such an if-test tree. This is a recipe for disaster: It is exceedingly easy to inadvertently add an if test in one function but forget to do so in another function, make a typo in one function on one if test but not in another function, etc.

In contrast, with objects, *adding another resource type requires no code changes or additions* to sort_entries_alpha. The new resource type just needs a write_bib_entry method defined for it. So, if we wanted to be able to list web pages in our bibliography, we would only need to define something like the following:

```
class WebPage(AuthoredResource):
    def __init__(self, authorlast, authorfirst, title, url):
        super().__init__(authorlast, authorfirst, title)
        self.url = url

    def write_bib_entry(self):
        return self.authorlast \
            + ', ' + self.authorfirst + ', ' \
            + '"' + self.title + ',"  ' \
            + self.url + '.'
```

Objects of the class WebPage can now be included in the bibliography described by Bibliography, and the entries for web pages will be formatted the way web page entries should be formatted. *No changes need to be made to Bibliography.* And, if in the future we have even more resources types (e.g., DVDs, audio clips, personal communications, etc.), we would add them in as we did with WebPage. We would not have to change Bibliography to accommodate the additional resource types. So much easier!

Finally, we consider a more subtle point about the flexibility of OOP. In the Book and Article classes, we defined a method write_bib_entry that creates a bibliographic entry that is correctly formatted for the class' resource type. In write_bibliog_alpha, we make use of that method to generate the bibliography by looping through the self.entrieslist list and calling the write_bib_entry method attached to each object in the list (lines 62–64 in Figure 17.3).

Notice, though, that we can do this without putting any constraints on the objects in self.entrieslist besides the constraint that the objects have a write_bib_entry method. The objects do not have to be the same class or type, nor do they have to have a common ancestor (although in the present case both Book and Article have a common parent). In fact, because Python is dynamically typed, the type of ientry *changes each iteration* of the loop in line 62 of Figure 17.3. There is no assumption as to the type of ientry prior to entering the loop.

This flexibility arises because in OOP we carefully structure how a user of an object interacts or interfaces with an object. In this case, by defining a common method name for a task, and by *both* putting the details of how that task is accomplished into the method (which is like procedural programming) *and* customizing those details as necessary depending on the class the method is in (which is not something we can do in procedural programming), we have abstracted away the need to think about these details at the level of calling the method. Thus, we can loop through these objects the way we do, without needing to consider the types of the objects in any explicit way.

17.2.5 Automating Handling of Objects and Modules

In the `write_bibliog_alpha` method in Figure 17.3, we looped through a list of objects and called the `write_bib_entry` method bound to each object. In this example, we knew what method we wanted to execute, but what if we wanted the computer to pick the method to execute (or to pick the attribute to access)? The syntax we have learned thus far is unable to do this, because we have to type in the object's name, a period, then the attribute or method name to access the attribute or method.

Python has functions and methods we can use to get around this limitation. These routines enable us to access and manipulate attributes and methods using string references to the attributes and methods. Because strings can be generated by the computer, or can be read in from a file, this enables us to write code to direct the computer to access and/or call attributes and functions.

The following built-in functions are useful for this kind of "automated" manipulation of attributes and methods:[7]

- `callable(<obj>)` Returns `True` if the object *<obj>* is callable, `False` otherwise.
- `delattr(<obj>, <name>)` Deletes the attribute or method in object *<obj>* specified by the string *<name>*.
- `getattr(<obj>, <name>)` Returns the value of the attribute or method in object *<obj>* specified by the string *<name>*.
- `hasattr(<obj>, <name>)` Returns `True` if the attribute or method specified by the string *<name>* is in object *<obj>*, and `False` otherwise.
- `setattr(<obj>, <name>, <value>)` Assigns the attribute or method specified by the string *<name>* in object *<obj>* to object *<value>*.

Even though we have been referring to attributes and methods as two separate kinds of entities, in Python, the only difference between the two is that methods are callable and attributes are not.. We can think of methods, in fact, as "callable attributes," that is, as objects like any other attributes only these objects are callable. Thus, all the functions above that have "attr" in their names work on both attributes and methods. We can use `callable` to tell whether we are dealing with an attribute or methods (attributes and methods in Python are also objects).

Here is an example list showing the equivalence of typing in an attribute reference or method call and using one of the built-in functions above. The examples use the `wong` object defined in lines 74–75 in Figure 17.3:

Typing in	Built in functions
`del(wong.volume)`	`delattr(wong, 'volume')`
`wong.volume`	`getattr(wong, 'volume')`
`wong.volume = 5`	`setattr(wong, 'volume', 5)`

---

[7] https://docs.python.org/3/library/functions.html (accessed July 6, 2020).

Thus, the following:

```
callable(getattr(wong, 'volume'))
```

returns `False` because the `volume` attribute in the `wong` object is not callable.

Pretend we have a class `Counter` to count the number of species observed in a meadow. We can use objects of this method such as a hand-held counter used by ushers in a theater. The class has methods `add1`, `add2`, and `add3` that add one, two, or three, respectively, to the integer stored in the attribute `total`. The following code defines the class, creates an instance of `Counter`, and automatically calls each of the aforementioned methods through judicious use of Python's looping and string manipulation routines:

```
In [1]:  class Counter(object):
             def __init__(self, init):
                 self.total = init

             def add1(self):
                 self.total += 1

             def add2(self):
                 self.total += 2

             def add3(self):
                 self.total += 3

         count = Counter(0)
         for i in range(3):
             method_name = 'add' + str(i+1)
             getattr(count, method_name)()
             print(str(count.total))


         1
         3
         6
```

In the next-to-last line of code in cell `In [1]`, `getattr(count, method_name)` gets the `add1`, `add2`, and `add3` methods. The `()` afterwards calls the method.

In the above example, we do not first check `count` has a method called `method_name` before using `getattr`. This is, in general, bad programming practice, because if the method `method_name` does not exist, `getattr` will throw an exception, which, if unhandled, will cause the program to stop. It is better practice first to use `hasattr` to check if the attribute `method_name` exists, use `callable` to check if it is callable (i.e., is a method and not an attribute), and then use `getattr` to obtain the method.

What if we do not know the names of the methods begin with `'add'`? We just want to execute all the public methods in the object. In Section 9.2.2, we saw that the `dir` function will generate a list of all attributes and methods in an object. So, we can loop through the list of all attributes and methods, find the callable and nonprivate methods, and call those methods. Assuming `Counter` is defined as described above, the following code implements these steps:

```
In [1]:  count = Counter(0)
         for iattr in dir(count):
             if callable(getattr(count, iattr)) and (iattr[0] != '_'):
                 getattr(count, iattr)()
                 print(str(count.total))
```

```
1
3
6
```

The `dir(count)` call returns a list of strings, so `iattr` is the string name of each attribute and method in `count` (including both public and private attributes and methods).

We can apply this same automated handling of attributes within objects to the handling of functions within modules. In the following, we look through all attributes and functions in the NumPy module, look for those functions that start with `'cos'`, and call them using `angle` as input (these criteria select the cosine and hyperbolic cosine functions):

```
In [1]:  import numpy as np

         angle = 0.7
         for iattr in dir(np):
             if callable(getattr(np, iattr)) and \
                (iattr[0] != '_') and \
                ('cos' == iattr[:3]):
                 temp = getattr(np, iattr)(angle)
                 print(iattr + ': ' + str(temp))
```

```
cos: 0.7648421872844885
cosh: 1.255169005630943
```

In Figure 16.4, we stored statistical functions in a dictionary and looped through and called those functions without having to type in each call manually. But, we still needed to add those functions manually to the `metrics` dictionary. Using similar syntax as above, we can now automatically add whatever functions we want to that dictionary.

In the above examples, we have looked at the attributes and methods of objects and modules. What if we want to manipulate a variable or function defined in the current scope? The built-in `locals` function (called with no input arguments) returns the dictionary of what is defined locally. The names of what are defined are strings that are the keys of the dictionary, and the values are the objects themselves. So, in the dictionary, the string name of a function connects to the function object it refers to. The `globals` function does the same for the global scope, that is, for the entire module.[8]

---

[8] https://docs.python.org/3/library/functions.html (accessed August 19, 2020).

These examples illustrate a deeper truth: In Python, nearly everything is (or we can treat as) an object. Lists, arrays, dictionaries, and objects of built-in or user-written classes, are, as we would expect, objects. But, so too are most variables, functions, attributes, methods, classes, modules, and packages. The more comfortable we are with using Python in an object-oriented way, the more we can take advantage of its powerful tools.

17.3 Try This!

The examples in this section address the main topics of Section 17.2. We create our own classes, practice extending a class using inheritance, and override a method. We also practice using user-defined classes to accomplish more complex sorting and automated handling of objects.

Try This! 17-1 Creating a Class: Information on an Element

Create a class Element that holds the following information about an element as attributes:

- Elemental symbol.
- Number of protons.
- Number of neutrons (in its most common isotope).

Set the attributes using a constructor. Include a method write_summary that writes out a summary message using the three pieces of information listed above. Create an instance of the class for an element and output the summary message.

Try This! Answer 17-1

This code solves the problem:

```
In [1]:  class Element(object):
             def __init__(self, symbol, num_protons, num_neutrons):
                 self.symbol = symbol
                 self.num_protons = num_protons
                 self.num_neutrons = num_neutrons

             def write_summary(self):
                 return self.symbol + ': ' + str(self.num_protons) \
                        + ' proton(s), ' + str(self.num_neutrons) \
                        + ' neutron(s).'

         hydrogen = Element('H', 1, 0)
         print(hydrogen.write_summary())

         H: 1 proton(s), 0 neutron(s).
```

This class is not a child class of another parent class besides the topmost parent `object`, so in the `class` statement, `object` is placed inside the parentheses. Remember that `write_summary` has no input parameters because all the information it needs to act on is stored in the object attributes. A string variable is returned by `write_summary`.

Try This! 17-2 Private Attributes: Information on an Element

How would we change the class `Element` in Try This! 17-1 so that the numbers of protons and neutrons are private? Change the class definition appropriately.

Try This! Answer 17-2

We change `Element` so `self.num_protons` and `self.num_neutrons` have one (or two) underscores in front of the name (after the `self` and the period). Thus, the class definition would become:

```
1   class Element(object):
2       def __init__(self, symbol, num_protons, num_neutrons):
3           self.symbol = symbol
4           self._num_protons = num_protons
5           self._num_neutrons = num_neutrons
6
7       def write_summary(self):
8           return self.symbol + ': ' + str(self._num_protons) \
9                       + ' proton(s), ' + str(self._num_neutrons) \
10                      + ' neutron(s).'
```

The prepended underscore is part of the attribute name and so must be included when referring to the attribute in the class definition. By adding a single prepended underscore, the attributes are set to "normal privacy." For more privacy, we would add two prepended underscores.

Try This! 17-3 Extending a Class: Scientific Bibliography

For the `Book` and `Article` classes in Figure 17.3, create an attribute `self.natural_name` that stores the first and last names as if the person was being addressed by name. Thus, for the `jones` object in lines 72–73 of Figure 17.3, `jones.natural_name` would be set to `'Timothy Jones'`. Have this attribute set by the constructor of the classes.

Try This! Answer 17-3

The easiest way to solve the problem is to put the following line in the `AuthoredResource` constructor after line 12 of Figure 17.3:

```
self.natural_name = self.authorfirst + ' ' + self.authorlast
```

We could also put that line in each of the constructors for `Book` and `Article`, but because of inheritance, by putting the code in `AuthoredResource`, the attribute is automatically created in `Book` and `Article`.

Try This! 17-4 Overriding: Information on an Element

Consider the class `Element` from Try This! 17-1. Create a class `AbbrevElement` that is a child of `Element` but overrides `write_summary` so that it writes the summary using symbols. Thus:

```
In [1]:  hydrogen = AbbrevElement('H', 1, 0)
         print(hydrogen.write_summary())

         H: 1 p, 0 n.
```

Try This! Answer 17-4

This class definition solves the problem:

```
1   class AbbrevElement(Element):
2       def write_summary(self):
3           return self.symbol + ': ' + str(self.num_protons) \
4                              + ' p, ' + str(self.num_neutrons) \
5                              + ' n.'
```

Try This! 17-5 Custom Exception Classes: File Input/Output

In Section 9.2.7, we learned to use exception classes to raise and handle error conditions. One use of inheritance is to create custom exception classes to use for specific error situations. Let us pretend we want to create a custom version of the `FileNotFoundError` exception for handling the case where a specific climate data file called *climate_data.txt* does not exist. In the code example below, we call a function `access_climate_data` that raises our custom exception if the *climate_data.txt* file does not exist. In that case, a default version is generated and filled with default values before we try to access the file again:

```
1   try:
2       access_climate_data()
3   except ClimateDataNotFoundError:
4       create_default_file()
5       access_climate_data()
```

The function `create_default_file` creates a default version of the file so the second attempt at accessing the file object in line 5 will work. The exception class `ClimateDataNotFoundError` is the custom exception we want to create.

Try This! Answer 17-5

The custom exception class is defined as:

```
1   class ClimateDataNotFoundError(FileNotFoundError):
2       pass
```

We could, if we wished, change some of the attributes or override some of the methods of `FileNotFoundError`, but we do not have to. Although we have not added any additional functionality to the class definition, by naming a new class, we do have additional functionality: the ability to customize our handling of this kind of error. See the Python documentation discussion on errors and exceptions. We provide an up-to-date link to the page at www.cambridge.org/core/resources/pythonforscientists/refs/, ref. 13. The Python documentation also contains a list of built-in exceptions; an up-to-date link to the page is at www.cambridge.org/core/resources/pythonforscientists/refs/, ref. 12.

Try This! 17-6 Sorting Using `sorted`: Scientific Bibliography

Change the `Bibliography` class in Figure 17.3 so that `sort_entries_alpha` will sort the contents of `self.entrieslist` by the `title` attribute of the objects stored in `self.entrieslist`.

Try This! Answer 17-6

We only have to change lines 53–55 in Figure 17.3 to:

```
        tmp = sorted(self.entrieslist,
                key=operator.attrgetter('title'))
```

Try This! 17-7 Automating Calling of Functions: Data Analysis

Pretend we have a one-dimensional array of numerical data called `data`. We have a file *metrics.txt* that consists of a single file of the text names of NumPy statisical functions, such as:

```
mean
median
std
```

All functions whose names are listed in that file are called with a single parameter. Write code that calculates and prints out the results of all the statistical functions given in *metrics.txt*, whether there are 3, 30, or more. Assume the list of functions in that file is properly constructed (e.g., no typos) and that NumPy has been imported as np.

Try This! Answer 17-7

This code will solve the problem (assuming `data` is already defined):

```
1   fobj = open('metrics.txt', 'r')
2   metrics = fobj.readlines()
3   fobj.close()
4
5   metrics = [ival[:-1] for ival in metrics]
6
7   for iattr in metrics:
8       temp = getattr(np, iattr)(data)
9       print(iattr + ': ' + str(temp))
```

We use the list comprehension in line 5 to strip off the newline character from each element in the list read in from the file. In front of each result (`temp`), the code will print the name of the statistical function (`iattr`).

17.4 More Discipline-Specific Practice

The Discipline-Specific Jupyter notebooks for this chapter cover the following topics:

- Defining and using a class. Attributes and methods, constructors and creating and using objects, and private attributes and methods.
- Inheritance.
- More sophisticated sorting using `sorted`.
- Automating handling of objects and modules.

17.5 Chapter Review

17.5.1 Self-Test Questions

Try to do these without looking at the book or any other resources or using the Python interpreter. Answers to these Self-Test Questions are found at the end of the chapter.

Self-Test Question 17-1

Describe what is included in a `class` statement.

Self-Test Question 17-2

What does `self` represent in a class definition? Is it used outside a class definition?

Self-Test Question 17-3

What is an attribute and a method in OOP?

Self-Test Question 17-4

If we have a class `RandomWalk1D` that describes a one-dimensional random walk model, what would be the code to create a 100-element array of *x*-values initialized to zeros?

Self-Test Question 17-5

Pretend we are defining a class `Blueprint` that has a method `normalize_units`. This method takes a single input parameter, a string, that has the value `'si'` or `'us'`, which selects Système International (SI) units or US Customary units (i.e., feet, etc.), respectively. The method goes through all the components of the blueprint and converts, if needed, all measurements into the unit system specified through the positional input parameter. It produces no return value. What is the code to call this method if we are in another method, `render_in_si` (the latter displays the blueprint to the screen in the named system of units)? How would this differ from calling the `normalize_units` method from outside the class definition, if we have a `Blueprint` object called `building`?

Self-Test Question 17-6

What is the name of the constructor method in a class definition in Python?

Self-Test Question 17-7

Pretend we are defining a class `Engine`. The following is an example of creating an instance of this class:

```
civic = Engine('gasoline', cyl=8)
```

What would be a possible signature for the constructor of this class?

Self-Test Question 17-8

How does Python denote private attributes and methods?

Self-Test Question 17-9

How is a child class related to a parent class?

Self-Test Question 17-10

If we create a class `ClimateModel` that is a child of `Model` in Figure 17.1, and we decide to override the `run` method, what is the syntax to call the `Model` version of `run` while in the `ClimateModel` definition?

Self-Test Question 17-11

Pretend we have a list of objects `labels` that consists of a variety of different classes of objects that store labeling information about a lab containing various experimental apparatus. Assuming the module operator is imported, describe what the following line of code does:

```
post_sort = sorted(labels,
                   key=operator.attrgetter('style', 'weight'))
```

Self-Test Question 17-12

Pretend we have an object of class `GrowthModel` called `model`. Write code that would check to see whether the object has a method called `run`, and, if so, then executes the method. Do not use syntax that has a period in it. Assume `run` takes no input parameters and has no return value.

17.5.2 Chapter Summary

In this chapter, we discovered how to create and use our own classes and how to derive one class from another using inheritance. With the ability to design our own classes, we found we have capabilities that are much more difficult to come by using procedural programming. This chapter, in that sense, is something of a paean to object-oriented programming: this style of programming enables us to write programs that are *both* more complex and more reliable. Specific topics we covered in this chapter include:

- Defining and using a class

  - Classes are defined by a `class` statement that gives the class name and any parent classes.
  - The `self` token is used in class statements to refer to the instance of the class the class definition is defining. In that way, variables whose names are prepended by `self` and a period refer to attributes and methods of that instance.
  - Attributes are set through regular variable assignment.
  - Methods are defined as if there were functions inside the class (i.e., starting with the `def` statement).
  - To call a method, we start with `self`, add a period, then the method name, and add parentheses to hold the argument list (which could be empty, depending on the method).
  - Instances of user-written classes are created by giving the class name and a pair of parentheses. Inside the parentheses, we put any arguments required by the constructor.
  - The constructor is the method in the class definition named `__init__`.
  - Private attributes and methods are denoted by name: there are one or two underscores prepending the alphabetical portion of the name, and there are either zero or one underscores after the alphabetical portion of the name. If the name has two underscores before and after the alphabetical portion, these are special attributes and methods, which are discussed in Section 9.2.2.

- Single-underscore prepended private attributes and methods are private by convention. Double-underscore prepended private attributes and methods are "very private," and the Python interpreter will do "name mangling" to offer some protection against inadvertent access to these attributes and methods.

- Inheritance
  - A child class, if built off a parent class, starts with the attributes and methods defined in the parent class. This is known as inheritance.
  - Inheritance is best used when the child classes are, in some way, "subsets" or "specific cases" of the parent class. Often, child classes add extra attributes or methods to what the child already has obtained from the parent.
  - A parent class is also known as a base class or superclass. A child class is also known as a derived class or subclass.
  - Inheritance can occur over many generations. The parents, grandparents, etc., of a class are known as ancestors. The children, grandchildren, etc., of a class are known as descendants.
  - We can check the specific type of an object by specifically comparing the output of the `type` function with a class. If we want to check whether an object is a particular class or has a particular class as its ancestor, we can use the `isinstance` function.
  - A child class can override a method it has inherited from its parent. In those cases, the child can still refer to the parent's version of the method by using the `super` built-in function.
  - Python supports multiple inheritance, where a child class has multiple parents.

- More sophisticated sorting using `sorted` and automating handling of objects and modules
  - The operator module has a function `attrgetter` that enables us to specify which attributes of each object to use for sorting a list of objects using the built-in `sorted` function.
  - OOP enables us to push customization decisions to the lowest level possible. In particular, it enables us to remove the decision of customization from the calling level of a function.
  - OOP makes programs more flexible and maintainable because, at any given level of function calls, we do not have to use `if` statements at that level to account for all the possible special cases.
  - Python has built-in functions such as `callable`, `delattr`, `getattr`, `hasattr`, and `setattr` that enable us to access and manipulate objects by referring to attributes and methods by string versions of their names. This enables us to use looping, string operations, and other programming constructs to tell the computer how to automatically process the attributes and methods.
  - The `dir` function returns a list of the string names of all attributes and methods in an object. We can loop through this list.

- The same features we use to automatically manage object attributes and methods can also be used to automatically manage module and package attributes and methods.
- Nearly everything in Python is an object!

17.5.3 Self-Test Answers

Self-Test Answer 17-1

The class statement begins with class, has a space, then gives the name of the class, an opening parenthesis, the name of the parent class (or object if there is no parent class), a closing parenthesis, and finally a colon. The body of the definition is under this statement, indented in one level.

Self-Test Answer 17-2

The self token represents an instance of the class being defined by the class statement. Thus, self plus a period and then an attribute or method name refers to the attribute or method of that instance. Outside a class definition, self is generally not used. Instead, the actual name of the instance is used. When referring to attributes or methods of that object, the name of the object is used, followed by a period, and then the attribute or method name.

Self-Test Answer 17-3

An attribute is a piece of data or information that is bound to an object. A method is a function that is bound to an object.

Self-Test Answer 17-4

In a method, preferably the constructor, include the following line (assuming NumPy has been imported as np:

```
self.x = np.zeros(100, dtype='d')
```

The code needs to be indented appropriately.

Self-Test Answer 17-5

This line of code calls the method while in the class definition:

```
self.normalize_units('si')
```

Outside the class definition, with the object building, we would call the method by:

```
building.normalize_units('si')
```

Self-Test Answer 17-6

The constructor is a special method in the class definition called __init__.

Self-Test Answer 17-7

This is one possible signature for the constructor:

```
def __init__(self, fuel_type, cyl=None):
```

A different default value can be chosen for the keyword input parameter cyl.

Self-Test Answer 17-8

An attribute or method with one or two underscores prepended in front of the alphabetic portion of the attribute or method's name denotes that the attribute or method is private. Zero or one underscores are permitted after the alphabetic portion of the attribute or method's name. A single prepended underscore means the attribute or method is private by convention only. Two prepended underscores means the attribute or method is very private, and the Python interpreter attempts to prevent access by doing some behind-the-scenes "name mangling."

Self-Test Answer 17-9

The child class has the attributes and methods the parent class has.

Self-Test Answer 17-10

```
super().run()
```

Self-Test Answer 17-11

The list post_sort is a list consisting of the objects in labels but ordered based upon the values of, first, the style attribute of the objects in labels, and, second, the weight attribute of the objects in labels.

Self-Test Answer 17-12

```
1  method = 'run'
2  if hasattr(model, method):
3      if callable(getattr(model, method)):
4          getattr(model, method)()
```

18 More Ways of Storing Information in Files

So far, we have seen how to use Python to access and/or store information in two different kinds of file formats: text files (Chapter 9) and image files (Chapter 13). The former hold numbers, letters, punctuation characters, and some special characters. The latter hold the image information and are encoded in formats such as the Portable Network Graphics (PNG) and Joint Photographic Experts Group (JPEG) formats. In the present chapter, we consider some additional file formats, the scientific and engineering applications they are useful for, and ways of using Python to interact with such files.

Similar to Chapter 14, the code and ideas in this chapter do not lend themselves as well to line-by-line footnoting and referencing as in other chapters. Many of the ideas and portions of code are from or built from those given in the pandas User Guide (see www.cambridge.org/core/resources/pythonforscientists/refs/, ref. 47, for an up-to-date link to the site) and Application Programming Interface (API) Reference (see www.cambridge.org/core/resources/pythonforscientists/refs/, ref. 44, for an up-to-date link to the site), the pickle module documentation (see www.cambridge.org/core/resources/pythonforscientists/refs/, ref. 61, for an up-to-date link to the site), the netCDF4 module documentation (see www.cambridge.org/core/resources/pythonforscientists/refs/, ref. 58, for an up-to-date link to the site),[1] and the documentation found in the Python interpreter. Additional references are given at www.cambridge.org/core/resources/pythonforscientists/refs18/.

Although in this chapter we explicitly create our file objects and explicitly use their `close` methods to close the files, we can also use the `with`/`as` notation as described in Section 9.2.6.

Files used in this chapter that are not generated by code in this chapter are available for download. See the "Supporting Resources and Updates to This Text book" section in the Preface for details.

18.1 Examples of Using Other File Formats

In Try This! 14-2, we considered a set of fictitious elevations (in meters above mean sea level) taken on a small hill.[2] The elevations are shown in Figure 14.14. Each location is 2 m away from each other (in North–South and East–West directions). The grid of elevations in Figure 14.14 is oriented so the northernmost row is the top row and the westernmost column is the left column. A contour plot of these data is shown in Figure 14.17. Assuming the data

---

[1] The reference for most of this chapter was https://unidata.github.io/netcdf4-python/netCDF4/index.html, which has since moved.

[2] Parts of this section are adapted from Chapter 5 of Lin (2012).

```
1    #- Module imports:
2    import numpy as np
3    import pandas as pd
4
5    #- Create labels for rows and columns:
6    nrows = np.shape(elev)[0]
7    ncols = np.shape(elev)[1]
8    rows = ['y = ' + str(i * 2.0) + ' m' for i in range(nrows)]
9    cols = ['x = ' + str(i * 2.0) + ' m' for i in range(ncols)]
10
11   #- Write out 2-D DataFrame to Excel file:
12   elev_df = pd.DataFrame(elev, index=rows, columns=cols)
13   elev_df.to_excel('hill.xlsx')
14
15   #- Read in 2-D DataFrame from Excel file and compare:
16   elev_df_in = pd.read_excel('hill.xlsx', index_col=0, header=0)
17   print(np.allclose(elev_df.values, elev_df_in.values))
18
19   #- Write out 1-D Series to Excel file:
20   elev0_ser = pd.Series(elev[:,0], index=rows, name=cols[0])
21   elev0_ser.to_excel('hill0.xlsx')
22
23   #- Read in 1-D Series from Excel file and compare:
24   elev0_df_in = pd.read_excel('hill0.xlsx', index_col=0, header=0)
25   print(np.allclose(elev0_ser.values, np.ravel(elev0_df_in.values)))
```

Figure 18.1 Code to write out `elev` from Figure 14.15 into the Excel files *hill.xlsx* (the entire array, Figure 18.2) and *hill0.xlsx* (the first column of the array, Figure 18.3).

are in the two-dimensional array `elev` as defined in Figure 14.15, in this section, we provide code to write out the data (and metadata) to an Excel file, a pickle file, and a netCDF file. Details about these file formats and Python tools for interacting with these files are given in Section 18.2.

Excel File Input/Output   Figure 18.1 shows code to write out data into cells in an Excel file, to read it back in, and to compare whether what was read back in matches the original data. This is done once using the full two-dimensional array `elev` (lines 11–17) and a second time using the first column of `elev` (lines 19–25). In both cases, we use classes and functions that are part of the pandas package. The result of writing out the two-dimensional array `elev` to the Excel file *hill.xlsx* is shown in Figure 18.2. The result of writing out the first column of `elev` to the Excel file *hill0.xlsx* is shown in Figure 18.3.

In lines 2–3 of Figure 18.1, we import NumPy and pandas. In lines 6–9, we extract the number of rows and columns in `elev` and create the row and column labels that will be written out to the Excel file. These labels give the *x* and *y* distance values relative to the northwest corner of the grid. Lines 8–9 use list comprehensions, which are described in Section 16.2.2.

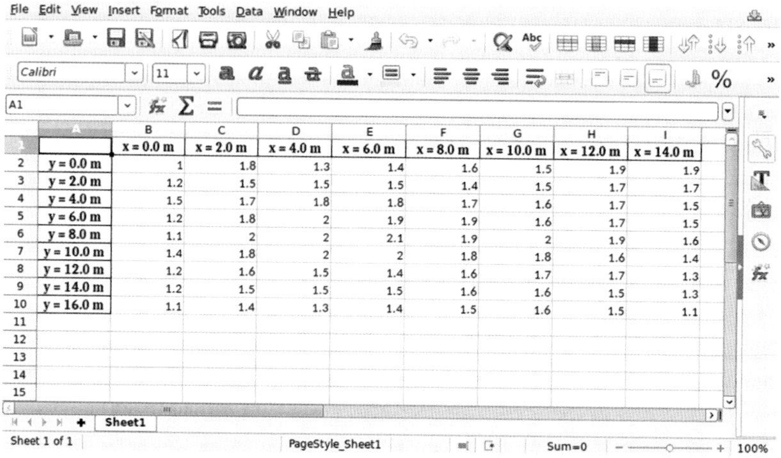

	x = 0.0 m	x = 2.0 m	x = 4.0 m	x = 6.0 m	x = 8.0 m	x = 10.0 m	x = 12.0 m	x = 14.0 m
y = 0.0 m	1	1.8	1.3	1.4	1.6	1.5	1.9	1.9
y = 2.0 m	1.2	1.5	1.5	1.5	1.4	1.5	1.7	1.7
y = 4.0 m	1.5	1.7	1.8	1.8	1.7	1.6	1.7	1.5
y = 6.0 m	1.2	1.8	2	1.9	1.9	1.6	1.7	1.5
y = 8.0 m	1.1	2	2	2.1	1.9	2	1.9	1.6
y = 10.0 m	1.4	1.8	2	2	1.8	1.8	1.6	1.4
y = 12.0 m	1.2	1.6	1.5	1.4	1.6	1.7	1.7	1.3
y = 14.0 m	1.2	1.5	1.5	1.5	1.6	1.6	1.5	1.3
y = 16.0 m	1.1	1.4	1.3	1.4	1.5	1.6	1.5	1.1

Figure 18.2 Screenshot of *hill.xlsx* generated by the code in Figure 18.1 (reformatted manually for readability). The file shown is opened in the spreadsheet application LibreOffice (as is also shown in Figures 18.3, 18.7, and 18.9).

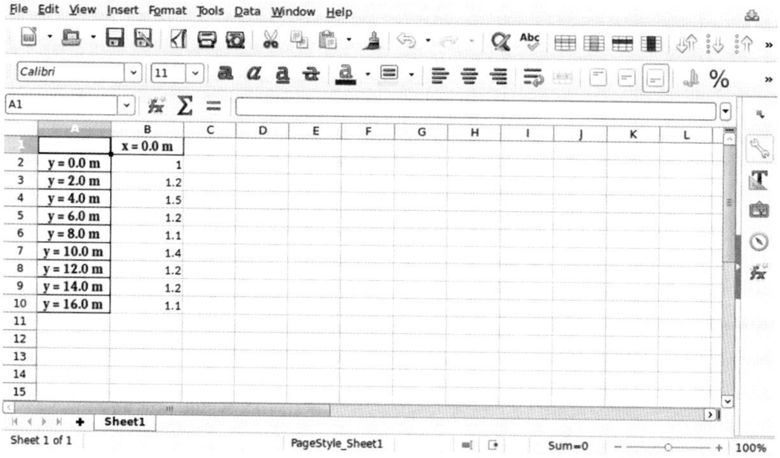

	x = 0.0 m
y = 0.0 m	1
y = 2.0 m	1.2
y = 4.0 m	1.5
y = 6.0 m	1.2
y = 8.0 m	1.1
y = 10.0 m	1.4
y = 12.0 m	1.2
y = 14.0 m	1.2
y = 16.0 m	1.1

Figure 18.3 Screenshot of *hill0.xlsx* generated by the code in Figure 18.1 (reformatted manually for readability).

In lines 11, we create a `DataFrame` object from the `elev` array, with the row labels set by `rows` and the column labels set to `cols`. Line 13 writes the `DataFrame` object `elev_df` to the Excel file *hill.xlsx*. In line 16, we read in the contents of *hill.xlsx* into a separate `DataFrame` object and compare those values with the original `elev_df` values. The value `True` is output to the screen, because both arrays are the same.

```
1    #- Module import:
2    import pickle
3
4    #- Make copy of elev:
5    copy_elev = elev.copy()
6    copy_elev_df = elev_df.copy()
7
8    #- Write objects to a pickle file:
9    fileobj = open('hill.pkl', 'wb')
10   pickle.dump([copy_elev, copy_elev_df], fileobj, protocol=4)
11   fileobj.close()
12
13   #- Delete copied variables and the file object:
14   del(copy_elev)
15   del(copy_elev_df)
16   del(fileobj)
17
18   #- Read in objects from the pickle file:
19   fileobj = open('hill.pkl', 'rb')
20   result = pickle.load(fileobj)
21   copy_elev = result[0]
22   copy_elev_df = result[1]
23   fileobj.close()
24
25   #- Check read in objects are the same as originals:
26   print(np.allclose(elev, copy_elev))
27   print(np.allclose(elev_df.values, copy_elev_df.values))
```

Figure 18.4 Code to write out `elev` from Figure 14.15 into the pickle file *hill.pkl*.

In line 20, we create a `Series` object from the first column of `elev`. The row labels are set by `rows` and the column label is the first element of `cols[0]`. We write out the `Series` object to *hill0.xlsx* in line 21. In line 24, we read in that Excel file and save the result in a new `Series` object, `elev0_df_in`. Line 25 compares the original values of `elev0_ser` and `elev0_df_in`. The flattened (i.e., `raveled`) version of the values of `elev0_df_in` is used in the comparison because the Excel file values come in as a column in a two-dimensional array. The value `True` is also output to the screen in line 25.

pickle File Input/Output   Figure 18.4 shows code to write out data into a pickle file, a special Python format that enables us to store entire Python objects in a file. That is, what we read in from the file is not merely data values but the complete object itself. In Figure 18.4, we assume the imports that were executed in Figure 18.1 have also been executed, and that `elev` and `elev_df` have already been defined as in Figure 18.1. We leave out the code that imported and defined those objects to highlight the pickle-related code.

We begin in line 2 of Figure 18.4 by importing the pickle module. In lines 5–6, we make copies of `elev` and `elev_df`. We do so to keep the originals unchanged as we execute the rest

```
1    #- Module import:
2    import netCDF4 as nc
3
4    #- Write data to a netCDF file:
5    fileobj = nc.Dataset('hill.nc', mode='w')
6    fileobj.createDimension('x', ncols)
7    fileobj.createDimension('y', nrows)
8
9    elev_var = fileobj.createVariable('elev', 'd', ('y','x'))
10   elev_var.units = "m above mean sea level"
11   elev_var[:,:] = elev
12
13   x_var = fileobj.createVariable('x', 'd', ('x'))
14   x_var.units = 'm'
15   x_var[:] = np.arange(ncols) * 2.0
16
17   y_var = fileobj.createVariable('y', 'd', ('y'))
18   y_var.units = 'm'
19   y_var[:] = np.arange(nrows) * 2.0
20
21   fileobj.title = "Fictitious elevations on a small hill"
22
23   fileobj.close()
24
25   #- Delete netCDF variable objects and the file object:
26   del(elev_var, x_var, y_var)
27   del(fileobj)
```

Figure 18.5 Code to write out `elev` from Figure 14.15 into the netCDF file *hill.nc*.

of the code. In lines 9–11, we write a list that contains the `copy_elev` and `copy_elev_df` objects into the pickle file *hill.pkl*. In lines 14–16, we delete the `copy_elev`, `copy_elev_df`, and `fileobj` objects. We do so to make sure they do not exist before we recreate these objects in lines 19–23. We want our comparison between our original and read-back objects to be honest. In lines 19–23, we read in the list we had written out to the pickle file in line 10 and save those objects in that list to the names `copy_elev` and `copy_elev_df`. In lines 26–27, we compare the values of the read-in versions of `copy_elev` and `copy_elev_df` with the originals of those objects and confirm they are the same.

netCDF File Input/Output   Figure 18.5 shows code to write out data into a netCDF file, a special portable binary format that can be opened on many different operating systems and computer hardware. In Figure 18.5, we assume the imports that were executed in Figure 18.1 have also been executed and that `elev`, `nrows`, and `ncols` have already been defined as in

```
28   #- Read in data from the netCDF file:
29   fileobj = nc.Dataset('hill.nc', mode='r')
30
31   elev_data = fileobj.variables['elev'][:]
32   elev_units = fileobj.variables['elev'].units
33   file_title = fileobj.title
34
35   x_data = fileobj.variables['x'][:]
36   x_units = fileobj.variables['x'].units
37
38   y_data = fileobj.variables['y'][:]
39   y_units = fileobj.variables['y'].units
40
41   fileobj.close()
42
43   #- Check read in objects are the same as originals:
44   print(np.allclose(elev, elev_data))
45   print(elev_units == "m above mean sea level")
46   print(file_title == "Fictitious elevations on a small hill")
47   print(np.allclose(x_data, np.arange(ncols) * 2.0))
48   print(x_units == 'm')
49   print(np.allclose(y_data, np.arange(nrows) * 2.0))
50   print(y_units == 'm')
```

Figure 18.5   (continued)

Figure 18.1. We leave out the code that imported and defined those objects to highlight the netCDF-related code.

In line 2 of Figure 18.5, we import the netCDF4 module. The capitalization is as shown, which is different than most Python modules. In lines 5–7, we create and open the *hill.nc* netCDF formatted file for writing and create netCDF dimension entities that describe the number of elements in the *x*- and *y*-directions. In line 9, we use the dimension entities to create a netCDF variable entity of the correct shape to hold the data from elev (we call this variable 'elev'). In line 10, we add metadata for the units of the elev data. We fill the variable entity with the data values in elev in line 11. Lines 13–15 and 17–19 create similar netCDF variable entities for the values in the *x*- and *y*-directions. The netCDF file is closed in line 21.

In lines 25–27, we delete the objects corresponding to the netCDF variable entities created earlier, as well as the netCDF file object. In lines 29–39, we read in the variables stored in *hill.nc*, saving the data values of the elevations and the *x*- and *y*-direction locations to elev_data, x_data, and y_data, along with the metadata for those variables. In lines 44–50, we check that the data and metadata values read in from the *hill.nc* file are the same as the values written out.

18.2   Python Programming Essentials

The three file formats we used in Section 18.1 – Excel, pickle, and netCDF – have different purposes and features, and Python interacts with them in different ways. In this section, we describe each format, the Python interface to those files, and the uses of the format. We provide only a brief treatment of these file formats. For details, please consult the documentation for the pertinent packages, described at the end of each subsection.

18.2.1 Excel Files

Excel spreadsheets are the bread-and-butter of small-to-medium-scale data analysis, for good reason. Graphical user-interface (GUI) spreadsheet programs such as Excel give us an intuitive and spatial sense of the cells of data, their relationships to each other, and how to manipulate them. There are times, however, when the calculations we want to do with the data in a spreadsheet are more involved than can be easily done using a point-and-click interface, or we want to automate tasks using a spreadsheet's data that are more sophisticated than Excel's **macro language** can easily carry out. In those cases, it can be more productive to read/write Excel files to/from Python, and manipulate the data in Python.

In the present section, we use pandas and pandas `Series` and `DataFrame` objects to write out and read in array data to and from an Excel spreadsheet. In Section 15.2.2, we said that `Series` objects are "enhanced" one-dimensional arrays and that `DataFrame` objects are enhanced two-dimensional arrays. Each row of a `Series` object, and each row and column in a `DataFrame` object, can be assigned labels beyond the integer indices NumPy arrays have. In that way, we can give meaningful, alphanumeric labels to each row or row and column. In Section 19.2.3, we further describe `DataFrame` objects and discuss how we can also use `DataFrame` labels to select columns, rows, and/or subranges from `Series` or `DataFrame` objects, which then can be manipulated.

When we create `DataFrame` objects, we assign the row labels using the `index` keyword input parameter and the column labels using the `columns` keyword input parameter. In Figure 18.1, line 12, we create a `DataFrame` from the two-dimensional array `elev` using the `rows` and `columns` list of strings we created in lines 8–9 for the labels. The resulting object, `elev_df`, looks like the following (only the first four columns are shown):

```
In [1]:  print(elev_df)


                     x = 0.0 m   x = 2.0 m   x = 4.0 m   x = 6.0 m ...
        y = 0.0 m        1.0         1.8         1.3         1.4 ...
        y = 2.0 m        1.2         1.5         1.5         1.5 ...
        y = 4.0 m        1.5         1.7         1.8         1.8 ...
        y = 6.0 m        1.2         1.8         2.0         1.9 ...
        y = 8.0 m        1.1         2.0         2.0         2.1 ...
        y = 10.0 m       1.4         1.8         2.0         2.0 ...
        y = 12.0 m       1.2         1.6         1.5         1.4 ...
        y = 14.0 m       1.2         1.5         1.5         1.5 ...
        y = 16.0 m       1.1         1.4         1.3         1.4 ...
```

`Series` objects work similarly for the one-dimensional case. When we create `Series` objects, we assign the row labels using the `index` keyword input parameter and the single column label using the `name` keyword input parameter. In Figure 18.1, line 20, we create a `Series` object using the first column of `elev` and its column label. The resulting object, `elev0_ser`, looks like the following:

```
In [1]:  print(elev0_ser)

         y = 0.0 m      1.0
         y = 2.0 m      1.2
         y = 4.0 m      1.5
         y = 6.0 m      1.2
         y = 8.0 m      1.1
         y = 10.0 m     1.4
         y = 12.0 m     1.2
         y = 14.0 m     1.2
         y = 16.0 m     1.1
         Name: x = 0.0 m, dtype: float64
```

The `to_excel` method of `Series` and `DataFrame` objects write the contents of the object to an Excel file, which can be specified as a single input parameter (lines 13 and 21 of Figure 18.1). This includes the row and column labels, as shown in Figures 18.2 and 18.3.

The pandas `read_excel` function reads an Excel file. Lines 16 and 24 of Figure 18.1 call this function, passing in the filename of the Excel file as well as specifying which column in the file (`index_col`) to use for the returned `DataFrame`'s index labels and which row in the file (`header`) to use for the returned `DataFrame`'s column labels. The row and column labels are stored as the `DataFrame` object's `index` and `columns` attributes, respectively.

The data (as a NumPy array) in the `DataFrame` object are stored in the `values` attribute. As we saw in Section 18.1, for the one-column array stored in *hill0.xlsx*, the values come in as a one-column `DataFrame` object rather than a `Series` object. Thus, to properly compare with the values of a `Series` object (line 25 in Figure 18.1), we have to use the flattened version of the values of `elev0_df_in` to properly compare with the values of `elev0_ser`.

Please see the pandas API Reference on the `DataFrame to_excel` method for more options we can use when calling the method. We provide an up-to-date link to the reference at www.cambridge.org/core/resources/pythonforscientists/refs/, ref. 48. The `Series to_excel` method documentation is similar. The pandas API Reference on the `read_excel` function also describes more options to use when reading in from an Excel file. We provide an up-to-date link to the reference at www.cambridge.org/core/resources/pythonforscientists/refs/, ref. 50. Multiple worksheet workbooks can also be handled by pandas (see Try This! 18-2 and 18-3 for examples).

For more low-level interaction with an Excel file, please see the openpyxl package, which allows us to customize Excel files, including formatting. An up-to-date link to a tutorial is at www.cambridge.org/core/resources/pythonforscientists/refs/, ref. 59. This package may not be installed in the Anaconda distribution by default. If it is not installed, first install it following the instructions in Section 1.3.

18.2.2 pickle Files

The pickle format essentially saves a "snapshot" of an object or objects to a binary file. Thus, when we load that snapshot from the pickle file (which can be named whatever we want, but the *.pkl* suffix works well), the object is ready to be used in Python. Its attributes and methods are all present. We do not have to convert the object into something else we are more interested in. When we **pickle** an object, we dump it into a pickle file. When we **unpickle** an object, we load it from a pickle file. The word "pickle" can be a verb or an adjective.

To create a pickle file, we use open as in the case of text files (see Section 9.2.6). However, the mode of the file has to be set to denote both "writing mode" and "binary mode." This is done by setting the second positional input parameter in open to 'wb' (as in line 9 in Figure 18.4). To read from a pickle file, we set the mode to 'rb', which denotes "reading mode" and "binary mode" (as in line 19 in Figure 18.4). We close a pickle object by calling the file object's close method.

To pickle an object, we use the pickle module's dump function. In line 10 in Figure 18.4, we call dump with two positional and one keyword input parameter (reproduced here for convenience):

```
pickle.dump([copy_elev, copy_elev_df], fileobj, protocol=4)
```

The two positional input parameters specify the object being pickled and the pickle file's file object, respectively. The keyword input parameter specifies the "protocol" to use; if absent, a default value is used. As the dump function pickles a single object, when we want to pickle multiple objects, it is probably easiest to put all the objects we want to pickle into a list and then pickle that list.[3] Our call to dump in line 10 takes this approach: The first positional input parameter is a list of two objects.

To unpickle an object, we use the pickle module's load function, passing in the pickle file's file object as the single input parameter (as in line 20 in Figure 18.4). The return value from calling load is the object that was pickled. Thus, in the line 20 example, what is unpickled is the list of two objects that was pickled. In order to obtain the objects that were elements of that list, we have to extract them from those elements (as in lines 21–22).

As of this writing, Python supports six different pickle file "protocols," all of which have different features and are compatible with certain versions of Python. Details on each protocol are found in the pickle module's documentation.[4] Protocol number 4 is the default protocol starting with Python 3.8 and has been available since Python 3.4. Protocols numbers are only used when pickling. When unpickling, the load function automatically detects which protocol the file uses.

There are some limitations on what types can and cannot be pickled. Please see the pickle module documentation for more options we can use when calling the methods for dumping

---

[3] https://stackoverflow.com/a/20725705/8430411 (accessed July 9, 2020).
[4] An up-to-date link to the pickle documentation is at www.cambridge.org/core/resources/pythonforscientists/refs/, ref. 61.

or loading data. An up-to-date link to the documentation is at www.cambridge.org/core/resources/pythonforscientists/refs/, ref. 61.

18.2.3 netCDF Files

Text files, Excel files, and pickle files all work fine for storing information.[5] But for larger datasets that we also want to be usable on nearly any computer and accessible by nearly any programming language, these options have limitations. To explain why, and why the netCDF format and similar formats exist, we need to step back and add some more background regarding how files are stored in a computer.

As we mentioned in Section 9.2.2, the most fundamental representation of data in a computer is as a binary digit or bit (a zero or one). Whether in the memory of a computer or as a file on a hard drive or other form of storage, this is how data are stored. Thus, the most basic way of storing a file is to store it directly in this binary form, as expressed by the operating system for the computer's hardware. Such files are called plain or flat binary files. Such files are the most compact and quickly accessed way of storing data in a computer. But, binary files are specific to the computer that created them. A binary file created on a Linux computer cannot be directly read on a Windows computer, and so on. If we store data as a binary file, we severely limit who else can use the data.

The other file formats we have seen do not have that limitation. Although these other formats also fundamentally store information in bits (i.e., in binary form), the formats follow standards that enable them to be readable on many different computers. Text files follow the Unicode (or similar) standard for representing characters, Excel follows the format for spreadsheets defined by Microsoft, and pickle files follow the format defined by Python. None of these formats, however, is an acceptable replacement for the binary format in encoding a large data file. For text files, the conversion to and from Unicode takes substantial amounts of processor time, and text files are substantially larger than their binary counterparts. Although we can manipulate Excel files in Python, Excel files are meant to be used by an office productivity program and thus do not work well when the dataset is more than hundreds of thousands of lines. Python directly loads a pickle file into memory, but programs written in other programming languages cannot, in general, read pickle files.

What we need is a data-file format that has the compactness, rapid access, and thus scalability for large datasets that binary files provide, coupled with the portability (i.e., the file can be used on nearly any computer) that a text file provides. This is the purpose of the netCDF format. It defines a format that, although binary, nonetheless is able to be read on many different computers, through the use of utility routines that can be run on many different operating systems and using many different languages. The format also gives tools to include metadata with the data, so both are bundled together in the same file. Finally, the format's structure easily handles regular n-dimensional arrays, so netCDF files work well for gridded datasets, where the data are located on a regular grid (for instance, a longitude–latitude grid). Because of these features, the format is often used for global-scale climate datasets.

---

[5] Parts of this section are adapted from Chapter 5 of Lin (2012).

Before we describe the netCDF4 Python interface, we briefly review the structure of the netCDF data format itself. There are four general types of parameters in a netCDF file: global attributes, variables, variable attributes, and dimensions. It is unfortunate that the netCDF format uses the term "attributes," because the term attributes has a very specific meaning in object-oriented languages. When we talk about object attributes in close proximity to netCDF attributes, we will try to add the modifier "netCDF" to denote a netCDF format attribute. Global attributes are usually strings that describe the file as a whole, e.g., a title, who created it, what standards it follows, etc. Variables are the entities that hold data. These include both the data proper (e.g., elevation, temperature, etc.), the domain the data are defined on (delineated by the dimensions), and metadata about the data (e.g., units). Variable attributes store the metadata associated with a variable. Dimensions define the extent of each axis in an array of data in the dataset (i.e., the number of elements in each dimension). Dimensions are used to define the shape of arrays of variable data in the netCDF file, but they also often have variable values of their own (e.g., latitude values, longitude values, etc.). Thus, we usually create variables for each dimension (to hold those values) that are the same name as the dimension (which hold the number of elements in that dimension of the domain).[6]

As an example of a set of variables and dimensions for a netCDF file, consider our Section 18.1 example. Because the `elev` array is a two-dimensional grid, with two dimensions in space (in the x- and y-directions), we want to define two netCDF dimensions and two corresponding netCDF variables, one for each dimension. In lines 6 and 7 in Figure 18.5, we define an `'x'` netCDF dimension and a `'y'` netCDF dimension to be equal to the number of columns in `elev` (`ncols`) and the number of rows in `elev` (`nrows`). In lines 13 and 17, we define an `'x'` netCDF variable and a `'y'` netCDF variable. Each of these netCDF variables is a one-dimensional arrays of the same size as the corresponding netCDF dimensions, and these netCDF variables hold the x- and y-location values of each `elev` grid point. Finally, we define the `'elev'` netCDF variable (line 9), to hold the elevation data, which has a shape where the number of rows is given by the `'y'` netCDF dimension and the number of columns is given by the `'x'` netCDF dimension.

Writing a netCDF File   In order to write out a netCDF file, we first create a file object set for writing, using the `Dataset` constructor in the netCDF4 module (line 5 in Figure 18.5). The `mode` keyword parameter is set to `'w'`. Once the file object exists, we use methods of the file object to create the netCDF dimensions and netCDF variable objects that will be in the file. We have to create the netCDF dimensions before the netCDF variable objects (because the latter depends on the former). And we have to create the netCDF variable objects first before we fill them with values and metadata.

The `createDimension` method of the `Dataset` object creates a netCDF dimension (lines 6–7 in Figure 18.5). This method creates the name of the dimension and sets the value (length or number of elements) of the dimension. The name of the dimension is a string and

---

[6] For more on the netCDF file format, see www.unidata.ucar.edu/software/netcdf/docs (accessed July 10, 2020).

is passed in as the first argument to the method, and the number of elements in the dimension is an integer and is passed in as the second argument to the method.

The `createVariable` method (lines 9, 13, and 17 in Figure 18.5) creates a netCDF variable object. It only creates the infrastructure for the netCDF variable (e.g., the shape of the netCDF variable) and does not fill the values of the variable, set any netCDF variable attributes, etc. A call to the method has three arguments: first, the name of the netCDF variable (as a string); second, the type of the netCDF variable elements as a string (e.g., `'d'` means double-precision floating point); and third, the shape of the array the netCDF variable holds. The shape is a tuple of strings where the elements are names of dimensions. Thus, in line 9 of Figure 18.5 (reproduced here for convenience):

```
elev_var = fileobj.createVariable('elev', 'd', ('y','x'))
```

the array being stored in `elev_var` has the shape given by (`nrows`, `ncols`), because the tuple being passed into the `createVariable` call is (`'y'`,`'x'`). If the value being stored in the netCDF variable object is not an array but a scalar, an empty tuple (i.e., `()`) is passed in as the third argument to the `createVariable` call. The return value of the `createVariable` call is a Python object (e.g., `elev_var` of line 9), so we use Python objects to represent a netCDF variable object.

The way we fill netCDF variable objects with data depends on whether the variable is an array or a scalar. If it is an array (as in the Figure 18.5 examples), we use the slicing syntax (i.e., the colon) with the variable object in an assignment operation (e.g., lines 11, 15, and 19 in Figure 18.5). Thus, line 11 (reproduced here for convenience):

```
elev_var[:,:] = elev
```

means "fill in the elements of the array the `elev_var` netCDF variable object stores with the corresponding elements in the NumPy array `elev`." The values of scalar variables are assigned to the variable object through the `assignValue` method of the *netCDF variable object* (not of the file object). For example, the following would create a double-precision floating-point scalar netCDF variable called `'value'` and fill it with the number `42.0`:

```
scalar_var = fileobj.createVariable('value', 'd', ())
scalar_var.assignValue(42.0)
```

Note that `'value'` is the name of the netCDF variable *in the netCDF file*. The name of the netCDF variable object *in the Python interpreter* is `scalar_var`.

Finally, netCDF variable attributes are set using Python's regular object assignment syntax, as applied to the netCDF variable object. For instance, line 10 (reproduced here for convenience):

```
elev_var.units = "m above mean sea level"
```

creates a netCDF variable attribute `units` that is attached to the netCDF variable object `elev_var`. The netCDF variable attribute is the string, `"m above mean sea level"`.

Besides netCDF dimensions and variables, a netCDF file can hold global attributes. These are usually strings that document the file as a whole. They can include a title, authorship, references to other documentation, example usage of the data, a description of the dataset, etc. In line 21 of Figure 18.5, we create a single netCDF global attribute called `title`. This is created using the object attribute assignment syntax, but in contrast to netCDF variable attributes, we use Python object attributes assigned to the *file object* to represent netCDF global attributes. In line 21, the `title` netCDF global attribute is assigned to the string, `"Fictitious elevations on a small hill"`. After we have finished writing to the netCDF file, we close the file object using the object's `close` method.

Reading a netCDF File   To read in a netCDF file, we first create a file object using the netCDF4 module's `Dataset` constructor, but with the `mode` keyword input parameter set to `'r'` for read (line 29 of Figure 18.5). When we are finished reading the file object, we close it using the `close` method (line 41).

When we created the netCDF global attribute `title` in line 21 of Figure 18.5, we saved it as an object attribute to the Python file object `fileobj` representing the netCDF file. When we read in a netCDF file, netCDF global attributes are represented the same way. Thus, in line 33, when we read the `title` netCDF global attribute, we use the Python syntax for referencing attributes in an object to reference the `title` attribute of `fileobj`.

The netCDF variables are stored in a netCDF file object attribute called `variables`. This object attribute is a dictionary-like data structure (technically a class `OrderedDict` object), where the keys of the dictionary are the names (as strings) of the netCDF variable and the values associated with those names are netCDF variable objects that contain the values of the array the variable holds and the netCDF variable attributes. Thus, for example, in lines 31 and 32, the expression:

```
fileobj.variables['elev']
```

refers to the netCDF variable object associated with the name `'elev'`. When we use array slicing to select elements from that object, we are accessing the array values. In line 31 (reproduced here for convenience):

```
elev_data = fileobj.variables['elev'][:]
```

all the two-dimensional array data in the `'elev'` netCDF variable are read out and saved to the Python variable `elev_data`. Even though the data are two-dimensional, it is enough to use the syntax `[:]`. That syntax works regardless of the number of dimensions of the array of data. Also, when the array of data is read in, it comes in as a masked array. If there are no missing values, the masked array has no mask, but the object is still of the masked-array type.

To obtain the netCDF variable attributes, we can refer to the netCDF variable attribute name via the syntax in Python to refer to object attributes. Recalling `fileobj.variables` `['elev']` is a Python object, the right-hand side of line 32 refers to the `units` attribute of that object. That value is a string which is stored as the Python variable `elev_units`.

For scalar netCDF variables, to obtain the value of the data, we use a method attached to the netCDF variable object called `getValue`. For a scalar netCDF variable named `'value'` in a netCDF file object `fileobj`, the following would return the value of the variable:

```
fileobj.variables['value'].getValue()
```

which could then be saved to a Python variable via assignment. NetCDF variable attributes for scalar netCDF variables are read in the same way as for array data netCDF variables.

The netCDF file object also has a dictionary-like attribute called `dimensions` that holds the netCDF dimensions. Our Figure 18.5 does not have any examples of using `dimensions`. The keys of `dimensions` are strings that are the names of the netCDF dimensions. The values are objects that hold information about the dimension, for instance the length of the netCDF dimension as the attribute `size`. Thus, if after opening the netCDF file in line 29 of Figure 18.5 we executed this expression:

```
fileobj.dimensions['x'].size
```

the value 8 would be returned, which is the number of columns of `elev`.

From this brief description of the netCDF4 interface, we can see that the format is less flexible, in terms of how the data are stored, than text files and pickle files. Compared with Excel files, netCDF files do not (as yet) support as flexible a graphical interface as the Excel format does. Still, there is no such thing as a perfect file format, just as there is no such thing as the perfect programming language. For its limitations, the netCDF format and netCDF4 module enable us to relatively easily store and access large amounts of data in a compact yet portable format. Please see the netCDF4 module documentation for more options we can use when reading and writing netCDF files. We provide an up-to-date link to the documentation at www.cambridge.org/core/resources/pythonforscientists/refs/, ref. 58.

18.3   Try This!

The examples in this section address the main topics of Section 18.2. We practice reading and writing Excel, pickle, and netCDF4 files.

For all the exercises in this section, assume NumPy has already been imported as np.

Try This! 18-1 Write to an Excel File: Air Quality

Write the air quality data in the NumPy array `pm25` (assume it is already defined as in Figure 7.7) to an Excel file *pm25.xlsx*. Provide appropriate row and column labels (given in the caption of Table 7.7).

Try This! Answer 18-1

```
1  import pandas as pd
2
3  rows = [str(i) for i in range(np.shape(pm25)[0])]
4  cols = ['Dongsi', 'Dongsihuan', 'Nongzhanguan', 'U.S. Embassy']
5
6  pm25_df = pd.DataFrame(pm25, index=rows, columns=cols)
7  pm25_df.to_excel('pm25.xlsx')
```

Figure 18.6 Code to write out NumPy array `pm25` from Figure 7.7, with appropriate row and column labels, into the Excel file *pm25.xlsx* (Figure 18.7).

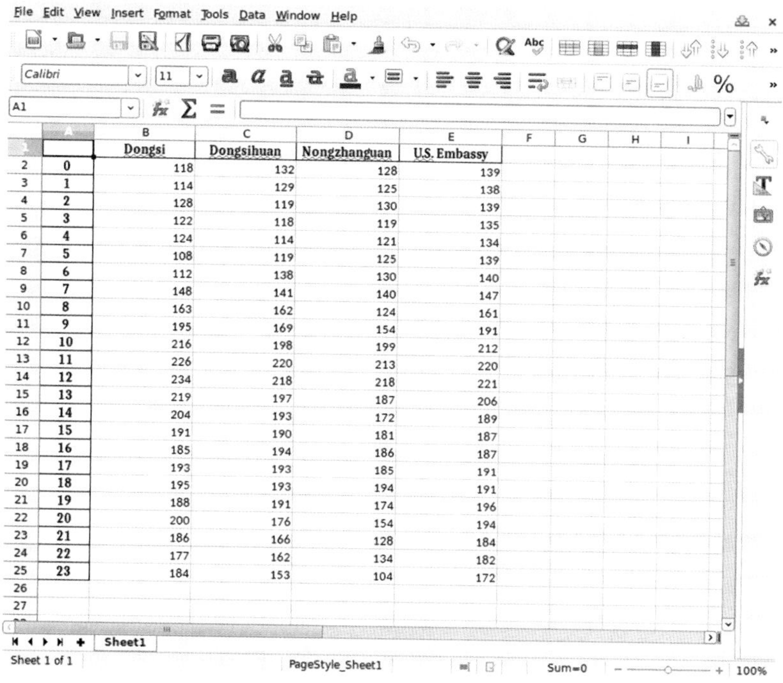

	Dongsi	Dongsihuan	Nongzhanguan	U.S. Embassy
0	118	132	128	139
1	114	129	125	138
2	128	119	130	139
3	122	118	119	135
4	124	114	121	134
5	108	119	125	139
6	112	138	130	140
7	148	141	140	147
8	163	162	124	161
9	195	169	154	191
10	216	198	199	212
11	226	220	213	220
12	234	218	218	221
13	219	197	187	206
14	204	193	172	189
15	191	190	181	187
16	185	194	186	187
17	193	193	185	191
18	195	193	194	191
19	188	191	174	196
20	200	176	154	194
21	186	166	128	184
22	177	162	134	182
23	184	153	104	172

Figure 18.7 Screenshot of *pm25.xlsx* generated by the code in Figure 18.6 (reformatted manually for readability). The file shown is opened in the spreadsheet application LibreOffice (as is also shown in Figures 18.8, 18.9, and 18.9).

Figure 18.6 shows the code that generates this file. Figure 18.7 shows the Excel file that is created. To construct the `rows` row labels in line 3 of Figure 18.6, we use list comprehensions, which are described in Section 16.2.2.

```
1    import pandas as pd
2
3    rows = [str(i) for i in range(np.shape(pm25)[0])]
4    cols = ['Dongsi', 'Dongsihuan', 'Nongzhanguan', 'U.S. Embassy']
5
6    fileobj = pd.ExcelWriter('pm25-1cols.xlsx', mode='w')
7    for i in range(len(cols)):
8        pm25_ser = pd.Series(pm25[:,i], index=rows, name=cols[i])
9        pm25_ser.to_excel(fileobj, sheet_name=cols[i])
10
11   fileobj.close()
```

Figure 18.8 Code to write out each column of NumPy array pm25 from Figure 7.7, with appropriate row and column labels, into separate worksheets of the Excel file *pm25-1cols.xlsx*. The second worksheet is shown in Figure 18.9.

Try This! 18-2 Write Multiple Worksheets to an Excel File: Air Quality

Do Try This! 18-1 again, but write each column to a separate worksheet in the Excel workbook. For each worksheet's name, use the column label. In pandas, multiple worksheet writing requires using the ExcelWriter class. Hint: Please see the ExcelWriter documentation for details. We provide an up-to-date link to the documentation at www.cambridge.org/core/resources/pythonforscientists/refs/, ref. 49.

Try This! Answer 18-2

Figure 18.8 shows the code that generates this file. Figure 18.9 shows the second worksheet of the Excel workbook that is created. In Figure 18.8, lines 1–4 are the same as in lines 1–4 of Figure 18.6.

In pandas, in order to write multiple worksheets to the same file, the DataFrame and Series to_excel methods have to have an ExcelWriter object to operate on. If we provide a string filename and use that filename for multiple to_excel calls (say, in an effort to write different worksheets to the same file), that file will be repeatedly overwritten each time the to_excel method evaluates. If we provide an ExcelWriter object to the to_excel call, the different worksheets will be added until the file object is closed.

In line 6 of Figure 18.8, we create an ExcelWriter object and set it to write mode. We loop through the indices of each column in pm25 in line 7, create a Series object from each column in line 8, and write that Series object to a worksheet in the workbook in line 9. The to_excel method has a keyword input parameter sheet_name that sets the name of the worksheet. We close the file object in line 11.

Try This! 18-3 Read Multiple Worksheets from an Excel File: Air Quality

Read in the multiple-sheet Excel workbook generated in Try This! 18-2 so separate DataFrame objects represent the worksheets in the workbook, and the collection of DataFrame objects is

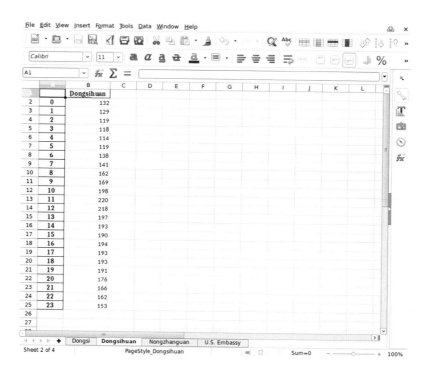

Figure 18.9 Screenshot of the second worksheet of the Excel workbook *pm25-1cols.xlsx* generated by the code in Figure 18.8 (reformatted manually for readability).

stored in a dictionary. Hint: Please see the `read_excel` documentation for details. We provide an up-to-date link to the documentation at www.cambridge.org/core/resources/pythonforscientists/refs/, ref. 50.

Try This! Answer 18-3

Assuming pandas is imported as `pd`, the following line of code solves the problem, saving the contents of *pm25-1cols.xlsx* in the dictionary-like object `contents`:

```
contents = pd.read_excel('pm25-1cols.xlsx', sheet_name=None)
```

When the `sheet_name` keyword input parameter is set to `None`, all the worksheets are read in from the Excel workbook and placed into a dictionary-like structure (technically, an `OrderedDict` object). The keys for the dictionary are the names of the worksheets and the values associated with the keys are the `DataFrame` object of that worksheet's information. The `read_excel` function generates `DataFrame` objects, even if there is only a single column of data in the worksheet. Thus, for `contents` defined above, `contents['Dongsihuan']` gives the two-dimensional, one-column `DataFrame` object holding the information from the second worksheet of the workbook, and `contents['Dongsihuan'].values` gives the array values of that data.

```
1    import pickle
2
3    fileobj = open('pm25.pkl', 'wb')
4    pickle.dump(pm25, fileobj, protocol=4)
5    fileobj.close()
6
7    fileobj = open('pm25.pkl', 'rb')
8    result = pickle.load(fileobj)
9    fileobj.close()
10
11   print(np.allclose(pm25, result))
```

Figure 18.10 Code to write out NumPy array pm25 from Figure 7.7 into the pickle file *pm25.pkl*, read in the object from that file, and compare the read in value with the original value.

Try This! 18-4 Writing to and Reading from a pickle File: Air Quality

Write the NumPy array of air quality data pm25 (assume it is already defined as in Figure 7.7) to a pickle file *pm25.pkl*. After doing so, read the array object from the file into an array of another name. Compare the read in array and the original array to confirm they are the same.

Try This! Answer 18-4

Figure 18.10 shows code that solves the problem. Line 11 in the figure outputs True, because the two arrays are the same.

Try This! 18-5 Write to a netCDF File: Surface Temperature

Consider the surface/near-surface air temperature data in Table 7.1 (as shown), for the locations whose longitude and latitude coordinates are given in Tables 7.2 and 7.3, respectively. Write the data out to a netCDF file, along with the longitude and latitude values of the grid, and appropriate global and variable attributes.

Try This! Answer 18-5

Figure 18.11 shows code that solves the problem. The *x*-direction netCDF dimension is named 'lon', and the *y*-direction netCDF dimension is named 'lat'. There are netCDF variables of the same names that hold the longitude and latitude values for each column and row in air_temp. There are two netCDF global attributes in this netCDF file, title and date_written (lines 26–28).

```
1    import netCDF4 as nc
2
3    air_temp = np.array( \
4        [[ 7.9,    7.1,    5.7,    3.6,    2.3,    1.7,    1.0],
5         [ 9.9,    8.5,    7.4,    5.7,    4.9,    4.5,    2.1],
6         [12.8,   10.6,    9.1,    8.1,    8.8,    8.3,    4.0],
7         [15.4,   13.5,   11.1,   10.5,   11.3,   10.1,    6.8],
8         [17.6,   17.0,   14.7,   15.3,   16.3,   15.1,   14.4]] )
9
10   fileobj = nc.Dataset('air_temp_mw.nc', mode='w')
11   fileobj.createDimension('lon', np.shape(air_temp)[1])
12   fileobj.createDimension('lat', np.shape(air_temp)[0])
13
14   air_temp_var = fileobj.createVariable('air_temp', 'd', ('lat','lon'))
15   air_temp_var.units = "deg C"
16   air_temp_var[:,:] = air_temp
17
18   lon_var = fileobj.createVariable('lon', 'd', ('lon'))
19   lon_var.units = 'deg E'
20   lon_var[:] = np.array([262.5, 265.0, 267.5, 270.0, 272.5, 275.0, 277.5])
21
22   lat_var = fileobj.createVariable('lat', 'd', ('lat'))
23   lat_var.units = 'deg N'
24   lat_var[:] = np.array([42.5, 40.0, 37.5, 35.0, 32.5])
25
26   fileobj.title = "Surface/near-surface air temperature over the " \
27                 + "U.S. Midwest at November 11, 2017, 18:00Z"
28   fileobj.date_written = "July 13, 2020"
29
30   fileobj.close()
```

Figure 18.11 Code to write out the data in Table 7.1 and related metadata, to the netCDF file *air_temp_mw.nc*.

Try This! 18-6 Read from a netCDF File: Surface Temperature

Read in the surface/near-surface air temperature data from the netCDF file we created in Try This! 18-5.

Try This! Answer 18-6

This code solves the problem (assuming the netCDF4 module is already imported in as nc):

```
1    fileobj = nc.Dataset('air_temp_mw.nc', mode='r')
2    air_temp_in = fileobj.variables['air_temp'][:]
3    fileobj.close()
```

To extract the array data from the netCDF variable object, we need to use array slicing on the netCDF variable object.

18.4   More Discipline-Specific Practice

The Discipline-Specific Jupyter notebooks for this chapter cover the following topics:

- Excel files.
- pickle files.
- netCDF files.

18.5   Chapter Review

18.5.1 Self-Test Questions

Try to do these without looking at the book or any other resources or using the Python interpreter. Answers to these Self-Test Questions are found at the end of the chapter.

Self-Test Question 18-1

Describe the differences between a pandas `Series` and `DataFrame` object.

Self-Test Question 18-2

Describe the basics of using the pandas package to read in and write out Excel files.

Self-Test Question 18-3

How do we use pandas to read in multiple worksheets from an Excel workbook?

Self-Test Question 18-4

How do we use pandas to write out multiple worksheets to an Excel workbook?

Self-Test Question 18-5

What is a pickle file?

Self-Test Question 18-6

When creating a file object for a pickle file, what access mode do we need to set the object to?

Self-Test Question 18-7

How do we write an object to a pickle file?

Self-Test Question 18-8

How do we read in an object from a pickle file?

Self-Test Question 18-9

What is a reason *not* to use a pickle file to store data?

Self-Test Question 18-10

Describe some reasons why a user would or would not want to save data in a netCDF formatted file.

Self-Test Question 18-11

Describe how a netCDF global attribute differs from a netCDF variable attribute, and how the netCDF4 module interface represents them.

Self-Test Question 18-12

What is a netCDF dimension and how do we create such a dimension in a netCDF file?

Self-Test Question 18-13

What is the `variables` attribute to a netCDF4 file object?

Self-Test Question 18-14

If we have a `Dataset` object `fobj` that has a scalar netCDF variable `energy` stored in it, what would be the code to read that value and store it as the Python variable `tot_energy`?

18.5.2 Chapter Summary

Often when using computers to do science or engineering work, we have to balance competing demands: memory used, performance, portability, flexibility, etc. The same is true for the storage of data and other information in a file. The text file of Chapter 9, although very flexible and portable, is unsuitable for large datasets because of the size of the files the format generates and its slow read/write access speeds. In this chapter, we looked at three other file formats, all of which overcome certain limitations of text files while having their own limitations. Our job, when using files, is to find out what capabilities really matter for the particular science or engineering use case we are trying to solve, and then choose the best format amongst the available options. Specific topics we covered in this chapter include reading and writing:

- Excel files

    - pandas `DataFrame` and `Series` objects have the method `to_excel` that can be used to write the data in those objects to an Excel file.
    - The pandas package's `read_excel` opens an Excel file and returns the contents of a worksheet as a `DataFrame` object or collection thereof.
    - Options to `read_excel` enable us to read in more than one worksheet or all the worksheets in an Excel workbook.
    - When paired with a pandas `ExcelWriter` object, we can write multiple worksheets out to a single Excel workbook using the `to_excel` method and looping (as needed).

- pickle files

- pickle files save a "snapshot" of a Python object to a file. When this object is read from the file, the entire object is recreated in the state and form it was saved.
- We call saving an object to a pickle file "pickling" the object and reading an object from a pickle file "unpickling" the object.
- We create or access pickle files by first creating a file object. If we want to write to the file, we set the access mode to 'wb'. If we want to read from the file, we set the access mode to 'rb'.
- The pickle module's dump function writes an object to a pickle file.
- To save multiple objects to a pickle file, we can place them into a list and then pickle the list.
- The pickle module's load function reads an object from a pickle file and returns that object.
- Not all Python objects can be pickled.

- netCDF files

  - netCDF files work well to store large amounts of gridded data and the metadata associated with the data.
  - netCDF files are in a special binary format that can be accessed on many different operating systems and computers.
  - netCDF files have global attributes, dimensions, variables, and variable attributes.
  - netCDF global attributes usually encode information about the file as a whole. This may include a title, authorship, description, etc.
  - netCDF dimensions define the shape of the arrays that store the data in netCDF variables.
  - netCDF variables hold the data for the variable as well as metadata about the variable. The metadata are called netCDF variable attributes and include such fields as the units of the variable, the long name of of the variable, etc.
  - We create a netCDF file object using the Dataset constructor of the netCDF4 module. The access mode is set to 'r' or 'w', depending on whether we are reading from or writing to the file, respectively.
  - The createDimension method of the netCDF file object is used to create a netCDF dimension. The createVariable method of the netCDF file object is used to create a netCDF variable.
  - To fill the data values of a netCDF variable that corresponds to an array (i.e., is not a scalar), we use the array slicing or selection notation (e.g., [:], [:, :], etc.). The same notation is used when reading the array data from a netCDF variable object.
  - To fill the values of a netCDF variable that corresponds to a scalar, we use the assignValue method of a netCDF variable object. To read a scalar value, we use the getValue method of a netCDF variable object.
  - netCDF global attributes are set by using Python object attribute assignment on a netCDF file object. We read netCDF global attributes using Python object attribute notation on a netCDF file object.

- netCDF variable attributes are set by using Python object attribute assignment on a netCDF variable object. We read netCDF variable attributes using Python object attribute notation on a netCDF variable object.
- The `dimensions` "dictionary" of a netCDF file object holds the netCDF dimensions in the file. The names of the netCDF dimensions are the keys and the corresponding netCDF dimension objects are the values.
- The `variables` "dictionary" of a netCDF file object holds the netCDF variables in the file. The names of the netCDF variables are the keys and the corresponding netCDF variable objects are the values.

There is much more that can be said about reading and writing files. There are many more file formats besides the ones we considered in Chapter 9 and the present chapter. Two important formats include the Structured Query Language (SQL) format,[7] which is used for general purpose databases, and the Hierarchical Data Format (HDF) format,[8] which can be used for collections of data and different formats of data. The HDF format is also a superset of the netCDF format. There are also many more Python packages that can be used to interact with files of various formats. Many of the principles we describe in the present work, however, are utilized in other formats and Python interfaces to those formats.

18.5.3 Self-Test Answers

Self-Test Answer 18-1

A `Series` object represents a single column of data and has row labels and a label for the single column itself. A `DataFrame` object represents a two-dimensional grid of data (i.e., multiple columns of data) and has row and column labels.

Self-Test Answer 18-2

The function `read_excel` can be used to read in one or more worksheets from an Excel file. The `to_excel` method of `Series` and `DataFrame` objects can write the object (including its row and column labels) to an Excel worksheet.

Self-Test Answer 18-3

The `read_excel` function has the keyword input parameter `sheet_name` that allows us to specify which worksheet in an Excel workbook to read in. When the parameter is set to `None`, all worksheets are read into a dictionary-like structure where the worksheet names are the keys to the `DataFrame` objects that represent the worksheet's information.

---

[7] www.iso.org/standard/63555.html (accessed July 14, 2020).
[8] www.hdfgroup.org/solutions/hdf5 (accessed July 14, 2020).

Self-Test Answer 18-4

We can use the `to_excel` method of `Series` and `DataFrame` objects to write a single worksheet, but we have to pass in an `ExcelWriter` object to these method calls in order for each call of the method to generate a worksheet in the same file (the file that the `ExcelWriter` object describes). When we call `to_excel`, the worksheet's name is passed in by the `sheet_name` keyword input parameter.

Self-Test Answer 18-5

A pickle file is a "snapshot" of a Python object that can be saved to a file and restored into memory at a later time.

Self-Test Answer 18-6

In both cases, the file object has to be in binary mode and thus `'b'` has to be part of the access mode argument. For reading a pickle file, we have to set the access mode to `'rb'`. For writing a pickle file, we have to set the access mode to `'wb'`.

Self-Test Answer 18-7

We use the pickle module's `dump` function. We pass in the object to be written out and the pickle file's file object. We can also (optionally) set the protocol number to be used to encode the object in the pickle file.

Self-Test Answer 18-8

The pickle module's `load` function returns the object in the pickle file. It takes one input parameter, the pickle file's file object.

Self-Test Answer 18-9

A pickle file can, generally, only be read by another Python program. This severely limits its usability, as a program written in another programming language is unable to access the data from the file.

Self-Test Answer 18-10

A netCDF file stores data in a format close to native binary but is portable across many different operating systems and computers. This makes a netCDF file suitable for large datasets. The format also enables us to store metadata with our data, enabling us to include a full range of documentation in our data file. At the same time, the netCDF format is not as flexible (in terms of how data are stored) as the text file and pickle file format.

Self-Test Answer 18-11

A netCDF global attribute is a string that usually documents something about the entire data file. In the netCDF4 interface, this is stored as a Python object attribute of the file object.

A netCDF variable attribute is a string that documents a characteristic of a netCDF variable, such as the units or a longer, more descriptive name of the variable. In the netCDF4 interface, netCDF variable attributes are stored as Python object attributes of the netCDF variable object.

Self-Test Answer 18-12

A netCDF dimension describes a number that can be used to specify the number of elements along an axis of an array of data in a netCDF variable. The `Dataset` object's `createDimension` method is used to create a netCDF dimension.

Self-Test Answer 18-13

The `variables` attribute is a dictionary-like structure that stores the netCDF variable objects defined in the netCDF file.

Self-Test Answer 18-14

```
tot_energy = fobj.variables['energy'].getValue()
```

19   Basic Searching and Sorting

The topics of **searching** (finding certain values in a collection of values) and **sorting** (putting values into an order) are not new to us. As we have examined various scientific and engineering computing use cases, we have encountered both topics. For searching, we have seen the use of `if` statements, boolean expressions, etc. For sorting, we have seen the use of the `sorted` function.

In this chapter, we summarize, in one place, different methods we have already seen (or are part of structures we have seen) of searching and sorting. The range of different ways to conduct searches and sorts in Python motivates us to ask if all methods are equal, and if not, what those differences look like. We will find that differences in how we conduct searches or sorts are fundamentally differences in the algorithm used to do the search or sort. Finally, we describe a "power tool" for searching and sorting: the pandas package. In Chapters 15, 16, and 18, we already introduced the `Series` and `DataFrame` classes. In the present chapter, we examine ways these classes make searching and sorting labeled arrays of data very convenient to do.

As an aside: In this chapter, we print the results of many expressions. Remember that the `print` function just formats what is in between the parentheses of the call, so that it displays the result of the inputted expression nicely on the screen. It is the expression inside the parentheses that actually does the tasks and creates the object we are interested in.

19.1   Examples of Searching and Sorting

Consider the fictitious beaver tracking scenario from Try This! 16-4, where we tagged five beavers with radio transmitters, and after the beaver has traveled one mile, its transmitter radios back information about the beaver. Pretend these are the data we received:[1]

Beaver ID	Nickname	Date of birth	Mother's ID	Father's ID
1243	Max	2018-04-17	2352	4722
9384	Polly	2018-04-17	2352	4722
2352	Fong	Empty string	−1	−1
4722	Edward	Empty string	−1	−1
1198	Jessie	Empty string	−1	−1

[1] Ideas for this table and related beaver data in the rest of the chapter are from https://animals.mom.me/fun-baby-beavers-7869.html, www.livescience.com/52460-beavers.html, and www.beaversolutions.com/beaver-facts-education/beaver-behavior-and-biology/ (all accessed August 4, 2020), but all data are fictitious.

```
 1    #- Import module:
 2
 3    import pandas as pd
 4
 5
 6    #- Define function to return minimum value and location:
 7
 8    def min_val_loc(values):
 9        current_min = values[0]
10        current_min_loc = 0
11        for i in range(len(values)):
12            if values[i] < current_min:
13                current_min = values[i]
14                current_min_loc = i
15        return current_min, current_min_loc
16
17
18    #- Define data as a dictionary, data as a list, and row and column
19    #  labels:
20
21    data = { 1243: ['Max', '2018-04-17', 2352, 4722],
22             9384: ['Polly', '2018-04-17', 2352, 4722],
23             2352: ['Fong', '', -1, -1],
24             4722: ['Edward', '', -1, -1],
25             1198: ['Jessie', '', -1, -1] }
26
27    rows = []
28    data_list = []
29    for ikey in data:
30        rows.append(ikey)
31        data_list.append(data[ikey])
32
33    cols = ["Nickname", "Date of Birth", "Mother's ID", "Father's ID"]
```

Figure 19.1 Code demonstrating searching and sorting through the fictitious beaver tracking data of Section 19.1.

where all identification numbers are four-digit integers and the nickname and date of birth are strings. Default values are empty strings for the nickname and date of birth or "−1" for identification numbers.

Figure 19.1 shows examples of searching and sorting using the data given above. The first block of code (lines 1–33) sets up the modules, functions, and data the searching and sorting examples will use in the second block of code (lines 34–56). Line 3 imports the pandas package. In lines 8–15, we define the function min_val_loc that will be used in line 46. This function takes a list and returns the minimum value in that list and the positional index of the first minimum value in that list.[2] The min_val_loc function and its call illustrate how

---

[2] See Section 19.2.3 for why we use the phrase "positional index" in this chapter to refer to the address of an element in a list, tuple, or NumPy array.

```python
34    #- Find nickname for beaver ID 1243:   Linear search:
35
36    for i in range(len(rows)):
37        if rows[i] == 1243:
38            nickname_linear = data_list[i][0]
39
40
41    #- Sort nicknames by ascending beaver ID:   Selection sort:
42
43    keys_unsorted = list(data.keys())
44    nicknames_sorted = []
45    for i in range(len(keys_unsorted)):
46        min_value, min_value_loc = min_val_loc(keys_unsorted)
47        nicknames_sorted.append(data[min_value][0])
48        temp = keys_unsorted.pop(min_value_loc)
49
50
51    #- Find nickname for beaver ID 1243 and sort nicknames by ascending
52    #  beaver ID using pandas:
53
54    data_df = pd.DataFrame(data_list, index=rows, columns=cols)
55    nickname_pd_ref = data_df.loc[1243, 'Nickname']
56    nicknames_df = data_df.sort_index()['Nickname']
```

Figure 19.1 (continued)

to return multiple values from a function and how to store those values from the call. The list object already has methods that can do what we do in min_val_loc, but as one of the goals of this chapter is to explore how algorithms work, we write our own function so we can see the steps that function's minimum finding algorithm takes.

In lines 21–25, we provide the data from above as a dictionary. The keys are the values in the "Beaver ID" column. The value associated with each key is a list that holds the rest of the information in the corresponding row in the table.

In lines 27–31 of Figure 19.1, we create a version of the data so that each of the values in data are elements in a list data_list, and each of the keys in data (i.e., the beaver identification numbers) are elements in a list rows. Later on, we use these lists in our **linear search** example and to create a DataFrame object. We also do this using loops because the values in data are of different types, so we cannot hold the data in a two-dimensional NumPy array and use slicing to get whatever subarrays we want. In line 33, we list the column labels (cols) for the elements in the lists stored as values in data and elements in data_list. Note that "Mother's ID" includes an apostrophe, so we use double quotation marks to delimit the column label. We do the same with "Father's ID".

In lines 34–56 of Figure 19.1, we conduct one search and one sort. In the first block of code, lines 36–38, we use a linear search – where we go through all items and ask whether we have found what we are looking for – to obtain the nickname of the beaver with the identification number 1243. In lines 43–48, we sort the data records by ascending beaver identification number and obtain the list of nicknames (`nicknames_sorted`) of the beavers given in that order. We use a sorting algorithm called **selection sort** to accomplish the sorting by beaver identification number. Finally, in lines 54–58, we do the same search and sort task as in lines 34–48 but using selection syntax and sorting methods provided by the pandas package.

19.2 Python Programming Essentials

In this section, we summarize ways of searching and sorting that we have seen (or are part of structures we have seen) in previous chapters, examine some different algorithms for searching and sorting, and describe the basic use of tools in the pandas package for searching and sorting.

19.2.1 Summary of Some Ways to Search and Sort

So far, we have seen a number of different ways of finding something in a collection of items and sorting items in Python.[3] This summary is in no particular order. Each searching or sorting method is illustrated using a brief (and closely similar) example.

Searching by Testing in a Loop   We can loop through a list of items and use an `if` statement to find what we are looking for. One place we used this strategy was in Figure 6.3. Below, we take some numbers and store the positional index in the list where the number 6 is found in the variable `loc`:

```
1   data = [3, 6, 1, -8, 9]
2   for i in range(len(data)):
3       if data[i] == 6:
4           loc = i
```

In the above code, `loc` is set to the positional index of the last occurrence of the number 6.

Searching Using the `index` Method of a List   In Section 9.2.3, we were introduced to using lists of objects. The list method `index` returns the positional index where the first occurrence of the search target is found. Below, we search the `data` list for the positional index where the first occurrence of the number 6 is located and store that positional index in `loc`:

```
1   data = [3, 6, 1, -8, 9]
2   loc = data.index(6)
```

---

[3] This section is adapted from Section 8.1 in Lin (2019).

Searching Using the count Method of a List As we saw in the answer to Self-Test Question 9.5, the count method of a list will count the number of occurrences in the list of a search target. Below, we search the data list to find how many times the number 6 occurs and store that number in count_six:

```
1   data = [3, 6, 1, -8, 9]
2   count_six = data.count(6)
```

Searching Using Array Syntax and Testing for Equality In Sections 13.2.2–13.2.3, we saw that array syntax automatically applies logical operators like equality element-wise across an array. Below, we look in data for where it equals 6:

```
1   import numpy as np
2   data = np.array([3, 6, 1, -8, 9])
3   mask = data == 6
```

The variable mask is a boolean array showing which elements of data equal 6 and which do not:

```
In [1]:   print(mask)

          [False   True False False False]
```

This works with NumPy arrays but not lists.

Searching Using the where Function on Arrays We saw this searching method in Section 13.2.3. Below, we search data for the number 6 and return an array where elements equal to 6 are set to its positional index and -999 otherwise:

```
1   import numpy as np
2   data = np.array([3, 6, 1, -8, 9])
3   mask = np.where(data == 6, np.arange(np.size(data)), -999)
```

The result of the where call is stored in mask, which is:

```
In [1]:   print(mask)

          [-999    1 -999 -999 -999]
```

The where method also has the single input parameter syntax which returns a tuple of indices in the condition array where the condition is True. We can then use that tuple to select the elements at those indices.

Searching Using the `isclose` Function on Arrays   Starting with Chapters 8 and 9, we saw how to use the NumPy functions `isclose` and `allclose` to do "equality" tests on floating-point numbers. Below, we look in `data` for where it is close to `6.0`:

```
1  import numpy as np
2  data = np.array([3., 6., 1., -8., 9.])
3  mask = np.isclose(data, 6.0)
```

and, as earlier, the variable `mask` is a boolean array showing which elements of `data` are close to `6.0` and which are not:

```
In [1]:  print(mask)


         [False   True False False False]
```

Searching Using Membership Testing   In the discussion of dictionaries in Section 16.2.1, we were introduced to membership testing, where we use the `in` operator to check whether a dictionary has a specific key. Below, we construct a dictionary `data` and ask whether the dictionary has a key named `'six'`:

```
In [1]:  data = {'three':3, 'six':6, 'one':1, 'neg_eight':-8, 'nine':9}
         print('six' in data)


         True
```

Membership testing also works on lists and strings. This does not tell us where a search target is in the list or string, but it does tell us if the target is present in the list or string. Below, we test whether 6 is in the list `data`:

```
In [1]:  data = [3, 6, 1, -8, 9]
         print(6 in data)


         True
```

To test whether a target is not in a list, string, etc., we use `not in`.

Searching Using the `glob` Function   In Section 10.2.5, we were introduced to the glob module's `glob` function. This function can search a directory for filenames matching a given pattern. The following searches the current working directory for all files that have the numeral "6" somewhere in their name and returns the list of those files as `files6`:

```
1  import glob
2  files6 = glob.glob("*6*")
```

Sorting Using the `sort` Method of a List   The list method `sort` will sort the list in place. Below, we sort the list `data` and print out the sorted version:

```
In [1]:  data = [3, 6, 1, -8, 9]
         data.sort()
         print(data)

         [-8, 1, 3, 6, 9]
```

Sorting Using the `sorted` Function   We were introduced to the use of `sorted` in the discussion of dictionaries in Section 16.2.1. In Section 17.2.3, we saw how `sorted` could be used for more sophisticated sorting. Below, we sort the list `data` and print out the sorted version:

```
In [1]:  data = [3, 6, 1, -8, 9]
         print(sorted(data))

         [-8, 1, 3, 6, 9]
```

For more on this and the previous sorting method, see the Python wiki page on sorting. An up-to-date link to this wiki page is at www.cambridge.org/core/resources/pythonforscientists/ refs/, ref. 97.

19.2.2 Searching and Sorting Algorithms

In Sections 19.1 and 19.2.1, we have seen a variety of different ways of doing searches and sorts. Some of these ways are different because they involve different Python operations, methods, or data structures, most of whose inner-workings are beyond the scope of this book. Some ways are different, however, because of differences in the choices and order of subtasks they include. That is, they are different because different algorithms are being used.

To illustrate these differences, below we consider two different search algorithms (linear search and **binary search**) and two different sorting algorithms (selection sort and **insertion sort**). The main point is to show how we can do the same task in different ways, and how our choice of algorithm can make a difference in the ease with which we can accomplish a task. Our discussion in the present chapter of the relative performance of algorithms only scratches the surface of the topic. See Chapter 22 for more on performance.

Linear and Binary Search

Lines 36–38 of Figure 19.1 provide an example of the linear search algorithm. In this example, we are looking for the nickname of the beaver whose identification number is 1243. In this kind of search, we loop through all items in a collection and ask whether or not the item we are examining is the item we are looking for. In line 37, we accomplish this by using an `if` statement that looks for the identification number 1243 in each element of `rows`. When we find it, we extract the nickname from the corresponding element in `data_list` and set the nickname to `nickname_linear`. The value saved is the string `'Max'`.

We might ask why we looped through `rows` and `data_list` instead of using the dictionary `data`. The reason is that dictionary lookups already conduct a kind of search, and so to reference a key:value pair in a dictionary does not illustrate how a linear search works. That is, because a key reference automatically "searches" for the key, this line:

```
nickname_linear = data[1243][0]
```

solves the problem. But that code does not show us what linear search does. By forcing ourselves to use lists in lines 36–38 of Figure 19.1, we can more clearly demonstrate the steps in linear search.

Linear search works fine and does not require that the collection of items is in any given order. But, if the collection of items is in some sort of order, linear search ignores the information that ordering gives. Thus, if we *know* the item we are looking for is near the end of a sorted list, linear search does nothing with that information. It starts at the beginning of the list and goes through the items one by one. We have to trudge through all the earlier items even though we know the item we are looking for is nowhere near the beginning of the list. This is a waste of computing resources.

Binary search, in contrast, takes advantage of the extra information that sorting gives us. In fact, binary search only works on already sorted collections. The algorithm has the following steps:

1. Consider the special cases where the sorted collection is one or two elements in length. In those cases, check to see whether the target is one of the elements in the sorted collection. If so, stop the search as we have found the value we are looking for. Otherwise, go on.
2. In the sorted collection, find the value in the middle of the collection.
3. Compare the search target to the middle value:
   (a) If the values are equal, stop the search: We have found the value we are looking for.
   (b) If the target is greater than the middle value: Go to step 2 using the right half of the collection as the sorted collection to search.[4] The right half includes the middle value as a lower bound.
   (c) If the target is less than the middle value: Go to step 2 using the left half of the collection as the sorted collection to search. The left half includes the middle value as an upper bound.

The algorithm assumes that the target is within the range bounded by the maximum and minimum values (both inclusively) in the sorted collection.

Figure 19.2 shows a binary search, following the above algorithm, through a sorted version of the `rows` list that was defined in lines 27–30 of Figure 19.1. The code searches for the value of `target` in `rows_sorted`. If the target is found, `target_found` is set to `True`.

---

[4] In this use of "half," we mean "approximate half," because the case of the middle value equaling the target removes one element from the consideration of halves. This use of half as usually being approximate continues throughout the chapter when referring to binary search.

```
1    rows_sorted = sorted(rows)
2    target = 1243
3
4    end_search = False
5    while not end_search:
6        if len(rows_sorted) == 1:
7            end_search = True
8            if target == rows_sorted[0]:
9                target_found = True
10           else:
11               target_found = False
12
13       elif len(rows_sorted) == 2:
14           end_search = True
15           if target == rows_sorted[0]:
16               target_found = True
17           elif target == rows_sorted[1]:
18               target_found = True
19           else:
20               target_found = False
21
22       else:
23           middle_loc = int(len(rows_sorted)/2)
24
25           if target == rows_sorted[middle_loc]:
26               end_search = True
27               target_found = True
28
29           elif target > rows_sorted[middle_loc]:
30               rows_sorted = rows_sorted[middle_loc:]
31
32           else:
33               rows_sorted = rows_sorted[:middle_loc+1]
34
35   print(target_found)
```

Figure 19.2 Code to implement a binary search to find the beaver identification number 1243 in rows.

Otherwise, target_found is set to False. In the last line of Figure 19.2, target_found is printed to the screen. The code here, and in Figure 19.3, does not test that the target is within the range bounded (inclusively) by the maximum and minimum values of rows, though it ought to.

The binary search algorithm of Figure 19.2 is, usually, substantially faster than a comparable linear search algorithm. This is because with each iteration of the algorithm, we decrease our search space – the total number of elements we have to consider – by half. The difference is the most dramatic when searching through a large set of values. With very small collections of items, however, a linear search is usually faster.

What are some drawbacks to the binary search algorithm? First, while the algorithm is faster, the algorithm only works on sorted collections. If the collection is not already sorted, the sorting process itself takes computational resources. Second, the implementation in code is more complex than the linear search code. There are more pieces of information to keep track of. Third, we have to be careful with binary search to keep any mapping between the items we are searching through and the data related to those items. In the case of the beaver tracking information, we have to be careful throughout the operations involved in the binary search (e.g., sorting, halving the domain) that we keep the connection between the beaver identification numbers and the list of information about that beaver. We can do this by arranging the beaver tracking information in the order of the identification number sort, and including that information when we halve the domain in the binary search. We take that approach in Figure 19.3, which revises the code from Figure 19.2 to obtain the nickname of the beaver with the identification number 1243.[5] In Figure 19.3, we also sort `data_list` (lines 2–4) and halve the sorted version of that list `data_list_sorted` each time we halve `rows_sorted` (lines 40 and 44). As a result, we are able to extract the information related to beaver identification number 1243 and print out the nickname of that beaver.

Selection and Insertion Sort

We can think of sorting algorithms as a sequence of steps where we transform an unsorted list of items into a sorted list of items. This transformation can be thought of as "moving" items from the first list (or **partition**) to the second.[6] Lines 43–48 of Figure 19.1 provide an example of the selection sort algorithm. In this example, we are sorting the beaver nicknames by ascending beaver identification number. That is, we sort the beaver identification numbers, and as we do so, we extract the beaver nicknames that correspond to each identification number.

This example, however, is not a clean demonstration of a selection sort, as it has the second step of extracting the nicknames. Figure 19.4 is a "stripped down" version (it actually has the same number of lines of code but it does fewer tasks) of our Figure 19.1 example that *only* sorts the beaver identification number into ascending order. Assume `data` and `min_val_loc` are defined as in Figure 19.1. Figure 19.5 shows the results of the first four iterations of the loop in Figure 19.4.

In selection sort, we iterate through a loop a number of times equal to the number of elements to be sorted. At each iteration, we find the minimum value of whatever elements remain in the unsorted partition, append it to the end of the sorted partition, then remove it from the unsorted partition. When the unsorted partition is exhausted, the sorted partition holds the original elements in ascending order. To sort in descending order, instead of finding

---

[5] Another approach with regards to the halving process is to conduct the halving on collections of positional indices (or ranges of positional indices, if the collection is already sorted). That way, we can keep the list of values to search through static while interacting with the positional indices of interest.

[6] Thanks to Rob Nash.

```
1    rows_sorted = sorted(rows)
2    data_list_sorted = []
3    for ikey in rows_sorted:
4        data_list_sorted.append(data[ikey])
5
6    target = 1243
7
8    end_search = False
9    while not end_search:
10       if len(rows_sorted) == 1:
11           end_search = True
12           if target == rows_sorted[0]:
13               target_found = True
14               nickname_binary = data_list_sorted[0][0]
15           else:
16               target_found = False
17               nickname_binary = ''
18
19       elif len(rows_sorted) == 2:
20           end_search = True
21           if target == rows_sorted[0]:
22               target_found = True
23               nickname_binary = data_list_sorted[0][0]
24           elif target == rows_sorted[1]:
25               target_found = True
26               nickname_binary = data_list_sorted[1][0]
27           else:
28               target_found = False
29               nickname_binary = ''
```

Figure 19.3 Code to implement a binary search solution to the task solved using linear search in lines 36–38 of Figure 19.1.

the minimum value of whatever elements remain in the unsorted partition, we find the maximum value of whatever elements remain in the unsorted partition.

As we see in the first iteration (where i is 0) in Figure 19.5, ids_unsorted starts out holding the beaver identification numbers in an unsorted order and ids_sorted is empty. Recall the keys of a dictionary, such as data, should be considered to be unordered. In line 4 of Figure 19.4, we call the min_val_loc function on the unsorted partition and obtain both the minimum value in that partition (min_value) and the positional index of that element in the list (min_value_loc). In line 5, we add that minimum value to the sorted partition, and in line 6, we remove that minimum value from the unsorted partition. We repeat this process for each subsequent iteration (as shown in Figure 19.5). At each iteration, ids_sorted grows by one element and ids_unsorted shrinks by one element. When ids_unsorted is out of elements, ids_sorted contains all the elements in ascending order.

```
30        else:
31            middle_loc = int(len(rows_sorted)/2)
32
33            if target == rows_sorted[middle_loc]:
34                end_search = True
35                target_found = True
36                nickname_binary = data_list_sorted[middle_loc][0]
37
38            elif target > rows_sorted[middle_loc]:
39                rows_sorted = rows_sorted[middle_loc:]
40                data_list_sorted = data_list_sorted[middle_loc:]
41
42            else:
43                rows_sorted = rows_sorted[:middle_loc+1]
44                data_list_sorted = data_list_sorted[:middle_loc+1]
45
46  print(target_found)
47  print(nickname_binary)
```

Figure 19.3 (continued)

```
1  ids_unsorted = list(data.keys())
2  ids_sorted = []
3  for i in range(len(ids_unsorted)):
4      min_value, min_value_loc = min_val_loc(ids_unsorted)
5      ids_sorted.append(min_value)
6      temp = ids_unsorted.pop(min_value_loc)
```

Figure 19.4 Code to sort beaver identification numbers in ascending order using selection sort. Assume all variables defined in lines 1–33 of Figure 19.1 already exist.

Note that the min_val_loc function provides two return values, given in line 15 of Figure 19.1. Thus, in our calls to min_val_loc (line 46 of Figure 19.1 and line 4 of Figure 19.4), we can assign two different variables to the two returns from the function. We also see that the list method pop has a form where we can pass in a specific positional index to remove that element from the list. Line 48 of Figure 19.1 and line 6 of Figure 19.4 show this use of pop. The temp variable holds the result of the call to pop and has no other use.

In the selection sort algorithm, all the comparisons are done on the unsorted partition (i.e., finding the minimum value). In the insertion sort algorithm, all the comparisons are done against the elements in the sorted partition. In each iteration of an insertion sort, we extract the first (i.e., next) remaining element in the unsorted partition and compare it with the last element of the sorted partition. If the extracted element is larger, we place the element after the element in the sorted partition. If the extracted element is smaller than the last element of the sorted partition, we compare the extracted element to the element before the

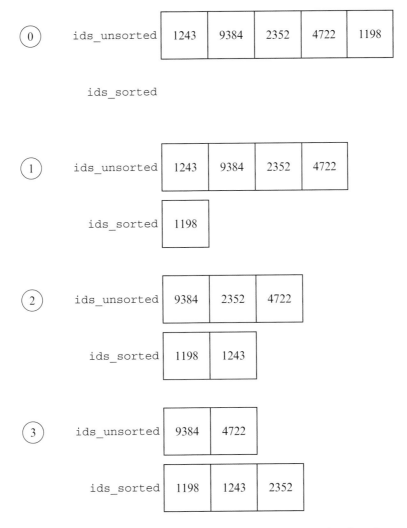

Figure 19.5 A schematic of the flow of execution of the first four iterations of the loop in the Figure 19.4 selection sort code. The number in the circle represents the value of i at the beginning of the iteration, i.e., the state of the ids_unsorted and ids_sorted right after line 3 of Figure 19.4, before the first line of the body of the loop.

last element of the sorted partition, and so on until we find where the extracted element goes in the sorted partition.

Figure 19.6 shows the code implementing an insertion sort for sorting the beaver identification numbers. The number of iterations of the outer for loop (line 3) equals the number of elements to be sorted. In lines 4–6, we handle the case where there are no elements in ids_sorted. In that case, we take the first element from ids_unsorted, place it in the ids_sorted array, and delete the element from ids_unsorted.

```
1    ids_unsorted = list(data.keys())
2    ids_sorted = []
3    for i in range(len(ids_unsorted)):
4        if len(ids_sorted) == 0:
5            ids_sorted.append(ids_unsorted[0])
6            temp = ids_unsorted.pop(0)
7
8        else:
9            element_placed = False
10           sorted_loc_check = len(ids_sorted) - 1
11
12           while not element_placed:
13               if ids_unsorted[0] >= ids_sorted[sorted_loc_check]:
14                   ids_sorted.insert(sorted_loc_check + 1, ids_unsorted[0])
15                   temp = ids_unsorted.pop(0)
16                   element_placed = True
17
18               elif (sorted_loc_check == 0) and \
19                   (ids_unsorted[0] < ids_sorted[0]):
20                   ids_sorted.insert(0, ids_unsorted[0])
21                   temp = ids_unsorted.pop(0)
22                   element_placed = True
23
24               else:
25                   sorted_loc_check -= 1
```

Figure 19.6 Code to sort beaver identification numbers in ascending order using insertion sort. Assume all variables defined in lines 1–33 of Figure 19.1 already exist.

For the cases where ids_sorted is not empty, we go through each of the elements in ids_sorted, starting from the end and working backward, to see where to place the first element from the remaining values of ids_unsorted (lines 9–25 in Figure 19.6). We use a while loop for these repeated comparisons. The logic of our comparison in lines 13–16 is to see whether the first element from ids_unsorted is greater than or equal to the element being compared to from ids_sorted. If so, we place the element from ids_unsorted *after* the element being compared to from ids_sorted (line 14). The list method insert (line 14) takes two positional input parameters (a positional index and an object) and places the object *before* the positional index given. Thus, in line 14, we place the element from ids_unsorted before positional index sorted_loc_check + 1. This works even if sorted_loc_check + 1 is beyond the positional index of the last element of ids_sorted.

The logic of lines 18–22 in Figure 19.6 covers the case when we have reached the first element of ids_sorted and the element from ids_unsorted is less than the first element of ids_sorted. In that case, the element from ids_unsorted is placed before the first element of ids_sorted. That is, insert is called with a positional index of 0 rather than $0 + 1 = 1$ (line 20).

Once the element from `ids_unsorted` is placed into the proper location in `ids_sorted`, the `while` loop ends (by setting `element_placed` to `True`). If the element from `ids_unsorted` is not yet placed, we subtract 1 from `sorted_loc_check` (line 25) so in the next iteration of the `while` loop we do the comparison with the element one to the left of the current element in `ids_sorted`.

Figure 19.7 shows the results of the first four iterations of the outer `for` loop in Figure 19.6. Figure 19.8 shows the comparison steps involved in moving between two of the iterations in Figure 19.7, namely from the ② to ③ iteration. The first element of `ids_unsorted` (2352) is compared with the last of `ids_sorted` (9384) to see if the `ids_unsorted` element is greater or equal to the `ids_sorted` element (the comparison is represented by the rightmost arrow and is done in line 13 of Figure 19.6). If so, the 2352 would be placed after the 9384, but as that is not the case, we increment `sorted_loc_check` (line 25 in Figure 19.6) to repeat the comparison with the 1243 in `ids_sorted`, in order to see if the 2352 should go between the 1243 and 9384 (the comparison is represented by the leftmost arrow in Figure 19.8). That is the case, so the 2352 is moved to between the 1243 and 9384 elements in `ids_sorted`.

From Figures 19.4 and 19.6, it may appear that selection sort is much simpler than insertion sort. Recall, though, in the selection sort code, we have placed some of the code that would have been in Figure 19.4 into the `min_val_loc` function. Also, in Figures 19.4 and 19.6, we implemented our unsorted and sorted partitions as separate regions of memory (i.e., as separate lists). In a real-world implementation of this algorithm, the sorted partition would be a subset of the memory holding the unsorted partition. If these sort algorithms are implemented using arrays, which have a fixed size, this would save memory as we would avoid creating two arrays of the same size. In our solution using lists, because lists grow and shrink in size, depending on the number of elements in the list, there is essentially no memory penalty for implementing our unsorted and sorted partitions as separate regions of memory.

Both selection and insertion sort perform similarly. Both use nested looping one level deep (the inner loop for our selection sort solution is in the `min_val_loc` function). Both sorting algorithms, in the worst-case scenario, loop through all the elements of the partition being considered in both levels of looping. As a result, both sorting algorithms take a similar number of operations. Other sorting algorithms have better performance characteristics, including the one used by the built-in function `sorted` (which uses an algorithm called Timsort, that is derived from the insertion sort and merge sort algorithms).[7] In Chapter 22, we address performance and how to measure performance in more detail.

The bottom line of our examination of different algorithms – selection and insertion sort and linear and binary search: There are different ways of accomplishing any given kind of task, and whereas some algorithms perform similarly, others do not.

---

[7] https://en.wikipedia.org/w/index.php?title=Timsort&oldid=963161039 and https://stackoverflow.com/a/10948946/8430411 (both accessed July 23, 2020).

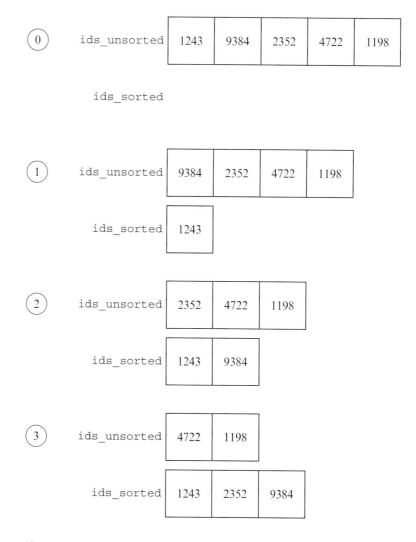

Figure 19.7 A schematic of the flow of execution of the first four iterations of the outer `for` loop in the Figure 19.6 insertion sort code. The number in the circle represents the value of `i` at the beginning of the iteration, i.e., the state of the `ids_unsorted` and `ids_sorted` right after line 3 of Figure 19.6, before the first line of the body of the loop.

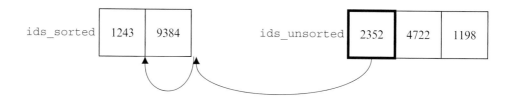

Figure 19.8 A schematic showing the comparison steps in moving from the ② to ③ iteration in Figure 19.7 (conducted by lines 12–25 of Figure 19.6).

19.2.3 Basic Searching and Sorting Using pandas

The pandas data objects are similar to NumPy arrays in that they represent gridded quantities, but they have additional features. First, pandas `Series` and `DataFrame` objects (hereafter in this chapter "pandas data objects") are labeled (like dictionary keys) with names (or integers, etc.) that have meanings besides the position of the element in a dimension. Second, pandas leverages this referencing system to define an interface that enables us to do a lot in very few lines of code. Third, unlike a NumPy array where each element has to be the same type, pandas has more flexibility to store different kinds of values. Thus, in the Figure 19.1 chapter example, the `DataFrame` object `data_df` (line 54) holds strings and integers.

So, although the ways we have used pandas data objects in prior sections – to handle missing values and to interface with Excel files – are useful, the additional features described above give pandas capabilities to select, manipulate, order, and group values in one- and two-dimensional datasets that go far beyond what NumPy arrays can do. In this section, we describe how to do basic selection and manipulation, boolean selection, sorting, and grouping using pandas. For a great, brief introduction to using pandas, see the "10 Minutes to Pandas" guide. (We provide an up-to-date link to the guide at www.cambridge.org/core/resources/pythonforscientists/refs/, ref. 45.) This guide, and the pandas interpreter help, also gave ideas for parts of the current section. The classic reference for pandas is McKinney (2017).

In the present section, assume pandas has been imported as `pd` and all variables defined in lines 21–33 of Figure 19.1 already exist.

More About `DataFrame` Objects

The `DataFrame` object `data_df` from Figure 19.1 is:

```
In [1]: print(data_df)

        Nickname Date of Birth  Mother's ID  Father's ID
1243       Max     2018-04-17         2352         4722
9384     Polly     2018-04-17         2352         4722
2352      Fong                          -1           -1
4722    Edward                          -1           -1
1198    Jessie                          -1           -1
```

This `DataFrame` object was created in line 54 of Figure 19.1. To create `data_df` we used three pieces of input. The list `data_list` is a list of lists, where each element of `data_list` is a list that contains the column data for a row in `data_df`. The column labels are defined by the list `cols` (line 33 in Figure 19.1) and the row names are defined by the list `rows` (defined in lines 27 and 30 in Figure 19.1). In line 54 of Figure 19.1, the row and column labels are passed into the `DataFrame` call via the `index` and `columns` keyword input parameters, respectively.

The row names on the left, while integers, are not positional indices, as in a two-dimensional NumPy array. Instead, they are the beaver identification numbers for the beavers in the data table in Section 19.1.

In a `DataFrame` object, the row and column names are stored in the attributes `index` and `columns`, respectively (output reformatted manually for readability):

```
In [1]:  print(data_df.index)
         print(data_df.columns)


         Int64Index([1243, 9384, 2352, 4722, 1198], dtype='int64')
         Index(['Nickname', 'Date of Birth', 'Mother's ID', 'Father's ID'],
             dtype='object')
```

Neither of the attributes is a list object, but the attributes behave like lists, with more features. They can be converted to lists using the built-in `list` function or the `tolist` method attached to each attribute. The `len` function, when applied to a `DataFrame` object, returns the number of rows in the object. This is the same value as when `len` is applied to the `index` attribute of the `DataFrame` object, because `index` lists the row names.

What can be confusing is that the collection of row names is often called the "index" (e.g., the keyword input parameter in line 54 of Figure 19.1), and so we might call the row names "indices." But, we also call the positional address for elements in lists, tuples, and NumPy arrays the "index" of an element. In the present chapter, we use "positional index" for the address based on the element's position in the data structure. We also refer to the "index" of a pandas data object as the collection of "row names" or "row labels," except when following the pandas convention of calling it the index is clearer.

Basic Selection and Manipulation

Because rows (and columns, for the two-dimensional case) are labeled in `Series` and `DataFrame` objects, we can refer to columns and rows using those labels, just as we refer to values in dictionaries using the keys associated with those values. To refer to a column of `data_df`, we use syntax similar to a dictionary:

```
In [1]:  print(data_df['Nickname'])


         1243        Max
         9384       Polly
         2352       Fong
         4722      Edward
         1198      Jessie
         Name: Nickname, dtype: object
```

The result of selecting a column from a `DataFrame` is a `Series` object:

```
In [1]:  print(type(data_df['Nickname']))
```

```
<class 'pandas.core.series.Series'>
```

We can select multiple columns by putting them in a list inside the square brackets selecting from the DataFrame:

```
In [1]:  print(data_df[["Nickname", "Mother's ID"]])
```

```
          Nickname   Mother's ID
1243           Max          2352
9384         Polly          2352
2352          Fong            -1
4722        Edward            -1
1198        Jessie            -1
```

The result of selecting multiple columns is a DataFrame object, rather than a Series object, because the result is a two-dimensional grid of data.

To select rows or to select ranges of rows and columns, we use the loc selection specifier. The specifier's syntax looks like a method call, in that loc comes after a period after the DataFrame object's name, but instead of parentheses after the loc, there are square brackets. For instance, the following selects the row of data_df with the row name 4722:

```
In [1]:  print(data_df.loc[4722, :])
```

```
Nickname              Edward
Date of Birth
Mother's ID              -1
Father's ID             -1
Name: 4722, dtype: object
```

The first element in the square backets is the row name (i.e., the beaver identification number for the beaver) and the second element is a colon specifying "use all columns." The result of this selection of a row is a Series object.

Using the loc selection mechanism, we can also select subranges of rows and columns. For instance:

```
In [1]:  print(data_df.loc[9384:4722, "Date of Birth":"Father's ID"])
```

```
          Date of Birth   Mother's ID   Father's ID
9384         2018-04-17          2352          4722
2352                              -1            -1
4722                              -1            -1
```

Using this selection mechanism, the range specifed by the lower- and upper-bound row or column names are **inclusive at both ends**. This is different from slicing or selection in lists, tuples, and NumPy arrays using positional indices where the upper bound is exclusive. Also, the row range specified by `9384:4722` means "start with the row whose name is 9384 and select rows through and including the row whose name is 4722." Unlike slicing using positional indices, if the row or column names are integers, the upper bound of the slice can be larger than the lower bound because the integer row or column names do not have to be in ascending order.

Besides specifying a subrange of columns for a row or subrange of rows, we can specify specific columns to select for a row or subrange of rows. For instance:

```
In [1]:  edward_row_partial = data_df.loc[4722, ["Nickname", "Mother's ID"]]
         print(edward_row_partial)
         print(edward_row_partial["Mother's ID"])
         print(edward_row_partial.loc["Mother's ID"])


         Nickname        Edward
         Mother's ID        -1
         Name: 4722, dtype: object
         -1
         -1
```

The variable `edward_row_partial` is a `Series` object that holds the `"Nickname"` and `"Mother's ID"` values for the beaver with beaver identification number 4722 (those column labels are now "row" labels in the `Series` object). As shown above, to extract a specific value from a `Series` object, we can use the row label with or without the `loc` notation; both give the same result.

In this example, we extract both multiple specific rows *and* multiple specific columns using the `loc` notation:

```
In [1]:  print(data_df.loc[[4722, 1198, 4722], ["Nickname", "Mother's ID"]])


                Nickname   Mother's ID
         4722     Edward            -1
         1198     Jessie            -1
         4722     Edward            -1
```

We select the row with beaver identification number 4722 two times and place the columns from that row in two different locations (as the first and last rows) in the resulting `DataFrame`.

We can also select an individual element of the `DataFrame` using row and column labels. For instance, the following provides the nickname for the beaver with identification number 4722:

```
In [1]:  print(data_df.loc[4722, 'Nickname'])


         Edward
```

This is the same syntax as line 55 of Figure 19.1.

Because row and column names are clearer and convey more information than positional indices, we usually manipulate pandas data objects using these names. Still, there are some cases where using positional indices make sense. In those cases, we can use the `iloc` notation. The notation works the same as list, tuple, and NumPy array slicing. We can also select specific rows and columns via lists in each dimension:

```
In [1]:  print(data_df.iloc[0:2, 1:3])

             Date of Birth   Mother's ID
      1243     2018-04-17          2352
      9384     2018-04-17          2352
```

```
In [2]:  print(data_df.iloc[[0, 2], [1, 3]])

             Date of Birth   Father's ID
      1243     2018-04-17          4722
      2352                           -1
```

```
In [3]:  print(data_df.iloc[2, 3])

      -1
```

When slicing ranges using `iloc` and positional indices, the upper bound is *exclusive*, just as slicing with lists, tuples, and NumPy arrays. For multiple specific rows and multiple specific columns, the specified row and column positional indices are used.

Finally, we can add and drop rows and columns (as appropriate) to pandas data objects, as we can to lists and dictionaries. Adding columns uses a syntax similar to dictionaries. In the example below, we make a more-or-less memory-independent copy of the first two columns (specified by column name) of the `data_df` object using the `DataFrame` object's `copy` method. We do this to make the table less wide, so after we add new columns, it will still fit on the page of the present book. Then, we add three columns of fictitious data, listing the color, the weight, and length of the beavers:

```
In [1]:  copy_df = data_df[['Nickname', 'Date of Birth']].copy()
         print(copy_df)

             Nickname Date of Birth
      1243       Max    2018-04-17
      9384     Polly    2018-04-17
      2352      Fong
      4722    Edward
      1198    Jessie
```

```
In [2]:  copy_df['Color'] = pd.Series(['brown', 'brown', 'dark brown',
                                       'dark brown', 'brown'], index=rows)
         copy_df['Weight'] = pd.Series([71, 65, 59, 60, 60], index=rows)
         copy_df['Length'] = pd.Series([26, 29, 32, 33, 30], index=rows)
         print(copy_df)
```

	Nickname	Date of Birth	Color	Weight	Length
1243	Max	2018-04-17	brown	71	26
9384	Polly	2018-04-17	brown	65	29
2352	Fong		dark brown	59	32
4722	Edward		dark brown	60	33
1198	Jessie		brown	60	30

Weight is in pounds and length is in inches. We specify the `index` keyword input parameter to equal the list `rows` (created in lines 27–30 of Figure 19.1). That way, pandas knows which element in the list passed into the `Series` constructor as a positional input parameter corresponds to which beaver identification number. If the right-hand side of the equal sign in the assignment is not a `Series` object but an array or list, the mapping to row names (indices) is based on the order of the elements and rows. Replacing existing columns in a `DataFrame` uses the same assignment syntax, just like with dictionaries.

The `drop` method drops specified rows and/or columns and returns the altered data object.

To add a row, we use the `loc` notation. The following will add data from a beaver with an identification number 7844 onto the `copy_df` object from above:

```
In  [1]:  copy_df.loc[7844, :] = pd.Series(['Lily', '', 'dark brown', 62, 34],
                                            index=copy_df.columns)
          print(copy_df)
```

	Nickname	Date of Birth	Color	Weight	Length
1243	Max	2018-04-17	brown	71.0	26.0
9384	Polly	2018-04-17	brown	65.0	29.0
2352	Fong		dark brown	59.0	32.0
4722	Edward		dark brown	60.0	33.0
1198	Jessie		brown	60.0	30.0
7844	Lily		dark brown	62.0	34.0

To add more than a single row to a `DataFrame` object, it makes more sense to make the new rows another `DataFrame` object and to use the pandas concatenation function `concat`.

Finally, in Section 15.2.2, we saw that the values of a `Series` object are stored in the attribute `values` as a NumPy array. Similarly, the values in a `DataFrame` object are stored in the attribute `values` as a two-dimensional NumPy array.

Boolean Selection

We saw with NumPy arrays that we could use array syntax and the `where` function, coupled with boolean arrays, to select certain elements in an array but not others. We can do something similar with pandas data objects.

Consider again the `DataFrame` object `data_df` from Figure 19.1. If we select a column from `data_df`, say `"Mother's ID"`, and check to see which values are greater than 0, we get the following:

```
In [1]:  print(data_df["Mother's ID"] > 0)

         1243      True
         9384      True
         2352      False
         4722      False
         1198      False
         Name: Mother's ID, dtype: bool
```

The operation returns a `Series` object where the greater than comparison is done element-wise down the column elements, and the boolean result is saved in the returned `Series` object. Because the return value is a `Series` object, the row labels (and the column name) are included.

We can use this kind of boolean "mask" to select specific rows of a `DataFrame` object. For instance, if we put this mask into square brackets selecting off of `data_df`, we obtain a `DataFrame` object consisting of the rows where the `"Mother's ID"` is greater than 0:

```
In [1]:  mask = data_df["Mother's ID"] > 0
         print(data_df[mask])

              Nickname Date of Birth  Mother's ID  Father's ID
         1243      Max    2018-04-17         2352         4722
         9384    Polly    2018-04-17         2352         4722
```

We can also accomplish the above in a single line:

```
print(data_df[data_df["Mother's ID"] > 0])
```

We can also conduct more complicated boolean expressions. The following returns the rows where `"Mother's ID"` is greater than 0 and the beaver identification number is greater than 5000:

```
In   [1]:  print(data_df[(data_df["Mother's ID"] > 0) & (data_df.index > 5000)])

              Nickname Date of Birth  Mother's ID  Father's ID
         9384    Polly    2018-04-17         2352         4722
```

In such pandas expressions, the logical "and" operator is the ampersand (`&`), and the logical "or" operator is the vertical strut (`|`).

Again, while we can do something similar with NumPy arrays, array syntax, and boolean arrays, the pandas selection mechanism and row and column labels make this kind of data subsetting more concise.

Sorting

`Series` and `DataFrame` objects include the `sort_index` and `sort_values` methods. The `sort_index` method sorts the elements (for `Series` objects) or rows (for `DataFrame` objects) based on the value of the index (row names). The `sort_values` method sorts the elements/rows based upon the values of the elements (for `Series` objects) or the values in one or more columns (for `DataFrame` objects).

Consider again the `DataFrame` object `data_df` from Figure 19.1. If we select the `'Nickname'` column to obtain a `Series` object, the `sort_index` method returns:

```
In [1]:  print(data_df['Nickname'].sort_index())

         1198     Jessie
         1243        Max
         2352       Fong
         4722     Edward
         9384      Polly
         Name: Nickname, dtype: object
```

Calling `sort_index` with the keyword input parameter `ascending` set to `False` will return the items in the `Series` object in order of descending index values (row names):

```
In [1]:  print(data_df['Nickname'].sort_index(ascending=False))

         9384      Polly
         4722     Edward
         2352       Fong
         1243        Max
         1198     Jessie
         Name: Nickname, dtype: object
```

The values of the `Series` object follow the reordering of the row names (indices) due to the sorting, because the values are referenced by those names.

The `sort_index` method attached to a `DataFrame` object works similarly to the `Series` object version. Applying the method to the `DataFrame` object `data_df` from Figure 19.1 gives us:

```
In [1]:  print(data_df.sort_index())

                 Nickname Date of Birth  Mother's ID  Father's ID
         1198     Jessie                          -1           -1
         1243        Max   2018-04-17           2352         4722
         2352       Fong                          -1           -1
         4722     Edward                          -1           -1
         9384      Polly   2018-04-17           2352         4722
```

When the row names (indices) are reordered due to the sorting, all the values of the entire row follow that reordering.

For a `Series` object, the `sort_values` method sorts based on the values of the elements of the object (not the row names). This method also has the boolean keyword input parameter `ascending` to control whether to sort the values in ascending or descending order:

```
In [1]:  print(data_df['Nickname'].sort_values(ascending=False))

         9384      Polly
         1243        Max
         1198     Jessie
         2352       Fong
         4722     Edward
         Name: Nickname, dtype: object
```

For `DataFrame` objects, when we use `sort_values`, we need to specify which column (or columns) of values to sort by. The `by` keyword input parameter accomplishes this. The following sorts by the `'Date of Birth'`:

```
In [1]:  print(data_df.sort_values(by='Date of Birth'))

                Nickname Date of Birth  Mother's ID  Father's ID
         2352       Fong                         -1           -1
         4722     Edward                         -1           -1
         1198     Jessie                         -1           -1
         1243        Max   2018-04-17           2352         4722
         9384      Polly   2018-04-17           2352         4722
```

To sort first by `'Date of Birth'`, then by `'Nickname'`, we can pass in a list of these columns names, in that order, to the `by` keyword input parameter:

```
In [1]:  print(data_df.sort_values(by=['Date of Birth','Nickname']))

                Nickname Date of Birth  Mother's ID  Father's ID
         4722     Edward                         -1           -1
         2352       Fong                         -1           -1
         1198     Jessie                         -1           -1
         1243        Max   2018-04-17           2352         4722
         9384      Polly   2018-04-17           2352         4722
```

Between rows with the same value for `'Date of Birth'`, the method sorts on the basis of the value of `'Nickname'`.

Both the `sort_index` and `sort_values` methods (for both `Series` and `DataFrame` objects) have the keyword input parameter `kind` that enables us to choose the sorting algorithm we wish to use. The default is an algorithm called Quicksort.[8]

Grouping

A typical operation using a set of columnar data is to group the data based upon one set of column values and then to perform an operation on the values of another column (or columns), for each grouping. The `groupby` method provides a handy way of accomplishing this task. Consider the final version of the `DataFrame` object `copy_df` from earlier in Section 19.2.3 (reproduced below for convenience):

```
In [1]:  print(copy_df)
```

```
         Nickname Date of Birth        Color   Weight  Length
1243         Max    2018-04-17         brown    71.0    26.0
9384       Polly    2018-04-17         brown    65.0    29.0
2352        Fong              dark brown        59.0    32.0
4722      Edward              dark brown        60.0    33.0
1198      Jessie                   brown        60.0    30.0
7844        Lily              dark brown        62.0    34.0
```

Let us say we want to group all rows with the same `'Color'` values and take the mean of the values of every column for which the operation can be applied, i.e., those with numerical values. For the above `DataFrame`, only the `'Weight'` and `'Length'` columns meet this criteria. The following will accomplish this task:

```
In [1]:  print(copy_df.groupby('Color').mean())
```

```
                Weight     Length
Color
brown         65.333333   28.333333
dark brown    60.333333   33.000000
```

The result is a `DataFrame` object where the columns are each column in `copy_df` the `mean` method can be applied to, and the row names (indices) are all the possible values of the `'Color'` column in `copy_df`. The `groupby('Color')` call groups the rows of `copy_df` based on the values in the column `'Color'` (which is `'brown'` and `'dark brown'`). That call returns an object of class `DataFrameGroupBy`. This object stores each grouping as a `DataFrame` and provides methods to do the calculation on each column of each grouping, as desired. In the above code, the `mean` method is applied to each *grouping*, and the result is the

---

[8] The Wikipedia article on Quicksort provides more information on the algorithm. An up-to-date link to the entry is at www.cambridge.org/core/resources/pythonforscientists/refs/, ref. 94.

value for each column for that group. Thus, in the `DataFrame` that results from calling `mean` on the `DataFrameGroupBy` object, the values corresponding to the row name `'brown'` are the mean of all the `'Weight'` values in the rows in `copy_df` with the `'Color'` `'brown'` and the mean of all the `'Length'` values in the rows in `copy_df` with the `'Color'` `'brown'`.

The `DataFrameGroupBy` has a host of other methods that can be applied to the columns in each group of values. Many of them are statistical methods and thus work only on numerical data. There is also the `apply` method that can be used to apply other functions on the groupings, including user-written functions.

To return a `DataFrame` of all the rows in a given group, we can use the `DataFrame GroupBy` object's `get_group` method. Thus, all the rows with `'Color'` `'brown'` can be obtained by:

```
In [1]:  print(copy_df.groupby('Color').get_group('brown'))

        Nickname Date of Birth  Color  Weight  Length
1243       Max    2018-04-17   brown    71.0    26.0
9384     Polly    2018-04-17   brown    65.0    29.0
1198    Jessie                 brown    60.0    30.0
```

This same `DataFrame` object can also be obtained by using this selection command:

```
copy_df[copy_df['Color'] == 'brown']
```

but what `groupby` does, in this example, is to do this kind of selection for *all* the possible values of `'Color'`, not just `'brown'`. The row names for all these groups are stored in the `groups` dictionary (reformatted for better display):

```
In [1]:  print(copy_df.groupby('Color').groups)

       {'brown': Int64Index([1243, 9384, 1198], dtype='int64'),
        'dark brown': Int64Index([2352, 4722, 7844], dtype='int64')}
```

The keys of the `groups` dictionary are all the possible values of `'Color'`. We do not have to manipulate the groupings directly, however. Applying the `mean` method, another calculation method attached to the `DataFrameGroupBy` object, or the `apply` method reduces the data in each and all groupings for all columns the method will work on. The bottom line: The `groupby` method enables us to do group-and-apply tasks on a table of data in one expression.

There are a lot of things we can do with pandas! We have just scratched the surface. We have seen, however, for tabular data, pandas can search (select and group) and sort a lot of complex categorical data in very few lines of code. See these resources mentioned earlier in this section for details on the capabilities of pandas.

19.3 Try This!

The examples in this section address the main topics of Section 19.2. We practice using various searching and sorting algorithms and routines. This includes using the powerful capabilities of pandas as well as non-pandas alternatives.

For all the exercises in this section, assume NumPy has already been imported as np and pandas as pd.

Try This! 19-1 Using Non-pandas Ways to Search: Monthly Precipitation

For the Seattle-area precipitation data in Table 6.1, use two of the methods from Section 19.2.1 to list the monthly precipitation values greater than four inches. Assume the precipitation data are already in a one-dimensional NumPy array `precip`.

Try This! Answer 19-1

The following presents a solution using loops and a solution using the `where` function:

```
In [1]:  for iprecip in precip:
             if iprecip > 4.0:
                 print(iprecip)

         5.58
         4.54
         4.31
         6.5
         5.41
```

```
In [2]:  print(precip[np.where(precip > 4.0)])

         [5.58 4.54 4.31 6.5  5.41]
```

Try This! 19-2 Binary Search: Equipment Inventory

Pretend we have the following inventory of laboratory equipment:

Part no.	Equipment	Quantity
3822	Safety goggles	16
2938	Test tubes	34
1123	Beakers	22
8733	Petri dishes	54
4389	Test strips	25
2311	Microscope slides	89
9320	Hot plates	11

All part numbers are four-digit integers. Write code using binary search that prints out whether beakers are listed in the inventory.

Try This! Answer 19-2

The problem is a search for whether part number 1123 is in the list of part numbers from the inventory. Our solution is similar to the code in Figure 19.2 and is given in Figure 19.9. After running the code, `True` is printed to the screen.

```python
part_num = np.array([3822, 2938, 1123, 8733, 4389, 2311, 9320])
part_num = sorted(part_num)
target = 1123

end_search = False
while not end_search:
    if len(part_num) == 1:
        end_search = True
        if target == part_num[0]:
            target_found = True
        else:
            target_found = False

    elif len(part_num) == 2:
        end_search = True
        if target == part_num[0]:
            target_found = True
        elif target == part_num[1]:
            target_found = True
        else:
            target_found = False

    else:
        middle_loc = int(len(part_num)/2)

        if target == part_num[middle_loc]:
            end_search = True
            target_found = True

        elif target > part_num[middle_loc]:
            part_num = part_num[middle_loc:]

        else:
            part_num = part_num[:middle_loc+1]

print(target_found)
```

Figure 19.9 Code to execute a binary search for Try This! 19-2.

Try This! 19-3 Selection Sort: Equipment Inventory

For the laboratory inventory data in Try This! 19-2, write code that uses selection sort to sort the part number in ascending order. Assume the function `min_val_loc`, as defined in Figure 19.1, is *not* available for use. We can, however, duplicate the functionality of that function in our own code.

Try This! Answer 19-3

This solution accomplishes the task:

```
In [1]:   part_num = [3822, 2938, 1123, 8733, 4389, 2311, 9320]
          part_num_sorted = []
          for i in range(len(part_num)):

              unsorted_min = part_num[0]
              unsorted_min_loc = 0
              for j in range(len(part_num)):
                  if part_num[j] < unsorted_min:
                      unsorted_min = part_num[j]
                      unsorted_min_loc = j

              part_num_sorted.append(part_num[unsorted_min_loc])
              temp = part_num.pop(unsorted_min_loc)

          print(part_num_sorted)


          [1123, 2311, 2938, 3822, 4389, 8733, 9320]
```

We use a list version of `part_num` since we want to use the `pop` method.

Try This! 19-4 Basic Selection and Manipulation Using pandas: Equipment Inventory

Create a pandas `DataFrame` object `inventory` using all the data in Try This! 19-2. Make the part numbers the row names (index). Extract the "Quantity" column as a separate `Series` object and display its values to the screen.

Try This! Answer 19-4

This solution accomplishes the tasks:

```
In [1]:  inventory = \
           pd.DataFrame([['Safety goggles', 16],
                         ['Test tubes', 34],
                         ['Beakers', 22],
                         ['Petri dishes', 54],
                         ['Test strips', 25],
                         ['Microscope slides', 89],
                         ['Hot plates', 11]],
                        index=[3822, 2938, 1123, 8733, 4389, 2311, 9320],
                        columns=['Equipment', 'Quantity'])
         print(inventory['Quantity'])
```

```
         3822    16
         2938    34
         1123    22
         8733    54
         4389    25
         2311    89
         9320    11
         Name: Quantity, dtype: int64
```

We can also select a column using the `loc` notation. Given the `inventory` object above, the following will also select the `'Quantity'` column and return a `Series` object:

```
inventory.loc[:,'Quantity']
```

Pretend all the quantities are off by a factor of 10 (maybe someone forgot to write down the "0" when they were taking the inventory). We can multiply all the current values by 10 and replace the `'Quantity'` column using the following:

```
In [1]:  inventory['Quantity'] = inventory['Quantity'] * 10
         print(inventory)
```

```
                       Equipment  Quantity
         3822      Safety goggles       160
         2938          Test tubes       340
         1123             Beakers       220
         8733        Petri dishes       540
         4389         Test strips       250
         2311   Microscope slides       890
         9320          Hot plates       110
```

Pandas data objects also support array syntax for calculations, as long as the operation is defined for the values in the object. Such array syntax calculations are not applied to the row names (indices) or column names.

Try This! 19-5 Boolean Selection and Sorting Using pandas: Equipment Inventory

For the original `inventory` object created in Try This! 19-4 (i.e., the object with the data as presented in the table in Try This! 19-2), select only those rows whose quantities of items is greater than 30, sort those rows by part number, and print the resulting `DataFrame` object to the screen.

Try This! Answer 19-5

This code solves the problem:

```
In [1]:  print(inventory[inventory['Quantity'] > 30].sort_index())


               Equipment   Quantity
      2311   Microscope slides        89
      2938          Test tubes        34
      8733         Petri dishes        54
```

The `inventory['Quantity'] > 30` portion creates a boolean `Series` object showing what rows have quantities greater than 30. When that is put into the square brackets after the first `inventory`, only those rows are selected. We can sort the resulting `DataFrame` by row names using the `sort_index` method.

Try This! 19-6 Grouping Using pandas: Equipment Inventory

Pretend that we want to print out labels for the equipment in our original laboratory inventory from Try This! 19-2. We have two sizes of labels (long and short) and we want to find out how many labels of each we need to print. The short label cannot hold more than 10 characters. That is, for the original `inventory` object created in Try This! 19-4, we need to:

• Add a column `'Name Length'` that gives the number of characters in the name listed under the `'Equipment'` column.
• Add a column `'Label Size'` that has the value `'long'` or `'short'`, depending on the label size that is appropriate for the name of the equipment.
• Group the rows by `'Label Size'` and add up all the quantities for each label size.

Some hints and helps:

- `Series` objects have an attribute `str` that holds methods that operate on string values. One of the methods is `len`, which returns the length of the string values for each value in the `Series` object. The return value is itself a `Series` object.
- `DataFrameGroupBy` objects also have the method `sum` that adds up all the values in each column for the `DataFrameGroupBy` object.
- The NumPy `where` function also operates on a boolean `Series` object as the condition. The resulting array can also be saved as a column in a `DataFrame` object.

Try This! Answer 19-6

This code solves the problem. After creating the `'Name Length'` and `'Label Size'` columns, we print out `inventory` to show what it looks like. We then print the result of doing a `sum` on our grouping over `'Label Size'` (after printing a blank line):

```
In [1]:   inventory['Name Length'] = inventory['Equipment'].str.len()
          inventory['Label Size'] = np.where(inventory['Name Length'] > 10,
                                             'long', 'short')
          print(inventory)
          print('\n')
          print(inventory.groupby('Label Size').sum())
```

	Equipment	Quantity	Name Length	Label Size
3822	Safety goggles	16	14	long
2938	Test tubes	34	10	short
1123	Beakers	22	7	short
8733	Petri dishes	54	12	long
4389	Test strips	25	11	long
2311	Microscope slides	89	17	long
9320	Hot plates	11	10	short

	Quantity	Name Length
Label Size		
long	184	54
short	67	27

Thus, we need 184 long labels and 67 short labels. The `sum` method is also applied to the `'Name Length'` column, but those values do not help us.

The `str` attribute of pandas `Series` objects is really useful! It enables us to do many different kinds of string operations over all values in the `Series` object. Versions of many of the regular Python string methods exist, including `upper`, `lower`, `split`, `strip`, etc. The `slice` method implements Python's square-bracket string-slicing capabilities for the `Series` object.

19.4 More Discipline-Specific Practice

The Discipline-Specific Jupyter notebooks for this chapter cover the following topics:

- Ways to search and sort in Python without using pandas.
- Searching and sorting algorithms.
- Basic searching and sorting using pandas.

19.5 Chapter Review

19.5.1 Self-Test Questions

Try to do these without looking at the book or any other resources or using the Python interpreter. Answers to these Self-Test Questions are found at the end of the chapter. If not otherwise stated, assume NumPy has been imported earlier as `np`, and pandas has been imported earlier as `pd`.

Self-Test Question 19-1

What do the list methods `index` and `count` do?

Self-Test Question 19-2

What is "membership testing" and how does Python implement the operation?

Self-Test Question 19-3

Describe the difference between a linear and binary search algorithm. What restrictions and constraints exist in the use of each algorithm?

Self-Test Question 19-4

Describe some differences between selection and insertion sort.

Self-Test Question 19-5

Pretend we have a `DataFrame` object `data`. How do we obtain the row and column names?

Self-Test Question 19-6

Pretend we have a `DataFrame` object `data`. How would we extract the column whose name is `'Mass'`? Assume that column name is valid in `data`.

Self-Test Question 19-7

Pretend we have a `DataFrame` object `data`. How would we extract the row whose name is 48324? Assume that row name is valid in `data`.

Self-Test Question 19-8

Pretend we have a `DataFrame` object `data`. How would we extract the sub-`DataFrame` object that consists of the first three rows and the first four columns? Assume this is a valid operation for `data`. Do this once using `loc` and once using `iloc`.

Self-Test Question 19-9

Pretend we have a `DataFrame` object `data`. Add a column to `data` called `'Row Number'` of integers that gives the row number, starting with one.

Self-Test Question 19-10

Pretend we have a `DataFrame` object `birds` describing data regarding a collection of birds. The object includes a column `'Count'` and another column `'Weight'` (the latter in pounds). How would we obtain the `DataFrame` consisting of those birds that weigh over 2 lb and have a count of over 15?

Self-Test Question 19-11

Pretend we have the `DataFrame` object `birds` from Self-Test Question 19-10. How would we sort the rows based upon putting the values in the column `'Count'` in descending order?

Self-Test Question 19-12

Pretend we have a `DataFrame` object `fish` that has two columns (besides the row names): `'Species'` and `'Count'`. The `'Count'` column gives the number of fish of the species given in the `'Species'` column. Each row represents observations taken at different times (the times are encoded in the `DataFrame` index of row names). How would we calculate the total number of each species of fish?

19.5.2 Chapter Summary

In this chapter, we described different ways of searching and sorting data in Python. Along the way, we described different algorithms for searching and sorting, noting that different algorithms have different requirements as well as different performance characteristics. Finally, we looked at some ways of using the methods attached to pandas data objects to select, manipulate, search, sort, and group data. The pandas tools and interface enable us to do a lot with data using very few lines of code. Specific topics we covered in this chapter include:

- Ways to search and sort in Python without using pandas

  - Searching by testing in a loop.
  - Searching using the `index` method or `count` method of a list.
  - Searching using array syntax and testing for equality.
  - Searching using the `where` function on arrays.

- Searching using the `isclose` function on arrays.
- Searching using membership testing.
- Searching using the `glob` function.
- Sorting using the `sort` method of a list.
- Sorting using the `sorted` function.

• Searching and sorting algorithms

- We can search and sort using different algorithms. Different algorithms have different dependencies and performance characteristics.
- Linear search involves looping through the items in a collection one-by-one and checking if the item is the target we are looking for.
- Binary search involves asking whether an item is greater than, equal to, or less than the midpoint of a sorted collection of items and repeating the inquiry on the half of the remaining collection the item is in.
- Linear search works on any collection of items. Binary search only works on a sorted collection of items.
- Sorting involves "moving" elements from an unsorted partition to a sorted partition until the unsorted partition is empty and all the elements are in the correct order in the sorted partition.
- Selection sort works by repeatedly finding the minimum value in the remaining items in the unsorted partition and "moving" the item to the end of the sorted partition.
- Insertion sort works by taking the next item in the unsorted partition, checking whether the item from the unsorted partition is larger than the last item in the sorted partition. If it is not, we repeat the check with the next-to-last item in the sorted partition, and so on with the items in the sorted partition until we find the place where the item from the unsorted partition should be placed (or inserted). The process then repeats for the next item of the unsorted partition.
- Both selection and insertion sort have similar performance characteristics.

• Basic searching and sorting using pandas

- We can refer to rows (and columns, if they exist) in `Series` and `DataFrame` objects by the row and column names.
- The row names are called the indices of the pandas data object and are held in the attribute `index`. `DataFrame` columns are listed in the attribute `columns`.
- pandas enables us to select one or more columns as well as one or more rows of a pandas data object. The rows and columns may be in a continuous range or not. The `loc` syntax uses the row and column names to do the selection. The `iloc` syntax uses the positional indices to do the selection.
- Single columns of `DataFrame` objects are `Series` objects. A multirow and column selection from a `DataFrame` object is also a `DataFrame` object.
- We can use boolean selection to select specific rows meeting given criteria. The ampersand (`&`) is the "and" operator and the vertical strut (`|`) is the "or" operator in these pandas boolean expressions.

- The `sort_index` method sorts elements or rows by the row names (index values). The `sort_values` method sorts elements or rows by the values in one or more column.
- The `groupby` method enables us to group rows in a `DataFrame` based upon values in one or more columns and then to do operations on other column values for each of those groups.

19.5.3 Self-Test Answers

Self-Test Answer 19-1

The `index` method returns the positional index where the first occurrence of what we are looking for is located in the list. The `count` method returns the number of occurrences in the list of what we are looking for.

Self-Test Answer 19-2

In membership testing, we are trying to see whether a collection has at least one occurrence of some value. In Python, this is implemented using the `in` operator. In an expression using the `in` operator, the value to the left of `in` is checked against the contents of the collection given to the right of the `in`. A boolean value is returned. Note, a `for` loop is *not* implementing membership testing with the inclusion of the `in` token.

Self-Test Answer 19-3

In linear search, we examine each item in a collection, one at a time, and see whether this is the item we are looking for. If so, we stop our examination and declare our search a success. In binary search, we find the midpoint of the collection we are searching through and ask whether the item we are looking for is greater than, less than, or equal to the midpoint. If it is greater or less than the midpoint, we repeat this question for the half of the collection the target is in, and so on until we either find the target or run out of items in the collection to compare with.

The linear search algorithm can be used with any collection of items. This makes the algorithm very flexible but slow, if the target is consistently located near the end of the collection. The binary search algorithm is generally faster than a linear search, but binary search can only be used on sorted collections. If a collection is unsorted, we either have to sort it before using binary search or use a linear (or other) search.

Self-Test Answer 19-4

In selection sort, our comparisons mainly occur on the items in the unsorted partition. In insertion sort, our comparisons mainly occur on the items in the sorted partition. Both algorithms have similar performance characteristics.

Self-Test Answer 19-5

The `data.index` attribute holds the row names (indices) while the `data.columns` attribute holds the column names.

Self-Test Answer 19-6

Either:

```
data['Mass']
```

or:

```
data.loc[:,'Mass']
```

works.

Self-Test Answer 19-7

```
data.loc[48324, :]
```

Self-Test Answer 19-8

This solves the task using `loc`:

```
data.loc[data.index[0:3], data.columns[0:4]]
```

This solves the task using `iloc`:

```
data.iloc[0:3, 0:4]
```

Self-Test Answer 19-9

This code will add the row:

```
data['Row Number'] = np.arange(len(data)) + 1
```

As mentioned in Section 19.2.3, the right-hand side does not have to be a `Series` object but can be an array or list.

Self-Test Answer 19-10

```
birds[(birds['Count'] > 15) & (birds['Weight'] > 2)]
```

Self-Test Answer 19-11

```
birds.sort_values(by='Count', ascending=False)
```

Self-Test Answer 19-12

This code solves the problem:

```
fish.groupby('Species').sum()
```

We do not have to know how the times for each row are encoded in the `DataFrame` row names index to solve this problem.

20 Recursion

In Section 10.2.4, we examined copying and deleting files and directories. The tools we used – `copytree`, `rmtree`, and `move` – automatically handled directory trees. For instance, a copy operation using `copytree` on a directory copies not just the topmost directory but all files and subdirectories underneath the topmost directory, and so on to every file and subdirectory. In the present chapter, we explore how a programming language implements the idea of "do-this-action-over-nested-levels" through the principle of recursion.[1]

Because we are fresh from Chapter 19, where we examined searching, in this chapter, we start with the problem of searching for items in a directory tree. This problem is an inherently recursive problem, because directories are made up of files and subdirectories, and those subdirectories in turn are made up of files and subdirectories, and so on. Python gives us tools to solve the problem using regular iteration (i.e., a `for` loop), and we look at that methodology first in Section 20.1. In our second solution to the directory search problem, we explicitly use recursion, and this motivates a deeper look at the construct. Like most programming structures and concepts which can be applied to multiple problems, the idea and usefulness of recursion goes beyond searching directory trees. Any problem that has the characteristic of "nested levels" is a candidate for recursion, and there are cases where writing a program using recursion is much easier than not using recursion.

20.1 Example of Recursion

In Chapter 10, we described how to use Python to manage files and directories. Amongst these were directory trees, collections of files and subdirectories, nesting down multiple levels, such as the directory tree of Figure 10.2. The `copytree`, `rmtree`, and `move` commands we looked at, however, operate on these trees as a whole. In this chapter's primary example, we look at code that, while also acting on the directory tree as a whole, allows us to do different tasks at different levels of the tree and on different items (files, subdirectories) in the tree. Specifically, we look at code that enables us to search for a specific file in a directory tree and, if found, prints out the location(s) of that file (because we can have files of the same name if they are located in different directories). In our solutions below, we are specifically searching for all occurrences of the file *Temp.txt* in the directory tree shown in Figure 10.2 that begins with the directory *MyFiles*.

---

[1] Portions of this chapter are adapted from Sections 8.2–8.3 in Lin (2019).

```
1   import os
2
3   file_occurrence_locations = []
4   for dirpath, dirnames, filenames in os.walk("MyFiles"):
5       if 'Temp.txt' in filenames:
6           file_occurrence_locations.append(dirpath)
```

Figure 20.1 Code to search for the locations of the file *Temp.txt* using the os module's `walk` generator.

In our first solution (Figure 20.1), we make use of the os module's `walk` generator[2] that enables us to use a loop to descend into a directory tree, providing the full path for every directory and subdirectory in the tree (as the variable `dirpath`), as well as the names of all files in each directory and subdirectory. This loop does not explicitly implement recursion, but it is recursive in its behavior. That is, its behavior relies on the fact that a directory tree is a construct that nests similar items, where every subdirectory is just like its parent directory (i.e., it consists of files and subdirectories), except for being nested within its parent directory. Each iteration of the loop visits each of these nested directories.

After running the code in Figure 20.1, the list `file_occurrence_locations` has the following contents:

```
In [1]:  print(file_occurrence_locations)

         ['MyFiles/StationData/July', 'MyFiles/StationData/June']
```

In Figure 20.1 and the `print` call above, we assume the directory *MyFiles* is found in the current working directory and is the directory tree given in Figure 10.2.

In our second solution, we make explicit use of recursion, a programming construct that expresses the idea of nesting similar items. The function `find_temp_txt`, defined in Figure 20.2, searches the directory specified in the input parameter `top_dir` and all subdirectories within `top_dir` and all subsequent subdirectories. Thus, the following call to this function:

```
In [1]:  print(find_temp_txt("MyFiles"))

         ['MyFiles/StationData/July', 'MyFiles/StationData/June']
```

returns a list of the paths of the directories that contain the file *Temp.txt*. In the call to `find_temp_txt` above, we assume that *MyFiles* is in the current working directory and is the directory tree given by Figure 10.2.

---

[2] See https://docs.python.org/3/library/os.html and http://stackoverflow.com/a/1724723 (both accessed August 21, 2020).

```
1   import os
2
3   def find_temp_txt(top_dir):
4       list_items = os.listdir(top_dir)
5       list_txt = []
6       for item in list_items:
7           item_path = os.path.join(top_dir, item)
8           if os.path.isdir(item_path):
9               list_txt = list_txt + find_temp_txt(item_path)
10          else:
11              if item == 'Temp.txt':
12                  list_txt = list_txt + [top_dir]
13      return list_txt
```

Figure 20.2 Code to define a function `find_temp_txt` that searches for the locations of the file *Temp.txt* using recursion.

20.2 Python Programming Essentials

We begin this section by describing the behavior of the os module's `walk` generator which, although not an example of recursive programming, enables us to search a directory tree recursively through a `for` loop. This enables us to explain how Figure 20.1 works. We then describe recursion and how to write recursive functions. This enables us to explain how Figure 20.2 works. We close by examining more applications of recursion, beyond searching directory trees.

20.2.1 Using the `walk` Generator

As we saw in Section 20.1, the os module's `walk` generator allows us to descend into a directory tree and visit each of the directories in that tree. However, although the `walk` generator looks like a function, it is not exactly a function. Most functions take an input and produce a return value. The return value is usually something we can save to a variable, such as a number, string, list, or array. What `walk` produces, however, is something called a generator. For our purposes, what a generator exactly is not that important.[3] Instead, we focus on using a generator. In short, a generator is an object whose purpose is to give us something to iterate through and return us something as we do the iteration. Thus, generators are often used in loops.

---

[3] See the Python wiki page on generators for more on the topic. An up-to-date link to this page is at www.cambridge.org/core/resources/pythonforscientists/refs/, ref. 15.

Consider the following example (assume the current working directory holds the folder *MyFiles*, which is the topmost directory of the Figure 10.2 directory tree):

```
In [1]:  import os

         for dirpath, dirnames, filenames in os.walk("MyFiles"):
             print( str(dirpath) + ":\n    " + str(dirnames) + \
                 "\n    " + str(filenames) )

         MyFiles:
             ['AudioClips', 'StationData', 'VideoClips']
             ['Report.pdf', 'Article.pdf']
         MyFiles/AudioClips:
             []
             ['Questions.mp3', 'Talk.mp3']
         MyFiles/StationData:
             ['July', 'June']
             ['Info.txt']
         MyFiles/StationData/July:
             []
             ['Temp.txt']
         MyFiles/StationData/June:
             []
             ['Humidity.txt', 'Wind.txt', 'Temp.txt']
         MyFiles/VideoClips:
             []
             ['Setup.mov', 'Poster.jpg']
```

What is happening? At each iteration of the `for` loop, the variables `dirpath`, `dirnames`, and `filenames` are filled. On the first iteration of the loop, `dirpath` is set to `'MyFiles'`, `dirnames` is set to the list `['AudioClips', 'StationData', 'VideoClips']`, and `filenames` is set to the list `['Report.pdf', 'Article.pdf']`. That is to say, on the first iteration, `dirpath` is set to the topmost directory in the tree (*MyFiles*) and `dirnames` and `filenames` are set to the names of the subdirectories and files that are in *MyFiles*.

On the second iteration, `dirpath` is set to *MyFiles/AudioClips*, that is, the *AudioClips* subdirectory in *MyFiles*. The `dirnames` variable is set to an empty list (because *AudioClips* contains no subdirectories), and the `filenames` variable is set to the list `['Questions.mp3', 'Talk.mp3']`.

The third iteration sets `dirpath` to *MyFiles/StationData*, `dirnames` to the list `['July', 'June']`, and `filenames` to the list `['Info.txt']`, and so on until all subdirectories in the *MyFiles* directory tree are visited. As a side note, we see from what the code prints out that when we reach a leaf directory (i.e., a directory that only contains files), the `dirnames` list will be empty (and thus have a length of zero). If we are interested in looking only at leaf directories, we can use `len(dirnames)` as a test to see whether we are in a leaf directory.

With this background, we can explain the code in Figure 20.1. In that code, the loop beginning in line 4 visits all directories in *MyFiles*. In each iteration of the loop, we visit another directory in the tree. At each iteration, the list of files in the directory being visited is `filenames`, and so we can check whether *Temp.txt* is in the directory being visited if it is in the `filenames` list (line 5). If that file exists in that directory, we append the path (starting with the argument to `walk`) of the directory that contains the files in `filenames` (i.e., `dirpath`) to the list `file_occurrence_locations`.

One final note about finding files: It is an unfortunate truth that different operating systems not only have different directory separation characters but also different rules about case sensitivity. On some operating systems (like Linux), filenames can be made from upper case or lowercase characters. On other operating systems (like Mac OS X), there is no real distinguishing between a file named *Temp.txt* and one named *temp.txt*.[4] This issue with case makes it possible that a plain string equality or membership test will fail, if we are working across operating systems. Python provides the fnmatch module to enable pattern matching for filenames that takes into account the case handling features of the operating system we are working on. See the fnmatch documentation for details. An up-to-date link to the fnmatch documentation is at www.cambridge.org/core/resources/pythonforscientists/refs/, ref. 54.

20.2.2 Recursion and Writing Recursive Code

Using the `walk` generator in a loop is a great way of searching through a directory tree. It is concise, flexible, and robust. However, although we do not need to discuss recursion to use `walk`, the directory tree searching problem very naturally lends itself to being solved by recursion. With `walk` we do not have to use recursion to traverse a directory tree, but it is worth using the tree traversal problem to learn about how recursion works, because recursion is a really useful concept that can help us write more concisely what would otherwise be a really complex function or program. We seldom *need* recursion, but when we do, we *really* need recursion.

Recursion is a repetition structure, and in that way it is like a loop. But, whereas a loop repeats any kind of task over and over (the task being defined in the body of the loop), a recursive function repeatedly executes *itself*. That is, recursive functions are functions whose definitions include one or more calls to itself. For instance, if we have a function `calculate`, within the definition of `calculate` there is at least one call to `calculate`.

That seems a little weird. How can we use a function while we are defining it? It sounds like an infinite regress: use a function we are defining using a function we are defining using a function we are defining, and so on without end.

---

[4] The real situation is a little more complex. The issue is with the file system format we choose and not with the Mac version of Unix itself. In addition, case-insensitive here does not mean it is not case-preserving. See http://apple. stackexchange.com/a/22304 for a nice discussion of the issue (accessed November 11, 2016).

A mathematical example might be helpful to describe both what recursion is and why it does not have to lead to infinite regress. Say we want to write a function that takes a list of numbers and adds the numbers up. Using a `for` loop, the function `add_up` would be:

```
def add_up(input_list):
    total = 0.0
    for i in input_list:
        total = total + i
    return total
```

When we go through the logic of this code in our head, we probably hear something like, "take one number, add it to `total`, take another number, add it to `total`, and so on until we are out of numbers."

But, we could think of the process of adding up these numbers in another way: "the sum of a list of numbers is the sum of the list of numbers *not including the last number* plus the last number." But how do we get the "sum of the list of numbers *not including the last number?*" Using the same logic, that sum is also "the sum of the list of numbers *not including the last number* plus the last number," where the "list of numbers" is not the original list of numbers but the original list without the last item on the list. The logic would continue until "the sum of the list of numbers *not including the last number*" is zero, because there are no numbers left in such a list. In Python, this would look like:

```
1  def add_up_recursive(input_list):
2      if len(input_list) == 0:
3          return 0.0
4      else:
5          return add_up_recursive(input_list[:-1]) + input_list[-1]
```

Some things to point out in this code: First, we see that our recursive call, i.e., our call to the `add_up_recursive` function inside the `add_up_recursive` definition, occurs in line 5. It is important to note, however, that while we are calling `add_up_recursive` in line 5, the input argument is *not* the same as the original, full list of numbers. If it were, we would get infinite regress. Instead, the input argument we are passing into the line 5 call to `add_up_recursive` is the current `input_list` without the last element. This is how we implement the "not including the last number" phrase in our above word description of our logic.

Second, our worries about infinite regress are solved by lines 2–3. The `if` statement checks to see whether the list of numbers `input_list` actually has any numbers in it and, if not, the function returns zero. This makes sense because the sum of nothing is zero. But from the standpoint of infinite regress, this means that there comes a point when we no longer make calls to `add_up_recursive`. The situation tested for and handled in lines 2–3 is called the **stopping case** or **base case**. As the name suggests, this condition terminates recursive calls to the function, preventing infinite regress. Every recursive function has to have at least one stopping case. The lack of one (or enough) stopping cases results in a recursive function eventually failing and throwing a `RecursionError` exception.[5]

In this example, the recursive function is not more concise than the one that uses a loop. For simpler problems, this is typical: the looping solution is more straightforward and shorter than the recursive version. For more complex problems, however, this is not the case. Consider again our Figure 20.2 recursive solution to the file search problem in Section 20.1. Let us first consider what is happening in each line of that code and compare that recursive solution with pieces of an iterative solution that does not have the benefit of the os module's `walk` generator.

In Figure 20.2, line 1, we import the os module. Line 3 defines the function. The function takes a single input string, `top_dir`, which is the name of the directory in which we look for all *Temp.txt* files. Line 4 saves a listing of all the items (files and directories) in `top_dir`. The `listdir` method only gives the names (i.e., the name without the path to the file or subdirectory) of the items. It does not provide the absolute path to the items. In line 5, we initialize an empty list to hold all the *Temp.txt* files in `top_dir` and any subdirectories (through to all leaf directories under them). For a given `top_dir`, this list will be initially empty because we have not processed the listing of the items (or sub-items) in `top_dir` yet.

In line 6, we loop through each item. For each item named *Temp.txt* (line 11), we add that to `list_txt` (line 12). We are assuming our operating system is case-sensitive. If an item in `top_dir` is a directory, we make a call to `find_temp_txt` (this is the recursive call) to obtain the list of the paths of *Temp.txt* files from that directory. The path we pass into the recursive `find_temp_txt` call (as well as in the line 8 check with `isdir`) has to start with the initial `top_dir`. That is, it has to start with the `top_dir` we passed in on the original call to `find_temp_txt`. In the example above, that was `find_temp_txt("MyFiles")`. The reason the path needs to begin with *MyFiles* is because the function does not change the current working directory. As a result, the current working directory is still the parent directory of *MyFiles*, and any item inside the directory tree will remain hidden if the path does not first specify *MyFiles*.

In successive recursive calls of the function, line 7 automatically adds whatever the current `item` is to the directory path that began with *MyFiles*, storing the result in `item_path`. This is because the line 9 `find_temp_txt` call's argument becomes the new `top_dir` in that recursive `find_temp_txt` call, and `join` adds on new subdirectories to that `top_dir` in line 7.

In lines 9 and 12, we use the concatenation operator to add items to `list_txt` rather than `append`, because `list_txt` might have zero, one, or more than one element. When the concatenation operator is used on two lists, it joins the two lists together. When one of the lists is an empty list, there is nothing to join, so the result is the other list, unaltered. If the lists have one or more elements, the second list extends the first. The method `append`, however, adds whatever is passed in as the argument as an element in the list. If that argument

---

[5] In other programming languages, a "stack overflow" results, which forces the program to terminate.

is an empty list, an empty list is added as an element of the existing list. If that argument is a list, that list becomes an element of the original list, not an extension of the original list. Here is an example:

```
In [1]:  a = [1, 2, 5]
         b = [3, -5, 9]
         c = a + b
         print(c)
         a.append(b)
         print(a)


         [1, 2, 5, 3, -5, 9]
         [1, 2, 5, [3, -5, 9]]
```

Finally, in line 13 of Figure 20.2, we return the output from `find_temp_txt`, which is the list of the paths (starting with the argument to the function) of all directories in the directory tree given by `top_dir` that have the file *Temp.txt* in them.

So, we see that recursion solves our problem in 13 lines of code. What if we took an iterative approach (i.e., used loops instead of recursion)? Now, in asking this, we have to add the constraint that we cannot use the os module's `walk` generator. With the `walk` generator, we can write a small loop that solves the problem (as Figure 20.1 shows).

The first issue we encounter when trying to write this using only loops (without `walk`) is how to deal with subdirectories. We could, for instance, write the following:

```
1    import os
2
3    def find_temp_txt(top_dir):
4        list_items = os.listdir(top_dir)
5        list_txt = []
6        for item in list_items:
7            item_path = os.path.join(top_dir, item)
8            if os.path.isdir(item_path):
9                list_items_2 = os.listdir(item_path)
10               for item_2 in list_items_2:
11                   item_2_path = os.path.join(top_dir, item_2)
12                   <... more code ...>
```

In lines 8–11 above, when we hit a subdirectory, we create another `for` loop to go through those items. If that list of items includes another directory, we would add another nested `for` loop to go through that subdirectory, and so on (line 12).

The problem with this approach is that we do not know, ahead of time, how many levels of directories there are. With two directory levels, we need two nested loops. With three directory levels, we need three nested loops, and so on. But, there could be one level of directories, two levels, or twenty levels. Do we write a twenty-level nested `for` loop structure in case there are that many directory levels? And how does one write a twenty-level nested `for` loop structure that is remotely readable? With recursion, though, *we do not have to know ahead of time* how many levels of directories there are. If we need to drop down another directory level, the recursive call does so, as many times as needed (within reason).[6]

Recursion works whenever we can identify some sort of "repeated nesting" structure of whatever we want to analyze. In the `add_up_recursive` example, it was the idea of a "list of numbers." In the directory tree searching example, it is the idea of a directory tree. Identifying that structure and describing all the possible kinds of input possibilities related to that structure is the key to these steps we generally follow in writing a recursive function:

1. Identify the "repeated nesting" structure in the recursive logic.
2. List out the possible "cases" of input into the function (or cases derived from the input into the function). So, in the `add_up_recursive` example above, the cases are when the input list `input_list` is empty (has length 0) and when it is not. In making this listing of possible cases, we are taking into account the nesting structure we have identified. In `add_up_recursive`, that structure is the idea of processing steadily decreasing "remaining" portions of the original input list.
3. Of the possible cases of input just identified, one or more will be a stopping case or condition. Identify all the possible stopping conditions and write the code for those.
4. The remaining possible cases of input will be recursive. Figure out what the recursive call(s) will be and write the code for these cases.

Recursive functions are tricky to follow, particularly when there are many recursive levels. Thus, it becomes imperative to walk through, line by line, what the function does given test input. When we do this analysis, we should consider test cases that cover all the possible cases of input, with some leading to stopping cases and the rest to recursive calls.

Because both are structures that implement the idea of repetition, we can, in general, write loops (i.e., iterative structures) as recursive functions and vice versa. When should we use recursion and when should we use looping? Loops are generally more readable and understandable, and the number of iterations we can make with loops is more or less unlimited. Recursive solutions, in contrast, can be confusing and the limit to the number of recursion levels available is generally much smaller than the number of elements of a medium-sized dataset (in the MB range). So, in the vast majority of cases, looping is the way to go. In the primary example of this chapter – the directory tree search – Python even gives us the `walk` generator to make sure we do not have to use recursion to solve this common problem. But, for certain problems, particularly those where we cannot know ahead of time how many "nesting levels" there are, recursion is the only way to solve the problem in a reasonable number of lines of code. We consider a few more such cases in the next section.

---

[6] The maximum number of recursive calls allowed is given by the sys module's `getrecursionlimit` function (https://docs.python.org/3/library/sys.html, accessed August 13, 2020). A typical value is 3000.

20.2.3 More Applications of Recursion

In this section, we mention a few more applications of recursion. In our discussion, keep in mind that recursion is a kind of programming structure or logic. It is not an algorithm. Thus, we can often write code implementing a given algorithm using recursion or not using recursion. In this section, we look at using recursion to implement a binary search and using recursion to traverse a "nested" list of function calls. Although divide-and-conquer approaches to sorting – such as the merge sort and quick sort algorithms – are classically implemented using recursion and are commonly discussed in introductory programming texts, we will not look at these algorithms in the present work.

Recursive Binary Search   In Figure 19.2, we wrote code to implement a binary search using a `while` loop. A binary search, however, also has a recursive logic: Chop the domain in half, find which half the target value would be in, then *do a binary search on the half the target value is in*, and so on. Figure 20.3 shows code defining a recursive binary search function `is_found_by_recursive_binary`, which we use below to duplicate the behavior of

```
1   def is_found_by_recursive_binary(input_list, input_target):
2       if len(input_list) == 1:
3           if input_target == input_list[0]:
4               return True
5           else:
6               return False
7
8       if len(input_list) == 2:
9           if input_target == input_list[0]:
10              return True
11          elif input_target == input_list[1]:
12              return True
13          else:
14              return False
15
16      middle_loc = int(len(input_list)/2)
17
18      if input_target == input_list[middle_loc]:
19          return True
20
21      if input_target > input_list[middle_loc]:
22          return is_found_by_recursive_binary( \
23              input_list[middle_loc:], input_target )
24      else:
25          return is_found_by_recursive_binary( \
26              input_list[:middle_loc+1], input_target )
```

Figure 20.3 Code to define a binary search function using recursion. This function duplicates the general logic of the iterative binary search code of Figure 19.2.

the iterative binary search code of Figure 19.2. Recall, in this example we are searching for whether beaver identification number 1243 is in the list `rows` (as defined in Figure 19.1):

```
In [1]:   rows_sorted = sorted(rows)
          target = 1243
          print(rows_sorted)
          print(is_found_by_recursive_binary(rows_sorted, target))


          [1198, 1243, 2352, 4722, 9384]
          True
```

Let us go through the logic of the recursive binary search function in Figure 20.3. The function stops in the following situations:

- When the input list has one or two elements.
- When the middle element of the input list equals the target.

The way we have written the code, if the input list has one or two elements, we manually check (i.e., through an explicit `if` statement) whether the target is equal to any of the elements in the input list. If so, we return `True` or `False`, as appropriate. In lines 2–6, we consider the stopping case for an input list of one element. In lines 8–14, we consider the stopping case for an input list of two elements. In lines 18–19, we consider the final stopping case, returning `True` when the middle element of the input list equals the target.

  The rest of the code in Figure 20.3 contains the recursive calls (i.e., the calls to the function that is being defined). Lines 21–23 call the function when the target is in the upper half of the input list, and lines 24–26 call the function when the target is in the lower half of the input list.[7] In the former case, the recursive call passes in only the upper half of the input list (lines 22–23), whereas in the latter case, the recursive call passes in only the lower half of the input list. This is the code expression of the "do a binary search on the half the target value is in" logic described earlier.

Recursive Traversal of a "Nested" List of Function Calls   In Section 17.2.5, we saw examples of using Python to automate handling of objects and modules by looping through attributes and methods. Let us take this one step further by looking at the case where we want to arbitrarily execute functions in any order desired, as specified in a list. In addition, we want the capability to save named lists of execution order (say in a dictionary), and then refer to those saved lists in our list of functions to execute. This kind of functionality, for instance, is implemented in the hybrid Python/Fortran intermediate-level tropical atmosphere circulation model of Lin (2009) to enable the user to set up "suites" specifying what methods to run. The

---

[7] In the current use of "half," we mean "approximate half," because the stopping case of the middle element removes one element from the consideration of halves. This use of half as usually approximate continues throughout the chapter when referring to binary search.

```
1    def run1():
2        print('bacteria')
3
4    def run2():
5        print('42')
6
7    def run3():
8        print('planet')
9
10   def run_execs(input_list, exec_lists):
11       for iname in input_list:
12           if iname in globals().keys():
13               globals()[iname]()
14
15           elif iname in exec_lists:
16               run_execs(exec_lists[iname], exec_lists)
17
18           else:
19               raise ValueError(iname + ' does not exist')
20
21   my_run = ['all', 'set3', 'run2']
22   my_exec_lists = {'all': ['run1', 'run2', 'run3'],
23                    'set1': ['run2', 'run2', 'run1', 'run1'],
24                    'set2': ['run3', 'run1'],
25                    'set3': ['run1', 'set2']}
26
27   run_execs(my_run, my_exec_lists)
```

Figure 20.4 Code to recursively traverse a "nested" list of function calls.

result is a model that can be easily configured to run with different capabilities and can even adapt and change what methods it uses as it runs.

As an example, consider Figure 20.4. Three functions, run1, run2, and run3 are defined that each print out a word (lines 1–8), and the function run_execs is defined to accomplish the arbitrary execution described above (lines 10–19). In this code, we want to execute the three functions or saved list of functions given in my_run (line 21). In that list, the last element is the name of one of the functions defined earlier ('run2'), but what are the first two elements in that list?

The first element, 'all', refers to the list stored with the 'all' key in the my_exec_lists dictionary. That list specifies the names of the three functions defined in the beginning of the code: 'run1', 'run2', and 'run3'. The second element of my_run, 'set3', refers to the list stored with the 'set3' key in the my_exec_lists dictionary. That list has two elements. The first is 'run1', referring to the run1 function, but the second element is 'set2', which refers to the list stored in the my_exec_lists dictionary under the 'set2' key. That is, the 'set3' list makes reference to another list. To execute what is described in the 'set3' list, we need to run the run1 function and then the functions listed in the 'set2' list. This is what we mean by these lists of function calls being "nested." A list element can refer to another saved list which in turn can have elements that refer to another saved list, and so on. Just as with the directory tree, we do not know ahead of time which lists use saved lists and which do not, nor whether a saved list has elements that refer to other saved lists (and how many levels of reference that extends to).

As a result, when we write the run_execs function to execute the list of functions and saved function lists given in my_run, it makes sense to use recursion. As specified in line 10 of Figure 20.4, we pass in two parameters to the function: a list of items to execute (input_list) and a dictionary of saved execution lists (exec_lists). The function loops through all items in the list of items (all strings). If the item refers to one of the run1, run2, or run3 functions defined in the module (line 12), the function is obtained from the symbol table and called (line 13). Recall from the discussion in Section 17.2.5 that globals returns the dictionary of everything defined in the module. Line 13 is the main stopping case. As to recursive calls, if the input_list item is one of the saved lists in exec_lists (line 15), we call run_execs using that saved list as input (line 16). Line 18 is also a stopping case, and catches situations where the item in input_list does not exist.

When we run the code in Figure 20.4, the following is printed to the console:

```
bacteria
42
planet
bacteria
planet
bacteria
42
```

The first three lines of console output come from executing the saved list 'all'. The next three lines come from executing the saved list 'set3', which itself executes the function run1 and the saved list 'set2', the latter which executes the functions run3 and run1. The final line of output comes from executing the function run2.

20.3   Try This!

The examples in this section address the main topics of Section 20.2. We practice manipulating a directory tree in different ways. We also practice writing recursive functions.

Try This! 20-1 Searching for Files in a Directory Tree Using walk

Consider the Figure 10.2 directory tree. Write a program, using the os module's walk generator, that lists the path relative to the current working directory and name of every file that ends in *.txt*. Assume the current working directory is the parent to *MyFiles*.

```
1    import os
2
3    file_occurrences = []
4    for dirpath, dirnames, filenames in os.walk("MyFiles"):
5        for ifile in filenames:
6            if os.path.splitext(ifile)[-1].lower() == '.txt':
7                file_occurrences.append(os.path.join(dirpath, ifile))
```

Figure 20.5 Code to list the path (starting with the argument to walk) and name of every file that ends in *.txt* for Try This! 20-1.

Try This! Answer 20-1

Figure 20.5 shows a solution to the problem. After running the code, the contents of the list file_occurrences are (output reformatted manually for readability):

```
In [1]:  print(file_occurrences)

         ['MyFiles/StationData/Info.txt',
          'MyFiles/StationData/July/Temp.txt',
          'MyFiles/StationData/June/Humidity.txt',
          'MyFiles/StationData/June/Wind.txt',
          'MyFiles/StationData/June/Temp.txt']
```

The os module's path submodule has two useful functions that we use in Figure 20.5: splitext and join.[8] Both are described in Section 10.2.1. In line 6, we use splitext to return the extension of the file ifile. We use the string method lower to convert the extension into lowercase, before we compare it with *.txt*. This way, we make sure the program works across different operating systems which may have different rules regarding case sensitivity (e.g., some operating systems do not distinguish between lower and upper case). In line 7, we use join to connect the path (starting with the argument to walk) of the directory the file ifile is in with the filename ifile. By using join, the two components are joined using the directory separation character for the operating system the program is running on.

Try This! 20-2 Manipulating a Directory Tree Using walk

Consider the Figure 10.2 directory tree. Write a program, using the os module's walk generator, that makes a copy of the contents of the *June* and *July* directories in *StationData* and puts those copies (called *June_copy* and *July_copy*, respectively) also in *StationData*. Hint: Recall the copytree function described in Section 10.2.4.

Try This! Answer 20-2

Figure 20.6 shows a solution to the problem. A few notes about this solution: In line 5, we use basename to return just the last part of dirpath to see whether the directory dirpath refers

---

[8] https://docs.python.org/3/library/os.path.html (accessed August 19, 2020).

```
1   import os
2   import shutil
3
4   for dirpath, dirnames, filenames in os.walk('MyFiles'):
5       if os.path.basename(dirpath) == 'StationData':
6           if 'June' in dirnames:
7               shutil.copytree(os.path.join(dirpath, 'June'),
8                               os.path.join(dirpath, 'June_copy'))
9           if 'July' in dirnames:
10              shutil.copytree(os.path.join(dirpath, 'July'),
11                              os.path.join(dirpath, 'July_copy'))
```

Figure 20.6 Code to make copies of *June* and *July* for Try This! 20-2.

to is *StationData*. If so, we check to see whether *June* and *July* are in *StationData* (lines 6 and 9), using membership testing on the list `dirnames` (which lists all the subdirectories in the directory specified by `dirpath`). Lastly, in lines 7–8 and 10–11, in our calls to `copytree`, we have to use `join` to make sure we are describing the path to *June*, *June_copy*, *July*, and *July_copy* relative to the current working directory, which is the parent directory of *MyFiles*. The `walk` generator, as it descends through the directory tree, does *not* change the current working directory. Because we pass in `'MyFiles'` to the call to `walk` (line 4), the current working directory has to be the parent directory to *MyFiles*, or Python would not be able to execute the `walk` call.

Although it makes no difference to the current problem's solution, if we want to alter `dirnames`, it may matter whether we effectively traverse the directory from the top-down or from the bottom-up. That is controlled by the `topdown` keyword input parameter to `walk`. See the `walk` documentation for details: https://docs.python.org/3/library/os.html#os.walk.[9]

Try This! 20-3 Recursive Function: Experiment Title

Create a recursive function `concat_title` that accepts a list of strings and produces a single string that is a concatenation of each element of the list, separated by a space. Thus, if the list of the strings are terms in an experiment title:

```
['Experiment', '3:', 'Determining', 'the', 'Melting', 'Point']
```

running `concat_title` on that list will return:

```
Experiment 3: Determining the Melting Point
```

This function behaves similarly to the string method `join`.

---

[9] Accessed December 19, 2020.

Try This! Answer 20-3

This function solves the problem:

```
1  def concat_title(input_list):
2      if len(input_list) == 0:
3          return ''
4      else:
5          return input_list[0] + ' ' + concat_title(input_list[1:])
```

The logic of the routine is "take the first element of the input list, concatenate a blank space to it, then concatenate the concatenated title of the second element of the input list all the way through the end of that list." There is one stopping case, when the input list is empty, in which case the function returns an empty string. The recursive call is part of the expression returned in the last line of the function. Note, for a one-element list, the slice [1:] returns the empty list.

Try This! 20-4 Searching for Files in a Directory Tree Using Recursion

Redo Try This! 20-1, but instead of using the `walk` generator, write a recursive function `find_all_txt` that, when called as:

    find_all_txt("MyFiles")

returns a list of the path (starting with the argument to the function) and name of every file in the Figure 10.2 directory tree that ends in *.txt*.

Try This! Answer 20-4

The function in Figure 20.7 solves the problem. When we concatenate two lists together (lines 9 and 12 in Figure 20.7), the second list continues after the first list.

```
1   import os
2
3   def find_all_txt(top_dir):
4       list_items = os.listdir(top_dir)
5       list_txt = []
6       for item in list_items:
7           item_path = os.path.join(top_dir, item)
8           if os.path.isdir(item_path):
9               list_txt = list_txt + find_all_txt(item_path)
10          else:
11              if os.path.splitext(item)[-1].lower() == '.txt':
12                  list_txt = list_txt + [item_path]
13      return list_txt
```

Figure 20.7 Code solving Try This! 20-4, defining a recursive function that can be used to list the path (starting with the argument to the function) and name of every file that ends in *.txt*.

20.4   More Discipline-Specific Practice

The Discipline-Specific Jupyter notebooks for this chapter cover the following topics:

- Using the `walk` generator.
- Recursion and writing recursive code.

20.5   Chapter Review

20.5.1 Self-Test Questions

Try to do these without looking at the book or any other resources or using the Python interpreter. Answers to these Self-Test Questions are found at the end of the chapter.

Self-Test Question 20-1

If we are given a loop defined by:

```
for dirpath, dirnames, filenames in os.walk("Documents"):
```

what can we say about the loop and the variables `dirpath`, `dirnames`, and `filenames`?

Self-Test Question 20-2

If we are given the loop defined in Self-Test Question 20-1, what does it mean if in an iteration the variable `filenames` is an empty list?

Self-Test Question 20-3

What does it mean that a function is recursive?

Self-Test Question 20-4

What is a stopping case or base case in a recursive function?

Self-Test Question 20-5

What happens if there is no stopping case or base case in a recursive function? Provide an example of a recursive function that has no stopping case.

Self-Test Question 20-6

Describe a situation where a recursive solution may be preferred over an iterative solution.

Self-Test Question 20-7

Consider the recursive function of Figure 20.2. What would happen if line 9 was removed and replaced with `pass`?

Self-Test Question 20-8

Explain why the recursive call input arguments are as specified in lines 23 and 26 of Figure 20.3.

Self-Test Question 20-9

If we changed line 21 of Figure 20.4 to:

```
my_run = ['set2', 'set3']
```

and executed the program, what is printed to the console?

Self-Test Question 20-10

If before line 27 of Figure 20.4 we added this line (to change the value of the `'set3'` entry in `my_exec_lists`):

```
my_exec_lists['set3'] = ['run1', 'set3']
```

what happens when the code file is executed?

20.5.2 Chapter Summary

Recursion is a programming construct that can be somewhat confusing. However, there are certain kinds of problems – those with repetition over an indeterminate number of nested levels – that lend themselves to the use of recursion. The task of visiting the various directories in a directory tree is one of these problems (although in Python the os module's `walk` generator enables us to go through the directories in a tree by using a single `for` loop). Whereas most of the time we use loops to accomplish repetitive tasks, every now and then there are problems that have to be done using recursion. Specific topics we covered in this chapter include:

* Using the `walk` generator

    - The os module's `walk` generator enables us to use a `for` loop to iterate through all directories in a directory tree and list all files and subdirectories in those directories.
    - Each iteration of the loop using the `walk` generator visits a different directory in the directory tree the generator accesses.
    - The generator provides three return values for each iteration of the loop which give the following information for the directory being visited in that iteration: the path (starting with the argument to `walk`) of the directory, the list of subdirectories in that directory, and the list of files in that directory.

- Recursion and writing recursive code:
  - Recursion works when there is some sort of "repeated nesting" logic to the problem we are looking to solve.
  - A function is recursive if somewhere in the function definition there is a call to itself.
  - All recursive functions have to have at least one case where the function terminates and returns without making a call to itself. This is called a stopping or base case.
  - A recursive function without any base case eventually throws a `RecursionError` exception.
  - The maximum number of recursive calls is given by the sys module's `getrecursion limit` function. A typical value is 3000.
  - Most problems involving repetition are better solved using loops. But problems that involve a modest but unknown (or varying) number of repeated nesting levels may be difficult-to-impossible to solve using plain iteration (looping). In those cases, recursion may be the only way to solve the problem concisely.
  - Binary search can be implemented using recursion.
  - We can also use recursion to implement the traversal of a "nested" list of function calls. This may be used in a modeling framework where we want to change what methods are executed as the model runs.

20.5.3 Self-Test Answers

Self-Test Answer 20-1

The loop will visit every directory and subdirectory in *Documents* once. Each iteration is a visit to a different directory. At each iteration:

- The variable `dirpath` is the path to the directory being visited at that iteration, starting with *Documents*.
- The variable `dirnames` is a list of all subdirectories in the `dirpath` directory.
- The variable `filenames` is a list of all files in the `dirpath` directory.

In both the `dirnames` and `filenames` lists, the names are relative to the `dirpath` directory.

Self-Test Answer 20-2

If `filenames` is an empty list, it means that for the directory given by `dirpath` at that iteration, there are no files in that directory.

Self-Test Answer 20-3

A recursive function is a function that, somewhere in the function definition, calls itself.

Self-Test Answer 20-4

A stopping case or base case is a situation that leads to the function ending or returning without a call to itself.

Self-Test Answer 20-5

The function continuously calls itself until a `RecursionError` is thrown. Here is an example of a recursive function, based upon a function in Section 20.2.2, that when called results in such an error:

```
def add_up_error(input_list):
    return add_up_error(input_list[:-1]) + input_list[-1]
```

Self-Test Answer 20-6

If the problem logic has the idea of repeatedly visiting "nesting levels," and the number of nesting levels is relatively small and unknown or varying as the program runs, this is a situation where a recursive solution may be preferred over an iterative solution.

Self-Test Answer 20-7

Line 9 of Figure 20.2 is the function's recursive call. If this were removed and replaced by `pass`, if `item` were a subdirectory of `top_dir`, nothing would happen. Thus, the list that is returned by the function only lists an occurrence of *Temp.txt* in `top_dir` at the initial calling of the function (if it exists in that directory), and none of the (possible) occurrences in subdirectories in the directory tree. The function would never descend into any of the subdirectories in the tree, because there is no recursive call into those directories.

Self-Test Answer 20-8

In the logic of a recursive binary search, after we figure out which half of the input list the target is in, we conduct another binary search on that half of the input list. As the recursive binary search function is called `is_found_by_recursive_binary`, the first argument passed into the call of the function in the case of the target being in the latter half of the list is that latter half of the list (line 23). In the case of the target being in the first half of the list, the first argument passed into the call of `is_found_by_recursive_binary` is the first half of the list (line 26). In both lines, the second argument is the target, which is the value we are searching the list for.

Self-Test Answer 20-9

The following is printed to the console:

```
planet
bacteria
bacteria
planet
bacteria
```

Self-Test Answer 20-10

A RecursionError exception is thrown. The 'set3' saved list makes reference to itself, and so that saved list is run over and over again until the recursion limit is reached and the exception is thrown.

Part IV

Going from a Program Working to Working Well

In Parts I–III, we have covered much of what would be covered in a two-quarter introductory programming sequence. Although there are many more topics in programming and scientific computing that we do not address – trees, numerical methods, parallel processing and cloud computing, data and database management, more about algorithms, to name a few – the topics covered in the first three parts are enough for us to write programs for most everyday scientific and engineering programming applications.

There is more to using a computer to do science and engineering, however, than just writing the program itself. Just because the Python interpreter runs our program and produces an answer does not mean the answer is right. Nor does it mean that anyone else is able to use our program. We might not understand our own program if we came back to it a month from now! And, just because our program works does not mean that it runs efficiently, is easily extended or adapted, or can interact with other programs written in the same or other languages.

There is a big difference between getting a program to work versus working well. Software engineers, however, have developed a suite of tools and practices that enable programmers of any type (scientific, engineering, or not) to write code that is flexible, extensible, useful to others, and efficient. In this part, we briefly consider some of these tools and practices. The descriptions will be concise and do not include a number of features from Parts I–III, such as the Try This!, More Discipline-Specific Practice, and Chapter Review sections. But the chapters will be enough to enable those new to these tools and practices to take the first steps in integrating them into their scientific and engineering programming workflows. For an in-depth treatment of these topics, we encourage readers to consult the many fine references available, some of which are listed in the following chapters.

Many of the tools described in this part are usually operated using a command-line interface. Section 2.2.3 introduces terminal windows and using a command-line interface, but a full treatment of the topic is beyond the scope of the current text.

21 Make It Usable to Others: Documentation and Sphinx

21.1 Introduction

In Section 4.2.4 we introduced code comments and Jupyter markdown, and in Section 6.2.6, we introduced docstrings. Although these constructs and practices are vital for the programs we write, there is more to documentation. Documentation is often unsung, but it plays a vital role in making our programs understandable and usable by ourselves and others. Python has a number of tools to streamline the process of creating documentation. In this chapter, we discuss principles of documenting, a more detailed general convention for docstrings than was described in Section 6.2.6, and the Sphinx documentation-generation program.

21.2 Principles of Documenting

When writing any document, it is imperative to understand our audience. Different audiences have difference goals and thus require different kinds of documents. Things are no different for software documentation. For software, there are at least four different kinds of documentation:[1]

- Tutorials. These teach a user how to program or use a program and are usually written in long-form prose. Often, tutorials assume the user has limited background in the kind of program or programming being discussed.
- Topic Guides. These address a specific task or theme such as creating plots, numerical integration, or manipulating images. Usually, topic guides assume users already have the background knowledge tutorials provide.
- How-To Guides. These are usually more specific and concise than topic guides. Code snippet galleries, cookbooks showing how to accomplish a specific task, and discussion/blog posts (such as provided by the Stack Overflow question-and-answer site) are examples of such guides.
- Reference Guides. These usually describe specific routines rather than topics. Application Programming Interface (API) documentation and function and class listings are examples of reference guides.

---

[1] These categories come from the Django documentation: https://docs.djangoproject.com/en/2.2/ (accessed August 24, 2020).

Although the document's audience (and thus purpose) is the most important consideration, there are other topics to consider when creating and using documentation. First, we need to recognize there is a distinction between official documentation and user-contributed documentation. Because official documentation is (in theory) the authoritative documentation regarding a program or package, we expect its information to be the most up to date and accurate. Unfortunately, this is not always the case, and revisions in documentation can lag behind new versions of the code. User-contributed documentation may be as or more accurate than the official documentation, and because anyone can be an author, it may also be more timely. Multiple authorship, however, may result in an inconsistent voice in the documentation as well as uneven quality.

Second, we need to consider how the documentation we are writing or reading will be or has been maintained. If we are writing documentation, we should consider how to ensure our documentation changes as the code changes. Version control, described in Chapter 24, can help keep the two in sync. As readers of documentation, we need to be aware whether the documentation describes deprecated features. Python has built-in exception classes that help developers communicate that certain features are deprecated or slated for deprecation in the future, such as `DeprecationWarning` and `PendingDeprecationWarning`.[2]

Third, we need to consider the kind of routines that are being documented. If we are documenting public functions and classes – those that users are supposed to make use of – the documentation for these features help define a "contract" with the user of the program. As a result, the documentation needs to be accurate, reliable, and unlikely to change. The interface to the functions and classes needs to remain as static as possible, because (possibly) many people will use the interface as documented. With private functions and classes, we have more latitude in the documentation and interface, because users are not expected to access these structures. Also, of all the possible kinds of structures that can be documented, classes and functions are the priority.

Finally, when making documentation, we need to consider the strengths and weaknesses of the different kinds of tools we can use for doing the documenting. Inline comments are tightly connected to the code they document. These comments, however, are only readable by someone who has the source code and requires the reader understand the source code in order to understand the comments. Word-processing applications, such as Word, are ubiquitous, easily changed, and can be used for tutorials, topic guides, how-to guides, and reference guides. These applications, however, are unconnected to the source code they are documenting, and so the translation from code to guide (and the editing and updating of documents) has to be done manually. Discussion and blog posts work well for brief discussions, and because they are live on the Internet are widely distributed. Such posts, however, are also unconnected to the source code they are documenting.

Documentation-generation programs that interface with source code, particularly code with standard documentation structures like docstrings, can be used for tutorials, topic

---

[2] https://docs.python.org/3/library/warnings.html (accessed August 24, 2020).

guides, and how-to guides. Because they can parse source code, they are able to generate reference guides "automagically" using docstrings and the function and class structure of the program. Documentation-generation programs, however, are more difficult to use than the tools described earlier. For any nontrivial documentation, however, a documentation generator greatly simplifies the process of documenting code.

21.3 General Convention for Docstrings: The NumPy Format

Our Figure 6.8 docstring example shows a nice, basic docstring that adequately describes the function it is attached to. For a more complex docstring, one that has additional sections and kinds of information, we would benefit from following a more detailed convention for docstrings, one that defines how to include these additional kinds of information. One such convention is the "NumPy or numpydoc format." We provide an up-to-date link to the format's documentation at www.cambridge.org/core/resources/pythonforscientists/refs/, ref. 63. We provide an up-to-date link to an example of a docstring following the NumPy format at www .cambridge.org/core/resources/pythonforscientists/refs/, ref. 62. Both the documentation and example are references for the present section.

If we were to rewrite our Figure 6.8 docstring example following the NumPy format for docstrings, it would look like Figure 21.1. From this example, we can identify a few core elements of the NumPy format (see the full documentation of the format for details):

- Section headers (e.g., "Parameters") are set off by a line of hyphens underneath the headers. That line is the same length as the header above it.
- When listing parameters, we list the name, a colon, then the type of the parameter.
- Variable and parameter names (e.g., `input_percent`) when found in the text of the docstring (as opposed to the parameters or returns listing) are placed between single backward tick marks (`).
- Types when found in the text of the docstring (as opposed to the parameters or returns listing) are placed between double backward tick marks (``).

As we describe in Section 21.4, text following these conventions is converted by a documentation generator like Sphinx into text with the appropriate formatting (colors, font size, etc.) in a document (e.g., HTML, PDF) for the kind of element that text is.

The meaning of these symbols is specified by the reStructuredText standard. The basic purpose of the standard is to enable a person to write a plain text file and specify formatting for that text in such a way that a reader of only the file can see both what is being written and how it should be formatted. For instance, the double backward tick marks tell a document generator following the reStructuredText standard that the text between this delimiter should be set in a "code-like" monospace font, while at the same time not unduly distracting the reader of the text file itself. The result is a code file that is both readable and, when processed by the right program, yields attractive documentation.

```
1   def percent_to_decimal(input_percent):
2       """Convert a percent to its decimal equivalent.
3
4       The decimal equivalent is computed by dividing `input_percent` by
5       100, and this value is returned by the function.  The result will be
6       ``float``, regardless of whether or not `input_percent` is ``float``
7       or ``int``.  `Input_percent` is assumed to be scalar; no testing is
8       done to see whether this is true.
9
10      Parameters
11      ----------
12      input_percent : float or int
13          Numerical value of a percent.  Scalar.
14
15      Returns
16      -------
17      float
18          Decimal equivalent.
19
20      Examples
21      --------
22      >>> print(percent_to_decimal(42))
23      0.42
24      """
25      return input_percent / 100.0
```

Figure 21.1 Figure 6.8 converted so the docstring follows the NumPy format.

There are many other formatting codes specified by the reStructuredText standard. We provide an up-to-date link to a quick reference to the standard (this reference is also a reference for the present section) at www.cambridge.org/core/resources/pythonforscientists/ refs/, ref. 66. The Sphinx documentation also contains a primer on reStructuredText. An up-to-date link to this primer is at www.cambridge.org/core/resources/pythonforscientists/refs/, ref. 65.

There are other docstring formats beside the NumPy format, with their own strengths and weaknesses. More important than the format we choose, however, is sticking to a single format. A program whose docstrings follow different formats is a program whose documentation will be more difficult to read and maintain.

21.4 The Sphinx Documentation Generator

The Sphinx documentation generator is a utility that substantially simplifies the creation of all four kinds of documentation listed in Section 21.2, and in a maintainable way. There are basically two main kinds of files Sphinx makes use of to generate documentation: reStructuredText files containing the content of tutorials, topic guides, etc., and Python source code. Sphinx is able to parse both kinds of files to generate the various kinds of documentation, in particular automatically generating API documentation from the source code docstrings.

To create a new Sphinx documentation project, we go to the directory of the project and execute the following in a command-line interface to the operating system:

```
sphinx-quickstart
```

This launches the quickstart utility, which sets up configuration files and other standard structures. For the "My Utilities" example, which we show in Figures 21.2–21.4, we use the default settings except for the project and author names, and we indicate that we want to use autodoc, the Sphinx extension that enables the automatic parsing of docstrings from Python code and conversion into API documentation.

After running quickstart, our documentation project directory consists of the following files and subdirectories:

_build
conf.py
index.rst
make.bat
Makefile
_static
_templates

The *make.bat* and *Makefile* files are used with the make utility, a tool that manages the compilation of the documentation, even if there are many different files and modules involved in the documentation. The *index.rst* is the reStructuredText source code for the main page of the documentation. The subdirectories whose names begin with an underscore are used in the documentation generation process.

The *conf.py* file contains settings for Sphinx. In order for Sphinx to understand the NumPy docstring format (Section 21.3), we have to manually add the string `'sphinx.ext.napoleon'` to the `extensions` list. After that addition, this will be the `extensions` line in *conf.py*:

```
extensions = ['sphinx.ext.autodoc', 'sphinx.ext.napoleon']
```

We can change the *index.rst* to suit our needs, as well as add *.rst* files to the current documentation project directory. We can also add Python modules into the current documentation project directory. There are ways of specifying alternate locations for our files, but for the example in this section, we put the documentation and code in the same directory, the project directory.

Once we have our reStructuredText files and Python modules the way we want, we create the HTML documentation based on those files by executing the following operating system command-line command:

```
make html
```

```
1   Welcome to My Utilities's documentation!
2   ========================================
3
4   This manual describes modules of utilities functions that I have written.
5   Currently, there is only a single module.
6
7   .. toctree::
8      :maxdepth: 2
9      :caption: Contents:
10
11     api
12
13
14  Indices and tables
15  ==================
16
17  * :ref:`genindex`
18  * :ref:`modindex`
19  * :ref:`search`
```

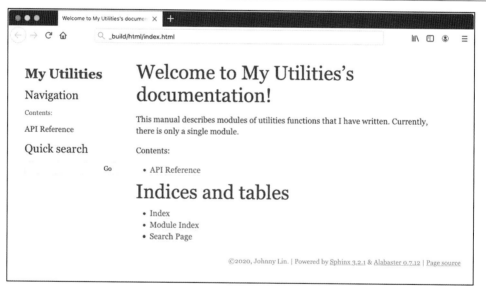

Figure 21.2 The My Utilities documentation *index.rst* file (top, with comment lines autogenerated by Sphinx manually removed) and the web page (*index.html*) generated from that file (bottom, with the front of the URL address manually truncated).

The HTML version of the documentation thus created is found in the *_build/html* sub-directory.

For our "My Utilities" example, we create two hand-edited documentation pages. The main page is specified by *index.rst*, and we add an additional *.rst* file, *api.rst*. The package has a single module, *util.py*, which consists of the code in Figure 21.1. All three files are placed in the current documentation project directory.

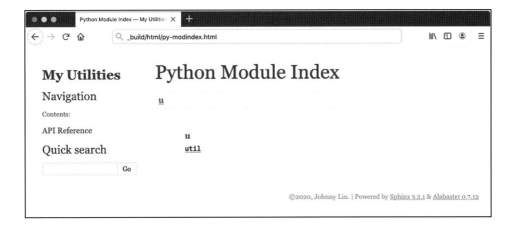

Figure 21.3 The module index (*py-modindex.html*) autogenerated by the Sphinx command in line 18 of Figure 21.2. The front of the URL address is manually truncated.

Figure 21.2 shows the documentation's main page, both the reStructuredText source code and the web page (*index.html*) that Sphinx generates from that source code. The reStructuredText code at the top of the figure include the following elements:

- Titles. These are denoted by equal signs under the title and extending the length of the title (lines 1–2 and 14–15). This is similar to how, in the docstrings of the Section 21.3 example, the sectioning was denoted by dashes under the section title. The title is rendered in HTML in a font, etc., as appropriate for a title.
- Paragraphs. Paragraphs of text are on subsequent lines (lines 4–5). Line breaks in the code are not typeset in the HTML, but the text flows as per the web page's style definition. Blank lines separate paragraphs of text.
- Table of contents. The `.. toctree::` directive (line 7) creates a table of contents for items listed underneath the directive. Lines 8 and 9 specify characteristics for the table of contents as options to the `toctree` directive, and line 11 lists the one reStructuredText code file listed in this table of contents (*api.rst*, that defines the API Reference web page). The *.rst* extension does not need to be provided, and the table of contents directive pulls the title from *api.rst* and places it in this location in the *index.html* page.
- Indices and tables. The Sphinx `:ref:` role is used to create cross-references to pages. In lines 17–19, references are made to the general index, module index, and search page that Sphinx automatically creates. Lines 17–19 are bulleted list items, following the reStructuredText convention for specifying a bulleted list. This convention is the same as the Markdown syntax summarized in Table 4.1.

Directives begin with two periods (`..`) and end with two colons (`::`) before and after the directive name (respectively). Options to directives (e.g., `:caption:`) are put in between two colons, as are text roles (e.g., `:ref:`).

```
1    API Reference
2    =============
3
4    This describes the application programming interface (API) for the module
5    listed below.
6
7    .. automodule:: util
8       :members:
```

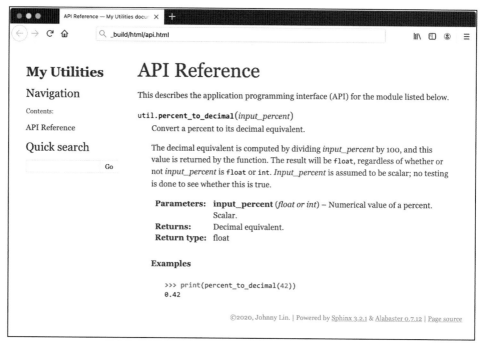

Figure 21.4 The My Utilities documentation *api.rst* file (top) and the web page (*api.html*) generated from that file (bottom, with the front of the URL address manually truncated).

The general index, module index, and search page Sphinx automatically creates use the source files and modules Sphinx is given, in response to a directive specifying the module (in this case line 7 of *api.rst*; see Figure 21.4). Figure 21.3 shows the module index web page (*py-modindex.html*) that is created by the Sphinx command :ref:`modindex`. If a module is present but the *.rst* files do not contain directives that access that module, that module is not included in the autogenerated indices. The left navigation bar including a search bar is also automatically generated.

Figure 21.4 shows the page describing the API for the package, both the reStructuredText source code (*api.rst*) and the web page that Sphinx generates from that source code. The key lines in *api.rst* are lines 7–8, which utilize the automodule directive from autodoc, the Sphinx extension which enables autogeneration of API documentation from Python module docstrings. Line 7 tells Sphinx to document the util module, and line 8 tells Sphinx to document all members of that module. If we do not want to document all members of a module, we can specify which members to document using the `members` option.

The web page that *api.rst* generates is shown in the bottom panel of Figure 21.4. The docstring in *util.py* for the `percent_to_decimal` function is formatted to look like a software documentation web page should. Headers are written in bold, code is typeset in a fixed-width font, and the example is typeset with fonts, colors, and shading that show it is an example.

In our example, we showed just a fraction of the capabilities of Sphinx, but from this small peek, we see that Sphinx enables us to create beautiful documentation with relatively little effort. The Sphinx documentation (and main reference[3] for this section) has more information. We provide an up-to-date link to the documentation at www.cambridge.org/core/resources/pythonforscientists/refs/, ref. 67. The Sphinx documentation includes more information on directives and autodoc. Up-to-date links to these pages are at www.cambridge.org/core/resources/pythonforscientists/refs/, refs. 69 and 68 respectively.

---

[3] Additional references include the source code for the Notecards documentation (an unpublished application by Johnny Lin), the sphinx-apidoc manual page, and the reStructuredText summary linked to at www.cambridge.org/core/resources/pythonforscientists/refs/, ref. 66.

22 Make It Fast: Performance

22.1 Introduction

Compared with their predecessors, modern computers are fast. Even wearable computing devices today boast processing power that only a few human generations ago would have placed them in the pantheon of the fastest computers on the planet. No amount of computational power, however, will do us any good if the program we write makes poor use of processor cycles and memory. Throughout the present work, we have focused on trying to "get the job done" and have written our programs with a focus on accomplishing the science or engineering task at hand rather than the speed at which the program accomplishes the task. Most of the time, that is the right focus. Sometimes, though, we have to pay attention to the performance of our programs if we are to finish the computational task at hand in a timely manner.

We begin in Section 22.2 by discussing two preliminary topics: ways of describing the **complexity** or efficiency of code and some coding practices that can result in inefficient programs. Section 22.3 discusses how to find bottlenecks in a program – places where the program runs particularly slowly or uses more memory than it ought – through the use of profilers, tools that report how much time and memory is spent in the various parts of the program. Once we know where the most resource-consuming parts of the program are, we can consider ways of fixing those bottlenecks (Section 22.4). Finally, in Section 22.5, we consider possible pitfalls when trying to improve the performance of a program. It is easy to think that "faster is always better," but as with many things in life, increased performance comes with a cost. We describe when it is not worth trying to make a program run faster.

22.2 Preliminaries

Before we address how to find bottlenecks and fix them, we need to discuss how to describe these bottlenecks. This is the topic of code or algorithmic complexity. After that, to motivate our later discussion on profiling, bottlenecks, and pitfalls, we describe some of the ways we can write inefficient code.

22.2.1 Describing the Complexity of Code

Our previous discussions on code performance (e.g., Section 13.2.4) used the amount of time it took for the code to run, based on the perception of the person running the code.

This is known as **wall clock time**. Wall clock time is a practical measure, as it describes how long it takes for the person to get the desired result, but as an analytical tool, wall clock time has limitations. For instance, when describing the complexity of code, we are often interested in comparing one implementation with another implementation. But wall clock time includes aspects of running a program – operating system performance, application and memory management, etc. – that have nothing to do with the code implementations we are examining. For the comparison of algorithms, another measure of code complexity is often more helpful. In this section, we briefly describe a measure of code complexity known as "**asymptotic algorithmic complexity**" or "**big-$\mathcal{O}$ notation**."

Asymptotic algorithmic complexity is an approximate measure of the number of operations conducted by the code when the number of items of input to the code is very large. As a result, we can ignore all contributions to the operation count except from the leading-order terms. To illustrate this, consider the following code, which doubles each element of a one-dimensional array `data` (assume this array is defined elsewhere and NumPy is already imported as `np`):

```
for i in range(np.size(data)):
    data[i] = data[i] * 2
```

How many operations does this code execute? To answer this, we could try to figure out how much time each part of this code takes to execute, and define every kind of "operation" (multiplication, assignment, etc.) in units of time. But, with big-$\mathcal{O}$ notation, we are interested in an approximate count, so we consider an "operation" as any arithmetic calculation, the storing of a variable, the calling of a function (that does not take longer to run if the number of input items increases), the setting up of a loop, and the comparison in an `if` statement. Based on this definition of operation, the number of operations in the above code is approximately $3n + 3$, where n is the number of elements in `data`. We count one operation for setting up the `for` loop, one for the call to `size`, and one for the call to `range`. For each iteration of the loop, we count one operation for assigning `i`, one for multiplying by 2, and one for assigning `data[i]`. Because there are n iterations, the looping contributes $3n$ to the operation count. We could also count the reading of the `data[i]` value to the right of the equal sign in line 2, which would make the contribution $4n$, but below we will see this makes no difference to the final answer.

To convert this polynomial expressing the operation count in terms of the number of input items (i.e., the n elements in `data`) into a big-$\mathcal{O}$ notation, we apply the following rules (which flow from our focus on large n) to the operation count expression:

- Ignore all terms except those of the largest power of n.
- Ignore any constant coefficients to terms in the operation count.

Thus, the $3n + 3$ operation count simplifies to n. In big-$\mathcal{O}$ notation, we say that the code has "order-n big-$\mathcal{O}$ complexity" or that the code is "$\mathcal{O}(n)$." This means that as the number of

Table 22.1 Typical big-$\mathcal{O}$ values with common names (if applicable).

$\mathcal{O}(\ldots)$	Common name
1	Constant
$\log n$	Logarithmic
n	Linear
$n \log n$	
n^2	Quadratic
n^3	Cubic
$n!$	

input items increases, the number of operations the code executes increases approximately proportionally. If n doubles, the operation count approximately doubles.

How is this useful? From a code's big-$\mathcal{O}$, we can see whether the code will scale to larger datasets and can compare the performance of that code with another code. For instance, in our $\mathcal{O}(n)$ case above, we see that if we want to handle a dataset approximately m times larger than n, it will take m times longer to run. In contrast, for code that is $\mathcal{O}(n^2)$, to handle a dataset approximately m times larger than n, it will take m^2 times longer to run.

Table 22.1 lists typical big-$\mathcal{O}$ values with common names for some of the values. The table is listed in order of ascending power of n. Code with constant big-$\mathcal{O}$ (i.e., $\mathcal{O}(1)$) is the most computationally efficient, as the algorithm conducts approximately the same number of operations regardless of the number of input items. On the other hand, $\mathcal{O}(n!)$ is bad news.

22.2.2 Practices That Can Result in Inefficient Code

Here is a brief summary of coding practices that can result in less performant code:

- Explicit looping. Sometimes we have to write explicit loops, but there are times when the operations in each iteration of the loop are independent of one another. There can thus be more efficient ways to handle the operation. In Section 13.2.4, we compared code using array syntax and code using explicit looping and found array syntax to be substantially faster.
- Inefficient access of memory. As we saw in Section 12.2.6, the order we loop through the dimensions of an n-dimensional array can impact our performance.
- Holding unnecessary copies of data in memory. Especially for larger datasets, it is often more efficient to do operations in place (i.e., modifying the original data) than to work on a copy.
- Writing extra operations (especially those involving loops). Sometimes we write code that can be condensed. For instance, we might have written operations using two loops that can

be done in one. Because the inclusion of a loop often results in many additional operations (owing to the inherently repetitive nature of a loop), extra loops can really slow down a program.

- Large amounts of file input/output. Reading from and writing to files can be expensive because file input/output requires going out to the hard drive or other storage device to get content. This is in contrast to the manipulation of data in memory, which is much faster.
- Not using code by others when we should. There is great educational value in writing our own functions, but when performance matters, we are often better off using a version written by someone else who has spent the time to tune the function so it runs as fast as possible.
- Language choice. Some programming languages produce faster code. Compiled languages (e.g., Fortran, C++), for instance, tend to result in highly performant code as compared with interpreted languages (of which Python is an example). Increasingly, however, this is becoming less and less of an issue because there are many libraries that compensate for the inefficiencies in various ways.

Let us look at an example task and how we could accomplish the task using (or not using) some of the above coding practices that result in less performant code. We are interested in taking the cube of each element in an array. We have written the function powb which returns the result of x^b, where x is given by the input parameter x and b is given by the input parameter exponent. We also define the array arr of values:

```
1   import numpy as np
2
3   arr = np.arange(1, 10)
4
5   def powb(x, exponent):
6       p = 1
7       for _ in range(exponent):
8           p *= x
9       return p
```

The function powb repeatedly multiplies x an exponent number of times. The use of _ as the iterator in line 7 is a Python convention for an iterator that is actually not used in the loop it is defined for. The lone underscore is a valid variable name, but the fact that leading underscore variable names are understood as private, coupled with the absence of alphanumeric characters in the name, indicates this iterator variable should not be used later in the program.

Our first method of accomplishing the task of cubing each element in arr and saving the result in an array is to use our powb function in an explicit loop set in a list comprehension:

```
cube_arr = np.array([powb(a, 3) for a in arr])
```

where `cube_arr` is:

```
In [1]:  cube_arr

Out[1]:  array([  1,    8,   27,   64, 125, 216, 343, 512, 729])
```

But, because the power operator is built in to Python, it will be faster to use that operator than to use our `powb` function, So, our second method of accomplishing the task is to use the power operator instead of `powb`:

```
cube_arr = np.array([a**3 for a in arr])
```

A third method of accomplishing the task is to use array syntax, which we have already seen outperforms explicit looping. By using array syntax, we can eliminate the loop in the list comprehension entirely:

```
cube_arr = arr**3
```

Finally, a fourth method of accomplishing the task is to use NumPy's function version of the exponentiation operator that does in-place operations. This enables us to avoid making extra copies of the array:

```
cube_arr = arr.copy()
np.power(cube_arr, 3, out=cube_arr)
```

In the above example, we make a copy of `arr` so the result of the exponentiation will be named `cube_arr` and to keep the `arr` array unchanged. Normally, when doing an in-place operation, we do not make such a copy, since avoiding extra copying is why an in-place operation is faster.

The bottom line: All the above code examples produce the same result, but some will take more processor time and memory than others. Once our code works and we want it to run faster, our task is to find out where it is running slow and change the way we do our calculations to make it run faster.

22.3 Finding the Bottlenecks Using Profilers

Those areas of the program using more resources than they ought to are the bottlenecks or logjams slowing down the execution of the program and are prime candidates for trying out the fixes described in Section 22.4. A tool called a profiler helps us find which parts of a program are the bottlenecks. The timing technique of Section 13.2.4 tells us how long a snippet of code takes, but it does not break down and analyze the code any further. A profiler analyzes units of the program (often the functions or methods) and produces a report on how much processing time and memory is used in each unit.

In this section, we describe four profilers:

- timeit. By far the easiest and suitable for timing small pieces of code (rather than fully profiling the code).
- cProfile. The recommended profiler for a more thorough breakdown of code operation.
- line-profiler. Provides easier to read line execution information.
- memory-profiler. Measures how much memory is used in a function.

The first two profilers are part of the Python standard library whereas the last two are external libraries that need to be installed into our Python distribution. Note that the last two are also referred to as, respectively, line_profiler and memory_profiler, using underscores instead of hyphens. In some contexts, they will be listed in one form, while in other contexts, the other. Inside Python code, a hyphen means subtraction. In our discussion of each profiler, we describe how to use them in their Jupyter notebook inline versions, through a more traditional Interactive Development Environment (IDE) like Spyder, and from the operating system's command line.

In our Jupyter notebook examples below, we run the profilers using "line magic," which can be thought of as shortcuts that enable us to do common development and execution tasks in the notebook.[1] Magic commands begin with a percentage character. The magic commands for each of the profilers above are:

- timeit `%timeit`
- cProfile `%prun`
- line-profiler `%lprun`
- memory-profiler `%memit`

Magic that has a single percentage character before the command applies to a single line of code in the cell. When we put two percentage signs before the magic command (`%%`), the command applies to the entire notebook cell.

A great text that includes discussion on timings and profiling (and is a reference for this section) is VanderPlas (2016).

22.3.1 timeit

The timeit profiler is by far the easiest to use and is suitable for timing small pieces of code (rather than fully profiling the code). The timeit profiler times how long it takes to run multiple times the code we give it. After dividing, timeit gives an estimate of how long the line of code will take to run. We apply timeit to single statements.

The three examples below show timeit timings for cases from Section 22.2.2.[2] The first shows using a list comprehension with calls to `powb`, the second shows using a list

---

[1] https://ipython.readthedocs.io/en/stable/interactive/magics.html (accessed August 25, 2020).

[2] As mentioned in the Preface, these timings should not be considered screenshots but rather model scenarios that are similar but not exact.

comprehension with the use of the built-in exponentiation operator (and also demonstrates the use of timeit magic on an entire notebook cell), and the final example shows the use of array syntax:

```
In [1]:  %timeit cube_arr = np.array([powb(a, 3) for a in arr])

         45.8 μs ± 2.01 ns per loop (mean ± std. dev. of 7 runs, 10000 loops each)

In [2]:  %%timeit
         cube_arr = np.array([a**3 for a in arr])

         24.3 μs ± 580 ns per loop (mean ± std. dev. of 7 runs, 10000 loops each)

In [3]:  %timeit arr**3

         5.6 μs ± 203 ns per loop (mean ± std. dev. of 7 runs, 100000 loops each)
```

The NumPy array syntax version is about four times faster than list comprehensions using the built-in exponentiation operator and eight times faster than list comprehensions using the powb function. Note that timeit benchmarks the expression by running it a number of times (in this case 10 000 or 100 000 times, as noted). So, we do not want to use timeit unless we are testing our algorithm with really small sample input.

Besides running timeit in a Jupyter notebook, we can also run it from inside a Python program (such as would be running in Spyder) and the operating system's command-line interface. The following would run timeit on the list comprehension case using the powb function, but from inside a Python program:

```
import timeit
timeit.timeit('np.array([powb(a, 3) for a in arr])', number=10000)
```

The following would do the same but from the operating system's command line:

```
python -m timeit 'np.array([powb(a, 3) for a in arr])'
```

The timeit documentation (and reference for this section) has more information. An up-to-date link to the documentation is at www.cambridge.org/core/resources/pythonforscientists/refs/, ref. 76.

22.3.2 cProfile

Although useful, the timeit profiler does not break down what is happening in the code being executed. The cProfile profiler can do that for us. Let us consider our list comprehension/powb example:

```
np.array([powb(a, 3) for a in arr])
```

What is happening in this code? We can think of programs or code as working like an onion, where each part of the program or code is a layer, and each layer has its own degree of complexity. In this case our layers are, starting from the outside going in:

1. The np.array function converts the final output into an array.
2. The [... for a in arr] loop in the list comprehension is applied to every element in arr.
3. The powb(a, 3) function call is executed for every element in arr.
4. The code inside the powb function. This runs the built-in multiplication operator in constant time, because the base variable is always predetermined (because it is a literal).

The cProfile profiler times these layers and reports how long each portion took. We can run cProfile on the code above in a Jupyter notebook using the following line of code:

```
%prun cube_arr = np.array([powb(a, 3) for a in arr])
```

In a Jupyter notebook, the result is placed in a separate window. The above generates a result similar to the following (reformatted manually for readability):

```
1          14 function calls in 0.000 seconds
2
3     Ordered by: internal time
4
5     ncalls  tottime  percall  cumtime  percall filename:lineno(function)
6         1    0.000    0.000    0.000    0.000 {built-in method builtins.exec}
7         9    0.000    0.000    0.000    0.000 <ipython-input-1-1e1f22924db5>:4
8                                                   (powb)
9         1    0.000    0.000    0.000    0.000 <string>:1(<listcomp>)
10        1    0.000    0.000    0.000    0.000 <string>:1(<module>)
11        1    0.000    0.000    0.000    0.000 {built-in method numpy.array}
12        1    0.000    0.000    0.000    0.000 {method 'disable' of
13                                                   '_lsprof.Profiler' objects}
```

Lines 7–8 refer to the powb(a, 3) call. Line 9 refers to the [... for a in arr] call. Line 11 refers to the NumPy array function call. Lines 6, 10, and 12–13 are artifacts of the profiler. The columns in the output refer to:[3]

Column title	Meaning
ncalls	Number of calls
tottime	Total time spent in the given function
percall	tottime divided by ncalls
cumtime	Time spent in this and all subfunctions
percall	cumtime divided by primitive calls
filename:lineno(function)	The respective input into each function

[3] Paraphrasing from https://docs.python.org/3/library/profile.html#module-pstats (accessed August 29, 2020).

A major limitation of cProfile is it gives times in units of **Central Processing Unit (CPU)** seconds, which is why all these report 0. The output of cProfile is ordered based on a sort of the statistics, in this case internal time, not by order of execution.

To run cProfile from inside a Python program, we can either use it as a context manager for short snippets of code:

```
1   import cProfile
2
3   with cProfile.Profile() as pr:
4       arr = np.arange(1, 10)
5       cube_arr = arr**3
6
7   pr.print_stats()
```

or larger blocks of code can be embedded in it:

```
1   import cProfile
2
3   pr = cProfile.Profile()
4   pr.enable()
5   arr = np.arange(1, 10)
6   cube_arr = arr**2
7   pr.disable()
8
9   pr.print_stats()
```

To run cProfile from the operating system's command line, the following works:

```
python -m cProfile myscript.py
```

where the module we want to profile is *myscript.py*.

The cProfile documentation (and reference for this section) has more information. An up-to-date link to the documentation is at www.cambridge.org/core/resources/pythonforscientists/refs/, ref. 75.

22.3.3 line-profiler

Because cProfile is not very detailed and does not go inside a function, we may find line-profiler to be useful. To run line-profiler in a Jupyter notebook, we have to first load it using the `%load_ext` magic command:

```
%load_ext line_profiler
```

After that, we can use `%lprun` to run line-profiler, such as:

```
%lprun -f powb np.array([powb(a, 3) for a in arr])
```

which produces, in a Jupyter notebook, something similar to the following in a separate window:

```
Total time: 0.000405 s
File: <ipython-input-1-e14f05e3fa35>
Function: powb at line 4

Line #      Hits         Time  Per Hit   % Time  Line Contents
==============================================================
     4                                           def powb(x, exponent):
     5          9         38.0      4.2      9.4      p = 1
     6         36        160.0      4.4     39.5      for _ in range(exponent):
     7         27        171.0      6.3     42.2          p *= x
     8          9         36.0      4.0      8.9      return p
```

Although line-profiler (and later memory-profiler) only profiles a specific function, it both analyzes how often the function is called in the expression in which it is called and goes into the lines of the function. In the example above, p = 1 is executed 9 times (i.e., Hits) because the list comprehension runs powb 9 times. The p *= x line is executed 27 times because the loop inside powb iterates 3 times while the function is called 9 times ($3 \times 9 = 27$).

Sometimes, the number of loop hits is one more than expected, because loops often technically iterate one more than their stopping condition (to check that they are done). For instance:

```
def loop():
    for _ in [1, 2, 3]:
        pass
%lprun -f loop loop()
```

results in this output:

```
Timer unit: 1e-06 s

Total time: 3.2e-05 s
File: <ipython-input-29-127f2d3e1b32>
Function: loop at line 1

Line #      Hits         Time  Per Hit   % Time  Line Contents
==============================================================
     1                                           def loop():
     2          4         20.0      5.0     62.5      for _ in [1, 2, 3]:
     3          3         12.0      4.0     37.5          pass
```

where the for _ in [1, 2, 3]: loop receives four hits.

In order to run line-profiler from inside a Python program, we make use of operating system executables that are accessed from the command line. In the Python script that contains the functions we want to profile, we add **decorators** to those functions we want to profile. A decorator is a function that takes another function as input. Consider the following example, where we want to profile the program *myscript.py*, consisting of the following lines of code:

```
1   @profile
2   def powb(x, exponent):
3       p = 1
4       for _ in range(exponent):
5           p *= x
6       return p
```

The `@profile` line is the decorator. It refers to a function in the line-profiler library called `profile`. The `profile` function takes as input the function being profiled, in this case the `powb` function. Because it is often cumbersome to write things like `profile(powb, x, exponent)` every time we want to use the `powb` function to profile it, instead we "attach" the `profile` function to `powb` via the decorator operator `@`. So, the `@profile` line tells Python, "every time the `powb` function is called, first call the `profile` function with `powb` as input."

Once we have added the decorator, on the operating system's command line, we run the following command to profile the script *myscript.py*:

```
kernprof -l myscript.py
```

The executable kerbprof is a program in the line-profiler module and is responsible for the profiling. The above call returns a binary file, which the line_profiler executable converts into a text output, when the following is executed:

```
python -m line_profiler myscript.py.lprof
```

There is also an Application Programming Interface (API) to embed the profiling in a tool. The line_profiler documentation contains more information about that embedding. An up-to-date link to that documentation is at www.cambridge.org/core/resources/pythonforscientists/refs/, ref. 70. The line-profiler documentation (and reference for this section) has more information about the tool in general. An up-to-date link to the line-profiler documentation is at www.cambridge.org/core/resources/pythonforscientists/refs/, ref. 71.

22.3.4 memory-profiler

Memory-profiler provides both a timeit-like interface to show how much memory a single statement uses and a line-profiler-like interface to show how much memory each line in a function uses. Here, we demonstrate `%memit` which is the `%timeit` equivalent. We do not demonstrate the `%lprun` equivalent, `%mprun`, because it must be run on a file.

```
In [1]:  %load_ext memory_profiler
```

```
In [2]:  %memit np.array([powb(a, 3) for a in arr])
```

```
         peak memory: 65.38 MiB, increment: 0.17 MiB
```

```
In [3]:  %memit np.array([a**3 for a in arr])
```

```
         peak memory: 65.40 MiB, increment: 0.00 MiB
```

```
In [4]:  %memit arr**2
```

```
         peak memory: 65.53 MiB, increment: 0.00 MiB
```

```
In [5]:  %memit arr
```

```
         peak memory: 65.55 MiB, increment: 0.01 MiB
```

These expressions all take the same amount of memory because the primary determinant of memory in all these operations is the size of arr, which is 65.55 MiB. A MiB is a "mebibyte," which equals 2^{20} bytes.[4] Note, these values describe the memory allocated by the process.

As with line-profiler, in order to run memory-profiler from inside a Python program, we make use of operating system executables that are accessed from the command line. In the Python script, we decorate any functions we wish to profile. For instance, if the following lines were in *myscript.py*:

```
1   from memory_profiler import profile
2
3   @profile
4   def powb(x, exponent):
5       p = 1
6       for _ in range(exponent):
7           p *= x
8       return p
```

then on the operating system's command line, to profile the script we run:

```
mprof run myscript.py
```

As with the other profilers, memory-profiler also provides an API to enable us to embed it in a larger tool. See the memory-profiler API documentation for details. An up-to-date link to

---

[4] https://en.wikipedia.org/w/index.php?title=Mebibyte&oldid=973476147 (accessed September 5, 2020).

the documentation is at www.cambridge.org/core/resources/pythonforscientists/refs/, ref. 73. The memory-profiler documentation (and reference for this section) has more information about the tool in general. An up-to-date link to the memory-profiler documentation is at www.cambridge.org/core/resources/pythonforscientists/refs/, ref. 72.

22.4   Fixing the Bottlenecks

After we have used a profiler to find a performance or memory bottleneck, we want to fix the bottlenecks. There are two general ways of doing so:

1. Use the most efficient algorithm for the circumstance.
2. Use the best programming practices for the circumstance.

As we have discussed in earlier chapters (e.g., Chapter 19) every algorithm has its own strengths and weaknesses. The choice of algorithm ultimately comes down to using the algorithm (or combination thereof) that works best for the use cases we want our program to address. In general, more efficient algorithms require more memory and vice versa. If we are limited in memory, we may need to use a slower algorithm. Likewise, if memory is not an issue, we may be able to use a faster algorithm.

We focus in this section on the second general way of fixing bottlenecks: the ways we do things with a computer (as opposed to the sequence of calculations we take to accomplish the task). In Section 22.2.2, we described some practices to avoid. We also described two practices to consider using: in-place operations and vectorization. When we use in-place operators, we replace the existing memory location that holds the old value with the new value the operator returns. In this way, the new value does not occupy additional memory. We saw that NumPy's function version of the exponentiation operator does in-place operations. In NumPy, the `+=`, `-=`, and similar operators also act in-place. Vectorization (or broadcasting) is when we use array-based math instead of loop-based math. NumPy array syntax is an example of vectorization. In addition to arithmetic operators in NumPy being vectorized when acting on NumPy arrays, NumPy also has the `vectorize` function to enable broadcasting (though this option is less efficient versus the built-in operators). In the rest of the present section, we describe two additional practices: the use of generators and just-in-time compilation.

22.4.1 Generators

Generators are objects that are designed to only provide the "next" value for a given task. For instance, the `walk` generator (Section 20.2.1) only gives the next directory in the directory tree as each iteration of the loop acts on the generator. Because it does so, rather than returning all the directories (and list of files and subdirectories for each directory) in the tree all at once, less memory is used. Thus, generators, in general, can help us save memory. They are particularly useful when each loop iteration returns something large (say a large array) that would require substantial amounts of memory to hold if we have many such entities. File input and output is a use case where generators can save us memory. For very large datasets, we want to avoid reading everything in, processing, and holding large results all at the same time. Generators are also useful when each iteration of the loop generates large amounts of intermediate data that, while needed, we do not need to keep once used.

We can create our own generators by using the `yield` command. Consider the following nongenerator function which takes each value in the list (or array) `arr`, raises it to the `exponent` power, and returns a list of the results:

```
def pow_arr(arr, exponent):
    values = []
    for x in arr:
        p = 1
        for _ in range(exponent):
            p *= x
        values.append(p)
    return values
```

This function has to retain all the values being calculated to return them as a block. This is because when a `return` statement is called, the function's scope ceases to exist, i.e., the function no longer knows about data that were passed into it or stored inside it. Thus, to work on an array, a nongenerator function has to work on the entire array and return the entire result.

In contrast, if we write a generator function that "returns" its results using a `yield` statement, even after the result is "returned," the current state of the function is retained, meaning the function knows what variables are defined, their values, and where it is in a loop if the function contains a loop. The next time the function is called, the `yield` statement updates the function's internal state (often by going to the next iteration of the loop if it is in one). This on-the-fly-only-as-needed generation of values to be returned by the function is an example of "lazy evaluation." A generator that behaves similarly to `pow_arr` is the following:

```
def pow_arr_g(arr, exponent):
    for x in arr:
        p = 1
        for _ in range(exponent):
            p *= x
        yield p
```

When `pow_arr_g` is evaluated, the `exponent` power of an element in `arr` is provided. But, the generator remembers where in `arr` the line 2 loop is, and so the next time `pow_arr_g` is evaluated, the next element in `arr` is acted on.

We can compare the use of the nongenerator function `pow_arr` and the generator `pow_arr_g`:

```
In [1]:  print(pow_arr(arr, 3))

         [1, 8, 27, 64, 125, 216, 343, 512, 729]

In [2]:  pow_arr_g(arr, 3)

Out[2]:  <generator object pow_arr_g at 0x7f8f247da9e8>

In [3]:  [a for a in pow_arr_g(arr, 3)]

Out[3]:  [1, 8, 27, 64, 125, 216, 343, 512, 729]
```

Just calling `pow_arr_g(arr, 3)` does not provide a value, as generators only provide values on evaluation (as `In [2]`/`Out[2]` shows). Thus, we obtain values only when the generator call is part of a list comprehension statement that creates a list (cell `In [3]`). In this particular example, because it is so small, the benchmarks for `pow_arr` and `pow_arr_g` are essentially the same (and thus are not shown). We can, however, use the principles behind `pow_arr_g` to write generators that help us save memory.

The `yield` documentation (and reference for this section) has more information. An up-to-date link to the documentation is at www.cambridge.org/core/resources/pythonforscientists/refs/, ref. 19.

22.4.2 Just-in-Time Compilation

Native Python can be slow, so there are tools that let us directly write parts of our Python code in another language or intelligently convert our Python code into an optimized version in a different language. The native Python **compiler** does not try to make specialized optimizations since the language aims to be domain independent.

There are two main tools to do just-in-time automated compilation: cython and numba. The cython tool basically allows us to write a subset of C directly in our Python code, and those portions written in C run faster than if they had been written in Python. The numba package compiles Python down to assembly and is optimized for NumPy. The advantage of numba is it mostly works by decorating existing functions, so we do not need to change our code to use its capabilities. In this section, we only look at an example of using numba.

Consider again an example similar to above, only in this case we put `arr**3` into a function:

```
1   import numpy as np
2
3   arr = np.arange(1, 10)
4
5   def pow_numpy(arr):
6       return arr**3
```

The timeit profiler gives us the following for a call to pow_numpy:

```
In  [1]:  %timeit pow_numpy(arr)

          5.78 μs ± 173 ns per loop (mean ± std. dev. of 7 runs, 100000 loops each)

In  [2]:  %timeit pow_numpy(arr)

          5.87 μs ± 128 ns per loop (mean ± std. dev. of 7 runs, 100000 loops each)
```

We have to run the timeit cell at least twice and take the second time, otherwise we will be including the compile time of the function. This is particularly noticeable below when we use numba.

If we import the jit function from numba, we can use it as a decorator for pow_numpy:

```
1   from numba import jit
2
3   @jit(nopython=True)
4   def pow_numpy(arr):
5       return arr**3
```

Setting the nopython keyword to True prevents the Python interpreter from being run with the function and thus results in better performance.[5] The timeit profiler will now give the following for a call to pow_numpy (output reformatted for better display):

```
In  [1]:  %timeit pow_numpy(arr)

          The slowest run took 4.29 times longer than the fastest.   This
          could mean that an intermediate result is being cached.
          13.5 μs ± 10.3 μs per loop (mean ± std. dev. of 7 runs, 1 loop each)

In  [2]:  %timeit pow_numpy(arr)

          2.54 μs ± 23 ns per loop (mean ± std. dev. of 7 runs, 100000 loops each)
```

The gains we obtain are not that substantial, because the exponentiation operation is already heavily optimized within NumPy. But the additional work we needed to expend to improve the performance of pow_numpy using numba is trivial! For a function that is not already optimized, as in our pow_numpy example, numba can make a real difference with little extra effort on our end.

The cython and numba documentation (and reference for this section) has more information. Up-to-date links to the documentation of each is at www.cambridge.org/core/resources/pythonforscientists/refs/, refs. 90 and 74 respectively.

---

[5] https://numba.pydata.org/numba-doc/0.41.0/user/5minguide.html (accessed September 19, 2020).

22.5 Pitfalls When Trying to Improve Performance

Improving performance seems like a no-brainer: Who does not want their code to run faster? However, the entire effort at improving performance comes at a cost. Rewriting algorithms and changing programming practices take time and effort, and optimization often comes at the expense of readability and understandability of our code. Before expending this effort to optimize, we have to ask if it is worth it. Here are a few questions to ask:

- How mission-critical is performance for our program? In some cases, such as short-range weather forecasting, if the computer model does not produce a result in under a day, it is useless. But, for some climate studies, it might be acceptable for a computer to take a month to finish its calculations.
- How well do we understand the tools for optimizing performance? The time needed to learn those tools also needs to be factored into the cost of optimizing.
- Is the program going to change a lot over the course of its lifetime? Optimizations are usually good for a specific configuration of the program. If the program is continuing to change – or we think we want to change it in the future – we might have to reoptimize after those changes are made, and time spent on optimizing now would be wasted.
- Do we have the resources to do the optimization? For instance, memory is usually limited, so if a faster algorithm is particularly memory-intensive, we might not be able to use it.
- Do the optimization tools span all the important use cases for our program? Some algorithms are more efficient in certain cases but not in others. Just because an algorithm has good big-$\mathcal{O}$ does not mean it is the best choice for all circumstances. We do not want to expend time and energy optimizing our code for certain use cases but find out later that our code now performs unacceptably for other important use cases.

Two final rules of thumb regarding optimization pitfalls: First, as computer scientist Donald Knuth has said, "… premature optimization is the root of all evil (or at least most of it) in programming."[6] We need to first make our program work, then make it fast. Second, if we have decided to put in the energy to optimize our program, we should do small changes that have a substantial impact first, then consider larger changes. We might find that the small changes may be enough to get the performance we need and avoid unnecessary extra work.

[6] Knuth (1974, p. 671).

23 Make It Correct: Linting and Unit Testing

23.1 Introduction

In Section 11.2, we discussed the importance of testing our programs in order for us to have confidence they are working the way they should. We discussed basic tests and the use of a debugger to help us step through our program slowly to find bugs. As our programs become larger, however, we need a more formalized framework and set of tools to help us in our testing. In this chapter, we introduce two such tools: linting or static checking, and formal unit testing frameworks. We conclude by describing the "test-driven development" mindset, which uses all these tools. That mindset can help us to write bulletproof programs.

23.2 Linting

Code linting is "static analysis," meaning these are tests that we run on the source code before it is executed. Linting checks for programming errors, can help enforce coding style standards (see Section 11.3), and can offer refactoring suggestions.

For example, say instead of defining the `powb` function as we did in Section 22.2.2, we defined `powb` in the file *mymath.py* as shown in Figure 23.1. Note the typo in the code and some poor formatting. We can run the linter pylint on that file using the operating system's command-line interface by typing in:

```
pylint mymath.py
```

This generates a report that reads something like given in Figure 23.2. There are slight variations depending on the editor and programming environment we use.

The default pylint message format is, quoting from the documentation:[1]

```
{path}:{line}:{column}: {msg_id}: {msg} ({symbol})
```

So, taking the second line of the sample output above, the path to the module being analyzed is `mymath.py`, the line number being referenced is 2, the column being referenced is 0, the message ID or code is `W0311`, the message is `Bad indentation. Found 2 spaces, expected 4`, and the symbolic message name is `bad-indentation`.

---

[1] http://pylint.pycqa.org/en/latest/user_guide/output.html (accessed December 9, 2020).

```
1   def powb(x, exponent):
2     p=1
3     for i in range(expnt):
4       p*=x
5     return p
```

Figure 23.1 Contents of *mymath.py*, showing code with a typo and some poor formatting.

```
************ Module mymath
mymath.py:2:0: W0311: Bad indentation. Found 2 spaces, expected 4
    (bad-indentation)
mymath.py:3:0: W0311: Bad indentation. Found 2 spaces, expected 4
    (bad-indentation)
mymath.py:4:0: W0311: Bad indentation. Found 4 spaces, expected 8
    (bad-indentation)
mymath.py:5:0: W0311: Bad indentation. Found 2 spaces, expected 4
    (bad-indentation)
mymath.py:1:0: C0114: Missing module docstring (missing-module-docstring)
mymath.py:1:0: C0103: Argument name "x" doesn't conform to snake_case
    naming style (invalid-name)
mymath.py:1:0: C0116: Missing function or method docstring
    (missing-function-docstring)
mymath.py:2:2: C0103: Variable name "p" doesn't conform to snake_case
    naming style (invalid-name)
mymath.py:3:17: E0602: Undefined variable 'expnt' (undefined-variable)
mymath.py:4:4: C0103: Variable name "p" doesn't conform to snake_case
    naming style (invalid-name)
mymath.py:1:12: W0613: Unused argument 'exponent' (unused-argument)
mymath.py:3:6: W0612: Unused variable 'i' (unused-variable)


------------------------------------
Your code has been rated at -22.00/10
```

Figure 23.2 Example of report from running pylint on *mymath.py*. Extra line breaks and indent spacing are added for increased readability.

The first letter of the message code describes the message type. Here are the different types, quoting from the pylint documentation:[2]

- [I]nformational messages that pylint emits (do not contribute to your analysis score).
- [R]efactor for a "good practice" metric violation.
- [C]onvention for coding standard violation.

---

[2] http://pylint.pycqa.org/en/latest/user_guide/output.html#source-code-analysis-section (accessed December 9, 2020).

```
1   """Module of math functions."""
2
3   def powb(x, exponent):
4       """Calculate the x to the exponent.
5       """
6       p = 1
7       for _ in range(exponent):
8           p *= x
9       return p
```

Figure 23.3 Contents of *mymath2.py*. The nondocstring lines of the function are the same as given in Section 22.2.2.

- [W]arning for stylistic problems, or minor programming issues.
- [E]rror for important programming issues (i.e. most probably bug).
- [F]atal for errors which prevented further processing.

Let us look at some of the warnings and errors in more detail. This error:

```
mymath.py:3:17: E0602: Undefined variable 'expnt' (undefined-variable)
```

says that in line 3 of *mypath.py*, which is:

```
for i in range(expnt):
```

there is a variable named `expnt` that is being used before it was assigned a value (which is what undefined means). We get a hint that `expnt` is a typo from this error:

```
mymath.py:1:12: W0613: Unused argument 'exponent' (unused-argument)
```

which tells us that the parameter being passed in on line 1:

```
def powb(x, exponent):
```

is not used anywhere else in the code.

We can fix most of the issues with the revision shown in Figure 23.3, which we save in the file *mymath2.py*. The use of _ as the iterator in the loop is described in Section 22.2.2. The result of running pylint on this module is (extra line breaks and indent spacing added for increased readability):

```
************* Module mymath2
mymath2.py:3:0: C0103: Argument name "x" doesn't conform to snake_case
    naming style (invalid-name)
mymath2.py:6:4: C0103: Variable name "p" doesn't conform to snake_case
    naming style (invalid-name)
mymath2.py:8:8: C0103: Variable name "p" doesn't conform to snake_case
    naming style (invalid-name)

------------------------------------
Your code has been rated at 4.00/10
```

We ignore the `snake_case` messages – which reminds us multiword variable names should use underscores to separate the words – because our x and p variables are only one "word" long.

Because linting analyzes the code statically, by definition it cannot evaluate how well the code executes. What are the results from the program and are they what we want them to be? To answer those questions, we need to use debuggers and unit testing. Debuggers were introduced in Sections 6.2.7 and 11.2. Unit testing is discussed in Section 23.3.

23.3 Unit Testing

We briefly discussed the topic of testing in Sections 6.2.7 and 11.2. In that discussion, we focused on interactive tests, tests we execute and examine as we are writing our code. Debuggers, which enable us to step through our code line by line and examine the state of our variables, are very useful for interactive testing.

As programs grow in complexity, interactive testing is not enough. For instance, interactive testing cannot detect whether or not a change we make to some existing code inadvertently breaks the code somewhere else, because interactive testing is focused on the code we are currently working on. In order to have confidence our code is working, in addition to interactive testing, we need to create a suite of tests that operate on limited portions of code (about the size of a method or function) and can be repeatedly run, testing the results of the method or function using various input against known benchmarks. This process is called unit testing and the tests that are written and then run are called **unit tests**. Large scientific software projects can have many unit tests, hundreds of thousands even. Even if we do not have to write that many unit tests for our projects, writing at least some tests will help our code to be more reliable.

In this section, we create tests using two different unit testing packages – unittest and pytest – for our Figure 23.3 *mymath2.py* module. For both packages, we include two tests of the `powb` function: one that tests to see how `powb` behaves when computing $2^3 = 8$ by calling `powb(2, 3)` and one that tests to see what happens when computing $2^{-3} = \frac{1}{8} = 0.125$ by calling `powb(2, -3)`. In the first test, the function should return what is expected, eight. In the second test, the function will not return what is expected, 0.125. Instead, it will return 1 because `range` of a negative number essentially gives the `for` loop in line 7 of Figure 23.3 an empty list, and so the body of the loop never executes.

When a unit testing package runs a test that produces the expected result, the test passes. Otherwise, the test fails. Normally, we are more interested in which tests failed, because it tells us what parts of the program need to be fixed. We do not need detailed information about a test that passes, beyond that it passed. Thus, the reports of a unit testing package are verbose for failed tests and concise for passed tests.

```
1    import unittest
2    from mymath2 import powb
3
4    class Testpowb(unittest.TestCase):
5        def test_pos(self):
6            self.assertEqual(powb(2, 3), 8)
7        def test_neg(self):
8            self.assertEqual(powb(2, -3), 0.125)
9
10   if __name__ == '__main__':
11       unittest.main()
```

Figure 23.4 Contents of *test_mymath2_unittest.py*.

23.3.1 unittest

We write another Python program, *test_mymath2_unittest.py*, to hold our tests of the function in *mymath2.py*. This test program is shown in Figure 23.4.

Tests in the unittest framework are methods of a child class of the TestCase class in the unittest module. These methods have names beginning with test.[3] When the main function of unittest runs (line 11 of Figure 23.4), by default it looks for and runs all methods following this naming convention (and defined in the module that calls main).

The test_pos method (lines 5–6 in Figure 23.4) calls the assertEqual method, which is inherited from TestCase. The assertEqual method checks to see if the two arguments passed into it are equal. If they are, the test passes. Otherwise, the test fails. The utilities in unittest (such as main) then handle the passed or failed test accordingly. In test_pos, powb(2, 3) returns 8, so the test will pass. In test_neg, powb(2, -3) returns 1, so the test will fail.

Note, we put the call to main in an if statement (line 10 in Figure 23.4) that checks to see whether the special variable __name__ equals the string '__main__'. That condition is met when the code in the module is being executed as a main program, such as when it is being run from the operating system's command line, in the Spyder console, or in the cell of a Jupyter notebook. This differs from when a module is executed as part of an import statement. When a module is being imported, __name__ is set to the string name of the module. Thus, by putting in line 10, we make sure our test program only runs when executed from the command line, not when imported by another module.

We run the test module at the operating system's command line by typing in:

```
python test_mymath2_unittest.py
```

[3] https://docs.python.org/3/library/unittest.html#basic-example (accessed December 22, 2020).

at the operating system prompt in a terminal window. This results in output to the screen similar to the following:

```
F.
========================================================================
FAIL: test_neg (__main__.Testpowb)
------------------------------------------------------------------------
Traceback (most recent call last):
  File "test_mymath2_unittest.py", line 8, in test_neg
    self.assertEqual(powb(2, -3), 0.125)
AssertionError: 1 != 0.125


------------------------------------------------------------------------
Ran 2 tests in 0.003s

FAILED (failures=1)
```

In the line above the horizontal rule created by equal signs, the unittest package writes out a period to the screen whenever a test passes and an F when a test fails. Below the rule created by equal signs, the unittest package provides detailed information about the failed tests and summarizes at the end that two tests were run with one failure.

Our unit test suite has successfully detected that our implementation of exponentiation in powb does not work for negative exponents. With this information, we can go back and make changes to powb to deal with this input case.

Besides assertEqual, the TestCase class has other "assert methods" that pass or fail on other conditions.[4] There are also assert methods that can detect whether an exception is raised, thus enabling us to check whether our error handling code is working properly. For more on unittest, see the documentation for the module (and reference for this section). An up-to-date link to the documentation is at www.cambridge.org/core/resources/pythonforscientists/refs/, ref. 79.

23.3.2 pytest

The unittest module is built-in with Python and thus convenient. It is, however, a little verbose and requires the defining of a class. The pytest package, while not built in, is a popular alternative. The testing code becomes much shorter as we do not need classes and can parameterize the different input cases. Figure 23.5 shows the same tests of powb as was provided in Figure 23.4.

Instead of writing separate methods for different input values to powb and different expected values, the @pytest.mark.parametrize decorator enables us to define a list of input argument(s) and expected values. Each of those items are then passed into separate calls to test_eval, and the assert command acts on the condition powb(x, exp) == expected using those arguments. The assert command in Python raises an Assertion

---

[4] https://docs.python.org/3/library/unittest.html#unittest.TestCase (accessed December 22, 2020).

```
1    import pytest
2    from mymath2 import powb
3
4    @pytest.mark.parametrize(
5        "x, exp ,expected",
6        [(2, 3, 8), (2, -3, 0.125) ],)
7    def test_eval(x, exp, expected):
8        assert powb(x, exp) == expected
```

Figure 23.5 Contents of *test_mymath2_pytest.py*.

```
============================ test session starts ============================
platform linux -- Python 3.6.10, pytest-6.2.1, py-1.10.0, pluggy-0.13.1
rootdir: /home/jlin/scra1
collected 2 items

test_mymath2_pytest.py .F                                            [100%]

================================= FAILURES ==================================
_____ test_eval[2--3-0.125] _____

x = 2, exp = -3, expected = 0.125

    @pytest.mark.parametrize(
        "x, exp ,expected",
        [(2, 3, 8), (2, -3, 0.125) ],)
    def test_eval(x, exp, expected):
>       assert powb(x, exp) == expected
E       assert 1 == 0.125
E        +  where 1 = powb(2, -3)

test_mymath2_pytest.py:8: AssertionError
========================= short test summary info ==========================
FAILED test_mymath2_pytest.py::test_eval[2--3-0.125] - assert 1 == 0.125
========================= 1 failed, 1 passed in 0.32s ======================
```

Figure 23.6 Example of report from running pytest on *test_mymath2_pytest.py*.

`Error` exception if the program is being run in "debug" mode and the condition is `False`.[5]
The pytest module handles this behavior and reports on the test accordingly.

We run the test module at the operating system's command line by typing in:

```
python -m pytest test_mymath2_pytest.py
```

at the operating system prompt in a terminal window. Figure 23.6 shows the report that is output to the screen. The report is similar to the one unittest produces. Again, we receive a detailed report on the failed test and a concise report on the passed test.

---

[5] https://docs.python.org/3/reference/simple_stmts.html#the-assert-statement (accessed December 22, 2020).

For more on pytest, see the documentation for the module (and reference for this section). An up-to-date link to the documentation is at www.cambridge.org/core/resources/pythonforscientists/refs/, ref. 77.

Unit tests are one kind of test. There are multiple levels of tests, and one common division is unit, integration, system, and acceptance:

1. Unit. Test a function that only does one thing.
2. Integration. Test a function that uses other functions in the program.
3. System. Test the composition of multiple functions that talk to each other to do a task.
4. Acceptance. Test how the user interacts with the system.

We do not need to write all these kinds of tests for every program we create. Most small projects will have a mix of unit and integration tests. There might not be any acceptance tests for projects that are development oriented. The goal of writing tests is mostly to verify that our code is doing what we expect it to do in a repeatable and automated matter. In Section 23.4, we describe ways of incorporating software testing into our workflows as scientists and engineers.

23.4   The "Test-Driven Development" Process

Having described what tests look like and their importance, in this section we address how to incorporate testing into a scientific or engineering workflow. For many scientists and engineers, linting and testing are optional extras: they are nice to have but ultimately not worth spending much time on.

Software engineers, however, realize this is not the case. Testing, of all kinds, is critical to the task of writing useful and accurate programs. Software engineering considers testing so important that an entire programming methodology exists that places testing front and center. **Test-Driven Development (TDD)** is a software engineering design method where we *first* write the tests and *then* write the code that makes those tests pass. The benefit of TDD is that in writing the tests, we are defining how the interface should work – what should be the inputs and outputs and the expected values for both – separately and before we write the implementation. In doing so, we substantially decrease the places where bugs can hide and substantially increase the likelihood of detecting any bugs that do exist.

Strict-TDD is probably a development methodology that is too structured for most scientists and engineers (unless they are working on a program where errors can result in death, injury, or damage to property) to implement in most of their projects. However, scientists and engineers can still benefit from adopting a "test-driven development" *mindset* for their workflow.

Most beginner (and many not-so-beginner) programmers follow this order in their workflow:

Write and debug → Lint and test

When we have a problem we want to solve by writing a program, we often sit down, turn on our computer, and immediately start writing code. Periodically, we might run the code and run into some syntax or other errors. When we do, we debug the program and fix the errors. After we are done with the program, then we run the program through a linter and write a few tests for it.

This strategy can work for relatively short programs (a few hundred lines), but for longer programs it is a gold-lettered invitation for trouble. By waiting to write tests until the program is finished, our tests will likely be weaker because of:

- Lack of motivation. Once our program is finished and gives a result, the last thing we want to do is to write a test that demonstrates our program does not work. Confirmation bias that "our program runs so it must be right!" is difficult to overcome.
- Lack of awareness of how the program was put together. After we have finished the program, it is unlikely we remember the algorithms used in each of the methods and functions, let alone what the input and output edge cases are for each method and function. The tests we write when we are done with our program will probably just test the final result in a handful of cases. Those cases are unlikely to cover the breadth of possible input values, and we have no way of knowing whether a passed test is the result of correct code or multiple errors in the code fortuitously canceling each other out.

As a result, we end up writing just a few tests that might catch some of the most egregious errors, but nothing more. Unfortunately, because the likelihood of errors in a program grows exponentially rather than linearly with each additional line of code (as each additional line results in a larger web of interconnections with other lines of code), a half-hearted effort at testing is unlikely to catch more than a few of the errors in our code.

The test-driven development mindset changes the order to:

Lint and test → Write and debug

In strict-TDD, the tests are literally written before the body of the method or function are written. But even if we do not use strict-TDD, by adopting a test-driven development mindset that puts testing first, we are reminded that testing cannot wait until the program is finished. Instead, we have to test each line or so as we write it (as we mentioned in Sections 6.2.7 and 11.2). This can be done by using a debugger to inspect our variables or by executing a line in the interpreter or console and printing the results. Many Interactive Development Environments (IDEs) also provide static checking as we type. Spyder uses PyFlakes to provide such real-time linting. And, after writing a method or function, we need to write multiple unit tests for the method or function that fully explore all the possible kinds of input and output.

Eventually, we will assemble a large enough suite of tests that it will take a substantial amount of time (perhaps hours) for us to run all the tests. We then use the tests in the following way. After a day of coding, at the end of the workday, we start our testing suite to run over the evening. When we come back the next morning, we have a report of all the tests that have failed and begin our workday by addressing those errors. Once those are fixed, we can add new code and write tests for that new code. When we leave work for the day, we run our testing suite again overnight. In this way, as we develop new features in our program, we have confidence both that the new features work and that the new features have not broken the old features. And, by running our testing suite overnight, before the workday begins, the "test-first" mindset, in a sense, becomes literally true: we test first, then write.

24 Make It Manageable: Version Control and Build Management

24.1 Introduction

When we first learn to program, our programs are usually very short, limited in their functionality, and seldom more than a single file. For such short programs, written by a single person, we can keep track of the different versions of our program and describe to a new user how to install the program by writing out a description and instructions in a few files of documentation (perhaps a *README.txt* file or a short user's manual written in Word).

Once our programs become more complex, however, it becomes much more difficult to keep track of the different versions of code and other files (e.g., configuration, settings, data, etc., files) that are part of our program. As we make changes to our code, we can easily forget which version of which files has what functionality. When multiple people collaborate on a program, this becomes even more difficult to keep track of. And complex programs are more difficult for users to install and use, because of the numerous files and dependencies the user needs to understand.

Software engineers have tools to help keep programs – their different versions, development, distribution, and installation – manageable. Version-control software keeps track of the state of a software project's files, enabling multiple developers to communicate about the status of the project and to combine multiple changes. Build managers provide tools to connect the writing of code, testing, and deployment of a program. Finally, packaging tools enable developers to bundle up the various components of their program (including describing needed dependencies) and simplify the process of installing programs. In this chapter, we briefly describe these tools. Again, our goal is not to be exhaustive but to provide enough background to encourage readers to take a deeper dive into whatever technologies (and more are being developed all the time) are the most helpful for their needs.

24.2 Version Control

A version-control system keeps track of changes, over time, in a software project's files. This enable us to describe the capabilities of different versions of each file, create snapshots of the state of a program's development, and integrate the changes from different snapshots together. There are different kinds of version-control systems and different implementations of those kinds of systems. At the time of this writing, distributed version-control systems are the most popular. Such version-control systems enable each person who downloads a copy of the version-controlled files of the software project (called a **repository**) to have full access to the history of the project and communicate (e.g., add new files, merge file changes) with other versions of the project. These capabilities enable developers to easily create side-branches to a project, make additions and changes, and reintegrate those side-branches into the main development branch of the project. Mercurial and Git are two examples of distributed version-control systems.

In this section, we describe a basic workflow using Git. There are graphical user-interfaces (GUIs) for using Git. We provide an up-to-date link to a list of Git GUI clients at www .cambridge.org/core/resources/pythonforscientists/refs/, ref. 84. Many programmers, however, use Git from the operating system's command line, and our description in this section takes that approach. Unless otherwise stated, the commands we type in this section are done in a command-line interface.

24.2.1 Using Git as a Single User

Pretend we have a folder *project* that has a single file in it, *README.txt*. Here are the steps to set up version control for this folder and its contents:

1. On the command line, navigate to the folder *project*.
2. Inside the folder, type `git init` to create the hidden folders that tell the computer this is a Git-tracked folder. This Git-tracked folder is now a repository.

Now that we have a repository, we want to manage the contents of the folder (or repository) using Git. In our initial snapshot of the repository, we want Git to recognize the *README.txt* file and track it. The process of tracking a file (and, later on, its changes) involves two phases in Git:

1. Stage. Place a "bookmark" on one or more new, deleted, or changed files saying that we plan to keep track of this file(s) for the next snapshot.
2. Commit. Take a "snapshot" of the repository with the changes in the file(s).

Once we have a snapshot, we can always go back to that version of the files in the repository. So, if we are developing our code and discover changes we had made need to be undone, we do not have to go to each file we made those changes in, find the changes, and delete them. Instead, we can rollback the state of our working copy of the repository to the snapshot taken before the changes were made. We can do this with a single file or the entire repository.

The reason why there are two phases is that it allows finer-grained control of what gets included in every snapshot (commit). Let us say we are working on and changing two files, *File A* and *File B*. If we stage both *File A* and *File B*, the next time we make a commit, that snapshot or version of the repository will include the changes in both files. Alternately, we could stage *File A*, commit that change, then stage *File B* and commit that change. The former snapshot would only contain changes to *File A* and not *File B*. This enables us, if desired, to rollback to a snapshot which includes *File A* changes but not *File B* changes. If there was no staging phase, anytime we changed more than one file, the only kind of snapshot we could take would be one that included the changes in all those files.

Let us return to our example repository, *project*, with the file *README.txt* that we want to take an initial snapshot of. To stage the file, at the operating system's command line we type:

```
git add README.txt
```

Then, we commit the changes by typing:

```
git commit -m "<a descriptive message>"
```

where *<a descriptive message>* is a message describing the snapshot just taken. We can view that message and other commit messages in a log to get a summary of what changes were made from snapshot to snapshot.

Our initial commit of our *project* repository is now done! We can confirm this by typing:

```
git status
```

This command prints a report of which files have changed and which have not been changed. After our initial commit of our *project* repository, if we have not made any changes to the repository, the report looks something like this:

```
On branch master
Your branch is up to date with 'origin/master'.

Changes not staged for commit:
  (use "git add <file>..." to update what will be committed)
  (use "git restore <file>..." to discard changes in working directory)

Untracked files:
  (use "git add <file>..." to include in what will be committed)

no changes added to commit (use "git add" and/or "git commit -a")
```

Typically, Git is used in conjunction with a service such as GitHub or GitLab, which are websites that host repositories. (We provide up-to-date links to the GitHub and GitLab websites at www.cambridge.org/core/resources/pythonforscientists/refs/, refs. 85 and 86 respectively.) The repository online is typically called the origin. On origin is the main development trunk, master (in Section 24.2.3 we describe branching from the trunk). The git status command is used throughout the life of the repository to show us what files need to be tracked, etc.

As we develop our program, we create new files, delete some old files, and change still other files. When we want to create a new snapshot of our repository, we repeat the steps of our initial commit. To stage the files we want in the new snapshot, we type:

```
git add <file 1> <file 2> ... <file n>
```

where the arguments after add are each separated by a blank space and are the file(s) to stage. Then, we commit the changes by typing:

```
git commit -m "<another descriptive message>"
```

where *<another descriptive message>* is a message describing this snapshot.

24.2.2 Using Git as a User Who Is Part of a Collaboration

In our discussion thus far, we have been working with a repository we created on our own local computer. What if we are working on someone else's repository with a team of programmers? We will probably download a copy of the repository rather than start one using git init. And, in addition to taking snapshots of our own work, we want to be able to add our changes into the remote repository and incorporate into our local repository the changes others have made to the remote repository.

If we are part of a collaboration, we will probably download a copy of the repository rather than start one using git init. An online repository will have a **Universal Resource Locator (URL)**, and this command:

```
git clone <repository URL>
```

puts a copy of the repository of interest onto our local computer. We can then work on that copy as described above, staging and committing snapshots on our local computer.

However, before cloning another person's repository, we may want to fork it first and then clone the fork. Forking means making a copy of a repository, but one that is remote (as opposed to a local copy) and one that we have the rights or permissions to manage (as opposed to the other person's repository, which we generally do not have administrator rights on). When we make changes to a copy that is a clone of a repository, we cannot integrate any of our changes back into the original repository because we lack the privileges to do so on the original repository. By making a fork, we can integrate the changes we make to our clone into the fork, because it is our fork. After enough changes are made, we can submit a pull request to the original repository asking that repository to consider merging the changes from our fork. On an online repository hosting service, if we have forked a repository, typically it is called origin and is the default our local copy (clone) will synchronize to. The original repository is called upstream.

Once we have a clone, made changes, and committed our changes to our local repository, at a certain point we want to upload those changes into the remote repository. The command to do so is:

```
git push
```

When we want to incorporate into our local repository the changes others have made to the remote repository, we type the following:

```
git pull
```

The changes in `origin` (the remote repository) may conflict with our local changes. If so, we have to resolve the conflicts by choosing whether to keep the local or remote changes. Most modern editors have tools to do so and a full discussion is beyond the scope of the present work.

Remember, it is **push** changes **to** remote, **pull** changes **from** remote. In summary, a typical basic workflow using Git is:

1. Add, delete, and/or change files in our local repository.
2. Stage and commit snapshots of our local repository, using `git status` to tell what files need to be tracked, etc.
3. Pull changes from the remote repository `origin`, to incorporate changes our collaborators have made to the remote repository into our local repository.
4. Push changes from our local repository to the remote repository for our collaborators to integrate.
5. If our remote repository is a fork of another remote, original repository, submit a pull request to the original repository asking to integrate our fork into the original repository.

If we are doing all the development work on our own local repository, our workflow only includes the first two steps. If we are collaborating with others remotely on a remote repository we manage, our workflow includes the first four steps. If our remote repository is forked from another repository, our workflow includes all five steps.

24.2.3 Using Git with Branching

More complex Git repositories use branches. A branch, as the name suggests, is a side or offshoot development of a code base. A branch can be used as a "sandbox" for new experiments, and it is easier to delete branches than revert commits, especially if remote changes become intermixed in the commit log. A common workflow is to complete a feature on a branch before integrating remote changes.

The trunk or stable working copy of the code is called `master`. A branch off `master` at a snapshot in time copies the code in `master` to this new branch. The programmer can then continue making changes in this new branch. When development along that branch is completed, the changes on that branch can be merged back into the trunk or `master` branch. If the branch goes awry – for example, the code has gotten hopelessly tangled – we can just delete the branch and create a new branch off of `master`.

A typical command-line workflow with branches involves these steps (below, *<name>* is the name of a branch or the trunk):

1. Navigate inside the repository directory.
2. Check what branch we are on by typing `git branch`.
3. Change to the branch we want to be on by typing `git checkout` *<name>*.
4. If we want to create a new branch off the trunk/branch we are currently on, type:

    `git checkout -B` *<name>*

5. Work on the branch and stage and commit files as described earlier.
6. Once we are finished with the code on the branch, we can do either of the following.
   (a) Merge (that is, combine the code) into `master` by typing:

   ```
   git checkout master
   git merge <name>
   ```

   (b) Or, if we are working on a fork, we push the branch to `origin` and open a pull request.

We usually want to pull changes from the remote repository to the matching local branch – for example, pull remote `master` to local `master` – and then synchronize other branches using a process called rebasing. Then, when a pull request is accepted, it includes the entire commit history on that branch (but there are also techniques to combine those commits into one, if that is what is desired).

We can do a lot more with Git. Many of the above steps are also automated in development environments. For more information on Git, see the Git documentation. An up-to-date link to the documentation is at www.cambridge.org/core/resources/pythonforscientists/refs/, ref. 83.[1]

24.3   Packaging

Programming libraries get their name from their physical equivalent. Just like a physical library is a collection of books that provides knowledge to advance a research project, programming libraries provide code encapsulated as functions and objects to carry out tasks related to the domain of the programming library.

Libraries need to be installed into programming environments before they can be used. A software package is the bundle of code, documentation, scripts, and utilities that is put (packaged) together to make the library installable on a computer system. Although software packages can be manually installed – this is called "installing from source" – whenever possible, we should use a package manager to install a package or library. Anaconda Navigator, which we described in Section 1.3, is one such manager.

The default Python installer, however, is called pip, and the default packaging index is called the Python Package Index (PyPI). Both are maintained by the Python Packaging Authority (PyPA). Anyone in the Python community can list a package on PyPI. We provide up-to-date links to the pip, PyPI, and PyPA sites at www.cambridge.org/core/resources/pythonforscientists/refs/, refs. 6, 9 and 8 respectively.

To install a package from PyPI from the operating system's command line, type:

```
pip install <package name>
```

where *<package name>* is the name of the package, here and later on in this section.

---

[1] See also https://blog.red-badger.com/2016/11/29/gitgithub-in-plain-english and https://githowto.com/staging_and_committing for more on using Git. These sites, and the Git documentation, are also references for this section (all accessed January 8, 2021).

The conda manager is a very popular package management system in the scientific Python community. Anaconda, Inc. maintains the main `conda` channel of packages, and community members can create their own channels of packages to upload. A very popular community-maintained channel is `conda-forge`. To install a package from the command line using conda, type:

```
conda install <package name>
```

To use the external channel `conda-forge`, type:

```
conda install <package name> -c conda-forge
```

See the conda documentation for more information. An up-to-date link to the documentation is at www.cambridge.org/core/resources/pythonforscientists/refs/, ref. 81.

In a Jupyter notebook, we can install packages using pip by typing the following line-magic command in a cell:

```
%pip install <package name>
```

Or, we can use conda by typing this line magic in a cell:

```
%conda install <package name>
```

Note the percentage sign in front of the commands. Section 22.3 discusses line magic in more detail.

The Python Packaging User Guide provides tutorials and guides for packaging in Python. There are also templating libraries we can use to help. A popular library is Cookiecutter and its derivatives. Up-to-date links to information on the Python Packaging User Guide and Cookiecutter are at www.cambridge.org/core/resources/pythonforscientists/refs/, refs. 10 and 82 respectively.

24.4   Build Management and Continuous Integration

When writing large or complex systems, it can be very tedious to manually check that the project installs, the documentation builds, or the tests pass on every single change in the variety of environments that the code can be expected to run in. Build management and continuous integration (CI) systems allow developers to automate these processes, often on every commit to the repository. Build management and CI are also commonly used to test if the project is packaged correctly such that it installs and runs on a variety of operating systems and with different versions of core dependencies. Each build management or CI tool is configured in a different way, as specified in the documentation of the pertinent tools.

Make-based build automation tools (such as make, cmake, and autotools) are commonly used in scientific projects to build the non-Python (usually C/C++ and Fortran) parts of the project. These kinds of tools are relatively stable, having been around for decades. The popularity of specific CI tools, at the time of this writing, changes rapidly. There are narrowly

```
1    # content of: tox.ini , put in same dir as setup.py
2    [tox]
3    envlist = py27,py36
4
5    [testenv]
6    # install pytest in the virtualenv where commands will be executed
7    deps = pytest
8    commands =
9        # NOTE: you can run any command line tool here - not just tests
10       pytest
```

Figure 24.1 Example configuration file that could be part of a continuous integration workflow using tox. Copied from https://tox.readthedocs.io/en/latest/ (accessed January 11, 2021).

focused tools commonly used for one task (such as documentation or Apple builds) and very broad tools that integrate well with version-control systems (such as Git). For example, there are a number of cloud-based CI systems (such as AppVeyor and Azure) that create sandboxed (isolated) virtual environments. The configuration files for these tools generally specify:

1. What is installed in these environments. Commonly, which operating systems and which versions of dependencies.
2. Which parts of the build and deploy system are executed. Building optional command-line tools, building documentation, and pushing to production are all tasks that can be specified in CI tools.

How these continuous integration tools get triggered is itself a configuration. Commonly, they are triggered when a pull request is opened against the master branch. On this trigger, the CI tool spins-up a virtual environment in which the pull request branch is cloned. This branch is then built and the tests are run against it. Then, depending again on the configuration, the code will be merged if the tests pass, or a report will be generated so that someone managing merges can then choose to manually merge the branch.

One example of a CI tool that is not tied to a cloud provider is tox, which automates testing against multiple environments. Figure 24.1 shows a sample file from the tox documentation that specifies the versions of Python to test a library against. Tox will:

- Create virtual environments for each version of Python in the envlist. In line 3 of Figure 24.1, this is specified as environments running Python 2.7 and 3.6.
- Install the listed dependencies (deps) in the environment. In line 7 of Figure 24.1, this is pytest. Pytest is discussed in Section 23.3.
- Run the listed commands in each environment, executed by the specified test runner. In line 10 of Figure 24.1, the pytest testing library is used.

See the web documentation (and reference for this section) for details on tox. Efforts have been made to help scientists and engineers make use of build management and CI. Scientific Python Cookiecutter is one resource. Up-to-date links to information on autotools, tox, and Scientific Python Cookiecutter are at www.cambridge.org/core/resources/pythonforscientists/refs/, refs. 80, 78 and 87 respectively.

25 Make It Talk to Other Languages

25.1 Introduction

In Section 22.4.2, we mentioned just-in-time compilation tools such as cython and numba that enable us to write parts of our Python code in another language or convert our Python code into an optimized non-Python version. The goal with using cython and numba is to enable our Python code to run faster. There are times, however, when we have a collection of legacy Fortran or C++ routines – battle-tested and reliable – that we want to use "as-is" from inside Python. The code files already exist external to Python, so we want to compile and wrap them up and expose them to Python to use. In this chapter, we describe two packages – f2py and ctypes – that enable us, from Python, to call routines and access variables in Fortran and C++ programs, respectively.

25.2 Talking with Fortran Programs

Although Fortran is quite an old programming language, it has not only a venerable past but an active present. Decades of legacy scientific computing code are written in Fortran, ranging from libraries of linear algebra routines and solvers to comprehensive climate models. Many of these programs are well-tested and run fast. It is unlikely that these routines will be rewritten into a more modern programming language such as Python in the forseeable future. Thankfully, the NumPy package comes with a module f2py that "automagically" converts a Fortran 77/90/95 function into a version that can be called from within Python. In this section, we introduce one way of using f2py to do this conversion.

Figure 25.1 shows a Fortran 90 function `powb` that takes a number `x` and raises it to the power `expon` by repeatedly multiplying `x` an `expon` number of times. This is a Fortran version of the Python code we introduced in Section 22.2.2.[1] When the f2py package is installed, a command-line tool `f2py` is also installed. The following command typed in the operating system's command-line interface creates a module called mathf from the Fortran 90 code in *mathf.F90*:

```
f2py -c -m mathf mathf.F90
```

---

[1] In Section 22.2.2, the second positional input parameter was called `exponent`. In Figure 25.1, we call it `expon` because `exponent` is an intrinsic procedure in Fortran.

```
 1    Integer Function powb(x, expon)
 2        Implicit None
 3        Integer x, expon, p, i
 4
 5        p = 1
 6        Do i = 1, expon
 7            p = p * x
 8        Enddo
 9        powb = p
10        Return
11    End Function powb
```

Figure 25.1 Contents of file *mathf.F90* that defines the Fortran 90 function powb. This function takes a number x and raises it to the power expon.

The command assumes a Fortran compiler has already been installed on our computer. The result is a *.so* object file which can be imported into Python. The filename of the *.so* file begins with the module name (in this case mathf), but includes other terms that are related to the system the module was built on. An example filename is *mathf.cpython-36m-x86_64-linux-gnu.so*.

The *.so* object file generated by f2py can be imported in Python as a module. When we import the module, we leave out the terms in the object file's filename that come after the module's name (in this case, after mathf). Functions and other entities defined in the module are referred to using the standard Python "dot" notation, as in:

```
In [1]:  import mathf
         mathf.powb(2, 3)
```

```
Out[1]:  8
```

Using the timeit profiler (described in Section 22.3), we can compare how fast our wrapped Fortran powb function is compared to the pure-Python powb function of Section 22.2.2:

```
In  [1]:  %timeit mathf.powb(2, 3)

          531 ns ± 2.69 ns per loop (mean ± std. dev. of 7 runs, 1000000 loops each)
```

```
In  [2]:  def powb(x, exponent):
              p = 1
              for _ in range(exponent):
                p *= x
              return p
```

```
In  [3]:   %timeit powb(2, 3)

           1.3 μs ± 2.58 ns per loop (mean ± std. dev. of 7 runs, 1000000 loops each)
```

The wrapped Fortran version is over two times as fast as the pure-Python version. This example is a short one, so the performance improvement is not as dramatic as we would like. With more involved Fortran routines, we would expect a larger effect. Still, this short example illustrates how we can include Fortran routines we have already written into our Python programs.

The f2py documentation (and main reference[2] for this section) has more information about the tool. An up-to-date link to the documentation is at www.cambridge.org/core/resources/pythonforscientists/refs/, ref. 91.

25.3   Talking with C/C++ Programs

The C/C++ programming languages are foundational to computer science. It would be hard to understate how many programs and routines are currently written in C/C++. We can make use of C/C++ routines in our Python programs by using the built-in ctypes module.

Similar to the Fortran example in Section 25.2, Figure 25.2 shows a C function powb that takes a number x and raises it to the power expon by repeatedly multiplying x an expon number of times. We first have to compile the source code and create the object file *mathc.o*:[3]

```
gcc -c -Wall -Werror -fpic mathc.c
```

With the object file, we can created a shared object library (*.so* file):

```
gcc -shared -o libmathc.so mathc.o
```

The gcc utility has a number of options. Please see the documentation for details. We provide an up-to-date link to the documentation at www.cambridge.org/core/resources/pythonforscientists/refs/, ref. 92.

Once we have created the shared object library, we can import ctypes and create a CDLL class object that gives us access to the function in the shared object library. Unlike with f2py, we do not import the *.so* file directly:

```
In  [1]:   import ctypes
           mathc = ctypes.CDLL("./libmathc.so")

In  [2]:   mathc.powb(2,3)

Out[2]:   8
```

---

[2] Additional references include Lin (2008), the source code of version 2.0 of the Neelin-Zeng Quasi-Equilibrium Tropical Circulation Model (QTCM1) (http://research.atmos.ucla.edu/csi/QTCM/qtcm.html), and the GNU Fortran Compiler documentation (https://gcc.gnu.org/onlinedocs/gfortran/). Both sites accessed June 19, 2021.

[3] www.cprogramming.com/tutorial/shared-libraries-linux-gcc.html, https://gcc.gnu.org/onlinedocs/gcc/Warning-Options.html, and https://stackoverflow.com/questions/5311515/gcc-fpic-option (accessed October 26, 2020).

```
1    extern int powb(int, int);
2
3    int powb(int x, int expon){
4         int p = 1;
5         while(expon-->0){
6              p *= x;
7         }
8         return p;
9    }
```

Figure 25.2 Contents of file *mathc.c* that defines the C function powb. This function takes a number x and raises it to the power expon.

In the above `ctypes.CDLL` call, we explicitly specify that the shared object file is in the current working directory (via the addition of the `"."` string to the file path, here on a Linux system). Not every computer requires that to be explicitly stated, but sometimes one's computer does require this.

Below, we use the timeit profiler to compare how fast our wrapped C powb function is compared to the pure-Python powb function of Section 22.2.2:

```
In  [1]:  %timeit mathc.powb(2, 3)

          1.14 μs ± 1.1 ns per loop (mean ± std. dev. of 7 runs, 1000000 loops each)


In  [2]:  def powb(x, exponent):
              p = 1
              for _ in range(exponent):
                  p *= x
              return p


In  [3]:  %timeit powb(2, 3)

          1.33 μs ± 6.4 ns per loop (mean ± std. dev. of 7 runs, 1000000 loops each)
```

The wrapped C version is a little bit faster than the pure-Python version. In this case, even more than in the earlier Fortran example, we see how the brevity of the example limits the improvement in performance obtained from writing the code in C. With longer C programs and routines, we can expect a larger effect. In those cases, ctypes provides a straightforward way to incorporate previously written C/C++ code into our Python programs.

The main reference for this section is a ctypes tutorial by New Zealand eScience Infrastructure. See that tutorial and the official Python documentation for ctypes for more information. Up-to-date links to the tutorial and documentation are at www.cambridge.org/core/resources/pythonforscientists/refs/, refs. 88 and 89 respectively.

APPENDIX A

List of Units

This list of units describes common units found in the natural sciences and used in this text. Système International (SI) units are a standard in many natural science disciplines and is based upon the kg, m, and s. Listed alphabetically (with prefixes counted as significant), considering Greek letters to be represented by their English name. References: Dickerson et al. (1984, pp. 15–16, 83, A-2, A-3, inside endcover), Resnick and Halliday (1977, pp. 81–82, inside endcover), and Wallace and Hobbs (1977, p. xv).

Abbreviation	Name	Description
cal	calorie	A unit of energy; 1 cal = 4.184 J.
cm	centimeter	A unit of length; $\frac{1}{100}$ th of a meter.
fs	femtosecond	A unit of time; 1 fs = 10^{-15} s.
g	gram	A unit of mass; $\frac{1}{1000}$ th of a kilogram.
in	inch	A unit of length; 1 in = 2.54 cm.
J	joule	The SI unit for energy; 1 J = 1 N m = 1 kg m^2 s^{-2}.
K	kelvin	The SI unit for temperature; each K is equal to a degree Celsius.
kcal	kilocalorie	A unit of energy; 1 kcal = 1000 cal.
kg	kilogram	The SI unit for mass; 1 kg = 2.204623 lb$_m$.
kJ	kilojoule	Unit for energy; 1 kJ = 1000 J.
l	liter	A unit of volume; 1 l = $\frac{1}{1000}$ m^3.
lb	pound	A unit of mass in British/US customary systems. More precisely and less colloquially, this unit is known as the "pound-mass" (and abbreviated "lb$_m$") to distinguish it from the "pound-force" (abbreviated "lb$_f$"); 2.204623 lb$_m$ = 1 kg. A pound-force is the force needed to accelerate a pound-mass to the "standard" acceleration of gravity on the surface of the Earth (Resnick and Halliday, 1977, pp. 81–82), and on the Earth, an object of mass 1 pound-mass will have a force due to gravity of 1 pound-force.
m	meter	The SI unit for length.
ml	milliliter	A unit of volume; 1000 ml = 1 l.

Abbreviation	Name	Description
mol	mole	A quantity; 1 mol of an item $= 6.022 \times 10^{23}$ of those items.
μg	microgram	A unit of mass; 1 μg $= 10^{-6}$ g.
μm	micron	A unit of length; 1 μm $= 10^{-6}$ m.
N	newton	The SI unit for force; 1 N $= 1$ kg m s^{-2}.
nm	nanometer	A unit of length; 1 nm $= 10^{-9}$ m.
Pa	pascal	The SI unit for pressure; 1 Pa $= 1$ N m^{-2}. One atm of pressure is 101 325 Pa.
s	second	The SI unit for time.

For discussion on memory units, see the Index entries for: bit, byte, MB, and MiB.

APPENDIX B

Summary of Data Structures

The table below summarizes some characteristics of different data structures. This table does not include some of the characteristics that are the most meaningful for that data structure, such as the use of masked arrays to mask out missing data:

Kind	Ordered?	Mutable?	Duplicates?	Types?	Refer item?
Array (NumPy)	Yes	No	Yes	Fixed	Index
DataFrame	Yes	No	Yes	Any	Multiple
Dictionary	No	Yes	Yes	Any	Key
List	Yes	Yes	Yes	Any	Index
Queue	Yes	Yes	Yes	Any	Front/back
Series	Yes	No	Yes	Any	Multiple
Set	No	Yes	No	Any	Value
Stack	Yes	Yes	Yes	Any	Top
Masked arrays	Yes	No	Yes	Fixed	Index

The table column entries have the following meanings:

- Ordered. An ordered data structure is one where there is a first element, a second element, etc.
- Mutable. A mutable data structure is one that can grow or shrink as the program executes.
- Duplicates. Are duplicate values allowed? If not, each item is unique.
- Types. "Single" means the data structure requires all items be of the same type. "Fixed" means all items have to be of the same type and occupy the same amount of memory. "Any" means the items can be the same type or different types.
- Refer item. "Index" means each element can be referred to by an index. "Value" means each item is referred to by its value. "Key" means each value is referred to by its key. "Top" means we only can manipulate the item at the top of the data structure. "Front/back" means we can only interact with the data structure items at the front and/or back. "Multiple" means elements can be referred to by multiple ways.

APPENDIX C

Contents by Programming Topic

This list of topics is syntax focused. It is broken down based on the syntax of the Python programming language, rather than the workflow of a scientist or engineer. The list is organized into three sections. The first two sections list the syntax topics that might be covered in the first (introductory) and second (intermediate) quarters of a traditional first-year programming sequence. The order of topics is given in the order many traditional programming textbooks for such courses use. In the third section, we list syntax topics that are less frequently covered in traditional first-year introductory programming courses. Some topics, particularly science or engineering focused application topics, do not fit in the categories of this list of topics. This is particularly true of some topics in Parts II–IV.

The Location column lists the chapter or section number (both prepended by the "§" symbol), the Try This! number (prepended by "TT!"), or the figure, table, or appendix number/letter (prepended by their title) where the topic is described. The topic may be just one part of the section, Try This!, etc., listed. Items in the Chapter Review sections are not included in this listing.

C.1 Introductory Programming Topics

C.1.1 What Is a Program and General Elements of Python

Topic	Location
Description of a program	§1.1
Capabilities of a computer	§1.1
Procedural versus object-oriented programming	§9.2.1
Code files	§3.2.4
Comment lines	§4.2.4; TT! 4-12
Docstrings	§6.2.6, 21.3, 21.4; TT! 6-10
Line continuation	§4.2.3; TT! 4-1
Block indentation and `pass`	§3.2.3, 6.2.4; TT! 5-2
Using the `with`/`as` construct	§9.2.6
The current working directory	§10.2.1, 20.2.1, 20.2.2; TT! 10-1, 10-4, 10-6

C.1.2   Variables and Expressions

Topic	Location
What is a variable	§2.1, 2.2.2
Variable assignment and names	§2.2.2
Accumulator and deaccumulator variables	§2.2.2; TT! 2-4
Expressions	§2.2.1
Evaluation order and order of precedence	§2.2.1, 8.2.2; TT! 2-3
Operators	§2.2.1
Arithmetic operators	Table2.1
Integer division	§2.2.1, 3.2.3
Modulus operator	§8.2.4
Incrementing and decrementing operators	TT! 6-5
Deleting variables	§3.2.2
Copying variables, data, and objects and assignment by value and reference	§9.2.5; TT! 9-6
Local and global variable scope	§14.2.5, 17.2.5

C.1.3   Typing and Some Basic Types

Topic	Location
What is a type	§5.2.7, 9.2.2
Dynamic versus static typing	§5.2.7
Converting between basic types	§5.2.7; TT! 7-3
Numerical types and memory	9.2.2
Floating-point imprecision	§2.2.1, 8.2.7, 9.2.2, 9.2.3; TT! 2-3
None	TT! 9-14

C.1.4 Strings

Topic	Location
Defining strings	§4.2.3
Converting to string	§4.2.3, 5.2.7; TT! 4-10, 5-6
Concatenating strings	§4.2.3; TT! 4-10
Empty string	§4.2.3
Newline character	§4.2.3
Tab character	§4.2.3
Backslash character	§10.2.1
Slicing substrings	§4.2.3
String attributes and methods	§9.2.4
Formatting values in a string	§9.2.4; TT! 9-5

C.1.5 Functions

Topic	Location
What are functions	§3.2.1
Writing functions	§3.2.3; TT! 3-6
Calling functions	§3.2.2, 3.2.3; TT! 3-1, 3-2
Parameter lists	§3.2.3; TT! 4-4, 4-5
Positional input parameters	§4.2.1
Keyword input parameters	§5.1, 5.2.1
Parameters versus arguments	§4.2.1
Dummy variables in parameter lists	§4.2.1
Flexible parameter lists	§14.2.5; TT! 14-5
The `del` function	§3.2.2
Functions that act on arrays	§6.2.2
Return values	§3.2.1
Multiple return values	§19.1; TT! 14-2
lambda functions	§16.2.2; TT! 16-8, 16-9

C.1.6 Branching, Conditionals, and Booleans

Topic	Location
What is branching and summary of branching statements	§6.1, 6.2.4; Figure 6.6
The basic `if` statement	§6.2.4
The `if-else` statement	§6.2.4
The `if-elif` statement	§6.2.4
Basic comparison operators	Table 6.2
Basic boolean operators	Table 8.1
Nested branching	§8.2.3
Examples of nonnested and nested branching	§6.2.5, 8.2.3; TT! 6-4, 6-5, 8-2, 8-3, 8-4
Scalar boolean values, operators, and expressions	§8.2.2; TT! 8-5, 8-6
Truth tables	§8.2.2; TT! 8-3
Comparison of floating-point numbers	§8.2.7, 13.2.2; TT! 8-10, 13-3
Comparison of all corresponding values between two arrays are close	TT! 9-9
Boolean arrays and operations using boolean arrays	§13.2.2, 13.1, 13.2.3, 15.2.1; TT! 13-3–13-7, 15-1
Boolean selection using pandas	§19.2.3

C.1.7 Looping

Topic	Location
What are loops	§6.2.3, 13.2.6
The `for` loop	§6.2.3; Figure 6.5
The `while` loop	§8.2.4; Figure 8.8
Nested looping	§7.2.5, 8.2.6, 12.2.6; TT! 7-6, 12-8, 12-9
Referring to offset indices in a loop	§6.2.3; TT! 6-8
Iterators and iterables	§6.2.3
Looping through strings	§6.2.3, 10.2.4
Examples of one-dimensional loops	§6.2.5; TT! 6-4, 6-5, 8-7
Examples of two- and *n*-dimensional loops	§7.1, 7.2.5, 12.1, 12.2.6; TT! 7-6–7-9, 8-9, 12-8, 12-9
The `continue` and `break` statements	§8.2.4
List comprehensions	§16.2.2; TT! 16-8

C.1.8   Console Input and Output

Topic	Location
The `print` command	TT! 2-4
The `format` method and replacement field formatting	§9.2.4; TT! 9-5

C.1.9   Text File Input and Output

Topic	Location
What are text files	§9
Directory paths	§10.2.1; TT! 10-1, 10-6
The current working directory	§10.2.1; TT! 10-1, 10-4, 10-6
Reading and writing files	§9.2.6
Using string methods to process file contents	§9.2.6
Using `genfromtxt` and csv to read files	§9.2.6
Catching file opening errors	§9.2.7

C.1.10 Exceptions

Topic	Location
What are exceptions	§9.2.7
Throwing exceptions and `raise`	§9.2.7
The `try`/`except` structure	§9.2.7; TT! 9-14
Custom exception classes	TT! 17-5

C.1.11 Arrays

Topic	Location
What are arrays	§5.2.6, 7.2, 12.2
Creating arrays	§5.2.6, 6.2.1, 7.2.2
Creating an array of indices	§6.2.1
Base type and typecodes	§6.2.1, 9.2.2; Table 9.1
Shape of an array	§7.2.1, 12.2.1, 12.2.4
Indexing elements in an array	§5.2.6, 7.2.3, 12.2.1; TT! 5-4, 7-8
Slicing (or selecting) subarrays	§5.2.6, 7.2.3, 12.1, 12.2.1, 12.2.2; TT! 5-4, 6-6, 7-2, 7-4, 12-2
Relationship of a slice to the original array	TT! 9-7
Array inquiry	§6.2.1
Array syntax calculations	§5.2.6, 7.2.4, 12.2.3, 12.2.6; TT! 5-5, 6-4
Array syntax versus looping	§8.2.6, 13.2.4; TT! 7-7
Relationship of array syntax expression results to the original array	TT! 9-8
Arrays versus lists	§5.2.6, 9.2.3; Table 9.2
Subarray offsetting	§12.2.5; TT! 5-5, 7-4, 12-5, 12-6
Functions that act on arrays	§6.2.2, 7.2.4, 12.2.3, 12.2.7; TT! 7-5
Looping through arrays by element index versus element value	TT! 6-2
Array attributes and methods	§9.2.2
Memory locations of elements	§12.2.4
Reshaping an array	§12.2.4; TT! 12-4
Boolean arrays and operations using boolean arrays	§13.2.2, 13.1, 13.2.3, 15.2.1; TT! 13-3–13-7, 15-1
Comparison of all corresponding values between two arrays are close	TT! 13-1, 9-9

C.1.12 Classes

Topic	Location
What are objects	§9.2.1
Defining and using a class	§17.2.1, 17.2.4, 17.2.5; TT! 17-1, 17-7
Listing and naming attributes and methods	§9.2.2
Public and private attributes and methods	§9.2.2, 17.2.1; TT! 17-2
Constructors	§17.2.1
Arrays as objects	§9.2.2; TT! 9-2
Lists as objects	§9.2.3; TT! 9-3
Strings as objects	§9.2.4; TT! 9-4
The Matplotlib object Application Programming Interface (API)	§14.2.1; TT! 14-1
Integer and floating-point objects	TT! 9-1
Looping through objects	§10.2.4, 13.2.6; TT! 13-9, 14-1
Copying variables, data, and objects	§9.2.5
Storing objects in lists and dictionaries	§9.2.3, 16.1.4
Overloading	§14.2.5
Inheritance	§17.1.1, 17.1.2, 17.2.2; TT! 17-3
Overriding	§17.2.2; TT! 17-4

C.2   Intermediate Programming Topics

C.2.1   Abstract Data Types and Structures

Lists and Tuples

Topic	Location
What are lists and defining a list	§4.2.1, 4.2.2, 9.2.3
Defining a tuple	§4.2.2
Indexing elements in a list and tuple	§4.2.2
Slicing sublists	§4.2.2; TT! 4-6–4-8
Relationship of a slice to the original list	TT! 9-7
Converting between lists and tuples	§4.2.2
List attributes and methods	§9.2.3

Dictionaries and Sets

Topic	Location
What are dictionaries	§14.2.5
Adding, deleting, and accessing items in a dictionary	§14.2.5, 16.2.1; TT! 16-4–16-6
Examples of using dictionaries	§16.1.3, 16.1.4, 16.2.1; TT! 16-4–16-7
What are sets	§16.2.1
Example of using sets	§16.2.1; TT! 16-3

Other

Topic	Location
Stacks and queues	§16.2.1; TT! 16-1, 16-2
`Series` and `DataFrame` objects	§15.2.2, 18.2.1, 19.2.3
Masked array objects	§15.2.3
Summary of data structures	Appendix B
When to use different data structures	§16.2.3

C.2.2 Algorithm Analysis

Topic	Location
Describing the complexity of code	§22.2.1
Big-$\mathcal{O}$ notation	§22.2.1; Table 22.1
Practices that can result in inefficient code	§22.2.2

C.2.3 Searching and Sorting

Topic	Location
Summary of some ways to search and sort	§19.2.1
Using `sorted`	§16.2.1, 17.2.3, 19.2.1; TT! 17-6
Linear and binary search	§19.2.2, TT! 19-2
Selection and insertion sort	§19.2.2; TT! 19-3
Basic searching and sorting using pandas	§19.2.3; TT! 19-4–19-6

C.2.4   Recursion

Topic	Location
What is recursion	§20.2.2
Comparing recursion with looping	§20.2.2
Examples of recursion	§20.1, 20.2.3; TT! 20-3, 20-4

C.3   Other Topics

C.3.1   How to Program and Programming Style

Topic	Location
Advice on writing code	§6.2.7, 11.1
Debugging and testing	§6.2.7, 11.2, 23.3, 23.4
The DRY principle	§6.2.3, 8.2.3
Coding style conventions	§11.3
When to use different data and execution structures	§16.2.3

C.3.2   Distributions and Interactive Development Environments (IDEs)

Topic	Location
The Anaconda distribution	§1.3
Installing packages	§1.3
Default Python interpreter in a terminal window	§2.1, 2.2.3, 3.2.4
Jupyter notebooks	§2.2.3, 3.2.4
Jupyter Markdown	§4.2.4; Table 4.1
Spyder	§3.2.5, 11.2
Getting help	§3.2.2

C.3.3   Packages and Modules

Topic	Location
Importing modules and using module items	§3.2.2; TT! 3-3, 3-4
Writing and importing our own modules	§3.2.3, 3.2.4, 24.3; TT! 3-7

C.3.4   Calculation Functions and Modules

Topic	Location
Exponential and trigonometric functions	§3.1, 3.2.2; TT! 3-1, 3-2
Random number generator functions	§8.2.1; TT! 8-1

C.3.5   Visualization

Topic	Location
Line plot	§4.1, 4.2; TT! 15-3, 15-4
Scatter plot	§4.1, 4.2
Line and shaded contour plots	§14.2.2, 14.2.3; TT! 14-2, 14-3, 15-5
Basic animation	§14.2.4; TT! 14-4
Data coordinates	§5.1, 5.2.2; TT! 5-1
Customizing how a plot looks	§5.2.2; Tables 5.1, 5.2, 5.3
Legends	§5.2.3
Multiple figures	§5.2.3; TT! 5-2
Multiple curves	§5.2.3; TT! 5-3
Multiple subplots	§8.2.5; TT! 8-8
Plot size	§5.2.4
Saving figures to a file and image formats	§5.1, 5.2.5
Reading, displaying, and writing images	§13.2.1
Color encoding in images	§13.2.1; TT! 13-1, 13-1
The Matplotlib object API	§14.2.1; TT! 14-1

Glossary

absolute path the path to a directory or file starting from the root directory (on Linux or Mac OS X) or the drive letter (on Windows).

accumulator a variable whose previous value is added to; sometimes is used to mean a variable that can both accumulate and deaccumulate value.

algorithm the way or set of steps the computer uses to make a calculation or computation.

ancestor any of the parent, grandparent, etc., to a class.

argument an item passed into a function as input; there is a subtle distinction from a parameter, but the two have similar meanings.

array a collection of values where each element of the collection is of the same type.

array syntax expressions using arrays where the operators automatically operate on corresponding elements of the arrays.

assigning connecting a variable name with a value or setting.

assignment the act of assigning.

assignment by reference creating a variable as an alias to another variable.

assignment by value creating a variable as a reference to a copy of the value being assigned.

asymptotic algorithmic complexity a measure of code complexity that only considers large numbers of input and thus only leading-order contributions to the number of operations executed by the code.

attribute data bound to an object that are designed to be acted on by methods also bound to that object; sometimes attributes are called "instance variables."

axis coordinates a coordinate system for a plot where locations are specified by the fraction

of the x- and y-axes data ranges, where each fraction ranges from 0 to 1.

back the end of a queue where items are added.

base the class a derived class inherits attributes and methods from.

base case the scenario in a recursive function that ends the sequence of recursive calls and prevents infinite regress.

base type the type of the elements in a NumPy array.

big-$\mathcal{O}$ notation typical way to express asymptotic algorithmic complexity.

binary of the form of or related to "2"; being characterized by 0 or 1; the numbering system where each place is a 0 or 1; or the memory or storage format where data are represented as bits.

binary operator an operator that acts on two elements, one to the left and one to the right of the operator.

binary search a search algorithm where we narrow our search field by half each time we do the comparison; the algorithm requires the collection be sorted.

bit a binary digit in the memory of a digital computer.

boolean an entity whose value is either true or false.

boolean algebra the system of mathematics that governs the manipulation of booleans.

boolean expression an expression that evaluates to a boolean value.

branching controlling (usually using conditionals) whether a program executes one set of commands or another.

breakpoint in an interactive development environment, this is a marker that tells the interpreter to stop at the location of the breakpoint to enable the programmer to examine the states of the variables in the program at that line.

bug an error in a program.

byte an amount of memory equal to 8 bits.

call execute or run a function.

calling executing or running a function.

child the class that a parent class provides attributes and methods to.

class the template or "pattern" all instances of that class follow.

code the commands in a program; a file of instructions for a computer to run.

color bar a rectangular-ish graphic showing what colors in a shaded contour plot (or similar) connect to what values.

color map a schema to connect a collection of values to a collection of colors.

command-line interface a way of interacting with a computer where you type in commands for the computer to perform, usually typed into a terminal window.

comment explanatory text in a code file that is not executed by the Python interpreter.

compiler a program or utility that converts human-readable source code into a machine-readable representation (sometimes called object code or byte code).

complexity a description of the efficiency or performance of code, usually with regards to the algorithm the code implements.

concatenate join together.

conditional one thing depending upon another thing; often found in `if` statements and boolean expressions to control whether a program executes one set of commands versus another, depending on whether the condition is true or not.

constructor the special `__init__` method that is the first method called when you create an instance of a class.

counter a variable that keeps track of how many times a given event has happened.

current working directory the directory you are currently in and that Python will base all relative file and directory references from.

data coordinates a coordinate system for a plot where locations are specified by the values of the x- and y-axes data ranges.

data structure an abstraction for storing, organizing, and accessing items in a collection of data.

deaccumulator a variable whose previous value is subtracted from.

debugger a tool to help us find bugs in a program.

decorator a function that takes another function as input.

degenerate when an n-dimensional array has a dimension of length 1, and the array can remove that dimension without loss of information, that dimension is called a degenerate dimension.

delimit show where a sequence or collection begins and ends.

dequeue remove an item from the front of a queue.

derived the class that a base class provides attributes and methods to.

descendant any of the children, grandchildren, etc., to a class.

deterministic a system where the state at each later time is calculated solely from the system state at an earlier time.

diagnostic calculations that describe the data as they are given, for the time(s) they are given at (or in the past), as opposed to calculating or predicting the values of variables in the future.

dictionary an unordered collection of Python objects whose size can grow and shrink as the program is running and whose values are referenced by keys, rather than by position, as in a list.

directory tree the hierarchical contents of a directory, showing all files and subdirectories, at all levels, underneath the top-level directory.

distribution a collection of the packages and programs needed to run a Python program.

docstring a triple quote delimited string that goes right after the `def` statement (or similar construct) and which provides a "help"-like description of the function.

double-precision something, generally a number, that occupies 8 bytes of memory.

dynamically typed variables take on the type of whatever value they are set to when they are assigned.

element an item of a collection of items.

empty string a string with no contents, not even blank spaces.

enqueue add an item to the back of a queue.

environment the interface we use to interact with the Python interpreter.

epoch a specific date that acts as a "datum" for a computer; this is operating system dependent.

exception an error state in the program that cannot be processed by the current scope.

executable lines of code or a file of commands that a computer can run or fulfill.

execute run or carry out.

expression a combination of values, variables, operators, and/or function or method calls that is evaluated to produce a return value.

front the end of a queue where items are removed.

full path another term for "absolute path."

function a "black box" that takes in input, processes the input, and returns a value.

generator a function that creates a value, as in a random number generator; a special object that only gives the "next" value.

immutable a variable/object that cannot be changed.

import compile a module or package and make what is in the module or package accessible to the Python program that is doing the importing.

increment increase the value of a variable, usually by one; or the amount a variable will increase.

incrementing to increment.

index the same meaning as a positional index, i.e., the address of an element in a list, tuple, or NumPy array based upon where the element is in the list, etc. (the first element usually has a positional index of 0); the collection of row names or labels in a pandas `Series` or `DataFrame` object.

infinite loop a loop that never stops running.

inherit incorporate the attribute and method definitions of another class into a definition of a new class of objects.

inheritance dealing with inheriting attribute and method definitions of another class into a definition of a new class of objects.

insertion sort a sorting algorithm which is a repeated application of comparing the next element in the remaining unsorted partition with elements (working backward) in the sorted partition and then moving that item into the sorted partition to the proper place.

instance an object that is the specific realization of a class of objects.

instantiate create an instance of a class.

interpreter the execution environment for Python commands.

iterable a data structure that one can go through, one element at a time; in such a structure, after you've looked at one element of it, it will move you on to the "next" element, but an iterable collection does not have to be ordered.

iteration a specific time we have repeated a loop.

iterator used nearly interchangably with the noun form of "iterable"; also used to mean the variable that is set to each of the values in an iterable data structure, one in turn.

key in a dictionary, the label that refers to a value stored in the dictionary.

keyword input parameter a parameter set by reference to a name or keyword rather than by position in a list.

lambda function a small function, written usually in a single line.

leaf directory the final directory in a directory path.

limited-access a characteristic of a data structure; items in the structure can only be accessed if certain conditions are met.

linear search a search algorithm where we examine each item in a collection, one at a time, and ask whether that item is the one we are looking for.

list an ordered collection of Python objects whose size can grow and shrink as the program is running.

list comprehension an execution structure that combines looping and list creation.

local limited in scope.

loop a programming construct that enables one to repeat a set of calculations.

looping the act of using or going through a loop.

macro language a programming language that is part of a GUI application that enables a user to automatically accomplish tasks in that application that would otherwise have to be done manually.

masked array an object that contains an array with a mask that describes which elements in the accompanying array are missing or not.

membership testing checking to see if an item is in a collection.

metadata information about an entity, as distinguished from the entity itself.

method functions bound to an object that are designed to act on the data also bound to that object.

module an importable Python source code file that typically contains function, class, and variable object definitions.

multiple inheritance dealing with an inheritance structure that has more than one parent class.

mutable a variable/object that can be changed.

n-dimensional having n dimensions, where n is usually more than one; when used to describe loops, this means there are n nested loops; when used to describe an array of data, this means the elements of that array are of rank n (e.g., if $n = 3$, the elements are arranged as are a cube-like structure).

nested branching a construct where one branching statement is found in the body of another branching statement; the result is that the inner branching statement test will execute only if the outer branching statement condition is true.

nested loop a construct where one loop is found inside another loop; the result is that the inner loop will fully execute over all of its iterations for each one of the iterations of the outer loop.

newline character a special text code that specifies a new line; the specific code is operating system dependent.

nondeterministic a system where the state at each later time is calculated solely from the system state at an earlier time..

nonproportional font a font where the width of each character is the same; e.g., an "i" is exactly as wide as an "m").

normalized coordinates a coordinate system where locations are specified by the fraction of the distance from a starting location to an ending location, where each fraction ranges from 0 to 1; in a plot or graph frame/axes, these are also called axis coordinates.

object a "variable" that has attached to it both data (attributes) and functions designed to act on that data (methods).

one-dimensional having only one dimension; when used to describe a loop, this means there is only a single loop with no nested loop(s) inside of that single loop; when used to describe an array of data, this means the elements of that array are in a linear sequence rather than in a grid of values.

open source a program or package that is licensed under conditions that permit the access and use of the underlying source code.

operand the value(s) an operator operates on.

operation mathematical or logical operator applied to some value or entity.

operator a token attached to one or more values in an expression that applies a function or operation to that value(s).

ordinal dealing with order or position, i.e., the idea of "first," "second," etc.; in Python lists, the first element has an ordinal position of one but an index of zero.

overloading writing multiple versions of a method that differ depending on the number and kind of input parameters the method takes.

package a directory of importable Python source code files (and, potentially, subpackages) that typically contains function, class, and variable object definitions.

parameter an item passed into a function as input.

parameter list a list of items passed into a function as input.

parent the class a child class inherits attributes and methods from.

partition a theoretical portion of memory devoted to a collection of items, such as a list; this term is used in discussions of sorting to describe a hypothetical region of memory that has a certain purpose, which in reality may not be implemented as a distinct portion of memory.

path the listing of what directories you need to go through to reach the file or directory at the end of the path.

pickle dump an object into a pickle file.

pixel a single element in an image; the tiniest dot that is the building block of an image.

pop remove an item from the top of a stack.

positional index the address of an element in a list, tuple, or NumPy array based upon where the element is in the list, etc. (the first element usually has a positional index of 0).

positional input parameter a parameter set by reference to its position in a list of parameters.

private an attribute or method meant to be accessed, changed, and/or used by developers only, not by the user of an object.

procedural programming a programming paradigm where a program is broken up into discrete procedures or subroutines, each of which do a specified task and communicate with the rest of the program solely (ideally) through input and output variables that are passed in argument lists and/or return values.

prognostic calculations predicting the future values of variables that evolve in time.

program a set of instructions for a computer.

proportional font a font where the width of each character changes depending on the shape of the letter; e.g., an "i" is not as wide as an "m").

pseudocode code-like words and phrases that convey what a program is to do but do not exactly follow the syntax of an actual programming language.

public an attribute or method meant to be accessed, changed, and/or used by the user of an object.

push add an item to the top of a stack.

queue a limited-access data structure where elements are added at the back and removed from the front.

random-access a characteristic of a data structure; any item in the structure can be accessed at any time (with the correct address).

raveling to turn an array into its one-dimensional version.

recursion a realization of something recursive.

recursive related to the idea that we can define some tasks in terms of themselves; this is expressed in terms of a function partially being defined by calls to itself.

relative path the path to a directory or file starting from the current directory.

replacement field when using a string's `format` method, the replacement field is the portion in that string where a formatted value of an argument passed to the `format` call is placed.

repository a copy of the version-controlled files for a software project.

return value what a function returns when it is called.

routine a generic name for a function, method, procedure, or other construct that does some action in the computer.

run execute the commands in a program.

scalar a quantity with only magnitude; a plain number; rarely, but sometimes, we also call one-element arrays scalars.

script a file of Python commands or instructions.

scripting writing a script.

searching the task of looking for an item that meets certain criteria.

seed an integer that determines a sequence of values generated by a random number generator.

select when used in relation to extracting parts of an array, this means the same as "slice."

selection sort a sorting algorithm which is a repeated application of a search for the minimum (or maximum, if sorting descending) value in the remaining unsorted partition and the appending of that item to the sorted partition.

set a data structure where all elements are unique.

shape a tuple whose elements are the number of elements in each dimension of an array; in Python, the elements are arranged so the fastest varying dimension is the last element in the tuple and the slowest varying dimension is the first element in the tuple.

sheet in a three-dimensional array, the two-dimensional array selected when the first dimension is constant and the second (row) and third (column) dimensions include all rows and columns.

shell a text-based interface to an operating system, the software that manages your files, directories, etc.

signature the interface of a method, described by the method name, the number and kind of input parameters, and the kind of value the method returns.

single-precision something, generally a number, that occupies 4 bytes of memory.

slice a portion of a list, tuple, or array; the act of extracting only a portion of a list, tuple, or array.

sorting the task of putting a collection of items into an order.

stack a limited-access data structure where elements are added and removed only at the top.

statically typed variables declared as having a certain type and are defined by that type for the life of the variable.

stopping case the scenario in a recursive function that ends the sequence of recursive calls and prevents infinite regress.

string a data type that holds characters such as letters, numerals, and special characters such as the newline character.

sub the class that a superclass provides attributes and methods to.

subarray a portion of an array.

sublist a portion of a list.

substring a string that is a subset of another string.

suffix of a filename, this is the letters after the rightmost period in the name.

super the class a subclass inherits attributes and methods from.

terminal window a text window in which you can directly type in operating system and other commands.

text file a file that holds only letters, numerals, punctuation characters, and special characters such as an end-of-line character in a format that a computer can read but is also easily readable by a human being.

timestep in a prognostic model, the amount of time between the current state of the model and the next calculated future state of the model.

token a discrete entity – usually a variable, object, operator, or function call – in an expression.

top the place in a stack where items are added or removed.

truth table a table showing the input(s) into and result of a boolean expression for all possible combinations of the input(s).

tuple an ordered collection of Python objects whose size and contents are fixed and unmutable for the life of the collection.

two-dimensional having two dimensions; when used to describe loops, this means there are two nested loops, one inside the other; when used to describe an array of data, this means the elements of that array are in a grid-like arrangement.

type what "kind" of value a value is; all values of the same type share certain characteristics.

typecode a single character string that specifies the type of the elements of a NumPy array.

typing defining types, i.e., different kinds of values that all share certain characteristics.

ufunc a NumPy universal function.

unary operator an operator that acts on one element.

unit test a check to see whether a small unit of code, usually a function or method, produces the correct results.

unit testing the procedure of checking whether a small unit of code, usually a function or method, produces the correct results given a range of possible input.

universal function a NumPy function that acts element-wise on arrays.

unpickle load an object from a pickle file.

variable a name or label that holds (or points to) a value or object and can be used as a reference to that value or object.

wall clock time the amount of time it takes for a program to run, based on the perception of the user.

whitespace characters typically used as separators between words and other text; the space character is whitespace, but in many contexts so too are the tab character and the newline character.

Acronyms and Abbreviations

API Application Programming Interface.

CI Continuous Integration.
CPU Central Processing Unit.
CSV comma-separated values.

dpi dots per inch.
DRY don't repeat yourself.

FIFO first in, first out.
FILO first in, last out.

GMT Greenwich Mean Time.
GUI graphical user-interface.

HDF Hierarchical Data Format.

IDE Interactive Development Environment.
IEEE Institute of Electrical and Electronics Engineers.

JPEG Joint Photographic Experts Group.

LIFO last in, first out.
LILO last in, last out.

MAST Mikulski Archive for Space Telescopes.

NaN Not a Number.
NASA National Aeronautics and Space Administration.
NCAR National Center for Atmospheric Research.
NCEP National Centers for Environmental Prediction.
NN nearest-neighbor.
NOAA National Oceanic and Atmospheric Administration.

OOP object-oriented programming.

PEP Python Enhancement Proposal.
PNG Portable Network Graphics.

RGB Red–Green–Blue.
RGBA Red–Green–Blue–Alpha.

SI Système International.
SQL Structured Query Language.

TDD Test-Driven Development.

URL Universal Resource Locator.
UTC Coordinated Universal Time.

Z Zulu Time Zone..

Bibliography

E. Ben-Jacob, O. Schochet, A. Tenenbaum, et al. Generic modelling of cooperative growth patterns in bacterial colonies. *Nature*, **368**: 46–49, Mar. 3, 1994.

B. M. Bush. The perils of floating point, 1996. URL www.lahey.com/float.htm.

R. E. Dickerson, H. B. Gray, M. Y. Darensbourg, and D. J. Darensbourg. *Chemical Principles*. The Benjamin/Cummings Publishing Company, Inc., Menlo Park, CA, 4th ed., 1984. ISBN 0-8053-2422-4.

D. Dua and C. Graff. UCI machine learning repository, 2019. URL http://archive.ics.uci.edu/ml.

J. D. Hunter. Matplotlib: a 2d graphics environment. *Comput. Sci. Eng.*, **9**(3): 90–95, 2007. doi: 10.1109/MCSE.2007.55.

E. Kalnay, M. Kanamitsu, R. Kistler, et al. The NCEP/NCAR 40-year reanalysis project. *Bull. Amer. Meteor. Soc.*, **77**:437–472, 1996. doi: 10.1175/1520-0477(1996)077<0437: TNYRP>2.0.CO;2.

D. E. Knuth. Computer programming as an art. *Comm. ACM*, **17**(12): 667–673, Dec. 1974.

X. Liang, S. Li, S. Zhang, H. Huang, and S. X. Chen. $PM_{2.5}$ data reliability, consistency, and air quality assessment in five Chinese cities. *J. Geophys. Res.*, **121**: 10 220–10 236, 2016. doi: 10.1002/2016JD024877.

J. W.-B. Lin. *qtcm User's Guide*. July 30, 2008. URL www.johnny-lin.com/py_pkgs/qtcm/doc/manual.pdf.

J. W.-B. Lin. qtcm 0.1.2: a Python implementation of the Neelin–Zeng Quasi-Equilibrium Tropical Circulation Model. *Geosci. Model Dev.*, **2**: 1–11, doi:10.5194/gmd–2–1–2009, 2009.

J. W.-B. Lin. *A Hands-On Introduction to Using Python in the Atmospheric and Oceanic Sciences*. Chicago, IL, 2012.

J. W.-B. Lin. *Lecture Notes on Programming Theory for Management Information Systems*. Bellevue, WA, 2019.

W. McKinney. *Python for Data Analysis*. O'Reilly Media, Inc., Sebastopol, CA, 2nd ed., 2017.

Met Office. *Cartopy: a cartographic python library with a Matplotlib interface*. Exeter, Devon, 2010–2015. URL http://scitools.org.uk/cartopy.

K. Patel and L. Dauphin. Lights out after Cyclone Fani, May 2019. URL https://earthobservatory.nasa.gov/images/145017/lights-out-after-cyclone-fani.

J. C. Phillips, R. Braun, W. Wang, et al. Scalable molecular dynamics with NAMD. *J. Comput. Chem.*, **26**: 1781–1802, 2005.

W. H. Press, B. P. Flannery, S. A. Teukolsky, and W. T. Vetterling. *Numerical Recipes in Pascal*. Cambridge University Press, Cambridge, 1989. ISBN 0-521-37516-9.

R. Resnick and D. Halliday. *Physics, Part I*. John Wiley & Sons, New York, 3rd ed., 1977. ISBN 0-471-71716-9.

A. B. Shiflet and G. W. Shiflet. *Introduction to Computational Science*. Princeton University Press, Princeton, NJ, 2nd ed., 2014. ISBN 978-0-691-16071-9.

S. van der Walt, J. L. Schönberger, J. Nunez-Iglesias, et al., and the scikit-image contributors. scikit-image: image processing in Python. *PeerJ*, **2**: e453, 6, 2014. ISSN 2167-8359. doi: 10.7717/peerj.453. URL https://doi.org/10.7717/peerj.453.

J. VanderPlas. *Python Data Science Handbook*. O'Reilly Media, Inc., 2016. URL

https://jakevdp.github.io/
PythonDataScienceHandbook/.

Z. Vinícius, G. Barentsen, C. Hedges, M. Gully-Santiago, and A. M. Cody. Keplergo/lightkurve, Feb. 2018. URL http://doi.org/10.5281/zenodo .1181928.

H. E. Volkman, H. Clay, D. Beery, J. C. W. Chang, D. R. Sherman, and

L. Ramakrishnan. Tuberculous granuloma formation is enhanced by a *Mycobacterium* virulence determinant. *PLoS Biol*, **2**(11): e367, 2004. URL https://doi.org/10.1371/ journal.pbio.0020367.

J. M. Wallace and P. V. Hobbs. *Atmospheric Science: An Introductory Survey*. Academic Press, San Diego, CA, 1977. ISBN 0-12-732950-1.

Index